STRUCTURES AND PROPERTIES OF RUBBERLIKE NETWORKS

TOPICS IN POLYMER SCIENCE
A Series of Advanced Textbooks and Monographs

Series Editor
James E. Mark, University of Cincinnati

Monte Carlo and Molecular Dynamics Simulations in Polymer Science, K. Binder, editor
Structures and Properties of Rubberlike Networks, B. Erman and J. E. Mark

STRUCTURES AND PROPERTIES OF RUBBERLIKE NETWORKS

Burak Erman
& James E. Mark

New York Oxford
Oxford University Press
1997

Oxford University Press

Oxford New York
Athens   Auckland   Bangkok   Bogota   Bombay   Buenos Aires
Calcutta   Cape Town   Dar es Salaam   Delhi   Florence   Hong Kong
Istanbul   Karachi   Kuala Lumpur   Madras   Madrid   Melbourne
Mexico City   Nairobi   Paris   Singapore   Taipei   Tokyo   Toronto   Warsaw

and associated companies in
Berlin   Ibadan

Copyright © 1997 by Oxford University Press, Inc.

Published by Oxford University Press, Inc.
198 Madison Avenue, New York, New York 10016

Oxford is a registered trademark of Oxford University Press

All rights reserved. No part of this publication may be reproduced,
stored in a retrieval system, or transmitted, in any form or by any means,
electronic, mechanical, photocopying, recording, or otherwise,
without prior permission of Oxford University Press.

Library of Congress Cataloging-in-Publication Data
Erman, Burak.
  Structures and properties of rubberlike networks / Burak Erman,
  James E. Mark.
      p.   cm. — (Topics in Polymer Science)
  Includes bibliographical references and index.
  ISBN 0-19-508237-0
  1. Polymer networks.   2. Elastomers.   I. Mark, James E., 1934–
    II. Title.
  QD328.P67E75   1997
  547'.842044—dc21   96-48699

9 8 7 6 5 4 3 2 1

Printed in the United States of America
on acid-free paper

Preface

In a sense, this book is a sequel to our 1988 book, *Rubberlike Elasticity: A Molecular Primer*. As such, it has much in common with its predecessor, particularly its strong emphasis on molecular concepts and theories. Similarly, only equilibrium properties are covered in any detail. This is not to question the importance of the dynamic aspects of viscoelastic phenomenon, nor even the importance of ultimately developing a theory that covers both rubber elasticity and viscoelasticity. Unfortunately, no such encompassing theory exists, at present, and it will probably be some time before one becomes available in a form that is useful to experimentalists, as well as theorists. It is also our opinion that a thorough understanding of equilibrium properties must, of necessity, precede an understanding of dynamic phenomena, at least in the molecular area. This was the approach taken in our earlier book, and in the now-classic 1975 book by Treloar (*The Physics of Rubber Elasticity*).

As was specified in its title, *Rubberlike Elasticity: A Molecular Primer* was a brief, elementary presentation written in a tutorial, introductory style. Very little background was required of the reader. This present book treats much the same subject matter, but is meant to be a more comprehensive, somewhat more sophisticated, treatment. Because of its more comprehensive character, much more detail is given but, for the convenience of the reader, this has generally been placed into the appendixes. It will be useful if the reader already has some familiarity with the fundamentals, such as those discussed in the introductory book. For this reason, some of this material is presented in condensed form in the first chapter, and in the first sections of several of the other chapters. A number of the books and review articles cited in the Selected General Bibliography at the end of this book should also be useful in this regard.

Contents

1 Overview and Some Fundamental Information   3

2 Classical Theories of Rubber Elasticity   7
  2.1 The Kuhn–Treloar Theory   8
  2.2 The Phantom Network Theory of James and Guth   10
    2.2.1 General Aspects   10
    2.2.2 The Elastic Free Energy   13
  2.3 The Affine Network Theory of Wall and Flory   19
  2.4 The Edwards Approach and Other Theories   19
  References   20

3 Intermolecular Effects: I. The Constrained-Junction Model   22
  3.1 The Model and its Assumptions   23
  3.2 Probability Distribution of Fluctuations in the Deformed Network   27
  3.3 The Elastic Free Energy   30
    3.3.1 General Aspects   30
    3.3.2 The Elastic Free Energy Due to Distortion of ΔR 30
    3.3.3 The Elastic Free Energy Due to Distortion of Δs 31
    3.3.4 The Total Elastic Free Energy   31
  References   32

4 Intermolecular Effects: II. Constraints along Network Chains   33
  4.1 The Slip-Link Model   34
  4.2 The Constrained-Chain Model   38
  4.3 The Diffused-Constraints Model   40
  4.4 Other Treatments of Entanglements   42
  References   42

5 Relationships between Stress and Strain   44

5.1 General Relationships of Finite Elasticity Theory   44
5.2 Stress–Strain Relations for the Phantom and Affine Network Models under Uniaxial Stress   46
5.3 Stress–Strain Relations for the Constrained-Junction Model under Uniaxial Stress   47
5.4 Stress–Strain Relations for the Slip-Link and Constrained-Chain Models under Uniaxial Stress   48
    5.4.1 The Slip-Link Model   48
    5.4.2 The Constrained-Chain Model   49
5.5 Comparison of Stress–Strain Relations with Experimental Data   49
References   52

6 Swelling of Networks   53

6.1 Free Energy of a Swollen Network   53
6.2 The Solvent Chemical Potential for an Isotropically Swollen Network   55
6.3 Thermodynamics of a Network Uniaxially Stretched in Solvent   57
6.4 Elastic Activity of a Swollen Network   60
6.5 More Recent Treatments of Network Swelling   62
6.6 Sorption and Extraction of Diluents   63
    6.6.1 Linear Diluents   63
    6.6.2 Branched Diluents   65
    6.6.3 Cyclic Diluents   66
6.7 Trapping of Cyclics within Network Structures   66
    6.7.1 Experimental Results   66
    6.7.2 Theoretical Interpretations   67
    6.7.3 Olympic Networks   69
References   70

7 Critical Phenomena and Phase Transitions in Gels   73

7.1 Theory of Critical Phenomena and Phase Transitions   74
7.2 Thermoreversible Gels   79
References   83

8 Calculations and Simulations   87

8.1 Spatial Configurations of an Isolated Chain   87
8.2 Statistical Averages of Configurational Variables   92
8.3 Distributions for End-to-End Separations for Specific Types of Network Chains   94
8.4 Stress–Strain Isotherms Calculated from the Non-Gaussian Distributions   100
8.5 Molecular Dynamics Calculations   104
References   104

9 Thermoelasticity   107
9.1 Theory   108
9.2 Typical Stress–Temperature Data   110
9.3 Illustrative Thermoelastic Results   112
9.4 Relevant Calorimetric Studies of Elastic Deformations   121
9.5 Relevant Viscosity–Temperature Results on Dilute Polymer Solutions   122
9.6 Rotational Isomeric State Interpretation of Stress–Temperature Results   125
References   131

10 Model Elastomers   134
10.1 The Dependence of the Stress on Network Structure   134
   10.1.1 General Approach   134
   10.1.2 Effect of Junction Functionality   135
10.2 The Issue of Entanglements   136
10.3 Interpretation of Ultimate Properties   141
10.4 Some Other Unusual Networks   143
   10.4.1 Dangling-Chain Networks   143
   10.4.2 Networks Containing Reptating Chains   145
   10.4.3 Networks Prepared in Solution or in a State of Strain   145
   10.4.4 Networks Containing Unusual Diluents   146
References   147

11 Segmental Orientation   150
11.1 Molecular Deformation   150
11.2 Segmental Orientation in Network Chains: The Simple Picture   153
11.3 Higher-Order Approximation for Segmental Orientation   155
11.4 Experimental Determination of Segmental Orientation in Rubbery Networks   157
11.5 Theoretical Interpretation of Infrared Dichroism Measurements of Segmental Orientation in Rubbery Networks   159
References   163

12 Networks with Semiflexible Chains and Networks Exhibiting Strain-Induced Crystallization   165
12.1 Networks with Semiflexible Chains   165
   12.1.1 The Lattice Model for the Semiflexible Chain   166
   12.1.2 The Partition Function and the Free Energy of Mixing of Network Chains   168
   12.1.3 A Set of Nonlinear Equations for Evaluating the Orientational Distribution under Deformation   169
   12.1.4 The Linearized Closed-Form Solution   173
   12.1.5 The Relationship of Stress to Deformation and Orientation   174

12.1.6 Isotropic–Nematic Phase Transitions in Deformed Polymer Networks   175
12.2 Strain-Induced Crystallization   178
   12.2.1 General Features   178
   12.2.2 Models for Strain-Induced Crystallization in Stretched Networks   179
   12.2.3 Predictions of the Molecular Theories   180
   12.2.4 The Effects of Strain-Induced Crystallization on Mechanical Properties   182
References   185

13 Networks Having Multimodal Chain-Length Distributions   188

13.1 Ultimate Properties and Non-Gaussian Effects   188
13.2 Bimodal Networks   189
   13.2.1 Materials and Synthetic Techniques   189
   13.2.2 Testing of the Weakest-Link Theory   191
   13.2.3 Elongation Results   193
   13.2.4 Results in Other Mechanical Deformations   203
   13.2.5 Results on Nonmechanical Properties   209
13.3 Trimodal Networks   212
13.4 Networks of Very High Modality   214
13.5 Elastomers that May Have Been Inadvertently Bimodal   214
13.6 Other Materials in which Bimodality May Be Advantageous   215
References   216

14 Small-Angle Neutron Scattering   220

14.1 General Features of SANS   220
14.2 Experimental Studies   223
14.3 Theory of SANS from Networks   224
   14.3.1 The Scattering Law   224
   14.3.2 Scattering from a Phantom Network Chain   225
   14.3.3 Scattering from an Affine Network Chain   228
   14.3.4 Scattering from a Chain whose Individual Segments Deform Affinely   229
   14.3.5 Scattering from a Labeled Path in the Network   229
14.4 Typical Results from Experiments and Comparison with Theory   230
References   233

15 Bioelastomers   235

15.1 Some General Observations   235
15.2 Chemical Aspects of Protein Bioelastomers   237
   15.2.1 Overall Amino Acid Composition   237
   15.2.2 Amino Acid Sequencing   239
   15.2.3 Cross-Linking Chemistry   240

15.3 Network Thermoelasticity   241
  15.3.1 General Relevance   241
  15.3.2 Elastin   242
  15.3.3 Resilin   247
  15.3.4 Other Protein Elastomers   247
15.4 Stress–Strain Behavior   248
  15.4.1 General Results   248
  15.4.2 Elastin   249
  15.4.3 Resilin   252
  15.4.4 Spider-Web Silk   252
  15.4.5 Other Protein Elastomers   252
15.5 Dynamic-Mechanical Properties   253
  15.5.1 Viscoelastic Responses in General   253
  15.5.2 Effects of Dehydration   256
15.6 Some Other Bioelastomeric Gels   257
  15.6.1 General Properties   257
  15.6.2 Some Specific Systems   259
References   262

16 Multiphase Systems   265

16.1 Some Theoretical Approaches   267
16.2 In Situ Generation of Fillers in Elastomers   269
  16.2.1 General Comments   269
  16.2.2 Preparation   270
  16.2.3 Electron Microscopy   270
  16.2.4 Scattering Techniques   271
  16.2.5 Nuclear Magnetic Resonance   272
  16.2.6 Aging   275
  16.2.7 Densities   275
  16.2.8 Calorimetry   275
  16.2.9 Thermogravimetric Analysis   276
  16.2.10 Mechanical Properties and Equilibrium Swelling   276
  16.2.11 Comparisons among Various Silica-Based Fillers   287
  16.2.12 Other Polymers   288
  16.2.13 Other Ceramic-Type Fillers   288
16.3 Preparation of Bicontinuous Systems   289
16.4 In Situ Generation of Elastomers in Ceramics   290
16.5 In Situ Generation of Catalysts in Polymers   291
16.6 In Situ Polymerizations of Glassy Polymers   292
  16.6.1 Isotropic Systems   292
  16.6.2 Anisotropic Systems   295
16.7 Fillers Responding to Magnetic Fields   295
16.8 Fillers of Controlled Crystalline Structure   297
References   299

xii CONTENTS

Appendixes

A. Network Structural Parameters 307
   References 310

B. Definitions in the Area of Rubber Technology 311
   B.1 Basic Definitions 311
   References 314

C. Deformation and Stress 315
   C.1 Deformation 315
      C.1.1 General Aspects 315
      C.1.2 Isotropic Swelling 317
      C.1.3 Simple Tension 317
      C.1.4 Biaxial Extension 317
      C.1.5 Pure Shear 318
   C.2 Stress 318
   References 320

D. Summary of Thermodynamics and Statistical Mechanics 321
   References 323

E. Fluctuations in Phantom Networks 324
   E.1 The Matrix Γ and its Inverse 324
   E.2 Expressions for Various Fluctuations 326
      E.2.1 Two Junctions Joined by a Single Chain 326
      E.2.2 Two Junctions Separated by Several Chains 327
      E.2.3 Points on a Network Chain 327
      E.2.4 Points on Network Chains that are Separated by Several Junctions 328
   References 329

F. Distributions of the Chain End-to-End Vector 330
   F.1 Examples of Distribution Functions 330
      F.1.1 The Freely Jointed Chain 330
      F.1.2 More Realistic Pictures of the Network Chain 331
   F.2 Transformation of Distribution Functions under Deformation 333
   References 333

G. Fortran Program for Monte Carlo Calculations 335
   G.1 Program (Calculation of Persistence Lengths and Mean-Square End-to-End Distances) 335
   G.2 Form of the Data Set 339
   G.3 Output of the Program 339

H. Some Historical Aspects 341
   H.1 The Earliest History 341
   H.2 Natural Rubber in the Un-Cross-Linked State 342
   H.3 Cross-Linked Natural Rubber 345
   H.4 The Plantation Movement East 346
   H.5 Some Additional Scientific Developments 346
   H.6 The Effects of World War II 348
   H.7 The Postwar Period 348
   References 349

Selected General Bibliography 351

Author Index 355

Subject Index 361

STRUCTURES AND PROPERTIES OF RUBBERLIKE NETWORKS

multiphase systems, as opposed to those obtained by attempting to blend badly agglomerated carbon black or silica (formed in a separate reaction) into an elastomer which generally has a high molecular weight and a correspondingly high bulk viscosity. Again, a wide variety of deformations is included to illustrate the reinforcing effects obtained in some of these systems.

The first of the series of appendices is included to remedy the fact that almost all treatments of rubber elasticity assume that the network structure is well defined, without clear guidance about what is involved. For this reason, appendix A provides a detailed description of the network structure that forms the basis of essentially all investigations, experimental as well as theoretical. Characterization of such structures is particularly difficult when the network has been obtained by the usual cross-linking techniques, such as vulcanization, that are uncontrolled and highly random. These techniques are nonetheless important, and appendix B attempts to describe some of the concepts in this area, and several others involving rubber technology in general. Consistent with the stated objectives of this book, the aspects that are covered here are those about which there is at least some molecular understanding or insight.

Appendix C describes in formal detail the deformation and stress in an elastic body. For generality, isotropic swelling, biaxial extension, and pure shear are covered, as well as the usual simple tension.

A concise summary of concepts in the area of thermodynamics and statistical mechanics is presented in appendix D. As would be anticipated, the emphasis is on aspects of particular relevance to rubberlike elasticity.

Since fluctuations of chain dimensions are fundamental elements of modern elasticity theories, a detailed discussion of this topic is presented in appendix E.

Appendix F is about distributions of the chain end-to-end vector. The primary topics here are the Gaussian limit, non-Gaussian alternatives, and the transformation of distributions under deformation. Distributions are fundamental to any molecular theory of rubberlike elasticity, and their importance is underscored by this separate treatment of the subject.

A Monte Carlo Fortran program for calculating persistence lengths and mean-square end-to-end distances of polymer chains is given in appendix G. It is based on rotational isomeric state models, and is useful in a number of respects, including gauging the extent to which a given chain structure has the flexibility required for rubberlike elasticity.

Appendix H discusses historical events involving elastomers and rubberlike elasticity. We hope that this information will provide an overview of early work in this area, and also give the reader a better appreciation of the discipline. These developments go back to the very beginnings of quantitative polymer science and engineering. In fact, the first quantitative theories in all of polymer science were addressed to the phenomenon of rubber elasticity!

Finally, there are the usual author and subject indices. A bibliography of general references to books and review articles on rubber elasticity is also included. It is essentially an updated version of the one provided in our earlier book, and should be a useful source of references for additional information.

Nearly all of the remaining chapters are more specialized. For example, chapter 10 is about the already mentioned model elastomers obtained by carrying out network-formation reactions much more carefully and specifically than in the usual technology. The fundamental advantage of these elastomers is the fact that they have known structures, particularly network chains of known molecular weights and molecular weight distributions, and junctions of known functionality. Chapter 10 describes the preparation of such networks, and their use in clarifying a number of structure–property relationships. Ultimate properties are touched upon, and are then considered further in chapters 12 and 13.

Chapter 11 focuses on chain segmental orientation accompanying deformation of an elastomeric network. An example of its fundamental importance is investigation of the molecular deformation arising from an imposed macroscopic strain, the importance of which is stressed in chapters 2, 3, and 4. A more frequently applied example of its importance is its connection to strain-induced crystallization, and the reinforcement it provides in the case of networks that can undergo it.

The classical theories of rubber elasticity are not valid when the chains making up the network have semiflexible segments. This is the area of liquid-crystalline networks, and great advances have now been made in this field. The treatment of such networks requires the description of the packing entropy of semirigid chains in a deformed lattice. This is covered in chapter 12, as is strain-induced crystallization of networks in general.

Chapter 13 describes a particular type of model network, that is, one in which the distribution of network chains is intentionally made multimodal. Bimodal elastomers are the simplest and most important example of this type, and have been much studied because they have unusually good mechanical properties, specifically simultaneously large values of the stress and extensibility. This, in turn, gives large energies for rupture, the standard measure of the toughness of a material. The use of these networks in both fundamental and applied studies is covered, using this as an occasion to describe mechanical property results in a variety of other mechanical deformations: biaxial extension, shear, torsion, tear, and cyclic deformations.

The use of neutron scattering measurements in the area of elastomeric materials is described in chapter 14. The coverage consists of the general features of the technique, theoretical treatments, and comparisons between theory and experiment.

"Bioelastomers," or elastomers in which the network chains are biopolymeric and produced by nature, are the subject of chapter 15. Again, this topic is inherently interesting, but it can also provide practical information; for example, how synthetic chemists might mimic nature in the ways it produces elastomeric materials.

The final chapter discusses a subject on which there has been relatively little molecular interpretation, namely the reinforcement of elastomeric networks by fillers. Some theoretical approaches are described, but most of the information concerns experimental results on elastomers in which the reinforcing phase is generated in situ through a suitably chosen chemical reaction. The justification offered is that under these conditions it may be possible to obtain relatively simple tributions at scales below the chain-length level. These topics, within the Gaussian chain formalism, are discussed in chapters 3, 4, and 8.

One important prediction of the molecular theories, especially the more modern ones, is the dependence of the modulus on strain, and this is discussed in chapter 5. The successful description of the stress–strain relationship that has been achieved is one of the major achievements of the molecular theories, with good agreement between theory and experiment being achieved over a twentyfold region of tensile and compressive strains.

The type of deformation most widely used to characterize elastomers in the stress–strain experiments just mentioned is simple elongation. It is the easiest to impose, and seems to be the easiest to think about intuitively. It is for these reasons that it is discussed so frequently throughout the book. However, networks also deform by imbibing solvent, that is, by swelling, to yield fragile but nonetheless solid materials called "gels." As might be anticipated, for their characterization they require some of the methodology used to study polymer solutions. The swelling process and the properties of the resulting gels are discussed in chapters 6 and 7. More specifically, chapter 6 provides a detailed analysis of the chemical potential of swollen networks, paving the way to the study of their large-scale volume changes in phase transitions, as presented in chapter 7. Some additional, less common types of deformation, such as biaxial extension, torsion, and shear, are described elsewhere, particularly in chapters 13 and 16.

Departures from Gaussian behavior may become significant in networks having unusually short chains, such as those intentionally introduced into some of the well-characterized, end-linked model networks discussed in chapters 10 and 13. Also, studies of rubbers at the segmental or the subchain level require structural information beyond that of the Gaussian chain. For these reasons, chapter 8 goes beyond the Gaussian chain model to discuss the properties of the isolated chain from the viewpoint of the rotational isomeric state (RIS) formalism. This approach takes into account the usual structural features of interest in any molecule (i.e., bond lengths, bond angles, and the locations and energies of bond rotational states). Specific applications of the RIS scheme to problems of rubber elasticity are given later in the chapter, along with the expressions of the end-to-end vector distribution functions for shorter chains. Some of the molecular dynamics simulations mentioned earlier are discussed at the end of chapter 8. Several controversial issues of the 1970s and the 1980s may, in principle be, resolved by such simulations. This technique should therefore not be underestimated, even though the size of the networks that may be treated by molecular dynamics is, at present, very small. Some initial attempts to model the reinforcement of elastomers by the incorporation of fillers are also described in chapter 8.

Although the most important relationship in rubberlike elasticity is that between stress and strain, the relationships between stress, strain, and temperature are also of great interest. They can be used, for example, to resolve elastomeric quantities into entropic and energetic components, and to test some of the major postulates of the molecular theory described in chapter 2. This subject of network "thermoelasticity" is therefore covered separately, in chapter 9.

1

Overview and Some Fundamental Information

This chapter is a brief overview of the topics treated in the book. It is aimed, in particular, at providing some qualitative information on rubber elasticity theories and their relationships to experimental studies, and at putting this material into context.

The following chapter describes in detail the classical theories of rubber elasticity, that is, the phantom and affine network theories. The network chains in the phantom model are assumed not to experience the effects of the surrounding chains and entanglements, and thus to move as "phantoms." Although this seems to be a very severe approximation, many experimental results are not in startling disagreement with theories based on this highly idealized assumption. These theories associate the total Helmholtz free energy of a deformed network with the sum of the free energies of the individual chains—an important assumption adopted throughout the book. They treat the single chain in its maximum simplicity, as a Gaussian chain, which is a type of "structureless" chain (where the only chemical constitution specified is the number of bonds in the network chain). In this respect, the classical theories focus on *ideal* networks and, in fact, are also referred to as "kinetic" theories because of their resemblance to ideal gas theories. Chain flexibility and mobility are the essential features of these models, according to which the network chains can experience all possible conformations or spatial arrangements subject to the network's connectivity. One of the predictions of the classical theories is that the elastic modulus of the network is independent of strain. This results from the assumption that only the entropy at the chain level contributes to the Helmholtz free energy. Experimental evidence, on the other hand, indicates that the modulus decreases significantly with increasing tension or compression, implicating interchain interactions, such as entanglements of some type or other. This has led to the more modern theories of rubber elasticity, such as the constrained-junction or the slip-link theories, which go beyond the single-chain length scale and introduce additional entropy to the Helmholtz free energy at the subchain level. This approach is supported by recent computer simulations on networks, using molecular dynamics, which are in support of entropic con-

2

Classical Theories of Rubber Elasticity

This chapter describes three molecular theories of rubber elasticity. Section 2.1 outlines the elementary theory of Kuhn[1–3] and Treloar[4,5], which is of particular importance since it presents the basic elements of rubberlike elasticity in a very transparent way. Section 2.2 presents the phantom network model developed by James and Guth[6–13], and section 2.3 presents the affine network model developed by Wall and Flory[14–20]. Historical aspects of the theories have been given in an article by Guth and Mark[21], and in a book prepared as a memorial to Guth[22]. Finally, the major features of both theories are briefly summarized in a review[23].

Separately, the James-Guth theory has been reviewed by Guth[24] and by Flory[25], and the phantom network model of section 2.2 is based on the Flory treatment. The affine network model has been described in detail in Flory's 1953 book[26]. This model is described in section 2.3 by generalizing the phantom network model (as was done in one of Flory's subsequent studies)[25].

The simple, elementary statistical theory described in section 2.1 paved the way to the current understanding of rubber elasticity. Further progress in the understanding of rubberlike systems was possible, however, only as a result of the two more precise and accurate theories: the phantom network and the affine network theories. Despite their differences, these two theories and the corresponding molecular models have served as basic reference points in this area for more than four decades. They still serve this purpose for the interpretation and explanation of experimental data. The differences between the assumptions and the predictions of the two models have led to serious disagreements during their development, as may be seen from the original papers cited earlier. The main point of disagreement was the magnitude of the front factor that appeared in the expression for the elastic free energy and the stress. For tetrafunctional networks, the James-Guth phantom network theory predicts one-half the value of the front factor obtained by the Wall-Flory affine network theory. As will be discussed in greater detail, in later chapters as well as in this one, the picture has become clearer and it is now well established that the phantom network model is the appropriate one for swollen networks. For networks in the unswollen state, the problem is still not unambiguously settled, and there is disagreement among researchers as to the nature of intermolecular correlations contributing to the elastic free energy of the network at equilibrium. According to one group of researchers, the action of entanglements in the unswollen network can ultimately suppress fluctuations of the junctions, and the network then approaches the Wall-Flory model. Others maintain that entanglements further contribute as chemical junctions, and therefore increase the elastic free energy and the modulus of the network above that of the affine network. These questions will be addressed in further detail in chapters 3 and 4 in relation to different entanglement theories of rubber elasticity.

In the final section of this chapter, the basic features of the Dean-Edwards treatment[27] are qualitatively outlined, and several other theories related to the classical theories of elasticity are discussed. The theory of Dean and Edwards unifies the physical picture behind the James-Guth and Wall-Flory theories and may be regarded as constituting the starting point of some of the most modern theories of rubberlike elasticity.

2.1 The Kuhn-Treloar Theory

The theory of Kuhn, as improved subsequently by Treloar, is based on the following fundamental assumptions:

1. The network consists of ν freely-jointed Gaussian chains, where such a network chain is defined as a sequence of skeletal bonds lying between two junctions. The mean-square end-to-end dimensions of the ensemble of network chains in the undeformed network are the same as those for an ensemble of chains in the bulk, un-cross-linked state. The mean-square end-to-end dimensions of the latter, in turn, are equal to those of the single chain in the unperturbed state, as given in eq. (F.9) of appendix F. This assumption is supported by neutron scattering experiments[5,28] by comparisons of chain dimensions in the bulk un-cross-linked and cross-linked states, and of those of an isolated chain.
2. There is no change in volume upon deformation (so long as crystallization does not occur). This assumption is supported by a large body of experimental evidence[29,30].
3. The junctions move affinely with macroscopic deformation. This assumption played a central role in the theories of rubberlike elasticity until neutron scattering experiments showed that the junctions are not rigidly embedded in the network and that their departure from affine displacement is substantial[31-33]. It should be noted that the affine assumption suppresses the fluctuations of the chain end points, but does not impose any constraints at points along the chain contour. In this respect, the chains do not interact with their environments along their contours. In current jargon, the chains are "phantomlike," although this terminology is usually reserved for the James-Guth theory.
4. The total elastic energy of the network is the sum of the elastic energies of the individual chains. Due to the assumption that the chains are freely jointed, all spatial arrangements are of the same energy, the network deformation is purely entropic, and the relation $\Delta A_{el} = \Delta E - T\Delta S$ becomes $\Delta A_{el} = -T\Delta S$. The treatment may be generalized to non-freely-jointed chains, however. Following later discussions by Volkenstein[34] and Flory[25], the entropy in the present treatment is generalized to the Helmholtz free energy or the elastic free energy.

CLASSICAL THEORIES OF RUBBER ELASTICITY

The elastic free energy of an isolated deformed Gaussian chain with its two ends fixed at **r** is given by eq. (F.14). Summing this equation over all chains of the network, the change ΔA_{el} in the total elastic energy at constant temperature (relative to that of the undeformed state) is obtained as

$$\Delta A_{el} = \frac{3kT}{2\langle r^2\rangle_0}\sum_{\nu}(r^2 - \langle r^2\rangle_0)$$
$$= \frac{3}{2}\nu kT\left(\frac{\langle r^2\rangle}{\langle r^2\rangle_0} - 1\right) \quad (2.1)$$

In going from the first to the second equality in eq. (2.1), the relationship $\langle r^2\rangle = \sum_{\nu} r^2/n$ is used to represent the average of the squared end-to-end vectors. Writing the end-to-end vector in terms of the Cartesian components and averaging over the ensemble of chains leads to

$$\langle r^2\rangle = \langle x^2\rangle + \langle y^2\rangle + \langle z^2\rangle \quad (2.2)$$

Dividing both sides of this equation by $\langle r^2\rangle_0$, using the isotropy of the network chains in the undeformed state, and using the assumption (see section 11.2) that the chain ends are displaced in proportion to the macroscopic strain, leads to

$$\langle r^2\rangle/\langle r^2\rangle_0 = (\lambda_x^2 + \lambda_y^2 + \lambda_z^2)/3 \quad (2.3)$$

Here, λ_x, λ_y, and λ_z are the components of the deformation tensor λ, defined as the ratio of the final length to the initial length, in each coordinate direction. (More detailed discussion of the state of deformation is given in appendix C.) Substitution of eq. (2.3) into eq. (2.2) leads to

$$\Delta A_{el} = (1/2)\nu kT(\lambda_x^2 + \lambda_x^2 + \lambda_x^2 - 3) \quad (2.4)$$

It should be noted that intermolecular interactions are zero in this model; that is, the system is essentially like an ideal gas. The expression for the force f acting on a prismatic sample under uniaxial tension along the x direction is obtained from the thermodynamic expression [eq. (D.8)]:

$$f = \left(\frac{\partial \Delta A_{el}}{\partial L}\right)_{T,V} = L_0^{-1}\left(\frac{\partial \Delta A_{el}}{\partial \lambda}\right)_{T,V} \quad (2.5)$$

where $\lambda = \lambda_x = L/L_0$ (with L and L_0 denoting the lengths of the sample along the direction of stretch in the deformed and undeformed states, respectively). Following the assumption that the volume of the sample remains constant during deformation, the y and z components of the deformation are written as $\lambda_y = \lambda_z = \lambda^{-1/2}$. Substituting eq. (2.4) into eq. (2.5) and performing the differentiation indicated leads to the elastic equation of state for the force:

$$f = \left(\frac{\nu kT}{L_0}\right)(\lambda - 1/\lambda^2) \quad (2.6)$$

Equations (2.4) and (2.6) are the most important achievements of the elementary theory of rubber elasticity. It should be emphasized that although the model rests on assumptions such as the simple additivity of the free energies of the individual chains and the lack of intermolecular interactions, its predictions approximate experimental data within reasonable bounds, as will be outlined in more detail in later chapters.

2.2 The Phantom Network Theory of James and Guth

2.2.1 General Aspects

The James-Guth theory is based on the following assumptions:

1. The network chains are Gaussian.
2. Some of the junctions at the surface of the networks are fixed and deform affinely with the macroscopic strain.
3. The chains are subject only to constraints that arise directly from the connectivity of the network. The effects of junctions and chains on one another is of no consequence, and the effect of the macroscopic strain is transmitted to a chain through the junctions to which a chain is attached at its two ends. This characteristic of a phantom network holds at all deformations.

The second assumption, according to which the macroscopic constraints are introduced by fixing microscopic variables, has been criticized by Edwards and Freed[35], Freed[36], and Flory†. Placing the fixed junctions at the surface has no significance and they could be inside the body instead. Fixing a subset of microscopic variables, however, is not strictly a rigorous approach in statistical mechanics[25]. Nevertheless, the theory based on these assumptions leads to significant improvements in the understanding of the properties of networks, such as microscopic fluctuations and neutron scattering behavior.

As a result of the third assumption, the configurational partition function Z_N of the network may be written as the product of the configurational partition functions of its individual chains. As described elsewhere[25], each of the latter partition functions is fully determined by the end-to-end vector $\mathbf{r}_{ij}$ for the chain connecting junctions i and j. Thus,

$$Z_N = \prod_{i<j} \tilde{Z}_{\mathbf{r}_{ij}} = Z^\nu \prod_{i<j} W(\mathbf{r}_{ij}) \qquad (2.7)$$

where $\tilde{Z}_{\mathbf{r}_{ij}}$ is the partition function for the chain with end-to-end vector $\mathbf{r}_{ij}$, and Z is the partition function corresponding to the free chain, and W is defined in eq. (2.8). Further description is given in appendix D. The second equality in eq. (2.7) follows from the fact that the distribution function is equal to the ratio of the partition functions $\tilde{Z}_{\mathbf{r}_{ij}}$ and Z. The products in eq. (2.7) include all pairs of junctions connected by a chain. The distribution of the instantaneous chain vec-

---

†A manuscript very similar to the 1976 paper[25] was submitted for publication to the *Journal of Chemical Physics* by Flory in 1964 and was widely circulated at that time. It was subsequently withdrawn from consideration for publication. The popular term "phantom network" first appeared in that manuscript.

tors in the undeformed network is identical to that of a free chain. This distribution may be thought of by conceptually freezing the network instantaneously and counting the number of chains with end-to-end vectors of specified magnitude and direction. In reality, the instantaneous vectors for individual chains of the network are not the same as those of the un-cross-linked melt because they are altered during the process of cross-linking. However, as stated by Flory[26], these alterations occur at random and the overall distribution of the chains remains unaltered. Thus, the instantaneous density distribution $W(\mathbf{r}_{ij})$ of the network chains is given by the Gaussian function [see eq. (F.13)] as

$$W(\mathbf{r}_{ij}) = \left(\frac{\gamma}{\pi}\right)^{3/2} \exp(-\gamma r_{ij}^2) \tag{2.8}$$

where

$$\gamma = \frac{3}{2\langle r_{ij}^2\rangle_0} \tag{2.9}$$

Substituting eq. (2.8) into eq. (2.7), and denoting the positions of junctions i and j by $\mathbf{R}_i$ and $\mathbf{R}_j$, eq. (2.7) may be written

$$Z_N = C \prod_{i<j} \exp\left(-\frac{1}{2}\sum_i\sum_j \gamma_{ij}^* |\mathbf{R}_i - \mathbf{R}_j|^2\right) \tag{2.10}$$

Here, $|\mathbf{R}_i - \mathbf{R}_j|^2$ is the square of the magnitude of the chain vector $\mathbf{r}_{ij}$, and $\gamma_{ij}^* = 3/(2\langle r_{ij}^2\rangle_0)$ if i and j are connected by a chain, and zero otherwise. Lowercase symbols denote chain vectors and uppercase letters denote positions of points in the network. The specification of γ_{ij}^* thus introduces the structure and topology of the network into the statistical theory, an aspect that is missing in the Kuhn-Treloar theory. Equation (2.10) expresses the partition function in terms of the positions of junction points rather than in terms of the chain vectors.

The position of the network may be fixed in space by locating one of the junctions at the origin of a laboratory-fixed coordinate system and measuring all other $\mathbf{R}_i$ values relative to this system. The position vectors $\mathbf{R}_i$ (with i ranging from 1 to the total number of junctions, μ) may be arranged in column form, represented as $\{\mathbf{R}\}$. Equation (2.10) may then be written

$$Z_N = C\exp(-\{\mathbf{R}\}^T \Gamma \{\mathbf{R}\}) \tag{2.11}$$

where the superscript T denotes the transpose. The elements γ_{ij} of the symmetric matrix Γ are obtained by expanding the quadratic form given by eq. (2.11) as

$$\Gamma = \begin{cases} \gamma_{ij} = -\gamma_{ij}^*, & i \neq j \\ \gamma_{ii} = \sum_j \gamma_{ij}^* = \sum_j \gamma_{ji}^* \end{cases} \tag{2.12}$$

In expanding eq. (2.11), products such as $\mathbf{R}_i\mathbf{R}_j$ are to be treated as scalar products. The matrix Γ completely describes the connectivity of the network. If all chains of the network are identical and are characterized by the same value of $\langle r_{ij}^2\rangle_0$, then all nonzero elements of Γ are equal and the latter reduces to the Kirchoff adjacency matrix adopted by Eichinger[37] in the study of perfect phantom networks by graph theory.

A small set $\{\sigma\}$ of the μ network junctions are assumed to be fixed as required by the second assumption stated previously. These may be thought to be junctions at the surface of the network, describing the size and the shape of the network and directly subject to macroscopic manipulations. The remaining set $\{\tau\}$ of junctions are assumed to be free to fluctuate. The two sets of junctions may now be used to partition the quadratic form of eq. (2.11) as

$$\{R\}^T \Gamma \{R\} = \begin{pmatrix} \{R_\sigma\} \\ \{R_\tau\} \end{pmatrix}^T \begin{pmatrix} \Gamma_\sigma & \Gamma_{\sigma\tau} \\ \Gamma_{\tau\sigma} & \Gamma_\tau \end{pmatrix} \begin{pmatrix} \{R_\sigma\} \\ \{R_\tau\} \end{pmatrix} \qquad (2.13)$$

or equivalently as

$$\{R\}^T \Gamma \{R\} = \{R_\sigma\}^T \Gamma_\sigma \{R_\sigma\} + 2\{R_\tau\}^T \Gamma_{\tau\sigma} \{R_\sigma\} + \{R_\tau\}^T \Gamma_\tau \{R_\tau\} \qquad (2.14)$$

where $\{R_\sigma\}$ and $\{R_\tau\}$ are the column vectors for the position vectors of the fixed and free junctions, respectively. The quantity Γ_σ is a square matrix comprising rows and columns of Γ for fixed junctions. Analogously, Γ_τ is a square matrix characterizing free junctions, and $\Gamma_{\tau\sigma} = \Gamma_{\sigma\tau}^T$ is the rectangular matrix whose rows correspond to free junctions and its columns to fixed junctions. Equation (2.14) may be rearranged as a sum of two terms, one containing elements for the free junctions and the other for the fixed junctions:

$$\{R\}^T \Gamma \{R\} = \{R_\sigma\}^T G_\sigma \{R_\sigma\} + \{\Delta R_\tau\}^T \Gamma_\tau \{\Delta R_\tau\} \qquad (2.15)$$

where

$$G_\sigma = \Gamma_\sigma - \Gamma_{\sigma\tau} \Gamma_\tau^{-1} \Gamma_{\tau\sigma} \qquad (2.16)$$

$$\{\Delta R_\tau\} = \{R_\tau\} - \{\bar R_\tau\} \qquad (2.17)$$

$$\{\bar R_\tau\} = -\Gamma_\tau^{-1} \Gamma_{\tau\sigma} \{R_\sigma\} \qquad (2.18)$$

The set of variables defined by eq. (2.17) indicates the fluctuations of the free junctions from their equilibrium positions. Equation (2.18) states that the mean positions of the free junctions are linear functions of the positions of the fixed junctions, and therefore transform affinely with macroscopic strain.

In order to have a feeling for the partitioning given in eq. (2.15), one may treat the variables as scalars by writing the proportionality $\bar R_\tau = A R_\sigma$, substituting into eq. (2.14), collecting terms, and comparing the result with eq. (2.15). This will give $A = -\Gamma_{\tau\sigma}/\Gamma_\tau$ and $G_\sigma = \Gamma_\sigma - \Gamma_{\sigma\tau}^2/\Gamma_\tau$. Repeating the same operations, while observing noncommutativity of the matrices, results in eqs. (2.15)–(2.18).

Substitution of eq. (2.15) into (2.11) yields

$$Z_N = C \exp(-\{R_\sigma\}^T G_\sigma \{R_\sigma\}) \exp(-\{\Delta R_\tau\}^T \Gamma_\tau \{\Delta R_\tau\}) \qquad (2.19)$$

Equation (2.19) thus leads to the partition function where terms relating to fixed and free variables are separated. The two products of the exponentials in eq. (2.19) may be defined as

$$Z_{N\sigma} = C' \exp(-\{\mathbf{R}_\sigma\}^T \mathbf{G}_\sigma \{\mathbf{R}_\sigma\}) \qquad (2.20)$$

$$Z_{N\tau} = C'' \exp(-\{\Delta\mathbf{R}_\tau\}^T \mathbf{\Gamma}_\tau \{\Delta\mathbf{R}_\tau\}) \qquad (2.21)$$

The product of the fluctuations of two junctions i and j, averaged over the network, may be obtained from eq. (2.21) as[23,38,39]

$$\langle \Delta\mathbf{R}_i \cdot \Delta\mathbf{R}_j \rangle = \frac{\int \Delta\mathbf{R}_i \cdot \Delta\mathbf{R}_j \exp[-\{\Delta\mathbf{R}_\tau\}^T \mathbf{\Gamma}_\tau \{\Delta\mathbf{R}_\tau\}] d\{\Delta\mathbf{R}_\tau\}}{\int \exp[-\{\Delta\mathbf{R}_\tau\}^T \mathbf{\Gamma}_\tau \{\Delta\mathbf{R}_\tau\}] d\{\Delta\mathbf{R}_\tau\}} \qquad (2.22)$$

$$= -\frac{\partial}{\partial \gamma_{ij}} \ln Z_\tau \qquad (2.23)$$

where

$$d\{\Delta\mathbf{R}_\tau\} \equiv d\Delta\mathbf{R}_{1\tau} \, d\Delta\mathbf{R}_{2\tau} \cdots d\Delta\mathbf{R}_{\mu\tau} \qquad (2.24)$$

and

$$Z_\tau = \int \exp[-\{\Delta\mathbf{R}_\tau\}^T \mathbf{\Gamma}_\tau \{\Delta\mathbf{R}_\tau\}] d\{\Delta\mathbf{R}_\tau\} = \left(\frac{\pi^{\mu_\tau}}{\det \mathbf{\Gamma}_\tau}\right)^{3/2} \qquad (2.25)$$

Equation (2.23) is an identity obtained by noting that the denominator of eq. (2.22) is Z_τ, and the Gaussian property of Z_τ is used in obtaining the right-hand side of eq. (2.25). Using eq. (2.25) in eq. (2.23) leads to

$$\langle \Delta\mathbf{R}_i \cdot \Delta\mathbf{R}_j \rangle = \frac{3}{2} \frac{\partial}{\partial \gamma_{ij}} \ln |\det \mathbf{\Gamma}_\tau| = \frac{3}{2} (\mathbf{\Gamma}_\tau^{-1})_{ij} \qquad (2.26)$$

where $(\mathbf{\Gamma}_\tau^{-1})_{ij}$ denotes the ijth element of the inverse $\mathbf{\Gamma}_\tau^{-1}$ of $\mathbf{\Gamma}_\tau$. The matrix $\mathbf{\Gamma}_\tau$ involves only structural variables of the network and otherwise is independent of macroscopic deformation. This observation leads to the important conclusion that fluctuations of junctions from their mean positions in a phantom network are independent of macroscopic deformation. More information concerning the fluctuation of various quantities in the phantom network may be extracted from eqs. (2.20)–(2.26). These are summarized in appendix E.

The conclusions that the mean positions of junctions transform affinely with macroscopic strain and that their fluctuations are Gaussian and independent of macroscopic strain gives the following expression for the transformation of the chain vector $\mathbf{r}_{ij}$ connecting two junctions i and j:

$$\mathbf{r}_{ij} = \lambda \bar{\mathbf{r}}_{ij} + \Delta\mathbf{r}_{ij} \qquad (2.27)$$

Here, $\bar{\mathbf{r}}_{ij}$ is the mean chain vector and $\Delta\mathbf{r}_{ij}$ is its fluctuation. The transformations of chain vectors $\mathbf{r}_{ij}$ are therefore nonaffine.

2.2.2 The Elastic Free Energy

2.2.2.1 General Approach In order to relate the statistical properties of the network to mechanical and thermodynamic quantities such as the force, energy, and entropy, the partition function must be averaged over the fluctuations of the free junctions. Integration over fluctuations of the μ_τ free junctions in eq. (2.19) leads to the configuration partition function of the network, as determined by the positions of the set $\{\sigma\}$ of fixed junctions (μ_σ in number). The result is

$$Z_{N,\sigma} = C\pi^{3/2\mu_\tau} \det(\mathbf{\Gamma}_\tau)^{-3/2} \exp(-\{\mathbf{R}_\sigma\}^T \mathbf{G}_\sigma \{\mathbf{R}_\sigma\}) \quad (2.28)$$

The elastic free energy ΔA_{el} of the network is obtained by substituting eq. (2.28) into eq. (D.8) of appendix D, resulting in

$$\Delta A_{el} = \Delta A^*(T) + kT\{\mathbf{R}_\sigma\}^T \mathbf{G}_\sigma \{\mathbf{R}_\sigma\} \quad (2.29)$$

The set of vectors specified by $\{\mathbf{R}_\sigma\}$ represents the positions of the fixed junctions, which transform affinely with macroscopic strain. Denoting the position of the fixed junction i in the undeformed state by $\mathbf{R}^0_{\sigma,i}$ and using the relation $\mathbf{R}_{\sigma,i} = \lambda \mathbf{R}^0_{\sigma,i}$ for the position of the junction in the deformed state (which holds for all fixed junctions i), eq. (2.29) may be written as

$$\Delta A_{el} = (1/3)kT[\{\mathbf{R}^0_\sigma\}^T \mathbf{G}_\sigma \{\mathbf{R}^0_\sigma\}](\lambda_x^2 + \lambda_y^2 + \lambda_z^2 - 3) \quad (2.30)$$

The factor in the square brackets in eq. (2.30) depends only on network structure. Its evaluation in closed form seems to be difficult, but one may use micronetworks[40,41] together with computer Monte Carlo simulations for its evaluation. In the following paragraphs, the value of this factor for a unimodal phantom network will be obtained by considering the statistics of forming a network with ν chains and μ junctions.

The instantaneous configuration of a network chain between junctions i and j is shown in figure 2.1. The quantity $\mathbf{r}_{ij}$ is the instantaneous end-to-end vector, and $\bar{\mathbf{r}}_{ij}$ represents the time average of $\mathbf{r}_{ij}$ over all possible configurations that may be taken by the chain in the network. The points and $\bar{i}$ and $\bar{j}$ represent the time-averaged locations of the junctions i and j, respectively. The quantities $\Delta \mathbf{R}_i$ and $\Delta \mathbf{R}_j$ indicate the instantaneous fluctuations of the respective junctions from their average locations. The vector $\mathbf{r}_{ij}$ may be written in terms of its average and fluctuations from this average as

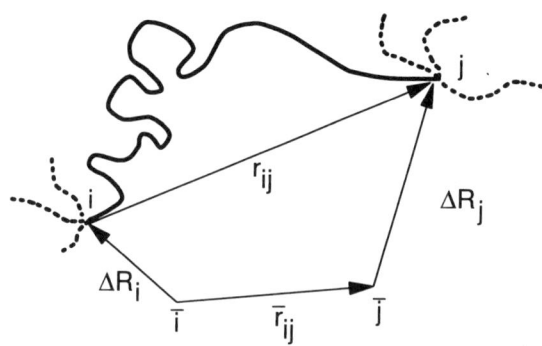

Figure 2.1 The instantaneous configuration of a network chain between junctions i and j. The variables shown are defined in the text.

$$\mathbf{r}_{ij} = \bar{\mathbf{r}}_{ij} + \Delta\mathbf{r}_{ij} = \bar{\mathbf{r}}_{ij} + \Delta\mathbf{R}_j - \Delta\mathbf{R}_i \tag{2.31}$$

where $\Delta\mathbf{r}_{ij} = \Delta\mathbf{R}_j - \Delta\mathbf{R}_i$ represents the instantaneous fluctuation in the chain vector $\mathbf{r}_{ij}$ from its average value $\bar{\mathbf{r}}_{ij}$.

The instantaneous distribution of the mean vectors $\bar{\mathbf{r}}_{ij}$ in the network and fluctuations $\Delta\mathbf{r}_{ij}$ are represented by $\chi(\bar{\mathbf{r}}_{ij})$ and $\Psi(\Delta\mathbf{r}_{ij})$, respectively. They are related to $W(\mathbf{r}_{ij})$ by the convolution expression

$$\begin{aligned} W(\mathbf{r}_{ij}) &= \chi(\bar{\mathbf{r}}_{ij}) * \Psi(\Delta\mathbf{r}_{ij}) \\ &\equiv \int \chi(\bar{\mathbf{r}}_{ij}) \Psi(\Delta\mathbf{r}_{ij}) d\bar{\mathbf{r}}_{ij} \end{aligned} \tag{2.32}$$

Here, the asterisk denotes the convolution expressed by integration over $d\bar{\mathbf{r}}_{ij}$. As will be shown below, the distribution $\Psi(\Delta\mathbf{r}_{ij})$, like $W(\mathbf{r}_{ij})$, is Gaussian for a phantom network of Gaussian chains. One may show, by straightforward integration, that if two of the distribution functions are Gaussian in eq. (2.32), then the third is also Gaussian. The distribution $\chi(\bar{\mathbf{r}}_{ij})$ of the mean vectors in the phantom network is therefore also Gaussian. The distributions for $\chi(\bar{\mathbf{r}}_{ij})$ and $\Psi(\Delta\mathbf{r}_{ij})$ may be expressed as

$$\Psi(\Delta\mathbf{r}_{ij}) = \left(\frac{\psi}{\pi}\right)^{3/2} \exp(-\psi(\Delta\mathbf{r}_{ij})^2)$$
$$\chi(\bar{\mathbf{r}}_{ij}) = \left(\frac{X}{\pi}\right)^{3/2} \exp(-\chi\bar{\mathbf{r}}_{ij}^2) \tag{2.33}$$

where

$$\psi = \frac{3}{2\langle(\Delta r_{ij})^2\rangle}$$
$$X = \frac{3}{2\langle(\bar{r}_{ij})^2\rangle} \tag{2.34}$$

Replacing (Δr_{ij}^2) in eq. (2.33) by the square of the magnitude of $\mathbf{r}_{ij} - \bar{\mathbf{r}}_{ij}$ and performing the integration leads to the following relationship between the parameters γ, χ, and ψ:

$$1/\psi + 1/X = 1/\gamma \tag{2.35}$$

which is equivalent to

$$\langle(r_{ij})^2\rangle = \langle(\bar{r}_{ij})^2\rangle + \langle(\Delta r_{ij})^2\rangle \tag{2.36}$$

Thus, the magnitudes of the instantaneous and average chain vectors and their fluctuations are not independent, but are related through eq. (2.36).

For a phantom network of junction functionality ϕ, the mean chain dimensions and fluctuations are related to network chain dimensions by the relations:

$$\langle\bar{r}_{ij}^2\rangle = \left(1 - \frac{2}{\phi}\right)\langle r^2\rangle_0 \tag{2.37}$$

16 STRUCTURES AND PROPERTIES OF RUBBERLIKE NETWORKS

and

$$\langle \Delta r_{ij}^2 \rangle = \frac{2}{\phi} \langle r^2 \rangle_0 \qquad (2.38)$$

Equations (2.37) and (2.38) follow from the work of Eichinger[37], Graessley[41], and Flory[25]. A detailed derivation is given in appendix E.

Formation of a network from individual polymer chains may take place in different ways, as described in chapter 1. Randomly cross-linking chains along their contours or end-linking chains with multifunctional junctions are the two techniques most frequently employed in practice. Conceptually, one may form a network by starting with a giant linear molecule and successively connecting it onto itself. This is the approach of Edwards[27] and was pursued in the later theories of Edwards and his collaborators. Alternatively, one may start with a giant acyclic tree, as is shown in figure A.1 of appendix A. Formation and deformation of a network in this manner is shown in figure 2.2. As shown in this figure, one may obtain the deformed phantom network following either the path 2, 3′, and 6, or the path 4, 5, and 6. In the latter case, the elastic free energy ΔA_{el} of the deformed phantom network relative to the undeformed network is

$$\Delta A_{el} = \Delta A_3 = \Delta A_4 + \Delta A_5 + \Delta A_6 - \Delta A_2 \qquad (2.39)$$

The elastic free energy for each of the stages will be described separately in this section.

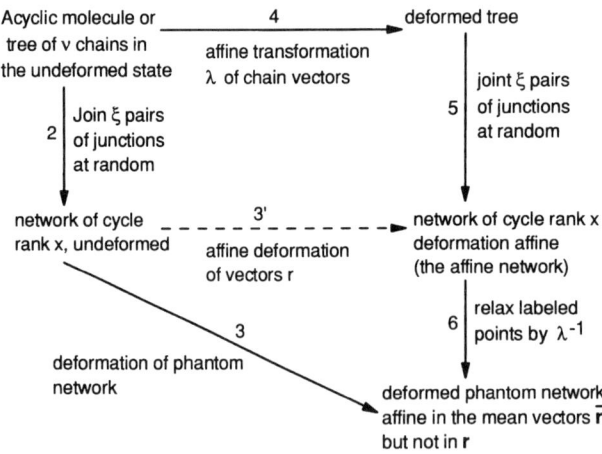

Figure 2.2 Formation and deformation of a network. One may obtain the deformed phantom network following either the path 2, 3′, and 6, or the path 4, 5, and 6. The affine network is obtained by following the path 2 and 3′ or 4 and 5

2.2.2.2 Calculation of ΔA_2

Two reactive groups must be within a volume element δV for reaction to be possible. Of primary interest is the probability that any given one of the ν labels has another label out of the $\nu - 1$ remaining ones that is situated within δV. This probability is proportional to $\delta V/V_0$, where V_0 is the volume of the reference state in which the network is formed. The probability of having ξ pairs of reactive sites that are situated within volume δV is proportional to $(\delta V/V^0)^\xi$. The number ξ of pairs of reactive sites is equal to the cycle rank discussed in appendix A. The constant of proportionality, not explicitly shown, is independent of deformation and is therefore of no importance in the present context. One may also ignore changes of free energy due to the chemical reaction taking place during the combination of the reactive centers, because they will be independent of deformation and cancel in calculating the total free energy change for deformation of the network. Thus, one may identify the probability $(\delta V/V^0)^\xi$ as the configurational partition function for converting the tree structure into a network of cycle rank ξ. The free energy ΔA_2 for this process follows then, as

$$\Delta A_2 = -\xi kT \ln(\delta V/V^0) \tag{2.40}$$

2.2.2.3 Calculation of ΔA_4

At this step, the network chains of the acyclic tree are transformed by the deformation λ. One first considers the group of chains having end-to-end vectors within the range $\mathbf{r}_i$ and $\mathbf{r}_i + d\mathbf{r}_i$ in the undeformed tree. Assume that there are ν_i chains in this ith group. In the deformed tree, the probability of occurrence of these chains in the stated interval is determined by the distribution function $(\det \lambda) W(\lambda \mathbf{r}_i) d\mathbf{r}_i$ and may be taken as Gaussian for sufficiently long chains. The probability of having all of the chains of the tree in the stated intervals is obtained as the product of the probabilities $(\det \lambda) W(\lambda \mathbf{r}_i) d\mathbf{r}_i$ for each group ν_i, or

$$\prod_j [(\det \lambda) W(\lambda \mathbf{r}_i) d\mathbf{r}_i]^{\nu_i} \tag{2.41}$$

However, there are $\nu!/\prod \nu_i!$ ways of combining the chains into the specified groups. The probability given by eq. (2.41) must therefore be multiplied by this number to yield the probability Ω_1 of having the stated distribution of chains in the deformed state:

$$\begin{aligned}\Omega_1 &= \nu! \prod [(\det \lambda) W(\lambda \mathbf{r}_i) d\mathbf{r}_i]^{\nu_i}/\nu_i! \\ &= \prod [(\nu/\nu_i)(\det \lambda) W(\lambda \mathbf{r}_i) d\mathbf{r}_i]^{\nu_i}\end{aligned} \tag{2.42}$$

The second line follows from use of Stirling's approximation of the factorial, that is, $\ln \nu! = \nu \ln \nu - \nu$. The ratio of Ω_1 in the deformed state to that in the undeformed state may be taken as the ratio of the configuration partition function $Z_\nu(\lambda)$ of the tree whose chains are deformed by λ to the configuration partition function Z_ν^0 of the undeformed one; that is,

18 STRUCTURES AND PROPERTIES OF RUBBERLIKE NETWORKS

$$Z_\nu(\lambda)/Z_\nu^0 = \prod [(\det \lambda) W(\lambda \mathbf{r}_i)/W(\mathbf{r}_i)]^{\nu_i}$$

$$= \exp\left\{-\sum \nu_i \gamma[(\lambda_x^2 x_i^2 + \lambda_y^2 y_i^2 + \lambda_z^2 z_i^2) - (x_i^2 + y_i^2 + z_i^2)]\right\}(V/V^0)^\nu$$

$$= \exp\{-\nu\gamma[(\lambda_x^2 - 1)\langle x^2\rangle_0 + (\lambda_y^2 - 1)\langle y^2\rangle_0 + (\lambda_z^2 - 1)\langle z^2\rangle_0]\}(V/V^0)^\nu \tag{2.43}$$

Here, the last line follows from the relationship of average chain dimensions in the deformed state to the undeformed ones, and $(\det \lambda)$ has been replaced by V/V^0. The elastic free energy ΔA_4 of deforming the tree follows then, from eq. (2.43), as

$$\Delta A_4 = (1/2)\nu k T (I_1 - 3) - \nu k T \ln(V/V^0) \tag{2.44}$$

where the first strain invariant I_1 is defined as

$$I_1 = \lambda_x^2 + \lambda_y^2 + \lambda_z^2 \tag{2.45}$$

2.2.2.4 Calculation of ΔA_5
Step 5 requires the joining of the ξ pairs of deformed chains. The change in free energy may be written in analogy with that in step 2, where now the reference volume is the deformed volume. Thus, one obtains, from eq. (2.40),

$$\Delta A_5 = -\xi k T \ln(\delta V/V) \tag{2.46}$$

2.2.2.5 Calculation of ΔA_6
In step 6, the fluctuations $\Delta \mathbf{R}$ of the $\nu - \xi$ labeled points are relaxed to their equilibrium distribution. This corresponds to the reversal of the part of the free energy change that took place during the affine deformation in step 4. Thus, eq. (2.43) must be written in inverted form for the fluctuations of the $\nu = \xi$ labeled points:

$$Z^0_{(\nu-\xi)}/Z_{(\nu-\xi)}(\lambda) = \prod [\Psi(\Delta \mathbf{r}_i)/(\det \lambda)\Psi(\lambda \Delta \mathbf{r}_i)]^{\nu_i} \tag{2.47}$$

where the denominator on the right-hand side indicates that the initial state of the junctions is the affinely deformed one. Performing the substitutions for the distribution $\Psi(\Delta \mathbf{r}_i)$, the elastic free energy is obtained as

$$\Delta A_6 = -(1/2)(\nu - \xi)kT(I_1 - 3) + (\nu - \xi)kT\ln(V/V^0) \tag{2.48}$$

2.2.2.6 Calculation of the Total Elastic Free Energy for the Phantom Network
Substituting eqs. (2.40), (2.44), (2.46), and (2.48) into eq. (2.39) leads to the elastic free energy $\Delta A_{el,ph}$ of the phantom network:

$$\Delta A_{el,ph} = (1/2)\xi k T (I_1 - 3) \tag{2.49}$$

One sees that the volume-dependent term $\ln(V/V_0)$ is not present in the final expression for the elastic free energy of the phantom network. This term was a point of controversy during the development of the affine and phantom theories since it appeared in the former but not the latter (see, e.g., Wall and Flory[20],

James and Guth[13]). The cancelation of this term takes place in adding the change in elastic free energy in the 6th step to the changes in steps 4 and 5. The elastic free energy given by eq. (2.49) is valid for imperfect as well as for perfect unimodal networks. The appearance of a single parameter (ξ) rather than two (such as the two Lame constants in classical linear elasticity theory) results from the assumption of incompressibility that is built into the theory in canceling the $\ln(V/V_0)$ term. For ordinary rubberlike materials, the changes in volume are indeed second order compared with changes in linear dimensions and may safely be neglected, as has been pointed out by Treloar (ref. 5, p. 67 and 68).

2.3 The Affine Network Theory of Wall and Flory

In the affine network model, it is assumed that the junctions of the network are embedded in the elastic continuum and transform affinely with macroscopic strain. These ideas follow from the earlier picture given in the pioneering work on networks by Kuhn[1-3]. No assumptions are imposed with regard to the behavior of the points between the junctions. Examination of figure 2.2 indicates that path $3'$ corresponds to the elastic free energy change ΔA_{af} of an affine network. One may therefore express the free energy as

$$\Delta A_{el,af} = \Delta A_{3'} = \Delta A_4 + \Delta A_5 - \Delta A_2 \qquad (2.50)$$

Substituting from eqs. (2.40), (2.44), and (2.46), leads to the expression

$$\Delta A_{el,af} = (1/2)\nu kT(I_1 - 3) - (2\nu/\phi)kT\ln(V/V^0) \qquad (2.51)$$

Equation (2.51) is the free energy expression corresponding to the Wall-Flory theory. The presence of the volume term shows that the material is not assumed to be incompressible. Neglecting the volume term and substituting for ν converts eq. (2.51) into

$$\Delta A_{el,af} = \frac{\phi}{2(\phi - 2)}\xi kT(I_1 - 3) \qquad (2.52)$$

Comparison of this expression with eq. (2.49) for the elastic free energy of a phantom network shows that the latter is always smaller than the elastic free energy of the affine network. For a tetrafunctional network (which is the case treated most extensively in the literature), the elastic free energy of the phantom network equals one-half of the elastic free energy of the affine network. Equation (2.52) is valid for imperfect as well as perfect networks.

2.4 The Edwards Approach and Other Theories

A unifying statistical mechanical treatment of rubber elasticity has been given by Deam and Edwards[27], in which the phantom and the affine network models emerge as limiting cases. The paper is based on the previous work of Edwards[42,43] on amorphous systems in general. The amorphous rubber is taken as a solid whose topology or connectivity is conserved during fabrication; namely, network formation. The theory recognizes different topologies, calculates their probabilities, and formulates the configurational partition function in the undeformed and the deformed states. The theory is more general than the James-Guth and the Wall-Flory theories because it includes the interactions between chains arising from the deformation of the topologically connected structure. These interactions arise from the discrete entanglements between chains and form the basis of entanglement theories, which are treated in chapters 3 and 4. In the absence of discrete entanglements, Edwards' theory reduces to either the phantom or the affine network models depending on the specific assumption made regarding the density of the polymer confined in a box in the model.

The result that the front factor of the elastic free energy should be $\nu kT/2$ for a tetrafunctional network has also been obtained by Duiser and Staverman[44], Eichinger[37], and Graessley[40,41]. On the other hand, the theory by Hermans[45,46] leads to a front factor of νkT for tetrafunctional networks. Further information on the other theories may be found in the review articles by Staverman[47], Eichinger[48], Edwards and Vilgis[49], Heinrich et al.[50], and Erman and Mark[51].

References

(1) Kuhn, W. *J. Polym. Sci.* 1946, 1, 380.
(2) Kuhn, W. *Kolloid Z.* 1936, 76, 258.
(3) Kuhn, W. *Angew. Chem., Int. Ed. Eng.* 1938, 51, 640.
(4) Treloar, L. R. G. *Trans. Faraday Soc.* 1946, 42, 77.
(5) Treloar, L. R. G. *The Physics of Rubber Elasticity*, 3rd ed, Clarendon Press: Oxford. 1975.
(6) James, H. M.; Guth, E. *Ind. Eng. Chem.* 1941, 33, 624.
(7) James, H. M.; Guth, E. *Ind. Eng. Chem.* 1942, 34, 1365.
(8) James, H. M.; Guth, E. *J. Chem. Phys.* 1943, 10, 455.
(9) James, H. M.; Guth, E. *J. Appl. Phys.* 1944, 15, 294.
(10) James, H. M. *J. Chem. Phys.* 1947, 15, 651.
(11) James, H. M.; Guth, E. *J. Chem. Phys.* 1947, 15, 669.
(12) James, H. M.; Guth, E. *J. Polym. Sci.* 1949, 4, 153.
(13) James, H. M.; Guth, E. *J. Chem. Phys.* 1953, 21, 1039.
(14) Wall, F. T. *J. Chem. Phys.* 1942, 10, 132.
(15) Wall, F. T. *J. Chem. Phys.* 1942, 10, 485.
(16) Wall, F. T. *J. Chem. Phys.* 1943, 11, 527.
(17) Flory, P. *J. Chem. Revs.* 1944, 35, 51.
(18) Flory, P. *J. Ind. Eng. Chem.* 1946, 38, 417.
(19) Flory, P. *J. J. Chem. Phys.* 1950, 18, 108.
(20) Wall, F. T.; Flory, P. J. *J. Chem. Phys.* 1951, 19, 1435.
(21) Guth, E.; Mark, H. F. *J. Polym. Sci., Polym. Phys. Ed.* 1991, 29, 627.
(22) Mark, J. E.; Erman, B., Eds. *Elastomeric Polymer Networks.* Prentice Hall: Englewood Cliffs, NJ. 1992.
(23) Kloczkowski, A. In *Physical Properties of Polymers Handbook*, J. E. Mark, Ed.; American Institute of Physics Press: Woodbury, NY. 1996; p. 61.
(24) Guth, E. *J. Polym. Sci.: Part C* 1970, 31, 267.
(25) Flory, P. J. *Proc. Roy. Soc. London, A* 1976, 351, 351.
(26) Flory, P. J. *Principles of Polymer Chemistry.* Cornell University Press: Ithaca, NY. 1953.

(27) Deam, R. T.; Edwards, S. F. *Phil. Trans. Roy. Soc. London* 1976, 280, 317.
(28) Cotton, J. P.; Decker, D.; Benoit, H.; Farnoux, B.; Higgins, J.; Jannink, G.; Ober, R.; Picot, C.; des Cloizeaux, J. *Macromolecules* 1974, 7, 863.
(29) Christensen, R. G.; Hoeve, C. A. J. *J. Polym. Sci.* 1970, A1, 8, 1503.
(30) Allen, G.; Kirkham, M. J.; Padget, J.; Price, C. *Trans. Faraday Soc.* 1971, 67, 1278.
(31) Hinckley, J. A.; Han, C. C.; Moser, B.; Yu, H. *Macromolecules* 1978, 11, 836.
(32) Beltzung, M.; Picot, C.; Herz, J. *Macromolecules* 1984, 17, 663.
(33) Yu, H.; Kitano, T.; Kim, C. Y.; Amis, E. J.; Chang, T.; Landry, M. R.; Wesson, J. A.; Han, C. C.; Lodge, T. J.; Glinka, C. J. In *Advances in Elastomers and Rubber Elasticity*; J. Lal and J. E. Mark, Eds. Plenum Press: New York. 1986; p. 407.
(34) Volkenstein, M. *Configurational Statistics of Polymer Chains*, Interscience: New York. 1963.
(35) Edwards, S. F.; Freed, K. *J. Phys.* 1970, C3, 739, 750.
(36) Freed, K. *J. Chem. Phys.* 1971, 55, 5588.
(37) Eichinger, B. E. *Macromolecules* 1972, 5, 496.
(38) Pearson, D. S. *Macromolecules* 1977, 10, 696.
(39) Kloczkowski, A.; Mark, J. E.; Erman, B. *Macromolecules* 1989, 22, 1423.
(40) Graessley, W. W. *Macromolecules* 1975, 8, 186.
(41) Graessley, W. W. *Macromolecules* 1975, 8, 865.
(42) Edwards, S. F. In *4th International Conference on Amorphous Materials*, R. W. Douglas and B. Ellis, Eds. Wiley, New York: 1970.
(43) Edwards, S. F. In *Polymer Networks: Structural and Mechanical Properties*, A. J. Chompff and S. Newman, Eds. Plenum Press: New York. 1971.
(44) Duiser, J.; Staverman, A. J.; *Physics of Non-Crystalline Solids*. North Holland: Amsterdam. 1965; p. 376.
(45) Hermans, J. J. *Kolloid Z.* 1943, 103, 209.
(46) Hermans, J. J. *J. Coll. Sci.* 1946, 1, 235.
(47) Staverman, A. J. *Adv. Polym. Sci.* 1982, 44, 73.
(48) Eichinger, B. E. *Ann. Rev. Phys. Chem.* 1983, 34, 359.
(49) Edwards, S. F.; Vilgis, T. A. *Rep. Prog. Phys.* 1988, 51, 243.
(50) Heinrich, G.; Straube, E.; Helmis, G. *Adv. Polym. Sci.* 1988, 85, 33.
(51) Erman, B.; Mark, J. E. *Ann. Rev. Phys. Chem.* 1989, 40, 351.

3

Intermolecular Effects
I. The Constrained-Junction Model

The classical theories of rubber elasticity presented in chapter 2 are based on a hypothetical chain which may pass freely through its neighbors as well as through itself. In a real chain, however, the volume of a segment is excluded to other segments belonging either to the same chain or to others in the network. Consequently, the uncrossability of chain contours by those occupying the same volume becomes an important factor. This chapter and the following one describe theoretical models treating departures from phantom-like behavior arising from the effect of entanglements, which result from this uncrossability of network chains.

The chains in the un-cross-linked bulk polymer are highly entangled. These entanglements are permanently fixed once the chains are joined during formation of the network. The degree of entanglement, or degree of interpenetration, in a network is proportional to the number of chains sharing the volume occupied by a given chain. This is quite important, since the observed differences between experimental results on real networks and predictions of the phantom network theory may frequently be attributed to the effects of entanglements. The decrease in network modulus with increasing tensile strain or swelling is the best-known effect arising from deformation-dependent contributions from entanglements. The constrained-junction model presented in this chapter and the slip-link model presented in chapter 4 are both based on the postulate that, upon stretching, the space available to a chain along the direction of stretch is increased, thus resulting in an increase in the freedom of the chain to fluctuate. Similarly, swelling with a suitable diluent separates the chains from one another, decreasing their correlations with neighboring chains. Experimental data presented in figure 3.1 show that the modulus of a network does indeed decrease with both swelling and elongation, finally becoming independent of deformation, as should be the case for the modulus of a phantom network. Rigorous derivation of the modulus of a network from the elastic free energy for this case will be given in chapter 5.

The starting point of the constrained-junction model presented in this chapter is the elastic free energy. In agreement with experimental observations, this model

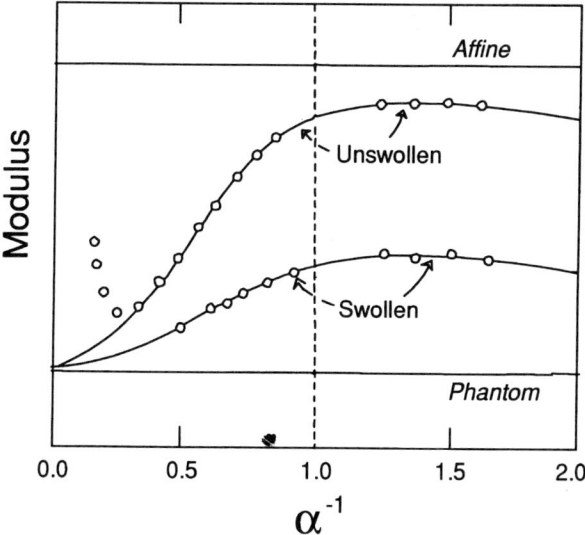

Figure 3.1 The effects of elongation or compression and swelling on the modulus of a network. The abscissa represents the inverse extension ratio as suggested by the semi-empirical Mooney-Rivlin representation [1-3] of the modulus: $2C_1 + 2C_2\alpha^{-1}$, where the C's are constants. The points represent experimental data, and comparisons between the two curves show the effects of swelling. The upturn in the modulus at high extensions could be due either to strain-induced crystallization or to limited chain extensibility. The upper and lower horizontal lines show the behavior predicted by the affine and phantom models, respectively.

takes the free energy of the deformed network as the sum of two contributions, one from the phantom network and the other from the entanglements.

The constrained-junction model was first proposed by Ronca and Allegra[1]. A more refined version of the theory was independently given by Flory[2], and this was subsequently modified by Erman and Flory[3-5]. The model has been further described in papers by Flory[6-8], as well as in review papers by Eichinger[9], Edwards and Vilgis[10], Heinrich et al.[11], and Erman and Mark[12].

3.1 The Model and its Assumptions

The constrained-junction model is, in a sense, intermediate to the James-Guth phantom model and the Flory-Wall affine network model discussed in chapter 2. The difference between the phantom and the affine network models is that the junctions are frozen in the latter, but may fluctuate over distances of the order of chain dimensions in the former. In neither of these models do the chains experience intermolecular interactions along their contours.

The mean-squared fluctuations $\langle(\Delta R)^2\rangle$ of junctions at each end of the chains in the phantom network is

$$\langle(\Delta R)^2\rangle = \frac{(\varphi-1)}{\varphi(\varphi-2)}\langle r^2\rangle_0 \qquad (3.1)$$

where φ is the average junction functionality. Thus, for a tetrafunctional network, $\langle(\Delta R)^2\rangle = (3/8)\langle(r)^2\rangle_0$. The derivation of this relation is given in appendix E. The corresponding fluctuations $\langle(\Delta r)^2\rangle$ in chain dimensions (from appendix E) are

$$\langle(\Delta r)^2\rangle = (2/\varphi)\langle r^2\rangle_0 \qquad (3.2)$$

In the constrained-junction model, the entanglement of network chains with their neighbors is assumed to diminish the size of the fluctuation domains of the junctions. An increase in the constraining effect of the surroundings results in a decrease in the magnitude of the fluctuations of the junctions, ultimately leading to the affine model, for which $\langle(\Delta R)^2\rangle = 0$. Conversely, a decrease in constraints from the environments of junctions, such as that obtained in stretching or swelling, results in an increase in the magnitude of the fluctuations, ultimately leading to the phantom model. It should be emphasized that consideration of the junctions as the centers being affected by the constraints, rather than points along the chains, is only a matter of convenience since the fluctuations of the two are proportional to one other, as acknowledged in eq. (3.1). A more recent treatment, by Erman and Monnerie[13,14], of constraints applied directly onto the chains rather than to the junctions has indeed shown that the results do not change appreciably. This model, referred to in the literature as the "constrained-chain model" is treated in some detail in chapter 4.

Figure 3.2 shows a network junction. The radius of the domain in which it may fluctuate in the phantom network is indicated by the dashed circle. The crosses show the other junctions of the network that share the space available to the central junction; these other junctions may be topologically close to or remote from the central junction under consideration. According to the constrained-junction model, the severity of the entanglements is measured by the degree of interpenetration. This interpenetration is measured by the average number ("Flory number")[5,11] N_F of junctions within the mean radius of the sphere occupied by the network chain, namely,

$$N_F = (4\pi/3)\langle r^2\rangle_0^{3/2}(\mu/V^0) \propto (\mu/V^0)^{-1/2} \qquad (3.3)$$

where V^0 is the volume in the state of reference (in which the network was formed). Since $\langle r^2\rangle_0$ varies inversely with the number of network junctions, N_F must be inversely proportional to the square root of the degree of cross-linking. This is shown by the proportionality part of eq. (3.3). In typical elastomeric networks, N_F is in the range of 25–100[8,15]. The significance of entanglements becomes immediately evident when one considers the space available to a given junction to be shared by 25–100 φ-functional junctions connected to different points in the network. The Flory number is discussed further in chapter 10.

Two types of forces act on a junction in a real network when it instantaneously fluctuates away from its mean position. (1) The restoring action of the phantom network pulls the junction toward its mean position. This effect results from the connectivity of the network. Since fluctuations of junctions are Gaussian in a

THE CONSTRAINED-JUNCTION MODEL 25

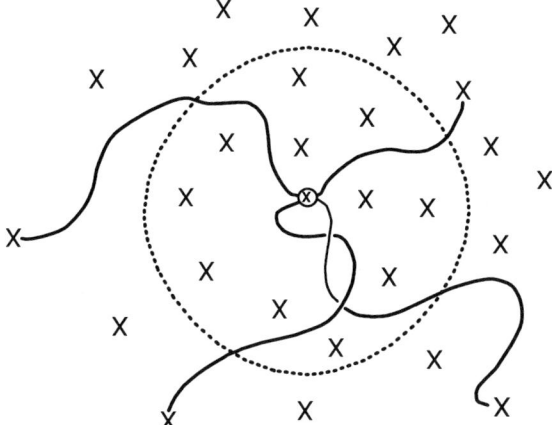

Figure 3.2 A central tetrafunctional junction surrounded by four other topologically neighboring junctions and several spatially neighboring junctions. The crosses show the other junctions of the ntwork that share the space available to the central junction. The fluctuations of the central junction are constrained by the presence of the spatially neighboring ones and their pendant chains.

phantom network, the restoring force obeys the linear-spring law. (2) Another spring-like force resulting from the constraining effect of the other junctions (and chains) present in the domain of the chosen junction. This elastic force is directed toward a center, called the "center of constraints," for the chosen junction. The location of this center is fixed during the formation of the network. One may imagine that in the absence of network connectivity, the junction would fluctuate around the center of constraints. In figure 3.3, various vectors describing the phantom network center and the center of constraints are shown relative to a laboratory-fixed coordinate system $0xyz$. One of the junctions of the network may

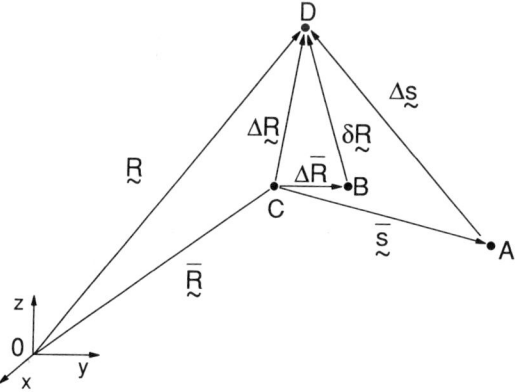

Figure 3.3 Various vectors describing a phantom network, and the center of constraints. See text for description of the vectors shown.

be assumed to be fixed at the origin, thereby fixing the position of the network in space. Point C at position $\bar{\mathbf{R}}_i$ locates the time-averaged position of the junction in the phantom network in the given state of strain. The vector $\bar{\mathbf{s}}_i$ locates the center of entanglements relative to the phantom network center. Point A at position $\bar{\mathbf{R}}_i + \bar{\mathbf{s}}_i$ is the center of constraints operating on the fluctuations of the junction i. (The subscript i identifying the junction i is not shown in the figure.) Point D at position $\mathbf{R}_i$ represents the instantaneous location of the junction in the real deformed network. Point B at position $\overline{\Delta \mathbf{R}_i}$ relative to the phantom network center locates the center of fluctuations of the junction obtained under the joint action of the phantom network and constraints. The vectors $\Delta \mathbf{R}_i$ and $\Delta \mathbf{s}_i$ are the instantaneous displacements of the junction from the phantom network center and the constraint center, respectively, and $\delta \mathbf{R}_i$ is the instantaneous fluctuation from the joint center of action of the network and constraints. Thus, the instantaneous position of the junction i is $\bar{\mathbf{R}}_i + \Delta \mathbf{R}_i$, where

$$\Delta \mathbf{R}_i = \overline{\Delta \mathbf{R}_i} + \delta \mathbf{R}_i = \bar{\mathbf{s}}_i + \Delta \mathbf{s}_i \tag{3.4}$$

The reference state of the network is prescribed by the temperature, volume, and shape prevailing during its formation. The volume V^0 may be adjusted as required by the temperature coefficient of $\langle r^2 \rangle_0$ if the experiments are conducted at a different temperature. The assumptions of the constrained-junction model are as follows.

1. The distribution of chains in the undeformed state is unaffected by entanglements and the cross-linking process. The instantaneous distribution of chain vectors in the network therefore equates to the Gaussian distribution of free chains:

$$W(\mathbf{r}) = \left(\frac{\gamma}{\pi}\right)^{3/2} \exp(-\gamma r^2) \tag{3.5}$$

where $\gamma = 3/(2\langle r^2 \rangle_0)$. The fluctuations of junctions in the phantom network are Gaussian and independent of macroscopic deformation, as follows from the theory of James and Guth. Thus, the distribution of fluctuations $\Delta \mathbf{R}_i$ of the ith junction of the phantom network model are expressed by

$$R(\Delta \mathbf{R}_i) = \left(\frac{\rho}{\pi}\right)^{3/2} \exp[-\rho(\Delta \mathbf{R}_i)^2] \tag{3.6}$$

where $\rho = 3/[2\langle(\Delta \mathbf{R}_i)^2\rangle]$, and $\rho = \varphi(\varphi - 2)(\varphi - 1)^{-1}\gamma$ [which follows from eq. ((3.1)].

2. The centers of entanglement $\bar{\mathbf{R}}_i + \bar{\mathbf{s}}_i$, like the mean positions $\bar{\mathbf{R}}_i$ of the junctions in the deformed phantom network, are affine in the strain. Thus, $\bar{\mathbf{s}}_i = \lambda \bar{\mathbf{s}}_i^0$, where $\bar{\mathbf{s}}_i^0$ is the mean position relative to $\bar{\mathbf{R}}_{i0}$ of junction i in the state of reference, and λ is the displacement gradient tensor. The distribution $H(\bar{\mathbf{s}})$ of positions of the entanglement centers relative to the phantom centers is assumed to be Gaussian and affine in the macroscopic deformation:

$$H(\bar{\mathbf{s}}) = \pi^{-3/2} (\det \boldsymbol{\eta}_\lambda)^{1/2} \exp[-(\bar{\mathbf{s}})^T \boldsymbol{\eta}_\lambda (\bar{\mathbf{s}})] \tag{3.7}$$

where η_λ is

$$\eta_\lambda = \eta_0(\lambda\lambda^T)^{-1} \tag{3.8}$$

with η_0 denoting a scalar parameter to be determined in the theory.

3. The a priori probability of the fluctuations $\Delta\mathbf{s}_i$ from the center of action $\bar{\mathbf{R}}_i + \bar{\mathbf{s}}_i$ of the entanglement constraints is Gaussian, from which it follows that the junctions fluctuate about the center of constraints according to the linear-spring law. It is also assumed that these fluctuations are affine in the macroscopic strain, that is, $\Delta\bar{\mathbf{s}}_i = \lambda \Delta\mathbf{s}_i^0$, where $\Delta\mathbf{s}_i^0$ is the fluctuation of the junction from the constraint center in the undeformed state. Thus, the probability is written as

$$S(\Delta\mathbf{s}_i) = \pi^{-3/2}(\det\sigma_\lambda)^{1/2}\exp[-(\Delta\mathbf{s}_i)^T\sigma_\lambda(\Delta\mathbf{s}_i)] \tag{3.9}$$

The parameter σ_λ is defined by

$$\sigma_\lambda = \sigma_0(\lambda\lambda^T)^{-1} \tag{3.10}$$

where σ_0 is a scalar characterizing the severity of the entanglement constraints in the reference state. Equation (3.10) is a statement of the basic postulate of the theory that the effect of constraints diminishes with increasing extension or swelling.

4. The network is of uniform structure.
5. The entanglement constraint about every junction is the same.

The formulation of the constrained-junction model may be simplified by separating the vectorial quantities into their x, y, and z components, and giving the expressions for the x components only. This is possible inasmuch as the distributions of chain dimensions and their fluctuations for the model are assumed to be Gaussians, and the various distribution functions may be factored into Gaussians for each principal axis of deformation represented by the three elements of the diagonal matrix λ. In the remaining sections of this chapter, the indices identifying a junction will be dropped, and only the x components of the equations will be given by replacing $\mathbf{R}$, $\Delta\mathbf{R}$, $\bar{\mathbf{R}}$, $\Delta\mathbf{s}$, and $\bar{\mathbf{s}}$ with X, ΔX, $\bar{X}$, x, and $\bar{x}$, respectively. According to this notation, the tensor σ_λ reduces to a scalar, $\sigma_\lambda = \sigma_0/\lambda_x^2$. Additionally, the deformation gradient tensor λ will be replaced by λ or λ_x, referring to the x component alone.

3.2 Probability Distribution of Fluctuations in the Deformed Network

A junction in a real network is under the joint action of network connectivity and entanglements. The probability $\rho(\Delta x)$ of having the junction at ΔX when its fluctuation from the constraint center is Δx is obtained as

$$P(\Delta X) = R(\Delta X) \times S(\Delta X - \bar{x})/\int R(\Delta X) \times S(\Delta X - \bar{x})d\Delta X \tag{3.11}$$

where eq. (3.4) is used for replacing Δs in the distribution function S. Stated in this manner, $P(\Delta X)$ is a conditional probability. The product in the numerator of eq. (3.11) is the joint probability of having the junction at ΔX, and eq. (3.11) is written on the basis of the Bayes theorem of probability. Substituting eqs. (3.6) and (3.9) into eq. ((3.11) and performing the integration results in

$$P(\delta X) = [(\rho + \sigma_\lambda)/\pi]^{1/2} \exp[-(\rho + \sigma_\lambda)(\delta X)^2] \tag{3.12}$$

where

$$\delta X = \Delta X - \Delta \bar{X}$$
$$\Delta \bar{X} = [\sigma_\lambda/(\rho + \sigma_\lambda)]\bar{x} \tag{3.13}$$

The argument of the function P on the left-hand side of eq. (3.12) is expressed in terms of δX since this is the variable appearing on the right-hand side of the same expression. Thus, P represents the probability distribution of fluctuations of the junction from the joint center of action of the network and entanglements. The second part of eq. (3.13) locates the position of the joint center relative to the average location of the junction in the phantom network. Inasmuch as the constraint action is spherically symmetric in the undeformed state, $\sigma_\lambda = \sigma_0$, and the vector $\Delta \bar{X}$ becomes parallel to $\bar{x}$. Solving the second part of eq. (3.13) for $\bar{x}$ and substituting into eq. (3.7) leads to the distribution $\Theta(\Delta \bar{X})$ of the variable as

$$\Theta(\Delta \bar{X}) = (\eta_\lambda/\pi)^{1/2}(1 + \rho/\sigma_\lambda) \exp[-\eta_\lambda(1 + \rho/\sigma_\lambda)^2(\Delta \bar{X})^2] \tag{3.14}$$

Having defined the probability of occurence of a given junction at a distance corresponding to the specified fluctuation δX, one may now consider the entire network and calculate the instantaneous distribution $R_*(\Delta X)$ of displacements ΔX of junctions in the real network from their mean positions in the phantom network in the deformed state. This distribution is defined (see appendix F) as the convolution of $P(\delta X)$ with the distribution $\Theta(\Delta \bar{X})$ of $\Delta \bar{X}$, that is,

$$R_*(\Delta X) = P(\delta X) * \Theta(\Delta \bar{X}) \equiv \int P(\delta X) \times \Theta(\Delta \bar{X}) d\Delta \bar{X} \tag{3.15}$$

The subscript asterisk in the distribution of ΔX differentiates it from the a priori distribution $P(\Delta X)$ of vectors ΔX obtained for the phantom network. In the reference state, the distribution $R_*(\Delta X)$ reduces to $P(\Delta X)$ and hence is Gaussian in this state. From eq. (3.15), it follows that $\Theta(\Delta \bar{X})$ is also Gaussian in the reference state. Finally, $H(\bar{x})$ is Gaussian at all levels of strain according to the assumption stated in eq. (3.7). Hence, eqs. (3.13) and (3.14) indicate that the distributions $\Theta(\Delta \bar{X})$ and $R_*(\Delta X)$, respectively, must both be Gaussian at all levels of strain. Substituting eqs. (3.12) and (3.14) into eq. (3.15) and performing the integration over $\Delta \bar{X}$ leads to

$$R_*(\Delta X) = (\rho_{*\lambda}/\pi)^{1/2} \exp[-\rho_{*\lambda}(\Delta X)^2] \tag{3.16}$$

where

$$\rho_{*\lambda} = (\rho + \sigma_\lambda)^2/(\rho + \sigma_\lambda + \sigma_\lambda^2/\eta_\lambda) \tag{3.17}$$

which can be rearranged to

$$\rho/\rho_{*\lambda} = 1 + (\sigma_\lambda/\rho)[(\sigma_\lambda/\rho)(\rho\eta_\lambda^{-1} - 1) - 1]/(1 + \sigma_\lambda/\rho)^2 \tag{3.18}$$

Compliance with the requirement that the actual distribution $R_*(\Delta X)$ should reduce to $R(\Delta X)$ for the phantom model in the state of reference results in

$$\eta_0^{-1} = \rho^{-1} + \sigma_0^{-1} \tag{3.19}$$

Replacement of $\rho\eta_\lambda^{-1}$ in eq. (3.18) by

$$\rho\eta_\lambda^{-1} = (\eta_0/\eta_\lambda)(1 + \rho/\sigma_0) \tag{3.20}$$

leads to

$$\rho/\rho_{*\lambda} = 1 + \kappa^2(\lambda_x^2 - 1)(\lambda_x^2 + \kappa)^{-2} \tag{3.21}$$

where the definition

$$\kappa = \sigma_0/\rho = \langle(\Delta R)^2\rangle_{ph}/\langle(\Delta s)^2\rangle_0 \tag{3.22}$$

has been adopted. In the case of a phantom network, no constraints operate on the junctions; thus, $\langle(\Delta s)^2\rangle_0$ is infinitely large and κ becomes zero. In the other limit, constituted by the affine model, the fluctuations of junctions are frozen and $\langle(\Delta s)^2\rangle_0$ equates to zero, leading to an infinitely large κ. The parameter κ, which represents a measure of entanglements of chains with their surroundings in the real network, is assumed[5] to be proportional to the Flory number, namely, the number of junctions in the volume occupied by a given junction [see eq. (3.3)]. Thus,

$$\kappa = I\langle r^2\rangle_0^{3/2}(\mu/V^0) \tag{3.23}$$

where I is the constant of proportionality, and μ is the number of junctions in the volume V^0 of the state of reference. For tetrafunctional junctions, κ is related to the cycle rank of the network by the relation[5]

$$\kappa = I(N_A d/2)^{3/2}(\langle r^2\rangle_0/M)^{3/2}(\xi/V^0)^{-1/2} \tag{3.24}$$

where N_A is Avogadro's number, d is the density, and M is the molecular weight of a chain whose mean-square end-to-end distance is $\langle r^2\rangle_0$. Equation (3.24) relates the constraint parameter κ to molecular parameters. Since, for a tetrafunctional network, the cycle rank is related to the molecular weight M_c of network chains by[5]

$$\xi = V^0 N_A d/2M_c \tag{3.25}$$

one sees from eq. (3.24) that the κ parameter increases with the square root of network chain length.

The distribution $S_*(\Delta x)$ of vector components Δx in the deformed network is given by the convolution of $P(\delta X)$ with the distribution of vectors joining point A to point B (see figure 3.3). Denoting the x components of these vectors by Δu, one obtains, using eq. (3.13),

$$\Delta u = -\frac{\rho}{\rho + \sigma_\lambda}\bar{x} \tag{3.26}$$

The distribution of Δu is the same as $H(\bar{x})$, where $\bar{x}$ is related to Δu by eq. (3.26). Thus,

$$S_*(\Delta x) - P(\delta X) * H(\bar{x}) \tag{3.27}$$

where $\delta X = \bar{x} + \Delta u$, from figure 3.3. Equations (3.12) and (3.7) furnish expressions for P and H, respectively. Replacing in $\bar{x}$ eq. (3.27) by Δu of eq. (3.26), completing the square in Δu, and integrating over $d\Delta u$, leads to

$$S_*(\Delta x) = (\sigma_{*x}/\pi)^{1/2} \exp[-\sigma_{*x}(\Delta x)^2] \tag{3.28}$$

where

$$\sigma_{*x}^{-1} = \sigma_0^{-1} \lambda_x^2 [1 + \kappa \lambda_x^2 (\lambda_x^2 - 1)(\lambda_x^2 + \kappa)^{-2}] \tag{3.29}$$

3.3 The Elastic Free Energy

3.3.1 General Aspects

The elastic free energy ΔA_{el} of the network is taken as the sum of the elastic free energy ΔA_{ph} of the phantom network and ΔA_c of the constraints:

$$\Delta A_{el} = \Delta A_{ph} + \Delta A_c \tag{3.30}$$

The expression for the elastic free energy of the phantom network model is given by eq. (2.49).

Contributions to the elastic free energy from constraints result from two effects. One is from the spring-like action of the constraints on the fluctuations $\Delta \mathbf{R}$ of junctions from their time-averaged positions in the network. The second contribution to the free energy results from distortion of the domains of constraint. According to the model, the domains of constraint are represented as regions having spring-like elasticity, and the distortion of the vectors $\Delta \mathbf{s}$ results in an additional change in the elastic free energy.

3.3.2 The Elastic Free Energy Due to Distortion of $\Delta \mathbf{R}$

The calculation of the probability of μ_j junctions having displacements in the range ΔX_j and $\Delta X_j + \delta(\Delta X_j)$ follows the same arguments given in chapter 2 for the phantom network. Thus, the number of configurations $\Omega_{\Delta X}$ consistent with the distribution $\{\mu_j\}$ of the μ_j junctions is given by

$$\Omega_{\Delta X} = \mu! \prod \omega_j^{\mu_j}/\mu_j! = \prod (\mu \omega_j/\mu_j)^{\mu_j} \tag{3.31}$$

Here, ω_j is the a priori probability of occurrence of a displacement ΔX_j in the range $\delta(\Delta X_j)$ and is given by $R(\Delta X_j)\delta(\Delta X_j)$. The relative incidence of junctions in the stated range in the deformed network is

$$\mu_j/\mu = R_*(\Delta X_j)\delta(\Delta X_j) \tag{3.32}$$

Using the expressions for $R(\Delta X_j)$ and $R_*(\Delta X_j)$ from eqs. (3.6) and (3.16), and taking the logarithm of eq. (3.28), leads to

$$\ln \Omega_{\Delta X} = (\mu/2) \ln(\rho/\rho_{*x}) - (\rho - \rho_{*x}) \sum \mu_j (\Delta X_j)^2$$
$$= (\mu/2)[\ln(\rho/\rho_{*x}) - (\rho/\rho_{*x} - 1)] \quad (3.33)$$

The second line of eq. (3.33) is obtained by using the expression for the average fluctuations:

$$\sum \mu_j (\Delta X_j)^2 = \mu \langle (\Delta X)^2 \rangle = \mu/2\rho_{*x} \quad (3.34)$$

Using eq. (3.21) for ρ/ρ_{*x} and the relation $\Delta A_{\Delta X} = -kT \ln \Omega_{\Delta X}$, the expression for the elastic free energy due to fluctuations of junctions from their mean locations in the network is obtained as

$$\Delta A_{\Delta R} = \frac{1}{2} \mu kT \sum_t [B_t - \ln(B_t + 1)] \quad (3.35)$$

where

$$B_t = (\rho/\rho_{*x}) - 1 = \kappa^2 (\lambda_t^2 - 1)(\lambda_t^2 + \kappa)^{-2} \quad (3.36)$$

3.3.3 The Elastic Free Energy Due to Distortion of Δs

The contributions to the elastic free energy due to distortion of the constraint domains is similarly obtained as

$$\Delta A_{\Delta s} = \frac{1}{2} \mu kT \sum_t [D_t - \ln(D_t + 1)] \quad (3.37)$$

where

$$D_t = (\sigma_{0t}/\sigma_{*x}) - 1 = (\sigma_0/\lambda_t^2 \sigma_{*x}) - 1$$
$$= \kappa \lambda_t^2 (\lambda_t^2 - 1)(\lambda_t^2 + \kappa)^{-2} = \lambda_t^2 \kappa^{-1} B_t \quad (3.38)$$

3.3.4 The Total Elastic Free Energy

Combining eqs. (3.35) and (3.37) leads to the total elastic free energy due to constraints. Combining the expression obtained in this manner with eq. (2.49), representing the elastic free energy of the phantom network model, leads to the total elastic free energy for the constrained-junction model:

$$\Delta A_{el} = \frac{1}{2} \xi kT \sum_t \{(\lambda_t^2 - 1) + (\mu/\xi)[B_t + D_t - \ln(B_t + 1) - \ln(D_t + 1)]\} \quad (3.39)$$

In the limit where the strength of the constraints vanishes, $\kappa \to 0$, B_t and D_t equate to zero, and the elastic free energy given by eq. (3.39) reduces to that of the phantom network model. When $\kappa \to \infty$,

$$B_t = \lambda_t^2 - 1, \quad D_t = 0 \quad (3.40)$$

and eq. (3.39) gives

$$\Delta A_{el} = \frac{1}{2} \nu kT(\lambda_1^2 + \lambda_2^2 + \lambda_3^2 - 3) - (2\nu/\phi)kT \ln(V/V^0) \qquad (3.41)$$

which is the elastic free energy of the affine network.

The elastic free energy of the constrained-junction model given by eq. (3.41) is used in chapter 5 for deriving the relationship between stress and strain. The effect of the κ parameter on the mechanical properties of the network is also illustrated in detail in that chapter.

References

(1) Ronca, G.; Allegra, G. *J. Chem. Phys.* 1975, 63, 4990.
(2) Flory, P. J. *J. Chem. Phys.* 1977, 66, 5720.
(3) Erman, B.; Flory, P. J. *J. Chem. Phys.* 1978, 68, 5363.
(4) Flory, P. J.; Erman, B. *Macromolecules* 1982, 15, 800.
(5) Erman, B.; Flory, P. J. *Macromolecules* 1982, 15, 806.
(6) Flory, P. J. *Polymer* 1979, 20, 1317.
(7) Flory, P. J. *Brit. Polym. J.* 1985, 17, 96.
(8) Flory, P. J. *Polym. J.* 1985, 17, 1.
(9) Eichinger, B. E. *Ann. Rev. Phys. Chem.* 1983, 34, 359.
(10) Edwards, S. F.; Vilgis, T. A. *Rep. Prog. Phys.* 1988, 51, 243.
(11) Heinrich, G.; Straube, E.; Helmis, G. *Adv. Polym. Sci.* 1988, 85, 33.
(12) Erman, B.; Mark, J. E. *Ann. Rev. Phys. Chem.* 1989, 40, 351.
(13) Erman, B.; Monnerie, L. *Macromolecules* 1989, 22, 3342.
(14) Erman, B.; Monnerie, L. *Macromolecules* 1992, 25, 4456.
(15) Flory, P. J. *Proc. Roy. Soc. London, A* 1976, 351, 351.

4

Intermolecular Effects
II. Constraints along Network Chains

In the constrained-junction model presented in chapter 3, intermolecular correlations were assumed to suppress the fluctuations of junctions. According to this model, the elastic free energy of a network varies between the free energies of the phantom and the affine networks. In a second group of models, to be introduced here, there is a constraining action of entanglements *along* the chains that may further contribute to the elastic free energy, as if they were additional (albeit temporary) junctions. Consequently, the upper bound of the elastic free energy of such networks may exceed that of an affine network. Since the entanglements along the chain contour are explicitly taken into account in the models, they are referred to as the constrained-*chain* models.

The idea of constrained-chain theories originates from the trapped-entanglement concept of Langley[1,2], and Graessley[3,4], stating that some fraction of the entanglements which are present in the bulk polymer before cross-linking become permanently trapped by the cross-linking and act as additional cross-links. These trapped entanglements, unlike the chemical cross-links, have some freedom, and the two chains forming the entanglement may slide relative to one other. The two chains may therefore be regarded as being attached to each other by means of a fictitious "slip-link," as is illustrated schematically in figure 4.1. The entangled system of chains representing the real situation is shown in part (a), and the representation of two entangled chains in this system joined together by a slip-link is shown in part (b). The slip-link may move along the chains by a distance a, which is inversely proportional to the severity of the entanglements. A model based on this picture of slip-links was first proposed by Graessley[3], and a more rigorous treatment of the slip-link model was given by Ball et al.[5] and subsequently simplified by Edwards and Vilgis[6]; section 4.1 describes this latter treatment in detail. In section 4.2, we present the extension of the Flory constrained-junction model to the constrained-chain model by including the effects of constraints along chains, following Erman and Monnerie[7,8]. One of the newest approaches, the diffused-constraints model, is then described briefly in section 4.3. Finally, section 4.4 describes various other treatments of entanglements.

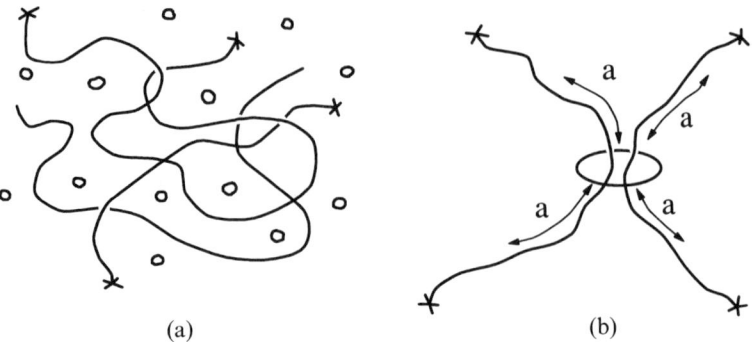

Figure 4.1 (a) An entangled system of chains. (b) The representation of two entangled chains joined by a slip-link. The slip-link may move along the chains by a distance a, which is inversely proportional to the severity of the entanglements.

Two excellent review articles on this topic, in general, are by Heinrich et al.[9], and by Edwards and Vilgis[10].

4.1 The Slip-Link Model

The probability $P[\mathbf{R}(s)]$ of the end-to-end vector $\mathbf{R}$ for a network chain is given by Edwards and collaborators[5,6,11] in the Wiener representation as

$$P[\mathbf{R}(s)] = \mathcal{N} \exp\left(-\frac{3}{2l}\int_0^L |\dot{\mathbf{R}}(s)|^2 ds\right) \quad (4.1)$$

Here, $\mathcal{N}$ is a normalization constant, $\mathbf{R}$ is the position vector in space, s is the arc length along the chain contour (varying between zero and the chain contour length L), $|\dot{\mathbf{R}}(s)| = \partial \mathbf{R}(s)/\partial s$, and l is the Kuhn length (the length of a bond in an equivalent freely jointed chain). Equation (4.1) gives the distribution function for a Gaussian chain.

A slip-link, as shown in figure 4.1(b), may be visualized as a sliding cross-link that joins a point of the ith network chain to a point of the jth network chain. This condition is expressed as

$$\mathbf{R}_i(s_i^j) = \mathbf{R}_j(s_j^i) \quad (4.2)$$

where $\mathbf{R}_i$ denotes the position vector of the slip-link on the ith chain. The quantity s_i^j locates the slip-link position along the ith chain contour, with the superscript j indicating that this point is common to the slip-link position on the jth chain. The constraint

$$\prod_{\text{cross-links}} \delta[\mathbf{R}_i(s_i^j) - \mathbf{R}_j(s_j^i)] \quad (4.3)$$

appears in the network partition function if the slip-links are frozen, as is the case for the permanent junctions or cross-links. Here, $\delta(x)$ represents the Dirac function, which is zero if $x \neq 0$, and ∞ if $x = 0$. If a link is not permanent but slides by a distance $\pm a$ along the contours of the two chains which it joins, then the constraint appearing in the product of eq. (4.3) becomes

$$\frac{1}{4a^2}\int_{-a}^{a}d\varepsilon_i\int_{-a}^{a}d\varepsilon_j\delta[\mathbf{R}_i(s_i^j+\varepsilon_i)-\mathbf{R}_j(s_j^i+\varepsilon_j)] \quad (4.4)$$

These constraints are built into the theory, by Deam and Edwards[11] for cross-links and by Ball et al.[5] for slip-links, by means of the replica formalism of statistical mechanics. As stated by the authors, to build these constraints into the theory is a difficult task and the replica method has serious algebraic complexity. Subsequent treatment of the problem by Edwards and Vilgis[6] bypassed these complexities by introducing the slip-link picture into the theory using the Flory-Wall affine model. The Gaussian distribution function given by eq. (2.8) is written for a freely jointed chain as

$$P(\mathbf{R},L)=\left(\frac{3}{2\pi lL}\right)^{3/2}\exp\left\{-\frac{3\mathbf{R}^2}{2lL}\right\} \quad (4.5)$$

Equation (4.5) is obtained by the substitution of the relation for a freely jointed chain:

$$\langle r^2\rangle_0 = nl^2 = lL \quad (4.6)$$

into eq. (2.8). Here, $L=nl$ denotes the contour length of the chain. The Gaussian expression given by eq. (4.5) is separable into contributions from three Cartesian coordinates as

$$P(\mathbf{R})=\prod_{i=1}^{3}P(X_i) \quad (4.7)$$

where

$$P(X_i)=\left(\frac{3}{2\pi lL}\right)^{1/2}\exp\left\{-\frac{3X_i^2}{2lL}\right\} \quad (4.8)$$

For brevity, the contour length L is not written in the argument of P in eq. (4.7). The slip-link process is modeled in the theory by the integral

$$P(X_i)=\int d\varepsilon P(\varepsilon)\frac{\exp\left\{-\dfrac{3X_i^2}{2l(L+\varepsilon)}\right\}}{\left[\dfrac{2l}{3}(L+\varepsilon)\right]^{1/2}} \quad (4.9)$$

Here, ε is the slippage length and $P(\varepsilon)$ is the probability of the arc length of the slippage. Assuming $P(\varepsilon)$ to be uniform along the slippage length, that is,

$$P(\varepsilon)=\frac{1}{2a} \quad (4.10)$$

eq. (4.9) may be written

$$P(X_i) = \int_{-a}^{a} \frac{d\varepsilon}{2a} \exp\left\{-\frac{3X_i^2}{2l(L+\varepsilon)} - \frac{1}{2}\ln(L+\varepsilon)\right\} \qquad (4.11)$$

This integral should be normalized for every value of ε. This is not easy when the complete expression given by eq. (4.11) is used. For further progress, the integrand is expanded into a Taylor series up to the second power of ε.

For a network subject to a principal state of deformation, with λ_i denoting the principal value of the deformation tensor along the ith coordinate direction, eq. (4.11) up to the second power of ε takes the form

$$P(\lambda_i X_i) = \mathcal{N} \int_{-a}^{a} \frac{d\varepsilon}{2a} \exp\left\{-\frac{3}{2lL}\sum_{i=1}^{3}\lambda_i^2 X_i^2\left[1 - \left(\frac{\varepsilon}{L}\right) + \left(\frac{\varepsilon}{L}\right)^2\right]\right.$$
$$\left. - \frac{3}{2}\left[\left(\frac{\varepsilon}{L}\right) - \frac{1}{2}\left(\frac{\varepsilon}{L}\right)^2\right]\right\} \qquad (4.12)$$

where $\mathcal{N}$ is the normalization constant. In order to perform the integration over ε, the exponent in eq. (4.12) is further expanded into a Taylor series up to the second power of ε. Performing the integration over ε yields

$$P(\lambda_i X_i) = \mathcal{N} \exp\left\{-\frac{3}{2lL}\sum_{i=1}^{3}\lambda_i^2 X_i^2\right\}\left\{1 + \frac{\overline{\varepsilon^2}}{L^2}\frac{1}{2}\left(\frac{3}{2lL}\right)^2\left(\sum_{i=1}^{3}\lambda_i^2 X_i^2\right)^2\right.$$
$$\left. - \frac{5}{2}\left(\frac{3}{2lL}\right)\sum_{i=1}^{3}\lambda_i^2 X_i^2 + \frac{15}{8}\right\} \qquad (4.13)$$

Here, $\overline{\varepsilon^2}$ denotes the second moment of ε defined by

$$\overline{\varepsilon^2} = \int_{-a}^{a} (d\varepsilon/2a)\varepsilon^2 \qquad (4.14)$$

The free energy $A(\lambda)$ of the network at the state of deformation λ is evaluated according to the expression

$$A(\lambda) = -kT \int d^3 R\, P(\mathbf{R}, L) \ln P(\lambda\mathbf{R}, L) \qquad (4.15)$$

The expression for the Helmholtz free energy given by eq. (4.15) results from the replica approach, which has been summarized by Vilgis[12]. The arguments are based on the difference between an un-cross-linked polymer melt and a cross-linked one. The former is treated as an "annealed," and the latter as a "quenched," thermodynamic system. Substituting from eq. (4.13) into eq. (4.15) and performing the integration leads to the expression for the Helmholtz free energy of each constraint as

$$A_c(\lambda)/kT = \frac{1}{2}\sum_{i=1}^{3}\lambda_i^2 - \frac{\overline{\varepsilon^2}}{8L^2}\left\{\left(\sum_{i=1}^{3}\lambda_i^2\right)^2 + 2\left(\sum_{i=1}^{3}\lambda_i^2\right)^4 - 10\sum_{i=1}^{3}\lambda_i^2 + 15\right\}$$

(4.16)

Here, the subscript c denotes constraints. This expression obtained by the simplified treatment of Edwards and Vilgis[6] is shown to fit the exact result of Ball et al.[5] to order $(\lambda_i^2 - 1)^2$. The latter is given for N_s slip-links as

$$A_c(\lambda) = \frac{1}{2}N_s kT \sum_{i=1}^{3}\left\{\frac{\lambda_i^2(1+\eta)}{(1+\eta\lambda_i^2)} + \ln(1+\eta\lambda_i^2)\right\}$$

(4.17)

where

$$\eta = \overline{\varepsilon^2}/L$$

(4.18)

As shown in chapter 2, the elastic free energy of the phantom network with a cycle rank of ξ is

$$\Delta A_{ph}(\lambda) = \frac{1}{2}\xi kT \sum_{i=1}^{3}(\lambda_i^2 - 1)$$

(4.19)

The total elastic elastic free energy is given as the sum of eqs. (4.17) and (4.19):

$$\Delta A = \Delta A_{ph} + A_c$$

$$= \frac{1}{2}\xi kT \sum_{i=1}^{3}\left\{(\lambda_i^2 - 1) + \frac{N_s}{\xi}\left[\frac{\lambda_i^2 - 1}{(1+\eta\lambda_i^2)} + \ln\frac{(1+\eta\lambda_i^2)}{(1+\eta)}\right]\right\}$$

(4.20)

Comparison of eq. (4.20) with eq. (3.39) shows that the deformation dependences of the two expressions are qualitatively very similar. If one identifies the η parameter of the slip-link model with κ^{-1}, eq. (4.20) may be written[13]

$$\Delta A = \frac{1}{2}\xi kT \sum_{i=1}^{3}\left\{(\lambda_i^2 - 1) + \frac{N_s}{\xi}[B_i + D_i - \ln(B_i + D_i + 1) - \ln\lambda_i^2]\right\}$$ (4.21)

where B_i and D_i are given by eqs. (3.36) and (3.38) with $\eta = \kappa^{-1}$, that is,

$$B_i = \eta^{-2}(\lambda_i^2 - 1)(\lambda_i^2 + \eta^{-1})^{-2}$$
$$D_i = \lambda_i^2 \eta B_i$$

(4.22)

Also, eq. (4.21) reduces to the phantom network result when N_s becomes zero. However, ΔA of eq. (4.21) may become indefinitely large as the number of slip-links increases, whereas the elastic free energy obtained by the constrained-junction model converges to that of the affine network model with increasing degree of constraints.

The reader is referred to the original papers of Edwards and collaborators[5,6,10,11,14–20] for the foundations of the theory leading to eq. (4.20).

4.2 The Constrained-Chain Model

In the constrained-junction model of rubber elasticity described in the previous chapter, intermolecular effects are assumed to act exclusively at the junctions. This assumption simplifies the way that constraints are introduced into the formulation. However, the magnitudes of junction fluctuations obtained from the neutron spin-echo scattering measurements described in chapter 14 suggest that it may be overdoing it to place the constraints entirely on the junctions themselves. Thus, a more general approach would be to assume that the constraints act instead along the entire chain contour, and to regard the constrained-junction model as a limiting case of this. However, this type of generalization immediately leads to several questions. The constrained-junction model assumes a constraint center. Can one introduce analogous constraint centers for points along the chain contour? If this can be done, how many of these constraint centers should there be? An upper bound for the number of constraint centers for a chain contour may be taken to be M_c/M_e, where M_e is the molecular weight between two entanglement points[21,22]. Unfortunately, this point has remained controversial ever since the introduction of the constraint or entanglement concept into the area of rubber elasticity. Edwards and Vilgis have taken the most constructive step in this direction by considering a single slip-link that may slide along the chain contour. On the other hand, the constrained-chain model introduced by Erman and Monnerie[7,8] proceeds along a different line, according to which the constraints operate on the mass center of each chain. This section outlines the Erman-Monnerie theory.

In figure 4.2, the instantaneous position of a network chain is shown and the variables used in the constrained-chain model are identified. The line AB denotes the instantaneous configuration of a network chain, and $\bar{A}$ and $\bar{B}$ are the time-averaged positions of the junctions A and B. Points G and D locate the instantaneous mass center of the chain and of the center of constraints operating on the

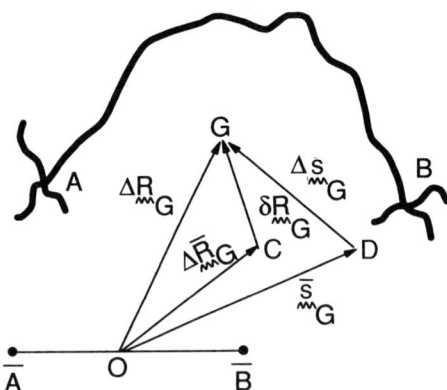

Figure 4.2 The instantaneous position of a network chain and the variables used in the constrained-chain model. See the text for the definitions of the variables shown.

chain, respectively. The time-averaged position of the mass center in the phantom network is at point O, and point C is the time-averaged position of the mass center under the joint action of the phantom network and the constraints. The quantities $\Delta \mathbf{R}_G$ and $\Delta \mathbf{s}_G$ are the instantaneous fluctuation vectors of the mass center from O and D, respectively, and $\delta \mathbf{R}_G$ is the fluctuation vector from the joint center C. The vectors $\Delta \bar{\mathbf{R}}_G$ and $\bar{\mathbf{s}}_G$ locate the positions of the joint center C and the constraint center D, respectively, from O.

The notation used in the derivation of the elastic free energy for the constrained-chain model follows that of the constrained-junction model described in the previous chapter. The distributions of the variables $\Delta \mathbf{R}_G$, $\bar{\mathbf{s}}_G$, $\Delta \mathbf{s}_G$, $\Delta \bar{\mathbf{R}}_G$, and $\delta \mathbf{R}_G$ are assumed to be Gaussian. In terms of x components, they are given by

$$R(\Delta X_G) = (\rho_\lambda/\pi)^{1/2} \exp[-\rho_\lambda(\Delta X_G)^2]$$
$$H(\bar{x}_G) = (\eta_\lambda/\pi)^{1/2} \exp[-\eta_\lambda(\bar{x}_G)^2]$$
$$S(\Delta x_G) = (\sigma_\lambda/\pi)^{1/2} \exp[-\sigma_\lambda(\Delta x_G)^2] \quad (4.23)$$
$$\Theta(\Delta \bar{X}_G) = (\theta_\lambda/\pi)^{1/2} \exp(-\theta_\lambda(\Delta \bar{X}_G)^2]$$
$$P(\delta X_G) = [(\rho_\lambda + \sigma_\lambda)/\pi]^{1/2} \exp[-(\rho_\lambda + \sigma_\lambda)(\delta X_G)^2]$$

where the parameters ρ_λ, η_λ, σ_λ, and θ_λ are given by

$$\rho_\lambda = 1/\{2\langle(\Delta X_G)^2\rangle\}$$
$$\eta_\lambda = 1/\{2\langle(\bar{x}_G)^2\rangle\}$$
$$\sigma_\lambda = 1/\{2\langle(\Delta x_G)^2\rangle\} \quad (4.24)$$
$$\theta_\lambda = 1/\{2\langle(\Delta \bar{X}_G)^2\rangle\}$$

The subscript λ implies that these variables are generally functions of deformation. Formulation of the elastic free energy follows essentially the same steps taken in the derivation of the constrained-junction model[23]. The parameter κ_G, reflecting the severity of constraints, is given by the relationship

$$\kappa_G = \sigma_0/\rho_0 = \langle(\Delta X_G)^2\rangle_0/\langle(\Delta x_G)^2\rangle_0 \quad (4.25)$$

which is similar to that of the constrained-junction model. Defined in this manner, κ_G becomes zero when the constraints do not operate along the chain contour and $\langle(\Delta x_G)^2\rangle_0$ is infinitely large. When the constraints are infinitely strong, $\langle(\Delta x_G)^2\rangle_0$ equates to zero and κ_G then becomes infinitely large.

The ratio $\sigma_\lambda/\eta_\lambda$ of the constrained-junction model[23] is obtained for the constrained-chain model as

$$\frac{\sigma_\lambda}{\rho_\lambda} = \lambda_x^{-2}\kappa_G[1 + (\lambda_x^2 - 1)\Phi] \equiv \lambda_x^{-2}h(\lambda_x) \quad (4.26)$$

where

$$h(\lambda_x) \equiv \kappa_G[1 + (\lambda_x^2 - 1)\Phi] \quad (4.27)$$

The parameter Φ in eq. (4.26) is given by Erman and Monnerie[7] as

$$\Phi = \left(1 - \frac{2}{\phi}\right)^2 \left(\frac{1}{3} + \frac{2}{n}\right) \tag{4.28}$$

where ϕ is the junction functionality and n is the number of equivalent Kuhn segments in the network chain.† The functions B_x and D_x characterizing the elastic free energy are given by

$$B_x \equiv \frac{\rho_\lambda}{\rho_{*\lambda}} - 1$$
$$= h(\lambda_x)\kappa_G(1 - \Phi)(\lambda_x^2 - 1)/[\lambda_x^2 + h(\lambda_x)]^2 \tag{4.29}$$

$$D_x = \lambda_x^2 B_x / h(\lambda_x) \tag{4.30}$$

Functions for the other two directions are similarly defined. Following the arguments in deriving the elastic free energy of the network in the constrained-junction model, one obtains the following expression for the total elastic free energy ΔA_{el} for the constrained-chain model:

$$\Delta A_{el} = \frac{1}{2}\xi kT \sum_t \{(\lambda_t^2 - 1) + (\nu/\xi)[B_t + D_t - \ln(B_t + 1) - \ln(D_t + 1)]\} \tag{4.31}$$

It should be noted that the contribution of constraints to ΔA_{el} is proportional to the number ν of chains, whereas in the constrained-junction model it is proportional to the number of junctions.

4.3 The Diffused-Constraints Model

This theory improves the constrained-chain model by distributing the constraints continuously along the chain contour, rather than at the chain mass center[24]. In its application to stress–strain isotherms in elongation[24], it has the advantage of having only a single constraint parameter, and the values it exhibits upon comparing theory and experiment seem more reasonable than in the earlier constraint models. Applications to strain birefringence[25], on the other hand, yield values of the birefringence that are much larger than those in the constrained-junction and constrained-chain theories.

The upper three graphs of figure 4.3 summarize the differences in the way the constraints are applied in the constrained-junction theory, constrained-chain theory, and diffused-constraints theory, respectively. Additional comparisons between theory and experiment for a variety of elastomeric properties should be very helpful. Also, neutron scattering measurements carried out on series of networks having different junction functionalities ϕ would be extremely useful in

---

†An alternate definition for the Φ parameter, based on a modified form of the constrained chain-model, is also given in reference 8.

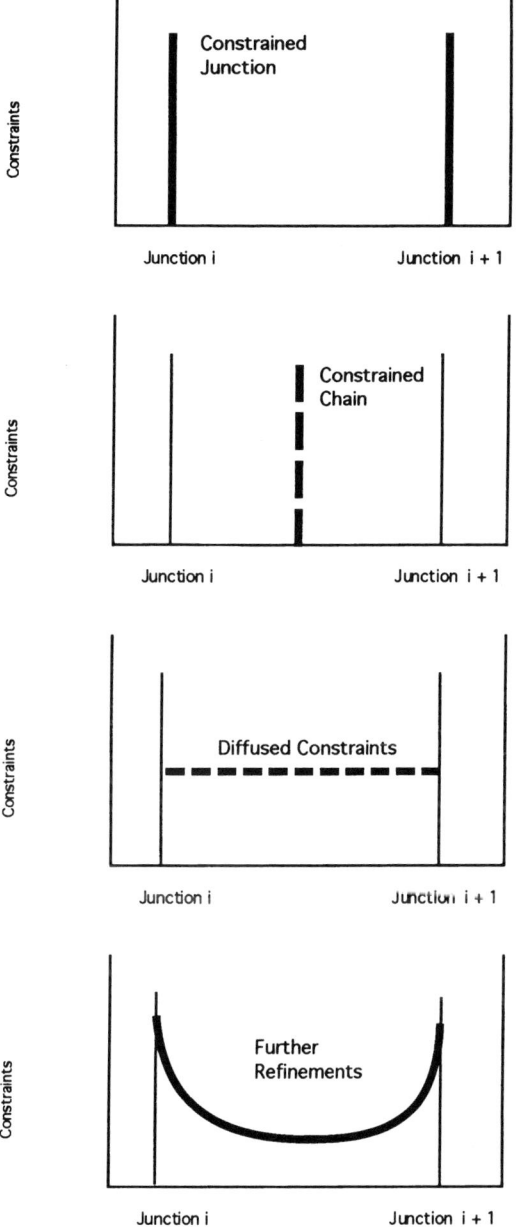

Figure 4.3 Some sketches showing differences in the way the constraints are applied in the constrained-junction theory, constrained-chain theory, and diffused-constraints theory, respectively. A possible further refinement in positioning, based on additional experimental evidence, is illustrated in the bottom graph

suggesting how to position the constraints along a chain in refining such models, since ϕ should have a pronounced effect on the magnitudes of the fluctuations of the junctions. A possible further refinement in positioning based on additional experimental evidence is illustrated in the lowest graph of figure 4.3.

4.4 Other Treatments of Entanglements

The slip-link model is essentially an extension of the more general tube model of entangled polymer systems. This tube model[10] is a convenient tool, especially for predicting the viscoelastic behavior of polymer melts[21,26], and it has been extended to the study of networks by several authors[9,12,27-29]. Heinrich et al.[9] for example, represented the tube as a harmonic pipe and derived the corresponding stress–strain relations for a rubber, and Graessley and Dossin[30] extended the Doi-Edwards theory to study the viscoelastic behavior of networks. Gaylord[31] and Marucci[32] used alternative forms of the tube model to derive the stress–strain behavior for networks.

References

(1) Langley, N. R. *Macromolecules* 1968, 1, 348.
(2) Langley, N. R.; Polmanteer, K. E. *J. Polym. Sci., Polym. Phys. Ed.* 1974, 12, 1023.
(3) Graessley, W. W. *Adv. Polym. Sci.* 1974, 16, 1.
(4) Graessley, W. W. *Adv. Polym. Sci.* 1982, 47, 67.
(5) Ball, R. C.; Doi, M.; Edwards, S. F.; Warner, M. *Polymer* 1981, 22, 1010.
(6) Edwards, S. F.; Vilgis, T. A. *Polymer* 1986, 27, 483.
(7) Erman, B.; Monnerie, L. *Macromolecules* 1989, 22, 3342.
(8) Erman, B.; Monnerie, L. *Macromolecules* 1992, 25, 4456.
(9) Heinrich, G.; Straube, E.; Helmis, G. *Adv. Polym. Sci.* 1988, 85, 33.
(10) Edwards, S. F.; Vilgis, T. A. *Rep. Prog. Phys.* 1988, 51, 243.
(11) Deam, R. T.; Edwards, S. F. *Phil. Trans. Roy. Soc. London* 1976, 280, 317.
(12) Vilgis, T. A. *Prog. Coll. Polym. Sci.* 1987, 75, 243.
(13) Vilgis, T. A., private correspondence.
(14) Edwards, S. F. *J. Phys. C* 1969, 2, 1.
(15) Edwards, S. F. *Proc. Phys. Soc. (London)* 1967, 92, 9.
(16) Edwards, S. F.; Freed, K. F. *J. Phys. A* 1969, 2, 145.
(17) Edwards, S. F.; Freed, K. F. *J. Phys. C* 1970, 3, 739.
(18) Edwards, S. F.; Freed, K. F. *J. Phys. C* 1970, 3, 750.
(19) Edwards, S. F.; Freed, K. F. *J. Phys. C* 1970, 3, 760.
(20) Edwards, S. F. *Discuss. Faraday Soc.* 1970, 49, 43.
(21) Ferry, J. D. *Viscoelastic Properties of Polymers*, 3rd ed. Wiley: New York. 1980.
(22) Fetters, L. J.; Lohse, D. J.; Colby, R. H. In *Physical Properties of Polymers Handbook*, J. E. Mark, Ed. American Institute of Physics Press: Woodbury, NY. 1996; p. 335.
(23) Flory, P. J.; Erman, B. *Macromolecules* 1982, 15, 800.
(24) Kloczkowski, A.; Mark, J. E.; Erman, B. *Macromolecules* 1995, 28, 5089.
(25) Kloczkowski, A.; Mark, J. E.; Erman, B. *Comput. Polym. Sci.* 1995, 5, 37.
(26) Ngai, K. L.; Plazek, D. J. In *Physical Properties of Polymers Handbook*, J. E. Mark, Ed. American Institute of Physics Press: Woodbury, NY. 1996; p. 341.

(27) Vilgis, T. A. *Prog. Coll. Polym. Sci.* 1987, 75, 4.
(28) Vilgis, T. A.; Heinrich, G. *Angew. Makromol. Chem.* 1992, 202/203, 243.
(29) Wagner, M. H.; Schaeffer, J. *J. Rheol.* 1993, 37, 643.
(30) Graessley, W. W.; Dossin, L. M. *Macromolecules* 1979, 12, 123.
(31) Gaylord, R. J. *Polym. Bull.* 1983, 9, 181.
(32) Marucci, G. *Macromolecules* 1981, 14, 434.

5

Relationships between Stress and Strain

In the first section of this chapter, the relationships between the Helmholtz free energy, the stress tensor, and the deformation tensor[1,2] are given for uniaxial stress. These relations follow from the general discussion of stress and strain given in appendix C, and the notation and approach closely follow the classic treatment of Flory[3]. The detailed forms of the stress–strain relations in simple tension (or compression) are given in the remaining sections of the chapter for the (1) phantom network, (2) affine network, (3) constrained-junction model, and (4) slip-link model. Results of theory are then compared with experiment. The effects of swelling on the stress–strain relations are also included in the discussion. It is to be noted that the stress–strain relations in this chapter are obtained by treating the swollen networks as closed systems. The conditions for such systems are fulfilled if solvent does not move in and out of the network during deformation. A network swollen with a nonvolatile solvent and subject to simple tension in air is an example of a closed system. The same network at swelling equilibrium and subjected to compression will exude some of the solvent under increased internal pressure[4], and is therefore not a closed system. For semiopen systems, such as those under compression, or, in general, networks stressed while immersed in solvent, a more general thermodynamic treatment is required. This situation will be taken up in the following chapter.

5.1 General Relationships of Finite Elasticity Theory

The state of stress on a network is obtained from the change in Helmholtz (or elastic) free energy ΔA, which is given[3] as the sum of two terms:

$$\Delta A = \Delta A^*(T, V) + \Delta A_{el}(T, \lambda) \tag{5.1}$$

where $\Delta A^*(T, V)$ is a function of temperature T and volume V, and denotes the contribution from intermolecular forces such as those in simple liquids. The quantity $\Delta A_{el}(T, \lambda)$ is the elastic free energy arising from the elasticity of the network chains. The component $\Delta A_{el}(T, \lambda)$ in eq. (5.1) was derived in chapters

2 through 4 for different models of elastomeric networks. In eq. (5.1), the argument λ represents the displacement (or deformation) gradient tensor, defined for the principal components of homogeneous deformation in a coordinate system $0x_1x_2x_3$ as[1,2]

$$\lambda = \begin{vmatrix} \lambda_1 & 0 & 0 \\ 0 & \lambda_2 & 0 \\ 0 & 0 & \lambda_3 \end{vmatrix} \quad (5.2)$$

Here, λ_i is the extension ratio along the x_i axis (specifically the ratio of the length of the side of a prism along x_i in the deformed state to that in the undeformed state).

In order to separate the effects of distortion from volume changes, λ may be expressed as the product of a pure dilation and a distortion at constant volume:

$$\lambda = (V/V^0)^{1/3}\alpha \quad (5.3)$$

Here, V^0 is the volume during the formation of the network, and $\det \alpha = 1$. The quantity α defined in this manner represents the distortion of the system and is referred to as the distortion tensor. In component form, eq. (5.3) simplifies to

$$\lambda_i = (V/V^0)^{1/3}\alpha_i = \left(\frac{v_{2c}}{v_2}\right)^{1/3}\alpha_i \quad (5.4)$$

where α_i denotes the ratio of the final length along the ith direction to the undistorted length measured at the final volume. The quantities v_{2c} and v_2 are the volume fraction of polymer during formation of network, and during the elasticity experiment, respectively.

The way the principal components of the stress tensor are related to the elastic free energy is derived in appendix C. The result is

$$t_i = p^* + \frac{2\alpha_i^2}{V}\left(\frac{\partial \Delta A_{el}}{\partial \alpha_i^2}\right)_{T,V} \quad (5.5)$$

where $p^* = -(\partial \Delta A^*/\partial V)_T$ is the contribution to the pressure from intermolecular interactions such as those obtained in simple liquids. One may eliminate p^* from eq. (5.5) by taking the difference of t_i along two principal directions. This leads to the "Treloar relations[4]," expressed as

$$t_i - t_k = \frac{2}{V}\left[\alpha_i^2\left(\frac{\partial \Delta A_{el}}{\partial \alpha_i^2}\right) - \alpha_k^2\left(\frac{\partial \Delta A_{el}}{\partial \alpha_k^2}\right)\right]_{T,V} \quad (5.6)$$

In order to simplify the discussion in the remainder of this chapter, we focus on the state of uniaxial tension (or compression). If x_1 is taken as the direction of the force in uniaxial loading, then the lateral directions x_2 and x_3 will experience atmospheric pressure only. One may thus define the stress τ along the x_1 direction relative to atmospheric pressure as the difference $t_1 - t_2$:

$$\tau \equiv t_1 - t_2 = \frac{2}{V}\left(\frac{v_{2c}}{v_2}\right)^{2/3}\left(\alpha^2\frac{\partial \Delta A}{\partial \lambda_x^2} - \alpha^{-1}\frac{\partial \Delta A}{\partial \lambda_y^2}\right) \quad (5.7)$$

Experimental data for stress–strain experiments are generally analyzed in terms of the reduced force $[f^*]$, which is defined as

$$[f^*] \equiv \frac{fv_2^{1/3}}{A_d(\alpha - \alpha^{-2})} = \frac{\tau v_2^{-1/3}}{\alpha^2 - \alpha^{-1}} \tag{5.8}$$

Here, A_d is the dry, undeformed cross-sectional area. The second equality on the right-hand side of eq. (5.8) is obtained by using the relation $A_d/A = v_2^{2/3}\alpha$, which follows from eq. (5.4).

In the limit of small α, the reduced force $[f^*]$ of eq. (5.8) can be related to the elastic modulus E or the shear modulus G. The elastic modulus can be obtained from Hooke's law, $\sigma = E\varepsilon$, where the engineering stress σ is conventionally defined as the force per unit area measured at the beginning of the experiment, and ε is the infinitesimal strain (defined as the change in length per unit undeformed length). The initial state of the sample is defined as the swollen but mechanically undeformed state. As stated earlier, elastomers are approximately incompressible during mechanical distortion, and therefore G is related to E by the relation $G = E/3$. For small deformations, τ equates to σ in eq. (5.8), and the term $\alpha^2 - \alpha^{-1}$ may be taken as 3ε. With these substitutions, eq. (5.8) may be written as

$$[f^*] = \frac{\tau v_2^{-1/3}}{3\varepsilon} = \frac{Ev_2^{-1/3}}{3} = Gv_2^{-1/3} \tag{5.9}$$

$$G = [f^*]v_2^{1/3}$$

5.2 Stress–Strain Relations for the Phantom and Affine Network Models under Uniaxial Stress

The elastic free energies for the phantom and affine network models are given by eqs. (2.49) and (2.51), respectively. Differentiating these expressions with respect to λ_i^2 and substituting into eq. (5.7) leads to the stress for the phantom and affine network models in simple tension or compression:

$$\tau = \begin{cases} \dfrac{\xi kT}{V^0}\left(\dfrac{v_2}{v_{2c}}\right)^{1/3}(\alpha^2 - \alpha^{-1}) & \text{Phantom} \\ \dfrac{\phi}{(\phi - 2)}\dfrac{\xi kT}{V^0}\left(\dfrac{v_2}{v_{2c}}\right)^{1/3}(\alpha^2 - \alpha^{-1}) & \text{Affine} \end{cases} \tag{5.10}$$

The reduced force is obtained by substituting eq. (5.10) into eq. (5.8):

$$[f^*] = \begin{cases} \dfrac{\xi kT}{V_d}v_{2c}^{2/3} & \text{Phantom} \\ \dfrac{\phi}{(\phi - 2)}\dfrac{\xi kT}{V_d}v_{2c}^{2/3} & \text{Affine} \end{cases} \tag{5.11}$$

It follows from eq. (5.11) that the reduced force, or equivalently the elastic modulus, is independent of deformation and degree of swelling for both the phantom and the affine network models. Substituting eq. (5.11) into eq. (5.9) shows that the shear modulus of the phantom and the affine networks decreases with increasing swelling.

5.3 Stress–Strain Relations for the Constrained-Junction Model under Uniaxial Stress

The total elastic free energy of the constrained-junction model is given by the sum of the phantom elastic free energy and the free energy due to constraints, $\Delta A_{el} = \Delta A_{ph} + \Delta A_c$. The component of the stress due to the phantom network is already given in eq. (5.4). The derivative of the elastic free energy of the constrained-junction model given by eq. (3.39) is

$$\frac{\partial \Delta A_{el}}{\partial \lambda_t^2} = \frac{1}{2} \xi k T \left[1 + \frac{\mu}{\xi} K(\lambda_t^2) \right] \tag{5.12}$$

where

$$K(\lambda_t^2) = [B_t \dot{B}_t (B_t + 1) + D_t \dot{D}_t (D_t + 1)] \tag{5.13}$$

with

$$B_t = (\rho/\rho_{*x}) - 1 = \kappa^2 (\lambda_t^2 - 1)(\lambda_t^2 + \kappa)^{-2}$$

$$D_t = \lambda_t^2 \kappa^{-1} B_t$$

$$\dot{B}_t \equiv \frac{\partial B_t}{\partial \lambda_t^2} = B_t [(\lambda_t^2 - 1)^{-1} - 2(\lambda_t^2 + \kappa)^{-1}] \tag{5.14}$$

$$\dot{D}_t \equiv \frac{\partial D_t}{\partial \lambda_t^2} = \kappa^{-1} (\lambda^2 \dot{B}_t + B_t)$$

Substituting eqs. (5.12) and (5.13) into eq. (5.7) leads to the following expression for the true stress:

$$\tau = \frac{\xi k T}{V^0} \left(\frac{v_2}{v_{2c}} \right)^{1/3} \left(\alpha^2 - \alpha^{-1} + \left(\frac{\mu}{\xi} \right) [\alpha^2 K(\lambda_x^2) - \alpha^{-1} K(\lambda_y^2)] \right) \tag{5.15}$$

The reduced force is obtained by using eq. (5.15):

$$[f^*] = \left(\frac{\xi k T}{V_d} \right) v_{2c}^{2/3} \left\{ 1 + \frac{\mu}{\xi} [\alpha K(\lambda_1^2) - \alpha^{-2} K(\lambda_2^2)](\alpha - \alpha^{-2})^{-1} \right\} \tag{5.16}$$

When the constraint parameter κ becomes zero, the function in the square brackets of eq. (5.16) becomes zero and the reduced force equates to that of the phantom network. In the other extreme, when κ approaches infinity, the term in the braces equates to $1 + \mu/\xi$ and the reduced force becomes equal to that of the affine network. The contribution $[f^*]_c$ of the constraints relative to that of the phantom network, $[f^*]_{ph}$, may be seen from the ratio

48 STRUCTURES AND PROPERTIES OF RUBBERLIKE NETWORKS

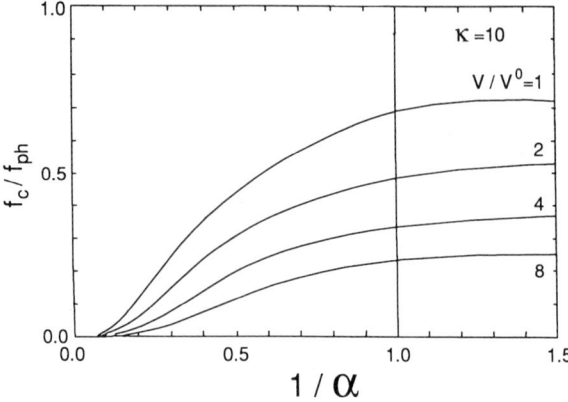

Figure 5.1 The dependence of the ratio given in eq. (5.17) on α and on V/V_0 for $\kappa = 10.^5$

$$\frac{[f^*]_c}{[f^*]_{ph}} = \frac{\mu}{\xi}[\alpha K(\lambda_1^2) - \alpha^{-2} K(\lambda_2^2)](\alpha - \alpha^{-2})^{-1} \tag{5.17}$$

In eq. (5.17), the effect of swelling is given in the arguments λ_1^2 and λ_2^2 of $K(\lambda_i^2)$. The dependence of the ratio given in eq. (5.17) on α and on V/V^0 is shown in figure 5.1 for $\kappa = 10.^5$ In this figure, the ordinate values of unity and zero correspond to the affine and phantom network models, respectively. The curves move toward the phantom network values with increasing values of V/V^0, in agreement with the postulate of the constrained-junction theory that swelling decreases the effects of constraints on junction fluctuations. Similarly, stretching the network decreases the constraints along that direction and the reduced force approaches that of the phantom network model (as observed from the $\alpha^{-1} \to 0$ intercepts of the curves).

5.4 Stress–Strain Relations for the Slip-Link and Constrained-Chain Models under Uniaxial Stress

5.4.1 The Slip-Link Model

The total elastic free energy of the slip-link model is given as the sum of the phantom elastic free energy and the free energy due to the slip-links, $\Delta A_{el} = \Delta A_{ph} + \Delta A_c$. Following the same procedure for obtaining the $K(\lambda_i^2)$ function outlined in the previous section, one obtains

$$K(\lambda_t^2) = \left[\frac{(B_t + D_t)(\dot{B}_t + \dot{D}_t)}{B_t + D_t + 1} - \lambda_t^{-2}\right] \tag{5.18}$$

with

$$\dot{B}_t \equiv \frac{\partial B_t}{\partial \lambda_t^2} = B_t[(\lambda_t^2 - 1)^{-1} - 2(\lambda_t^2 + \eta^{-1})^{-1}]$$
$$\dot{D}_t \equiv \frac{\partial D_t}{\partial \lambda_t^2} = \eta(\lambda^2 \dot{B}_t + B_t) \quad (5.19)$$

The stress, the reduced force, and the ratio $[f^*]_c/[f^*]_{ph}$ all follow from eqs. (5.15)–(5.17) by replacing μ with N_s.

5.4.2 The Constrained-Chain Model

The total elastic free energy for the constrained-chain model is defined in chapter 4. Differentiating the energy with respect to λ_t^2 yields the $K(\lambda_t^2)$ function as

$$K(\lambda_t^2) = \frac{B_t \dot{B}_t}{1 + B_t} + \frac{D_t \dot{D}_t}{1 + D_t} \quad (5.20)$$

where

$$\dot{B}_t \equiv \frac{\partial B_t}{\partial \lambda_t^2} = B_t \left\{ (\lambda_t^2 - 1)^{-1} - 2[\lambda_t^2 + h(\lambda_t)]^{-1} + \frac{\kappa_G}{h(\lambda_t)} \frac{[\lambda_t^2 - h(\lambda_t)]\Phi}{[\lambda_t^2 + h(\lambda_t)]} \right\} \quad (5.21)$$

$$\dot{D}_t \equiv \frac{\partial D_t}{\partial \lambda_t^2} = B_t \left[h(\lambda_t)^{-1} - \frac{\lambda_t^2 \kappa_G \Phi}{h(\lambda_t)^2} \right] + \frac{\lambda_t^2 \dot{B}}{h(\lambda_t)} \quad (5.22)$$

5.5 Comparison of Stress–Strain Relations with Experimental Data

The constrained-junction model has been very successful in predicting the stress–strain relations of elastomers. The slip-link model has not been tested extensively against experimental data but one should expect similar success for this model due to its resemblance to the constrained-junction model. The constrained-chain model, which is essentially a generalization of the constrained-junction model, also agrees well with experimental data. Some comparisons of the predictions of the constrained-junction and constrained-chain models with experimental data are given below.

Figure 5.2 presents some stress–strain results of Rivlin and Saunders[6] on natural rubber[7]. Results for compression were obtained by inflation of a sheet of the material, while those for extension were measured on a sample cut from the same specimen. Some of the results at high deformations may be artificially high because of strain-induced crystallization, and the results at small deformations may be susceptible to large errors. Given these complications, the theoretical curve gives a remarkably good account of the experimental results, over an impressively wide range of elongation and compression.

Some results emphasizing the effects of swelling on stress–strain isotherms in elongation are shown in figure 5.3[8]. All of the curves shown are for the set of

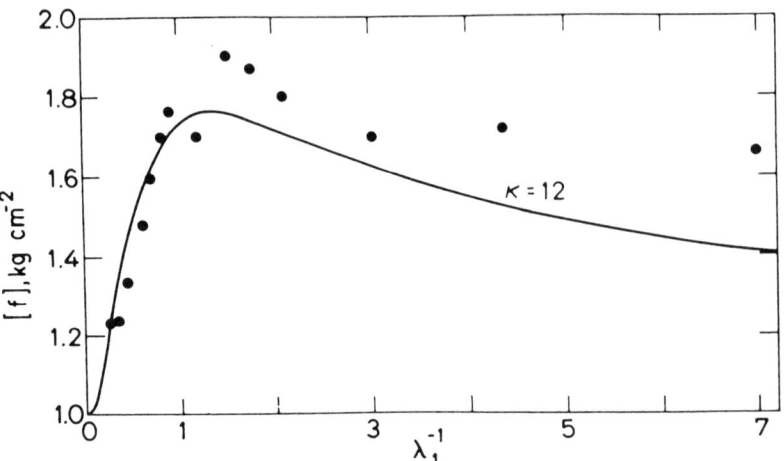

Figure 5.2 Results of Rivlin and Saunders[6] for natural rubber vulcanized with sulfur, shown by the data points, in comparison with the constrained-junction theoretical curve calculated with $\kappa = 12$ and ξ chosen to give $[f] = 1.05 \text{ kg cm}^{-2}$ at $\lambda_1^{-1} = 0$.[7]

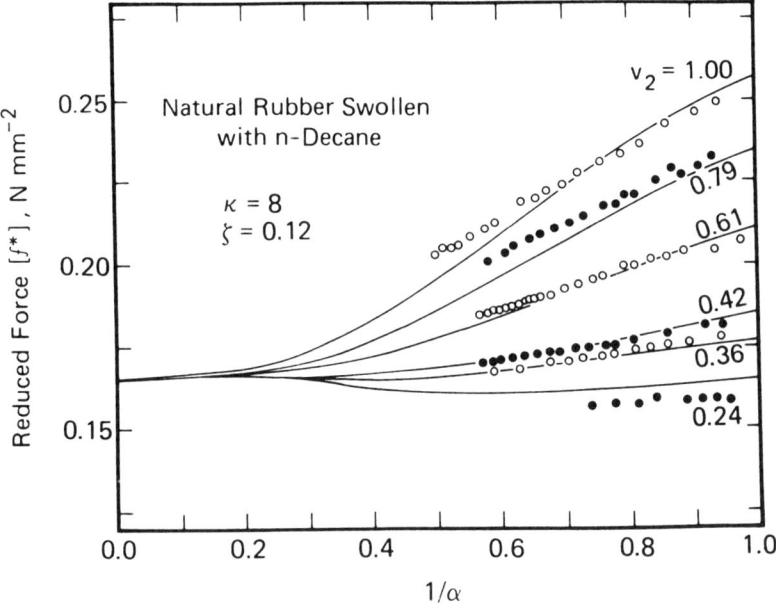

Figure 5.3 Reduced forces as a function of α^{-1} for natural rubber swollen to varying degrees with n-decane[8]. The points represent experimental results of Allen et al.[9] for specimens of their sample swollen to the extents indicated by the volume fractions v_2 of rubber included for each isotherm. The curves were calculated from the constrained-junction theory with $[f_{ph}^*] = 0.166 \text{ N mm}^{-2}$, $\kappa = 8$, and a value $\zeta = 0.12$ for a relatively unimportant parameter characterizing departures from affine transformations of some a priori probability functions in the theory[13]. Reprinted with permission from Erman, B. and Flory, P. J. (1982), *Macromolecules*, **15**. Copyright 1997 American Chemical Society.

parameters indicated, with the theory providing the changes in the isotherms with increased swelling (decreased v_2). Except at the lowest value of v_2, the dependence of stress on strain, and the effect of swelling on this dependence, are seen to be well reproduced by the theory.

Figure 5.4 compares results from the constrained-junction theory with experimental results[6] for multiaxial states of strain[9]. The theory is seen to closely approximate the experimental relationships among the specified elastic free energies and the principal extension ratios. Use of the Mooney-Rivlin relationship[4] would give the horizontal dashed line shown at the top of the figure. Phantom network results would also give a horizontal line[11] that would make them equally unsatisfactory for explaining the experimental observations.

Some comparisons involving the constrained-chain theory[12] are given in figure 5.5[13]. In this figure, the reduced force for a natural rubber network is shown as a function of the inverse extension ratio α^{-1}. The agreement between experiment and theory, similar to that for the constrained-junction model, is, in general, quite satisfactory.

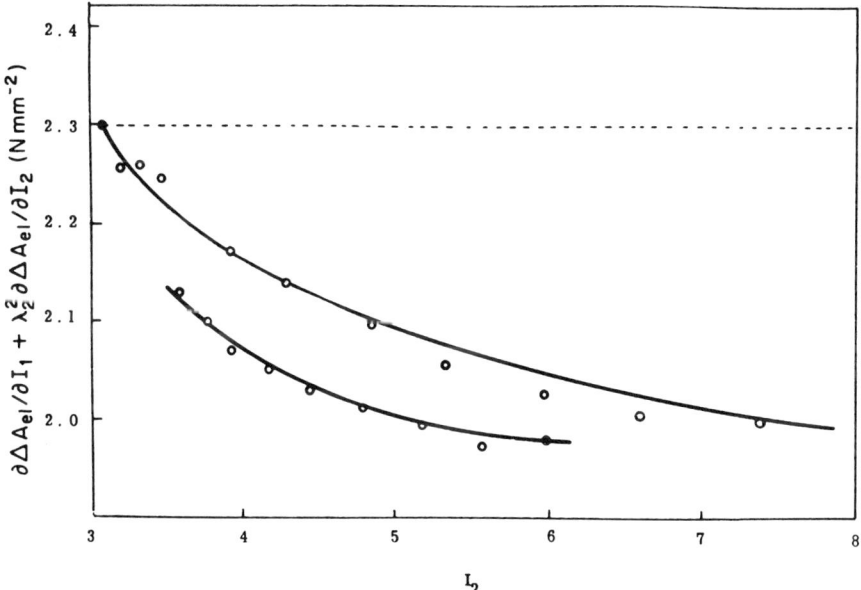

Figure 5.4 Comparison of experimental and theoretical results for natural rubber in pure shear and pure shear superposed on simple extension, where l_1 and l_2 are principal extension ratios[11]. Points show experimental data from Rivlin and Saunders[6], and the solid lines show results calculated from the constrained-junction theory with $\kappa = 3$. Values of the network structural parameter $\mu kT/2$ were arbitrarily chosen to match one experimental point (filled circle) with the theoretical result. The upper and lower solid lines represent results of calculations for pure shear and shear superposed on simple extension, respectively. The horizontal dashed line is for the Mooney-Rivlin relationship[4].

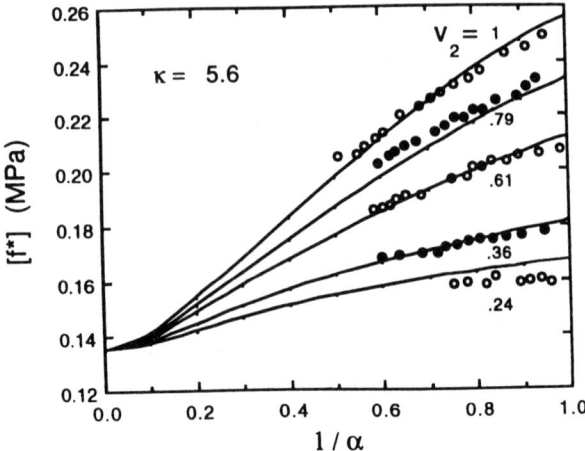

Figure 5.5 The reduced force for a natural rubber network[13,14] as a function of the inverse extension ratio α^{-1}. The circles show data points obtained for different degrees of swelling as identified on each curve. Calculations were performed according to the constrained-chain model[10] with $\kappa = 5.6$. Reprinted with permission from Erman, B. and Monnerie, L. (1992), *Macromolecules*, **25**. Copyright 1997 American Chemical Society.

References

(1) Murnaghan, F. D. *Finite Deformations of an Elastic Solid*. John Wiley & Sons: New York. 1951.
(2) Truesdell, C.; Toupin, R., Eds. *The Classical Field Theories*. Springer-Verlag: Berlin. 1960; Vol. III/1.
(3) Flory, P. J. *Trans. Faraday Soc.* 1961, 57, 829.
(4) Treloar, L. R. G. *The Physics of Rubber Elasticity*, 3rd ed. Clarendon Press: Oxford. 1975.
(5) Flory, P. J. *J. Chem. Phys.* 1977, 66, 5720.
(6) Rivlin, R. S.; Saunders, D. W. *Phil. Trans. Roy. Soc. London, Ser. A* 1951, 243, 251.
(7) Pak, H.; Flory, P. J. *J. Polym. Sci., Polym. Phys. Ed.* 1979, 17, 1845.
(8) Erman, B.; Flory, P. J. *Macromolecules* 1982, 15, 806.
(9) Allen, G.; Kirkham, M. J.; Padget, J.; Price, C. *Trans. Faraday Soc.* 1971, 67, 1278.
(10) Flory, P. J.; Erman, B. *Macromolecules* 1982, 15, 800.
(11) Erman, B. *J. Polym. Sci., Polym. Phys. Ed.* 1981, 19, 829.
(12) Erman, B.; Monnerie, L. *Macromolecules* 1989, 22, 3342.
(13) Erman, B.; Monnerie, L. *Macromolecules* 1992, 25, 4456 (correction to ref. 14).
(14) Fontaine, F.; Noel, C.; Monnerie, L.; Erman, B. *Macromolecules* 1989, 22, 3352 (see also ref. 13).

6

Swelling of Networks

In the preceding chapter, the stress–strain behavior of networks was described for the case where the swollen network was assumed to be a thermodynamically closed system (such that solvent molecules did not enter or leave the network during deformation). In this chapter, a more general thermodynamic analysis will be given for the case where the network–solvent system will be regarded as semi-open. In such systems, the solvent may enter or leave the network depending on the chemical potential of the solvent and the extent of deformation of the network. In the first section, we will describe general thermodynamic relations for network–solvent systems, followed by a discussion of isotropic swelling of networks in the second section. This topic has already been treated in classic books[1,2], and the reader is referred to them for background information. In this chapter, we discuss more recent improvements and approaches, and newer experiments in this field. In the third section of the chapter, we will describe the effects of an externally applied deformation on networks immersed in solvent. Again, the study of deformation of immersed networks goes back to approximately the mid-1940s[3,4], and only more recent developments will be included in this chapter. Also, we will consider in detail, in this chapter, the swelling of nonionic networks only. The effects of ionic groups on network chains are discussed in the following chapter in relation to critical phenomena and phase separation in swollen networks. We conclude with a discussion of sorption of linear and cyclic diluents into networks. Also covered is their extraction and, in the case of the cyclic molecules, their trapping within the network structure.

6.1 Free Energy of a Swollen Network

The free energy change ΔA from swelling an amorphous undeformed network with a solvent is given as the sum of two terms: one, the free energy of mixing ΔA_{mix}; and the other, the elastic free energy ΔA_{el} resulting from the dilation of the network:

$$\Delta A = \Delta A_{\text{mix}} + \Delta A_{el} \tag{6.1}$$

The expression for ΔA_{mix} is readily obtained from the lattice theory of polymer solutions. (Rigorously, it is the Gibbs free energy ΔG_{mix} that should be used for the mixing of the polymer, instead of the Helmholtz free energy. The difference is inconsequential, however.) The result is[5]

$$\begin{aligned}\Delta A_{\text{mix}} &= \Delta A_{\text{comb}} + A^R \\ &= RT(n_1 \ln v_1 + n_2 \ln v_2 + \bar{\chi} n_1 v_2)\end{aligned} \tag{6.2}$$

Here, ΔA_{comb} and A^R denote the combinatorial and the residual free energy, n_1 and n_2 are the mole numbers of solvent and polymer, respectively, and v_1 and v_2 are their volume fractions, respectively. In eq. (6.2), $\Delta A_{\text{comb}} = kT(n_1 \ln v_1 + n_2 \ln v_2)$ is obtained from the generalized ideal mixing law[6]. The residual free energy is $A^R = \bar{\chi} n_1 v_2$, where the quantity $\bar{\chi}$ is the dimensionless interaction parameter for the solvent–polymer system, and is, in general, a function of concentration. The quantity $kT\bar{\chi}$ is equal to the difference in energy of a solvent molecule immersed in pure polymer compared with a solvent molecule in the pure state. The residual free energy has acquired a central role in more recent studies of polymer–solvent or polymer–polymer mixtures because its concentration and temperature dependence has critical importance for the miscibility in such systems and for phase transitions. Its importance for critical phenomena and phase transitions in swollen networks will be discussed in chapter 7.

The elastic free energy in eq. (6.1) has been written for different models in the previous chapters. The discussion of swelling here will be directed to the constrained-junction model, in general, and will be given in terms of the functions B_t and D_t (described in chapter 3), which we write below for convenience:

$$\Delta A_{el} = \frac{1}{2} \xi kT \sum_t \{(\lambda_t^2 - 1) + (\mu/\xi)[B_t + D_t - \ln(B_t + 1) - \ln(D_t + 1)]\} \tag{6.3}$$

where t equates to x, y, or z. An advantage of adopting this general expression is that it simplifies readily into the expression for the phantom network by equating the terms in the square brackets to zero. By taking $B_t = \lambda_t^2 - 1$ and $D_t = 0$ in eq. (6.3), one obtains the expression for the affine model. The slip-link and the constrained-chain models may also be described with eq. (6.3) by making suitable adjustments, as was described in chapter 4. In the interest of studying the effects of isotropic dilation and anisotropic distortion separately, the components λ_t of the deformation tensor λ may be decomposed into two parts:

$$\lambda = (v_{2c}/v_2)^{1/3} \alpha = (V/V^0)^{1/3} \alpha \tag{6.4}$$

Here, v_{2c} is the volume fraction of polymer present during network formation, and λ and α are the extension ratios defined in appendix C. For swelling without distortion, α equates to the third-order unit tensor $\mathbf{E}$ and λ has three equal diagonal components, $\lambda = (v_{2c}/v_2)^{1/3}\mathbf{E}$. For uniaxial deformation along the x direction, the two lateral components of the extension ratio in eq. (6.4) become $\lambda_2 = \lambda_3 = [v_{2c}/v_2\lambda)]^{1/2} = (v_{2c}/v_2)^{1/2}\alpha^{-1/2}$, in which $\lambda_1 = \lambda$.

Once the free energy of the system is defined, several useful thermodynamic and mechanical relationships may be obtained for the swollen network. Below, we present a few of those that are often used in characterizing and studying such materials.

6.2 The Solvent Chemical Potential for an Isotropically Swollen Network

The chemical potential, $\Delta \mu_1 = \mu_1 - \mu_1^0$, of solvent in the swollen network is obtained by differentiating eq. (6.1) with respect to the number of solvent molecules at fixed temperature T and pressure p. A more convenient quantity to use is the reduced chemical potential, $\Delta \tilde{\mu}_1 = \Delta \mu_1 / RT$:

$$\begin{aligned} \Delta \tilde{\mu}_1 &= \ln a_1 = (RT)^{-1} \left(\frac{\partial \Delta A}{\partial n_1} \right)_{T,p} \\ &= (RT)^{-1} \left(\frac{\partial \Delta A_{\text{mix}}}{\partial n_1} \right)_{T,p} + (RT)^{-1} \left(\frac{\partial \Delta A_{el}}{\partial n_1} \right)_{T,p} \quad (6.5) \\ &= (\Delta \tilde{\mu}_1)_{\text{mix}} + (\Delta \tilde{\mu}_1)_{el} \end{aligned}$$

The quantity a_1 in the first line of eq. (6.5) represents the activity of the solvent in the network. The mixing component $(\Delta \tilde{\mu}_1)_{\text{mix}}$ of the reduced chemical potential defined in eq. (6.5) is obtained from eq. (6.2) as

$$(\Delta \tilde{\mu}_1)_{\text{mix}} = \ln(1 - v_2) + v_2 + \chi v_2^2 \quad (6.6)$$

The factor χ in eq. (6.6) is related to the interaction parameter $\bar{\chi}$ by the following relations:

$$\begin{aligned} \chi &\equiv \frac{1}{RT v_2^2} \frac{\partial A^R}{\partial n_1} \\ &= \frac{\partial (v_2 \bar{\chi})}{\partial v_2} - \frac{\partial \bar{\chi}}{\partial v_2} \quad (6.7) \\ &= \chi_1 + \chi_2 v_2 + \chi_3 v_2^2 + \cdots \end{aligned}$$

where the second line of eq. (6.7) is obtained by differentiating the third term of eq. (6.2). The third line of eq. (6.7) represents the Taylor series expansion for χ with only the first three terms shown. The parameters χ_1, χ_2 and χ_3 are now independent of concentration. The equation of state approach[7], random phase approximation[8], or the polymer reference interaction site model[9] are some of the techniques by which the interaction parameter may be related to molecular variables in a polymer–solvent system.

The elastic part of the reduced chemical potential is obtained by using the chain rule

$$(\Delta\tilde{\mu}_1)_{el} = \left(\frac{\partial \Delta A_{el}}{\partial \lambda}\right)_{T,p} \left(\frac{\partial \lambda}{\partial n_1}\right)_T$$
$$= \left(\frac{\partial \Delta A_{el}}{\partial \lambda}\right)_{T,p} \left(\frac{V_1}{3V^{2/3}V^{01/3}}\right) \tag{6.8}$$

The first factor on the right-hand side of eq. (6.8) is obtained by differentiating eq. (6.3) with respect to λ, after substituting $\lambda = \lambda_1 = \lambda_2 = \lambda_3$, and $\lambda^3 = V/V^0$ which corresponds to free swelling. The resulting expression is of the form

$$(\Delta\tilde{\mu}_1)_{el} = \left(\frac{\partial \Delta A_{el}/RT}{\partial \lambda}\right)_{T,p} = \frac{3\xi}{N_A}\left[1 + \frac{\mu}{\xi}K(\lambda)\right] \tag{6.9}$$

The second factor on the right-hand side of eq. (6.8) is obtained first by writing λ as

$$\lambda = \left(\frac{V}{V^0}\right)^{1/3} = \left(\frac{n_1 V_1 + xV_1 n_2}{V^0}\right)^{1/3} \tag{6.10}$$

where V_1 is the molar volume of solvent, and then differentiating with respect to n_1:

$$\left(\frac{\partial \lambda}{\partial n_1}\right)_T = \frac{1}{3}\frac{V_1}{V^{2/3}V^{01/3}} = \frac{1}{3}\frac{V_1}{V^0 \lambda^2} \tag{6.11}$$

Substituting eqs. (6.11) and eq. (6.10) into eq. (6.9) leads to the reduced solvent chemical potential due to the elastic activity of the network:

$$(\Delta\tilde{\mu}_1)_{el} = \frac{\beta}{\lambda}\left[1 + \frac{\mu}{\xi}K(\lambda)\right] \tag{6.12}$$

where $K(\lambda)$ is given by eq. (5.13), and

$$\beta = \frac{V_1}{RT}\frac{\xi kT}{V^0} = \frac{v_{2c}}{x_c}\left(1 - \frac{2}{\phi}\right) = \frac{v_{2c}\rho V_1}{M_c}\left(1 - \frac{2}{\phi}\right) \tag{6.13}$$

Here, M_c is the molecular weight of a network chain, ρ is the network density and the quantity x_c is the number of repeat units of the network chain, the volume of each being taken as equal to the volume of a solvent molecule.

Substituting eqs. (6.6), (6.7), and (6.12) into eq. (6.5) leads to the total reduced chemical potential of solvent in a swollen network:

$$\Delta\tilde{\mu}_1 = \ln(1 - v_2) + v_2 + \chi v_2^2 + \frac{\beta}{\lambda}\left[1 + \frac{\mu}{\xi}K(\lambda)\right] = \ln a_1 \tag{6.14}$$

When the network is immersed in solvent, it swells by taking up solvent until an equilibrium is reached between the entropic forces that encourage the solvent molecules to mix into the network, and the elastic forces of the stretched chains that tend to prevent further swelling. At conditions of equilibrium, the activity a_1 of solvent equates to unity and the solvent chemical potential equates to zero. The activity a_1 will thus be taken as unity in the following.

The solution of eq. (6.14) for v_2 gives the equilibrium degree of swelling. Alternatively, solving eq. (6.14) for β and using the expression for β given by eq. (6.13) leads to the molecular weight of a network chain between cross-links as

$$M_c = -\frac{\rho\left(1-\frac{2}{\phi}\right)V_1\left(\frac{v_2}{v_{2c}}\right)^{1/3}\left[1+\frac{\mu}{\xi}K(\lambda)\right]}{\ln(1-v_2)+v_2+\chi v_2^2} \quad (6.15)$$

where ρ denotes the network density during formation.

Equation (6.15) is a general expression for a perfect network of functionality ϕ. It is also general with regard to the function $K(\lambda)$ being left unspecified. By proper choice of this function, one may adopt any model ranging from the phantom network model to the constrained-junction model to the slip-link model, as described in chapter 5. Below, we give the explicit expression for two special cases: (1) a tetrafunctional phantom network, and (2) an affine network. For the phantom network model, $K(\lambda) = 0$ and eq. (6.15) takes the form

$$M_c = -\frac{\frac{1}{2}\rho V_1 \left(\frac{v_2}{v_{2c}}\right)^{1/3}}{\ln(1-v_2)+v_2+\chi v_2^2} \quad (6.16)$$

For the affine network model, $K(\lambda) = 1 - \lambda^{-2}$, and

$$M_c = -\frac{\rho V_1 v_{2c}^{-1/3}\left(v_2^{1/3} - \frac{1}{2}\frac{v_2}{v_{2c}}\right)}{\ln(1-v_2)+v_2+\chi v_2^2} \quad (6.17)$$

Equation (6.17) is the well-known Flory-Rehner equation[3], widely used for determining the cross-link density of a network. As specified, it is based on the affine network model. It should be noted, however, that the mechanical behavior of a swollen network is closer to that of the phantom network model, and therefore the use of eq. (6.16) rather than eq. (6.17) is more appropriate.

6.3 Thermodynamics of a Network Uniaxially Stretched in Solvent

When a network at swelling equilibrium is stretched uniaxially, in excess solvent, to a fixed extension ratio λ, equilibrium will be reestablished by the uptake of solvent by the network. Since the length is fixed, swelling will take place along the lateral directions. In this section, we give the derivation of some mechanical parameters, such as osmotic compressibility and bulk compressibility. For simplicity, we give the results for a network formed in the dry state, that is, with $v_{2c} = 1$. When the network is in equilibrium with the surrounding solvent at an extension ratio λ, the total volume will be V', the volume fraction of polymer will be v_2', and the lateral extension ratio will be related to the longitudinal one by[10]

$$\lambda = \alpha v_2^{-1/3}$$
$$\lambda_2 = \lambda_3 = v_2^{1/6}(v')^{-1/2}\alpha^{-1/2} \tag{6.18}$$

The reduced solvent chemical potential at constant length is

$$\Delta\tilde{\mu}_1 = (RT)^{-1}\left(\frac{\partial \Delta A}{\partial n_1}\right)_{T,p,\lambda} \tag{6.19}$$

leading to

$$\Delta\tilde{\mu}_1 = \ln(1 - v_2') + v_2' + \chi v_2'^2 + \beta\lambda_2^2 v_2'\left[1 + \frac{\mu}{\xi}K(\lambda_2)\right] = 0 \tag{6.20}$$

The change in volume of a network stretched in the immersed state is partly due to the solvent entering the system upon the application of the tensile force, and partly due to a decrease of the internal pressure that would take place if the system were closed. The coefficient of dilation η is defined as[11-13]

$$\begin{aligned}\eta &= \left(\frac{\partial \ln V}{\partial \ln \lambda}\right)_{T,p,\mu_1} \\ &= \left(\frac{\partial \ln V}{\partial \ln \lambda}\right)_{T,p,n_1} + \left(\frac{\partial \ln V}{\partial n_1}\right)_{T,p,\lambda}\left(\frac{\partial n_1}{\partial \ln \lambda}\right)_{T,p,\mu_1}\end{aligned} \tag{6.21}$$

The pressure p may be identified with the negative of the stress along the transverse directions. The first term in eq. (6.21), designated η_c, represents the coefficient for the volume change of the system at fixed composition. It may be expressed in terms of the pressure[12] as

$$\eta_c = \left(\frac{\partial \ln V}{\partial \ln \lambda}\right)_{T,p,n_1} = \lambda\kappa_L\left(\frac{\partial p}{\partial \lambda}\right)_{T,V,n_1} \tag{6.22}$$

where

$$\kappa_L = \frac{1}{V}\left(\frac{\partial V}{\partial p}\right)_{T,\lambda,n_1} \tag{6.23}$$

is the isothermal compressibility at fixed length. The second term in eq. (6.21) is the contribution from the change in composition and vanishes for closed systems. This term, denoted by η_s, is derived from the chemical potential of solvent as

$$\begin{aligned}\eta_s &= \left(\frac{\partial \ln V}{\partial n_1}\right)_{T,p,\lambda}\left(\frac{\partial n_1}{\partial \ln \lambda}\right)_{T,p,\mu_1} \\ &= \left(\frac{\partial \Delta\mu_1}{\partial \ln \lambda}\right)_{T,p,n_1} \bigg/ v_2'\left(\frac{\partial \Delta\mu_1}{\partial v_2'}\right)_{T,p,\lambda}\end{aligned} \tag{6.24}$$

The osmotic compressibility κ_{os} is simply the inverse of the bulk modulus K_v, and is defined as

$$K_v = \kappa_{os}^{-1} = v_2'\left(\frac{\partial \pi}{\partial v_2'}\right)_{T,p,\lambda} \tag{6.25}$$

where $\pi = -\Delta\mu_1/V_1$ is the osmotic pressure, namely, the extra pressure required inside the network (in the presence of the network chains which tend to lower μ_1) to give μ_1 the same value inside and outside the network[14]. Substituting eq. (6.25) into eq. (6.24) yields

$$\eta_s = \lambda \kappa_{os} \left(\frac{\partial \pi}{\partial \lambda}\right)_{T,p,n_1} \tag{6.26}$$

Equation (6.25) gives the bulk modulus at fixed length. Performing the differentiation in eq. (6.26) leads to the following expression for K_v:

$$K_v = \kappa_{os}^{-1} = \frac{RT}{V_1}[v_2'^2(1-v_2')^{-1} - 2\chi_1 v_2'^2 - 3\chi_2 v_2'^3 + \beta\left(\frac{v_2'}{v_{2c}}\right)\lambda_2^4\left(\frac{\mu}{\xi}\right)\dot{K}(\lambda_2^2)] \tag{6.27}$$

where $\dot{K}(\lambda_2^2) = \partial K(\lambda_2^2)/\partial \lambda_2^2$.

In most experimental work, the osmotic compressibility is measured in the swollen but undistorted state, that is, $\alpha = 1$. This is conveniently achieved by deswelling the network by bringing the swollen system into contact with a solution of known activity through a semipermeable membrane[15]. The linear dilation ratio λ then becomes $\lambda = \lambda_1 = \lambda_2 = \lambda_3 = (v_{2c}/v_2)^{1/3}$. The expression for the solvent chemical potential given by eq. (6.14) has to be used under these circumstances. The resulting expression is[13]

$$K_v = \kappa_{os}^{-1} = \frac{RT}{V_1}\left\{v_2^2(1-v_2)^{-1} - 2\chi_1 v_2^2 - 3\chi_2 v_2^3 \right.$$
$$\left. + \frac{\beta}{3}\left(\frac{v_2}{v_{2c}}\right)^{1/3}\left[1 + \left(\frac{\mu}{\xi}\right)K(\lambda^2) - 2\left(\frac{\mu}{\xi}\right)\left(\frac{v_2}{v_{2c}}\right)^{-2/3}\dot{K}(\lambda^2)\right]\right\} \tag{6.28}$$

The osmotic compressibility of a network depends strongly on the interaction parameter, as is seen in figure 6.1 (taken from the literature[13]), where the constrained-junction model was adopted with $\beta = 5 \times 10^{-6}$, $\kappa = 120$, and $RT/V_1 = 25\,\text{N mm}^{-2}$. The solid curve shows the result from eq. (6.28) for κ_{os} as a function of χ_1, where $\chi_2 = 0$. A crossover from high compressibilities for $\chi_1 < 0.5$ to very low compressibilities for $\chi_1 > 0.5$ is clearly seen. The dashed curve is obtained for $\chi_2 = 0.381$, a special value chosen to obtain a discontinuity in osmotic compressibility when $\chi_1 = 0.5$. Inasmuch as κ_{os}^{-1} is proportional to the second derivative of the elastic free energy, the point of discontinuity in figure 6.1 may be regarded as corresponding to a spinodal point. This will be discussed further in chapter 7.

The bulk compressibility of a swollen network is also sensitively dependent on the degree of swelling. The equilibrium degree of swelling of a network may vary because of changes in the χ parameter, as well as because of changes in the degree of cross-linking. Figure 6.2 shows results of experiments on poly(vinyl acetate) networks of different cross-link densities, swollen in acetone.

60 STRUCTURES AND PROPERTIES OF RUBBERLIKE NETWORKS

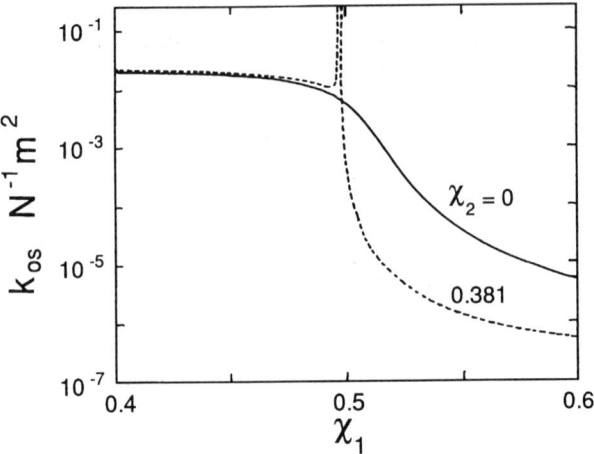

Figure 6.1 The osmotic compressibility of a network as a function of the interaction parameter. The results shown are taken from the literature[13].

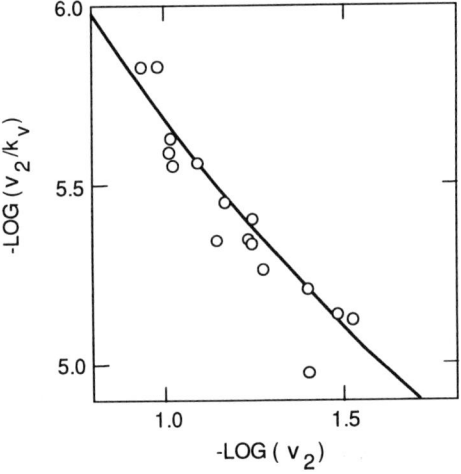

Figure 6.2 Results of experiments on poly(vinyl acetate) networks of different cross-link densities, swollen in acetone, and their comparison with theory. Circles represent data of Zrinyi and Horkay[15]. The solid curve is obtained from eq. (6.28) for a phantom network, with $\chi = 0.437 + 0.2v_2$.[13]

6.4 Elastic Activity of a Swollen Network

The difference between the chemical potential of a swollen network and an un-cross-linked bulk polymer at the same volume fraction v_2 shows the elastic contribution of the chains. The elastic contribution of the network to the chemical potential of the solvent is given by eq. (6.12). The activity a_1 of the solvent is given

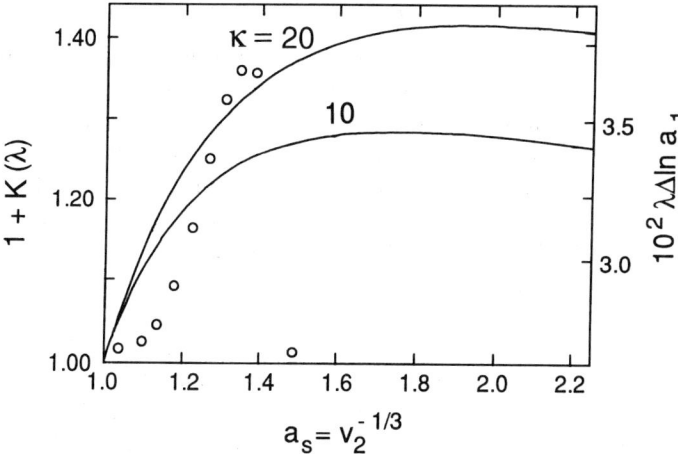

Figure 6.3 Activity differences $\Delta(\ln a_1)$ for networks and for the same polymer in the uncross-linked bulk of the same polymer. Open circles show results of measurements on the rubber–benzene system by Gee et al.[16]. All points except the right-most one were obtained by differential sorption measurements. This right-most point was obtained by equilibrium swelling measurements. The right-hand ordinate gives the experimentally determined values, $\Delta(\ln a_1)$ and the left-hand one gives the theoretical expression $1 + K(\lambda)$ for a tetrafunctional network. The solid curves were obtained by Flory[22] using the constrained-junction model for two values of κ, as specified in the figure.

through the relation $\ln a_1 = \Delta\mu_1$. If a_{1N} is the activity of the solvent in the network and a_{1L} is its activity in the un-cross-linked linear polymer at the same concentration, then eq. (6.14) may be written

$$\Delta(\ln a_1) \equiv \ln\left(\frac{a_{1N}}{a_{1L}}\right) = \frac{\beta}{\lambda}\left[1 + \frac{\mu}{\xi} K(\lambda)\right] \quad (6.29)$$

The activity differences $\Delta(\ln a_1)$ have been measured for the system rubber–benzene[16] and for poly(dimethylsiloxane) in various solvents[17–21]. Relevant experimental data from Gee et al.[16] are given in figure 6.3 by the open circles. The experimental points show a sharp maximum. This feature is also seen in the subsequent experiments of Eichinger et al.[17–21], performed with more accurate instruments. Although the theoretical curves show maxima, they are much less pronounced and at higher dilations than those given by experiment. The source of the discrepancy between experiment and theory in this matter is attributed to the nonseparability of the mixing and elastic free energies[17]. A statistical mechanical theory of network swelling obtained, using the replica formalism, by Ball and Edwards[23,24] results in the nonseparability of the mixing and elastic free energies. Whether the effects calculated by the Ball-Edwards theory correspond to experimental observations has not yet been established, however.

6.5 More Recent Treatments of Network Swelling

Interpretation of swelling data in terms of the constrained-junction model has led to the conclusion[25] that a highly swollen system may be treated essentially as a phantom network. Neutron scattering experiments by Bastide et al.[26] have shown that the transformation of chain dimensions in swollen networks gives results close to, and sometimes below, those predicted by the phantom network model. Comparison of neutron scattering data with predictions of the constrained-junction model has indeed shown[27] that the network chains deform less than is predicted for unswollen networks. In order to account for the low degrees of deformation of the chains, a theory of disinterspersion of networks has been proposed by Bastide et al.[28]. This approach is along the lines of the c^* theorem proposed by de Gennes[8], according to which the swollen coils exclude each other from a volume determined by the cross-link points, and the chains are essentially forced into contact at these points. The ideas have been discussed by Painter and Shenoy[29] in relation to the elastic activity of swollen networks. The effects of extractable components on the microscopic structure of a swollen network has also be investigated in some detail[30].

Extension of the theory of network swelling to ionic networks has been discussed by Flory[1]. However, a thorough understanding of the phenomenon of ionic gels was not possible until the 1990s. A treatment of positively ionized hydrogels has been presented by Hooper and coworkers[31,32], using the constrained-junction model for the elastic contribution, and the ideal Donnan theory for polyelectrolyte effects. Information on the synthesis, characterization, thermodynamics, mass transport, and applications of polyelectrolyte gels has been considered in the literature[33]. Properties of weakly charged polyelectrolytes have been investigated both theoretically and experimentally by Khokhlov and collaborators[34-36]. In the following chapter, we will study in more detail the effects resulting from the polyelectrolytic nature of the chains in relation to volume phase transitions in swollen networks.

The interesting problem of swelling a network of flexible chains in a nematogenic solvent has been studied theoretically by Brochard[37], who reached the conclusions that: (1) above the transition temperature of the nematic fluid, a uniaxial stress induces a nematic order within the network, and (2) below the transition temperature of the solvent, the gel collapses to a dense state and the solvent is expelled from the network. Predictions of the theory showed agreement with results of previous experiments by Rehage and collaborators[38] on networks swollen with nematogenic solvents.

The most thorough investigation of the thermodynamics of swollen networks has been carried out by Horkay, Zrinyi, Geissler, Hecht, and collaborators.[15,39-56]. Their work has mostly concentrated on studies of the mechanical properties (modulus and Poisson's ratio) of immersed and deformed networks, osmotic swelling and deswelling of networks, and small-angle neutron and x-ray scattering from swollen networks.

6.6 Sorption and Extraction of Diluents

The use of model networks has given a great deal of useful information leading to a better understanding of the properties of elastomeric materials at a molecular level. Much of this has to do with mechanical properties; for example, the evaluation of the molecular theories, the investigation of the effects of network chain-length distribution, and the effects of network imperfections such as dangling chains on ultimate strength and maximum extensibility. The present application is different in that it involves swelling as the elastomeric deformation, and specifically addresses questions related to the rate at which diluent absorbs into a network of known and controlled "pore size," and how rapidly it can subsequently be extracted[57–60]. The rate of absorption can be used to estimate diffusion coefficients, as can the rate of extraction. Extraction efficiencies can be used to obtain information on the extents of reaction in the end-linking procedure used to form the network, and the degree to which the extractable chains are entangled with the network chains that impede their removal from within the network structure. Of obvious interest is the dependence of these quantities on the molecular weight M_c of the network chains (as a measure of pore size), the molecular weight M_d of the diluent, the structure of the diluent (linear, branched, or cyclic), and whether or not the diluent is present during the end-linking process.

6.6.1 Linear Diluents

One way of obtaining a network swollen with diluent is to form the network in a first step, and then to absorb an unreactive diluent into it. Alternatively, the same diluent can be mixed into the reactive chains prior to their being end-linked into a network structure. In either case, oligomeric and polymeric diluents are of greatest interest, and they must be functionally inactive for them to "reptate" through the network (with snake-like motions) rather than being bonded to it. Both types of networks can then be extracted to determine the ease with which the various diluents can be removed, as a function of M_d and M_c.

Some typical results obtained as a function of M_c are shown in figure 6.4[58], and with respect to M_d in figure 6.5[57]. The ease with which a diluent could be removed from a network was found to decrease with increase in M_d and with decrease in M_c, as expected. High-molecular-weight diluents are extremely hard to remove at the values of M_c of interest in the preparation of model networks—a circumstance complicating the analysis of soluble polymer fractions in terms of degrees of perfection of the network structure. Figure 6.6 illustrates differences in the extraction efficiency observed for two types of networks[57]. As can readily be seen, the diluents added after the end-linking are the more easily removed, possibly because they are less entangled with the network structure, and this could correspond to differences in chain conformations of the diluent.

Some of these data can be used to estimate values of the diffusion coefficient D, either in sorption or in extraction. Such coefficients can also be estimated using

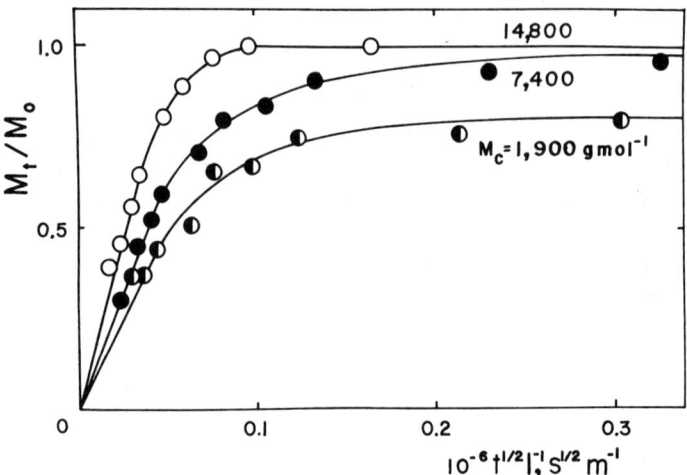

Figure 6.4 Amount of diluent extracted from PDMS networks, relative to the amount originally present, shown as a function of the square root of the extraction time (normalized by the thickness of the sample)[58]. In these illustrative results, at room temperature, the diluent was present during network formation and the solvent employed for the extraction was toluene. The diluent had a molecular weight M_d of 9600 g mol^{-1} and the network chains had the values of the molecular weight M_c between cross-links as specified in the figure.

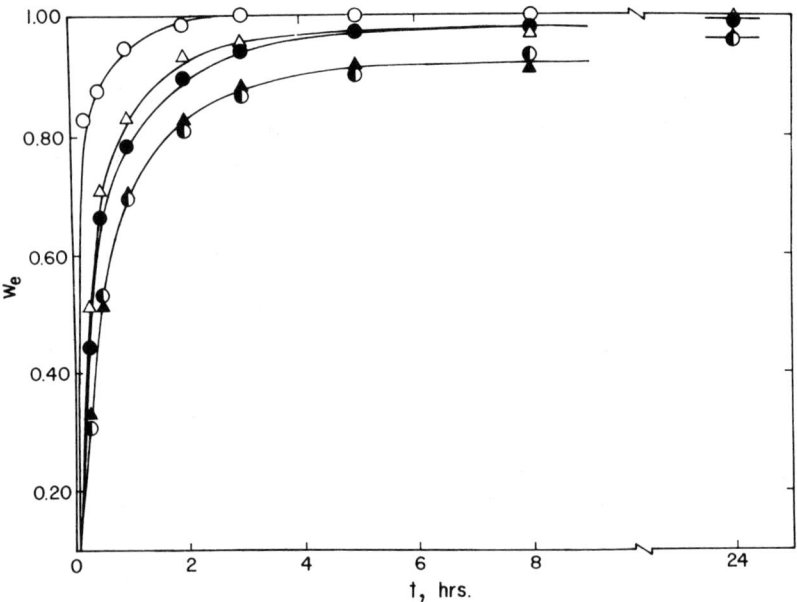

Figure 6.5 Time dependence of the weight fraction w_e of dimethylsiloxane diluent extracted from five PDMS networks ($v_2 \sim 0.65$) which were end-linked while diluted ($v_{2s} \sim 0.65$) with the linear PDMS diluents to be extracted[57]. The network had a value of $M_c = 11.3 \times 10^3$ g mol^{-1}, and the diluents had molecular weights corresponding to $10^{-3} M_d = 26.4$ (▲), 18.6 (●), 9.8 (◐), 6.7 (△), and 1.2 (○) g mol^{-1}.

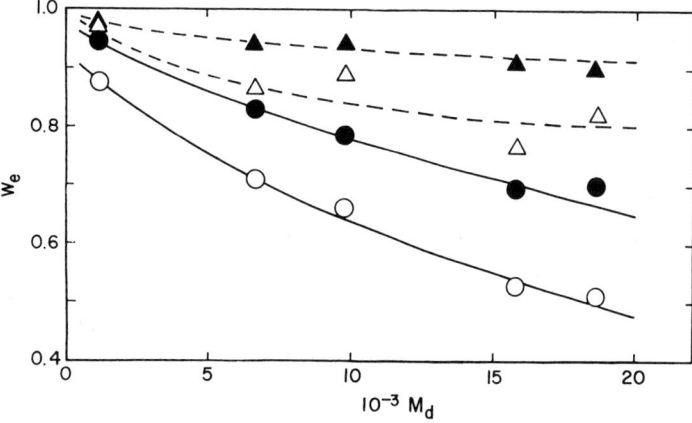

Figure 6.6 Differences in extraction efficiency for the two types of PDMS networks, as determined from results such as those shown in figure 6.5; polymer end-linked with diluent present, total extraction times of 0.5 (○) and 1.0 h (●); diluent added after end-linking: 0.5 (△) and 1.0 h (▲)[57].

networks at swelling equilibrium and either increasing the degree of swelling by stretching the network[61] or decreasing the degree of swelling by compressing it. It is also possible to use pulsed gradient nuclear magnetic resonance (NMR) to estimate diffusion coefficients[62]. In any case, there is particular interest in the differences in values obtained for linear chains and for the cyclics of the same molecular weight, as described in Section 6.6.3.

6.6.2 Branched Diluents

The extraction of diluents that are branched has considerable potential importance. For example, in the preparation of networks by radiation cross linking, there are presumably large numbers of soluble molecules formed that are highly branched (because of the essentially uncontrolled formation and coupling of free-radical species)[63]. Another example occurs in the area of the controlled release of drugs from cross-linked reservoirs[64]. Some of the drugs diffusing out of such delivery systems are assuredly "nonlinear," and the problems of understanding such diffusion are likely to become more important as interest shifts to larger and larger molecules (such as polypeptides). Finally, it should be mentioned that molecules with long branches can greatly affect the flow characteristics of a polymer[65], to the extent that they can be used as processing aids in polymeric systems that will subsequently be cross-linked. The ultimate in branched systems, the "dendrimers"[66,67], may eventually also find applications in network structures.

Unfortunately, however, such systems have apparently not been studied in this regard in any quantitative way. The effects of the branching are probably difficult to predict, since the increased compactness of branched molecules may be offset by the increased congestion and entangling around the branch points. Perhaps definitive experiments could be carried out using model branched molecules which have been prepared by controlled synthetic methods[68], and then purified and characterized by separation techniques such as gel permeation chromatography[69].

6.6.3 Cyclic Diluents

Diluents consisting of cyclic molecules have been extensively studied with regard to diffusion, in part because of the generation of such species in many polymerizations[70,71]. Again, such cyclics can be sorbed into the networks after the end-linking process, or they can be present during the process. In the latter case, some are permanently trapped, as will be described separately later, making difficult the calculation of diffusion coefficients D from the extraction data. In the former case, however, D is readily calculable. For both the linear and cyclic chains, D increases with decease in M_d or with increase in M_c, as expected. The cyclics have values of D larger than those of the corresponding linear chains, presumably because of their greater compactness. Quantitatively, they are found to be higher by a factor of 1.18, which is in excellent agreement with theory[72].

6.7 Trapping of Cyclics within Network Structures

6.7.1 Experimental Results

As already mentioned, if cyclic molecules are present during the end-linking of chains, some of them will be trapped because of having been threaded by the linear chains prior to the latter being chemically bonded into the network structure[73-76]. The fraction trapped is readily estimated from solvent extraction studies. Some typical results, in terms of the fraction trapped as a function of the degree of polymerization (DP) of the cyclic, are shown in figure 6.7[77], where the experimental results are represented by the open circles and the corresponding curve. The results were found to be independent of the amount of time given for the cyclics and linear chains to intermingle[76], thus demonstrating the very high mobility of the poly(dimethylsiloxane) (PDMS) chains. As expected, very small cyclics do not get trapped at all, but almost all of the largest cyclics do. The following section describes the interpretation of these results in terms of the configurational characteristics of PDMS chains.

These cyclics can change the properties of the network in which they are incarcerated. Since they restrict, to some extent, the motions of the network chains, they should increase the modulus of an elastomer. Some small but possibly significant increases in low-deformation modulus have, in fact, been observed[74]. Also, when PDMS cyclics are trapped in a thermoplastic material, they can act as a plasticizer that is, in a sense, intermediate to the usual external (dissolved) and internal (copolymerized) varieties. Interesting changes in mechanical properties have been observed in materials of this type[78,79].

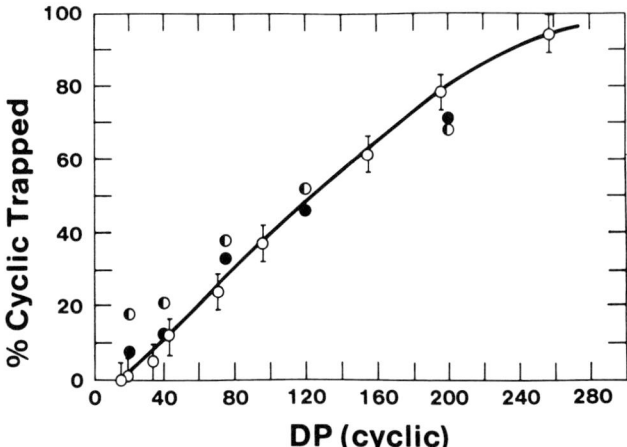

Figure 6.7 Experimental (○) and theoretical trapping efficiencies, using an overly simplified model[77] (◐) and a more realistic model (●), as described in the text for PDMS cyclics trapped in a model PDMS network[77]. Reprinted with permission from DeBolt, L. C. and Mark, J. E. (1984), *Macromolecules*, **20**. Copyright 1997 American Chemical Society.

6.7.2 Theoretical Interpretations

The trapping process has been simulated using Monte Carlo methods based on a rotational isomeric state model[80] for the cyclic chains[77]. The first step was generation of a sufficient number of cyclic chains having the known geometric features and conformational preferences, and the desired degree of polymerization. This was done for a large number of linear chains, using the methods of matrix multiplication standard to rotational isomeric state theory[80,81]. Up to this point, the method is identical with that used to generate distribution functions for a non-Gaussian approach to rubberlike elasticity, as described in chapter 8. In the present application, however, a chain having an end-to-end distance R less than a threshold value R_0 was considered to be a cyclic. The scheme is illustrated in figure 6.8[77] for a chain labeled 0 at one end and N at the other. The coordinates of each "cyclic" chain thus generated were stored for detailed examination of the chain's configurational characteristics, particularly the size of the "hole" it would present to a threading linear chain. Of particular interest is the size of this hole in comparison with the known diameter, 7–8 Å, of the PDMS chain.

The quality of the sets of cyclic conformations thus generated was tested by evaluating the mean-square unperturbed radius of gyration $\langle s^2 \rangle_0$, for each ring size[77], using the matrix multiplication methods[80,81]. The same quantity was calculated for linear PDMS chains of the same degree of polymerization. The results, in terms of the logarithmic dependence of $\langle s^2 \rangle_0$ on the degree of polymerization, are shown in figure 6.9. The linear relationships observed confirm previous results for shorter PDMS chains. The ratio of $\langle s^2 \rangle_0$ for the linear chains to that for the

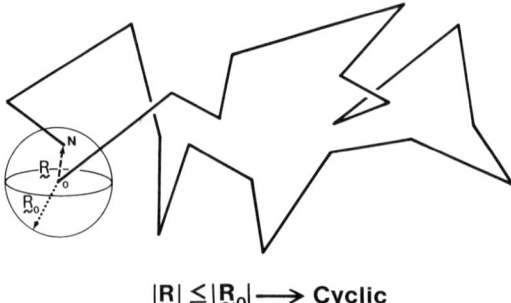

$|\underset{\sim}{R}| \leq |\underset{\sim}{R}_o| \longrightarrow$ **Cyclic**

Figure 6.8 Schematic of the criterion for "cyclization" of a generated Monte Carlo chain[77].

cyclics is shown in figure 6.10, and is seen to approach a value of 2 at large degrees of polymerization, as expected[82].

The trapping process was simulated using two models[77], the more realistic of which is described here. As shown schematically in figure 6.11, a torus was centered around each repeat unit in the cyclic. Any torus found to be "empty" was considered to provide a pathway for a chain of specified diameter to pass through, threading it and then "incarcerating" it once the end-linking process had been completed. The simulation results thus obtained are also shown in figure 6.7, and are seen to give a good representation of the experimental trapping efficiencies.

It is also possible to interpret these experimental results in terms of a power law for the trapping probabilities and fractal cross sections for the PDMS chains[83].

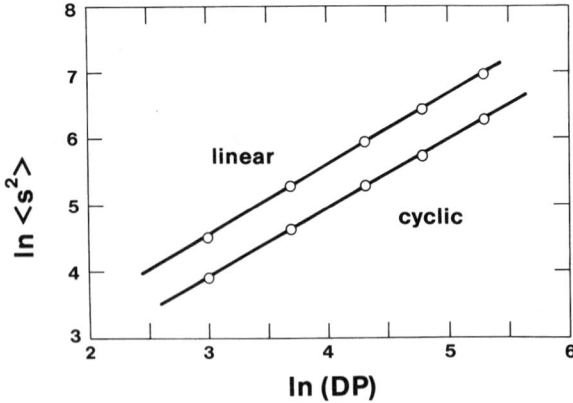

Figure 6.9 Logarithmic representation of the dependence of the mean-square unperturbed radius of gyration on the degree of polymerization for linear and cyclic PDMS[77]. Reprinted with permission from DeBolt, L. C. and Mark, J. E. (1984), *Macromolecules*, **20**. Copyright 1997 American Chemical Society.

SWELLING OF NETWORKS 69

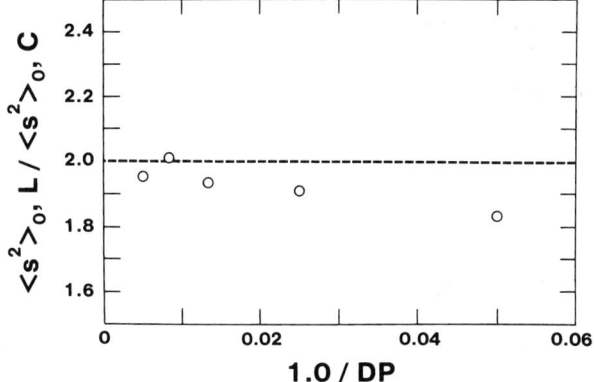

Figure 6.10 Ratio of the mean-square unperturbed radius of gyration of linear to cyclic PDMS as a function of the inverse degree of polymerization[77]. Reprinted with permission from DeBolt, L. C. and Mark, J. E. (1984), *Macromolecules*, **20**. Copyright 1997 American Chemical Society.

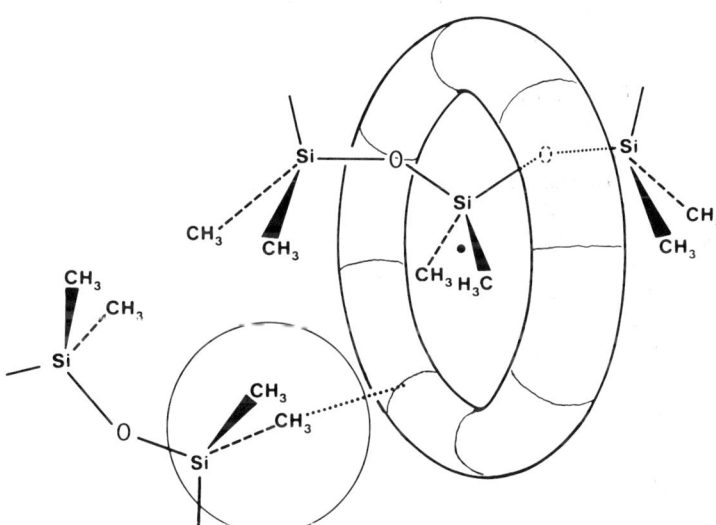

Figure 6.11 Schematic of the more realistic model, showing the criterion for deciding whether a cyclic has a unit free of conflicts, thus permitting the cyclic to be trapped[77]. Reprinted with permission from DeBolt, L. C. and Mark, J. E. (1984), *Macromolecules*, **20**. Copyright 1997 American Chemical Society.

6.7.3 Olympic Networks

It may also be possible to use this technique to prepare networks having no cross-links whatsoever[84]. Mixing linear chains with large amounts of cyclic and then *di*functionally end-linking them could give sufficient cyclic interlinking to yield an

"Olympic" or "chain-mail" network[8,73]. Such materials would be very interesting topologically, and would be similar, in some respects, to the catenanes and rotaxanes that have long been of interest to a variety of scientists and mathematicians[78,85–95]. Computer simulations[89], in particular, could be very useful with regard to establishing the conditions most likely to produce these novel structures.

References

(1) Flory, P. J. *Principles of Polymer Chemistry*. Cornell University Press: Ithaca, NY. 1953.
(2) Treloar, L. R. G. *The Physics of Rubber Elasticity*, 3rd ed. Clarendon Press: Oxford. 1975.
(3) Flory, P. J.; Rehner, J. *J. Chem. Phys.* 1944, 12, 412.
(4) Gee, G. *Trans. Faraday Soc.* 1946, 42B, 33.
(5) Erman, B.; Flory, P. J. *Polym. Commun.* 1984, 25, 132.
(6) Flory, P. J. *Principles of Polymer Chemistry*; Cornell University Press: Ithaca, NY. 1953; eq. (10) on p. 502.
(7) Flory, P. J. *Discuss. Faraday Soc.* 1970, 49, 7.
(8) de Gennes, P. G. *Scaling Concepts in Polymer Physics*. Cornell University Press: Ithaca, NY. 1979.
(9) Schweizer, K. S.; Curro, J. G. *Macromolecules* 1988, 21, 3070.
(10) Erman, B.; Flory, P. J. *Macromolecules* 1983, 16, 1607.
(11) Flory, P. J. *Trans. Faraday Soc.* 1961, 57, 829.
(12) Flory, P. J.; Tatara, Y. *J. Polym. Sci., Polym. Phys. Ed.* 1975, 13, 683.
(13) Bahar, I.; Erman, B. *Macromolecules* 1987, 20, 1696.
(14) Hill, T. L. *An Introduction to Statistical Thermodynamics*. Addison-Wesley: Reading, MA. 1960.
(15) Zrinyi, M.; Horkay, F. *J. Polym. Sci., Polym. Phys. Ed.* 1982, 20, 815.
(16) Gee, G.; Herbert, J. B. M.; Roberts, R. C. *Polymer* 1965, 6, 541.
(17) Yen, L. Y.; Eichinger, B. E. *J. Polym. Sci., Polym. Phys. Ed.* 1978, 16, 121.
(18) Brotzman, R. W.; Eichinger, B. E. *Macromolecules* 1981, 14, 1445.
(19) Brotzman, R. W.; Eichinger, B. E. *Macromolecules* 1982, 15, 531.
(20) Brotzman, R. W.; Eichinger, B. E. *Macromolecules* 1983, 16, 1131.
(21) Neuburger, N. A.; Eichinger, B. E. *Macromolecules* 1988, 21, 3060.
(22) Flory, P. J. *Macromolecules* 1979, 12, 119.
(23) Ball, R. C.; Edwards, S. F. *Polymer* 1979, 20, 1357.
(24) Ball, R. C.; Edwards, S. F. *Macromolecules* 1980, 13, 748.
(25) Erman, B.; Flory, P. J. *Macromolecules* 1982, 15, 806.
(26) Bastide, J.; Duplessix, R.; Picot, C.; Candau, S. *Macromolecules* 1984, 17, 83.
(27) Erman, B. *Macromolecules* 1987, 20, 1917.
(28) Bastide, J.; Picot, C.; Candau, S. *Macromol. Sci.* 1981, B19, 13.
(29) Painter, P. C.; Shenoy, S. L. *J. Chem. Phys.* 1993, 99, 1409.
(30) Falcao, A. N.; Pedersen, J. S.; Mortensen, K.; Boue, F. *Macromolecules* 1996, 29, 809.
(31) Hooper, H. H.; Baker, J. P.; Blanch, H. W.; Prausnitz, J. M. *Macromolecules* 1990, 23, 1096.
(32) Beltran, S.; Hooper, H. H.; Blanch, H. W.; Prausnitz, J. M. *J. Chem. Phys.* 1990, 92, 2061.

(33) Harland, R. S.; Prud'homme, R. K., Eds. *Polyelectrolyte Gels*. American Chemical Society: Washington, DC. 1992; Vol. 480.
(34) Khokhlov, A. R. *Polymer* 1980, 21, 376.
(35) Khokhlov, A. R.; Starodubtsev, S. G.; Vasilevskaya, V. V. *Adv. Polym. Sci.* 1993, 109, 123.
(36) Philippova, O. E.; Pieper, T. G.; Sitnikova, N. L.; Starodubtsev, S. G.; Khokhlov, A. R.; Kilian, H. G. *Macromolecules* 1995, 28, 3925.
(37) Brochard, F. *J. Physique* 1979, 40, 1049.
(38) Gebhard, G.; Rehage, G.; Schwarz, J. In *Proceedings of the 4th International Conference on the Physics of Non-Crystalline Solids*. Clausthal. 1976; p. 81.
(39) Hecht, A. M.; Geissler, E. *Polymer* 1980, 21, 1358.
(40) Geissler, E.; Hecht, A. M. *Macromolecules* 1980, 13, 1276.
(41) Geissler, E.; Hecht, A. M. *Macromolecules* 1981, 14, 185.
(42) Horkay, F.; Zrinyi, M. *J. Macromol. Sci.—Phys.* 1986, B25, 307.
(43) Horkay, F.; Zrinyi, M. *Macromolecules* 1988, 21, 3260.
(44) Horkay, F.; Geissler, E.; Hecht, A. M.; Zrinyi, M. *Macromolecules* 1988, 21, 2589.
(45) Geissler, E.; Hecht, A. M.; Horkay, F.; Zrinyi, M. *Macromolecules* 1988, 21, 2599.
(46) Horkay, F.; Geissler, E.; Hecht, A. M. *Macromolecules* 1989, 22, 2007.
(47) Horkay, F.; Hecht, A. M.; Geissler, E. *J. Chem. Phys.* 1989, 91, 2706.
(48) Horkay, F.; Zrinyi, M.; Geissler, E.; Hecht, A. M. *Makromol. Chem., Macromol. Symp.* 1990, 40, 195.
(49) Hecht, A. M.; Horkay, F.; Geissler, E.; Zrinyi, M. *Polym. Commun.* 1990, 31, 53.
(50) Horkay, F.; Zrinyi, M.; Geissler, E.; Hecht, A. M.; Pruvost, P. *Polymer* 1991, 32, 835.
(51) Horkay, F.; Hecht, A. M.; Mallam, S.; Geissler, E.; Rennie, A. R. *Macromolecules* 1991, 24, 2896.
(52) Zrinyi, M.; Rosta, J.; Horkay, F. *Macromolecules* 1993, 26, 3097.
(53) Horkay, F.; Burchard, W.; Geissler, E.; Hecht, A. M. *Macromolecules* 1993, 26, 1296.
(54) Horkay, F.; Burchard, W.; Hecht, A. M.; Geissler, E. *Macromolecules* 1993, 26, 3375.
(55) Hecht, A. M.; Stanley, H. B.; Geissler, E.; Horkay, F.; Zrinyi, M. *Polym. Commun.* 1993, 31, 2894.
(56) Geissler, E.; Horkay, F.; Hecht, A. M. *Phys. Rev. Lett.* 1993, 71, 645.
(57) Mark, J. E.; Zhang, Z.-M. *J. Polym. Sci., Polym. Phys. Ed.* 1983, 21, 1971.
(58) Garrido, L.; Mark, J. E. *J. Polym. Sci., Polym. Phys. Ed.* 1985, 23, 1933.
(59) Mark, J. E. *J. Appl. Polym. Sci., Symp.* 1989, 44, 209.
(60) Mazan, J.; Leclerc, B.; Galandrin, N.; Couarraze, G. *Eur. Polym. J.* 1995, 31, 803.
(61) Ozkul, M. H.; Onaran, K.; Erman, B. *J. Polym. Sci., Polym. Phys. Ed.* 1990, 28, 1781.
(62) Garrido, L.; Mark, J. E.; Ackerman, J. L.; Kinsey, R. A. *J. Polym. Sci., Polym. Phys. Ed.* 1988, 26, 2367.
(63) Dole, M., Ed. *The Radiation Chemistry of Macromolecules*. Academic Press: New York. 1973; Vols. 1 and 2.
(64) Peppas, N. A.; Langer, R. *Science* 1994, 263, 1715.
(65) Graessley, W. W. In *Physical Properties of Polymers*, J. E. Mark, A. Eisenberg, W. W. Graessley, L. Mandelkern, E. T. Samulski, J. L. Koenig, and G. D. Wignall, Eds. American Chemical Society: Washington, DC. 1993; p. 97.
(66) Frechet, J. M. J. *Science* 1994, 263, 1710.
(67) Tomalia, D. A. *Sci. Am.* 1995, 272(5), 62.

(68) Rempp, P.; Lutz, P. *Macromol. Symp.* 1992, 62, 213.
(69) Fried, J. R. *Polymer Science and Technology.* Prentice Hall: Englewood Cliffs, NJ. 1995.
(70) Odian, G. *Principles of Polymerization,* 3rd ed. Wiley-Interscience: New York. 1991.
(71) Semlyen, J. A. In *Siloxane Polymers,* S. J. Clarson and J. A. Semlyen, Eds. Prentice Hall: Englewood Cliffs, NJ. 1993; p. 135.
(72) Garrido, L.; Mark, J. E.; Clarson, S. J.; Semlyen, J. A. *Polym. Commun.* 1984, 25, 218.
(73) Garrido, L.; Mark, J. E.; Clarson, S. J.; Semlyen, J. A. *Polym. Commun.* 1985, 26, 53.
(74) Garrido, L.; Mark, J. E.; Clarson, S. J.; Semlyen, J. A. *Polym. Commun.* 1985, 26, 55.
(75) Clarson, S. J.; Mark, J. E.; Semlyen, J. A. *Polym. Commun.* 1986, 27, 244.
(76) Clarson, S. J.; Mark, J. E.; Semlyen, J. A. *Polym. Commun.* 1987, 28, 151.
(77) DeBolt, L. C.; Mark, J. E. *Macromolecules* 1984, 20, 2369.
(78) Fyvie, T. J.; Frisch, H. L.; Semlyen, J. A.; Clarson, S. J.; Mark, J. E. *J. Polym. Sci., Polym. Chem. Ed.* 1987, 25, 2503.
(79) Huang, W.; Frisch, H. L.; Hua, Y.; Semlyen, J. A. *J. Polym. Sci., Polym. Chem. Ed.* 1990, 26, 1807.
(80) Flory, P. J. *Statistical Mechanics of Chain Molecules.* Interscience: New York. 1969.
(81) Mattice, W. L.; Suter, U. W. *Conformational Theory of Large Molecules.* Wiley: New York. 1994.
(82) Casassa, E. F. *J. Polym. Sci., Part A* 1965, 3, 605.
(83) Galiatsatos, V.; Eichinger, B. E. *Polym. Commun.* 1987, 28, 182.
(84) Mark, J. E.; Erman, B. *Rubberlike Elasticity. A Molecular Primer.* Wiley-Interscience: New York. 1988.
(85) Frisch, H. L.; Wasserman, E. *J. Am. Chem. Soc.* 1961, 83, 3789.
(86) Callahan, D.; Frisch, H. L.; Klempner, D. *Polym. Eng. Sci.* 1975, 15, 70.
(87) Rigbi, Z.; Mark, J. E. *J. Polym. Sci., Polym. Phys. Ed.* 1986, 24, 443.
(88) Clarson, S. J.; Wang, Z.; Mark, J. E. *J. Inorg. Organomet. Polym.* 1991, 1, 223.
(89) Iwata, K.; Ohtsuki, T. *J. Polym. Sci., Polym. Phys. Ed.* 1993, 31, 441.
(90) Frisch, H. L. *New J. Chem.* 1993, 17, 697.
(91) Mark, J. E. *New J. Chem.* 1993, 17, 703.
(92) Wood, B. R.; Semlyen, J. A.; Hodge, P. *Polymer* 1994, 35, 1542.
(93) Gibson, H. W.; Bheda, M. C.; Engen, P. T. *Prog. Polym. Sci.* 1994, 19, 843.
(94) Yamaguchi, I.; Osakada, K.; Yamamoto, T. *J. Am. Chem. Soc.* 1996, 118, 1811.
(95) Gibson, H. W.; Liu, S. *Macromol. Symp.* 1996, 102, 55.

7

Critical Phenomena and Phase Transitions in Gels

The term "gel" has been used in a wide variety of contexts, and there have been difficulties in reaching an all-inclusive, workable definition for it[1,2]. Perhaps the simplest way to proceed is to list some of its most important characteristics[1]:

It is a solidlike material that when deformed responds in the manner of a typical elastic body, but generally with a very small modulus.

If it does show plastic flow, then this occurs above a threshold value of the stress, with full recoverability below this limit.

It typically consists of two or more components: one a liquid in substantial quantity, and the other generally a polymeric network.

One of the most direct ways of obtaining a gel is to place a network into a solvent known to be capable of dissolving the network chains in the absence of cross-links. In fact, a unique property of a highly extensible elastomer (resulting from a low degree of cross-linking) is its ability to swell greatly when exposed to a good solvent. A gel with less than 10^{-6} mol cm^{-3} of cross-links, for example, may increase its volume more than thousandfold when immersed in a suitable solvent. The extent to which such a network will swell depends specifically not only on the degree of cross-linking, but also on the interactions between the chains and the solvent[1,3,4]. While the degree of cross-linking is established during the preparation of a network, the extent of the interaction of chains and solvent may be modified as desired, and therefore the degree of swelling may be controlled.

A gel can be made to swell or shrink continuously by changing the quality of the solvent with which it is in contact. Alternatively, it may go through critical conditions and, in fact, can exhibit phase transitions, depending on the type of the polymer–solvent interaction and the extent of cross-linking. The discrete shrinkage of the gel, by changing the polymer–solvent interaction parameter, is a volume phase transition similar to the gas-liquid transition of a condensing gas. The possibility of such phase transitions was, notably, first discussed by Dusek and collaborators many years ago[5,6]. Their treatment was confined to nonionic networks. They showed that networks prepared to show this effect should be synthesized in the highly diluted state, and the swelling solvent should rather be poor and have an interaction parameter which is strongly concentration dependent. The possibility of first-order volume transitions in such systems was later shown by Tanaka[7] using the swelling expression for nonionic networks[3]. The general thermodynamic conditions for the realization of critical conditions and transitions were subsequently derived by Erman and Flory[8]. It was shown there that the presence of ionizable groups on the network chains would greatly facilitate the transition. The large volume of experimental literature on volume phase transitions that we will discuss in this chapter is almost entirely about networks with ionizable chains.

Advances in light scattering spectroscopy in the late 1970s and 1980s then resulted in a series of experiments (performed mostly on the system polyacrylamide–water–acetone) in which discrete volume changes were observed by changing the temperature, solvent composition, pH, and ionic composition, or by the application of a small electric field across the gel[7,9–18]. In later work, it was shown that such transitions are possible in networks of various types, such as gelatin, agarose, and DNA in mixtures of acetone and water as solvent[19]. Among gels that exhibit phase transitions, temperature-sensitive systems such as N-isopropylacrylamide and N, N'-diethylacrylamide gels in water attracted much attention, both theoretically and technologically[13,20–29]. Most of the literature on ionizable networks is covered in the two volumes *Responsive Gels: Volume Transitions I and II*[30].

The similarity of the transition phenomena in gels to the coil–globule transition of linear chains can be observed in the experiments of Tanaka et al. on linear polystyrene chains[31–34]. More recently, the classical lattice theory (on which the mixing part of the free energy is based) has been modified to take into account strong specific interactions that are encountered in aqueous solutions of nonelectrolytes[35]. The theory was extended[36] to polyelectrolyte gels and the results were compared with data from ionized gels. The effects of deformation on phase transitions in deformed gels have been discussed by Onuki[37], and a broad general review of phase transitions of gels is given by Li and Tanaka[38]. The literature on such systems is increasing at a rapid rate, especially that emphasizing their applicability in diverse fields of advanced technologies[39–74].

In the following section, we describe the theory of critical phenomena and phase transitions in gels based on the paper by Erman and Flory[8]. The theoretical treatment is followed by some examples of phase transitions observed on various systems. The related subject of thermoreversible gels makes up the second section of this chapter.

7.1 Theory of Critical Phenomena and Phase Transitions

The reduced chemical potential $\Delta\tilde{\mu}_1$ for solvent in an isotropically swollen network in equilibrium is given by eq. (6.14). Here, we generalize the potential to contain the effects of ionic groups on the chains. Thus,

$$\Delta\tilde{\mu}_1 = (\mu_1 - \mu_1^0)/RT = \ln(1-v_2) + v_2 + \chi v_2^2 + \frac{\beta}{\lambda}\left[1 + \frac{\mu}{\xi}K(\lambda)\right] + \frac{\Delta\mu_i}{RT} = \ln a_1 \tag{7.1}$$

where $\Delta\mu_i$ is the contribution to the total chemical potential by the presence of ionic groups on the chains.

As described in detail in chapter 6, χ depends, in general, on the composition, as well as on temperature and pressure. It is empirically expressed as a Taylor series in composition[75–77] as $\chi = \chi_1 + \chi_2 v_2 + \chi_3 v_2^2 + \cdots$. Results of experimental measurements of osmotic pressures and vapor pressures yield values of the coefficients χ_i. The dependence of the χ parameter on composition is of significant importance from the point of view of phase transitions, and this will be discussed in more detail later. For example, the equation for the system polystyrene in cyclohexane at 25°C is $\chi = 0.505 + 0.33 v_2 + 0.29 v_2^2$, and for the system polyisobutylene in benzene at 24.5°C it is $\chi = 0.500 + 0.30 v_2 + 0.30 v_2^2$.

The linear additivity of the ionic potential term in the expression for the chemical potential in eq. (7.1) is valid when the number of ionic groups on each network chain is small. In this case, the contribution of the ionizable groups to the chemical potential will be[8]

$$\Delta\mu_i/RT = -i\nu(V_1/V^0 N_A)(v_2/v_2^0)$$
$$= i\beta\left(1 - \frac{2}{\phi}\right)^{-1}\left(\frac{v_2}{v_2^0}\right) \tag{7.2}$$

This equation will be adopted in the following derivation of the theory.

The chemical potential and its first and second derivatives, taken with respect to composition, vanish at the critical point of a swollen gel for $a_1 = 1$. That is,

$$\Delta\tilde{\mu}_1 = 0 \tag{7.3}$$

$$\Delta\tilde{\mu}_1' \equiv \frac{\partial \Delta\tilde{\mu}_1}{\partial v_2} = -\frac{v_2}{(1-v_2)} + 2\chi_1 v_2 + 3\chi_2 v_2^2 + \cdots + \Delta\tilde{\mu}_{1,el}' - \frac{i}{x_c} \tag{7.4}$$

and

$$\Delta\tilde{\mu}_1'' \equiv \frac{\partial^2 \Delta\tilde{\mu}_1}{\partial v_2^2} = -\frac{1}{(1-v_2)^2} + 2\chi_1 + 6\chi_2 v_2 + \cdots + \Delta\tilde{\mu}_{1,el}'' \tag{7.5}$$

will equate to zero at the critical point. For a network with a specified structure, the critical conditions may be fulfilled at constant pressure only at a unique temperature and composition, corresponding to a unique value of the χ parameter. Thus, critical conditions obtain for a network when a specific value of χ is reached, imposing strict restrictions on it. Calculations based on eqs. (7.3)–(7.5), keeping only χ_1 and χ_2 in the χ-parameter expansion, show[8] that for nonionic networks, critical conditions may be reached only if $\chi_1 < 1/2$ and $\chi_2 > 1/3$. The fulfillment of these conditions for nonionic homopolymer systems seems difficult when one considers the generally observed values of the χ parameter. Presence of ionic groups on the network chains makes the fulfillment of these critical conditions easier, as shown by theoretical calculations[8], and systems with much lower χ_2 values may exhibit critical conditions. This conclusion explains the extensive study of ionizable networks reported in the literature.

Conditions for the coexistence of two swollen phases, one with composition v_2 and the other with v_2' is given at activity $a_1 = 1$, are

$$\Delta\tilde{\mu}_1(v_2) = \Delta\tilde{\mu}_1(v_2') = 0 \tag{7.6}$$

and

$$\Delta\tilde{\mu}_2(v_2) = \Delta\tilde{\mu}_2(v_2') \tag{7.7}$$

where $\Delta\tilde{\mu}_2$ is the reduced chemical potential of an arbitrary subunit of the network. When the solvent activity is unity, the condition given by eq. (7.7) is equivalent to the integral

$$\int_{v_2}^{v_2'} \Delta\tilde{\mu}_1 v_2^{-2} dv_2 = 0 \tag{7.8}$$

Solvent chemical potentials for $\ln a_1 = 0$ are plotted in figure 7.1 for three nonionic networks having different χ parameters. Curve a illustrates swelling with good solvent, curve b shows a critical point, and curve c shows a triphasic equilibrium between phases with $v_2 = 0$, 0.030, and 0.157.

Figure 7.2 shows results of calculations of phase transitions based on eqs. (7.6)–(7.8) for a constrained-junction model. The simultaneous solutions to eqs.

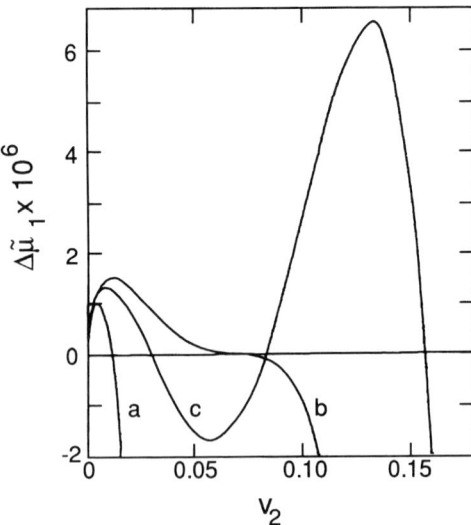

Figure 7.1 Solvent chemical potentials for for three nonionic networks having different χ parameters. A constrained-junction network model is adopted with $x_c = 10^5$ and $\kappa = 12$. Curve a is obtained with $\chi_1 = 0.49$, $\chi_2 = 0$; curve b with $\chi_1 = 0.4967$, $\chi_2 = 0.3386$; and curve c with $\chi_1 = 0.4950$, $\chi_2 = 0.4088$. Curve c shows a triphasic equilibrium between phases having $v_2 = 0$, 0.030, and 0.157.

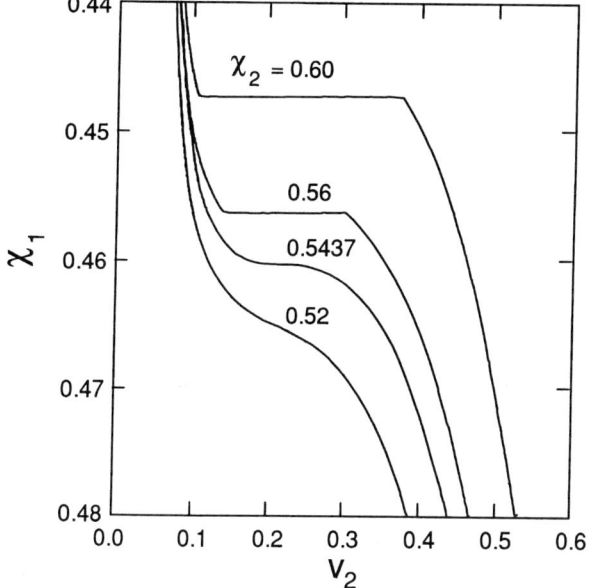

Figure 7.2 Results of calculations of phase transitions based on eqs. (7.6)–(7.8) for a constrained-junction network model with $x_c = 10^3$, $v_{2c} = 1.00$, and $\kappa = 12$. The curve for $\chi_2 = 0.5437$ passes through those for critical conditions. For values of χ_2 less than 0.5437, the volume change of the network does not result in a phase transition. For values of χ_2 larger than 0.5437, on the other hand, the system shows phase transitions at the indicated horizontal lines of χ_1.

(7.6)–(7.8) were obtained numerically by varying the values of χ_1 for a fixed value of χ_2. The values of v_2 at the extremities of the horizontal lines in the figure correspond to the concentrations of the coexisting dilute and concentrated phases. The third phase is the pure solvent phase. Thus, the phase equilibrium of the network corresponds to a triphasic equilibrium. Results of the solutions of eqs. (7.6)–(7.8) were also obtained for the same network in figure 7.2, but cross-linked in solution with $v_{2c} = 0.035$. The significant increase in the widths of the horizontal lines observed in the figure is the unique effect on phase transitions of cross-linking in solution. Figure 7.3 shows results of experiments by Tanaka[11] on a temperature-sensitive gel. The transition is reported to be fully reversible.

The phenomenon of phase transitions in swollen gels is especially of interest when the system has a low degree of cross-linking, such as in biological gels that are only slightly past the threshold point of gelation. In such cases, the κ parameter is large and the gel behavior is close to that of the affine model. In the limit $\kappa = \infty$, the function $K(\lambda)$ takes the simple form

$$K(\lambda) = 1 - \lambda^{-2} \qquad (7.9)$$

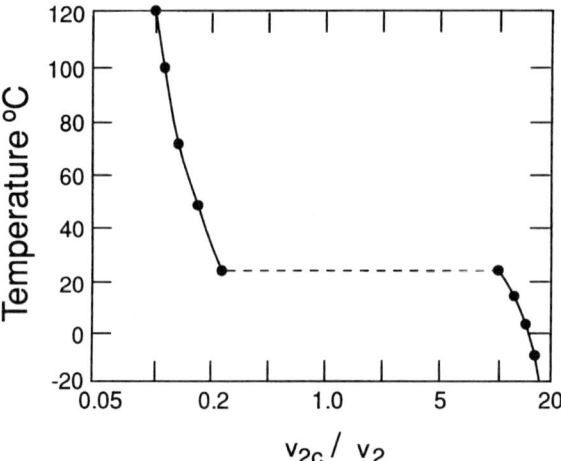

Figure 7.3 Results of experiments by Tanaka[11] on a temperature-sensitive gel. The gel was immersed in a 42 wt% acetone and 52 wt% water mixture of fixed composition and the temperature was varied. The left branch of the curve corresponds to the highly swollen state. The gel collapses upon cooling to 22°C, and the right portion of the curve corresponds to the collapsed state.

Expanding the logarithmic term up to the fourth order in eq. (7.1), setting $\chi_j = 0$ for $j > 2$, and using eqs. (7.3)–(7.5), one obtains the following set of equations in the affine limit:

$$\Delta\tilde{\mu}_1 = \left(\chi_1 - \frac{1}{2}\right)v_2^2 + \left(\chi_2 - \frac{1}{3}\right)v_2^3 - \frac{v_2^4}{4} + \frac{v_{2c}^{2/3}}{x_c}v_2^{1/3} - \left(i + \frac{1}{2}\right)\frac{v_2}{x_c} = 0$$

$$\Delta\tilde{\mu}_1' = 2\left(\chi_1 - \frac{1}{2}\right)v_2 + 3\left(\chi_2 - \frac{1}{3}\right)v_2^2 - v_2^3 + \frac{1}{3}\frac{v_{2c}^{2/3}}{x_c}v_2^{-2/3} - \left(i + \frac{1}{2}\right)/x_c = 0$$

$$\Delta\tilde{\mu}_1'' = 2\left(\chi_1 - \frac{1}{2}\right) + 6\left(\chi_2 - \frac{1}{3}\right)v_2 - 3v_2^2 - \frac{2}{9}\frac{v_{2c}^{2/3}}{x_c}v_2^{-5/3} = 0$$

(7.10)

Solution of eq. (7.10) for χ_1, χ_2, and v_2 for large x_c leads to

$$v_{2,\text{critical}} = \left[\frac{80}{9}v_{2c}^{2/3}\right]^{3/11}x_c^{-3/11}$$

$$\chi_1 \approx \frac{1}{2}$$

$$\chi_2 \approx \frac{1}{3}$$

(7.11)

These results, which become exact in the limit of infinite x_c, are independent of ion concentration. This implies that, in the case of chains with small ion concentration (which was the assumption of the model described above), ionic and nonionic networks behave identically in the limit of infinite chain length.

More recent theoretical analysis of weakly charged polyelectrolyte gels by Khokhlov and collaborators[78,79] direct attention to the possibility of formation of microdomain structures as a result of microphase separation.

7.2 Thermoreversible Gels

There is another type of gel that undergoes a phase transition, specifically those gels that are "thermoreversible." These materials are highly swollen networks in which the cross-links are temporary or physical; for example, crystallites or intersegmental aggregates[80-100]. These types of gels are thermoreversible in that they reliquefy upon increase in temperature sufficient to melt the crystallites or break up the aggregates, and then reform the gel upon cooling. The most commonplace example of such a material is collagen in the form of gelatin, which is in the gelled form with water in some food products. Some other examples of biopolymers that exhibit this property are described in chapter 15 on bioelastomers.

A variety of synthetic polymers also form thermoreversible gels, and some of them are described in this section. The ones that have been studied most extensively consist of polyethylene[80-84,86,90,93,99,100], and vinyl/vinylidene polymers, such as atactic polystyrene[87,94-96] and poly(methyl methacrylate)[97]. Most investigations of such gels have employed techniques for characterizing the gelation process (and subsequent liquefication), the nature of the aggregation (particularly in the case of noncrystallizable polymers), the corresponding phase diagrams, calorimetric effects accompanying the transitions, and structures and morphologies. Some of the techniques used to characterize these gels and the polymers from which they are formed are: analytical and preparative gel permeation chromatography, fractional precipitation, ^{13}C NMR and Raman spectroscopy, viscometry, scanning calorimetry, visual observations of the initiation and cessation of flow for the gelation process, small-angle x-ray and light scattering, polarized light microscopy, and scanning and transmission electron microscopy[93]. Relatively little has been done on the mechanical properties of the gels, and some of it focuses on dynamical mechanical properties, rather than equilibrium moduli[87,96].

It is possible, however, to obtain estimates of the moduli of these gels, in spite of the fact that their high degrees of dilution make them quite fragile. One direct way of doing this is by adding known weights to the surface of the gel and then measuring the change in height of the sample[93]. An alternative is to enclose the solution being gelled in a dilatometer-shaped glass cell, imposing a pressure above the surface of the gelling system, and then measuring the change in height of a mecury column contacting the lower portion of the gel[99,100]. This pressure-pulsing approach has the advantage of being able to follow relatively rapid gelations, and the transparency of the cell permits simultaneous scattering measurements for monitoring changes in structure, even prior to the gelation point. Some of the results obtained by these two methods are described below.

Figure 7.4[99] shows the effects of polymer concentration and temperature on the critical concentration c^*, below which gelation cannot occur. The results are for ethylene hexene-1 copolymers, where the comonomer is used with the ethylene to

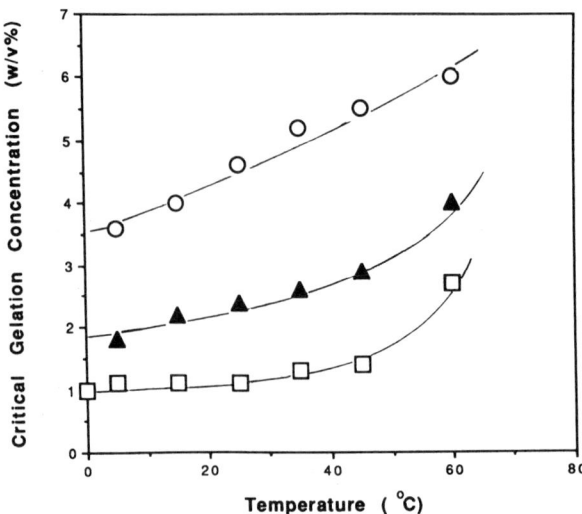

Figure 7.4 Critical gelation concentration for ethylene hexene-1 copolymers as a function of temperature[99]. The values of $10^{-3} M_n$ are 26.1 (○), 43.5 (▲), and 111 (□) in g mol^{-1}, and values of the mol% branches are 1.21, 1.21, and 1.47, respectively.

introduce known numbers of branches of known length and structure [($-$CH$_2$)$_3$CH$_3$]. The values of c^* are generally considerably higher than those of polyethylene homopolymers having the same molecular weight. This demonstrates the expected interference of the branch points with the crystallization process that provides the temporary cross-links required for gelation. At each molecular weight, c^* increases with increase in temperature, as was also found for polyethylene linear homopolymers[93]. At each temperature, c^* increases with decrease in molecular weight because more cross-links are required for network formation in the case of the shorter chains. For all three molecular weights, increases in temperature cause increases in modulus, presumably because of an increase in the number of crystallites.

This increase in modulus with increase in temperature can also be seen from the evolution of the modulus with time, as is shown for one of the same polyethylene copolymers in figure 7.5[100]. As expected, the modulus reaches an asymptotic limit as the crystallization approaches its equilibrium value.

The effects of concentration c on the low-deformation modulus have also been reported for some polyethylene homopolymers. Significant curvature is observed in the dependence of $[f^*]$ on concentration itself, but its dependence on c^2 is linear within experimental error, as is shown in figure 7.6[93]. The modulus is seen to increase significantly with increasing c, and the slopes of the $[f^*]$ vs. c^2 curves increase asymptotically with increase in molecular weight. This is presumably due to a decrease in the number of chain ends, which have to be relegated to the interlamellar regions when crystallization (cross-linking) occurs. Such cross-linking is obviously more efficient when the number of such chain ends is minimized.

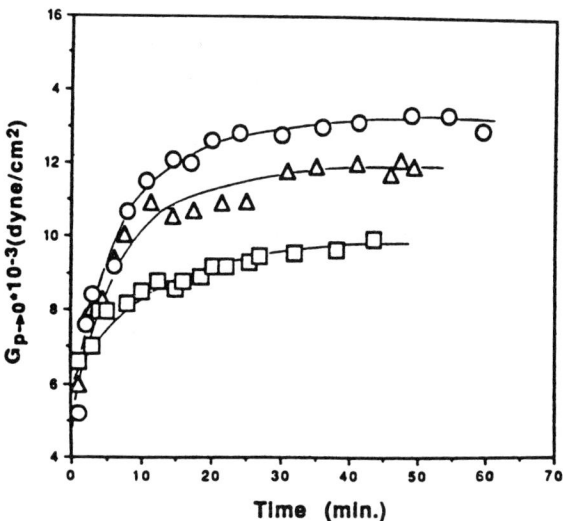

Figure 7.5 Values of the modulus in the limit of very small deformation pressure for gels prepared from ethylene hexene-1 copolymers having $10^{-3}\ M_n = 43.5\,\text{g}\,\text{mol}^{-1}$ as a function of time[100]. The temperatures were 15 (□), 25 (△), and 35 (○), and the concentration was approximately 50% higher than the critical concentration c^*.

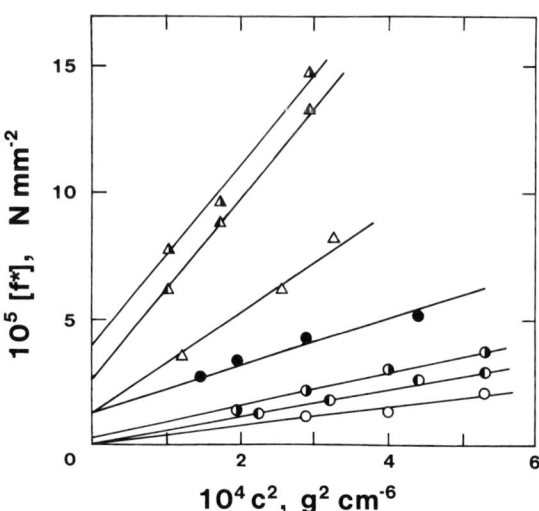

Figure 7.6 Moduli of gels prepared from polyethylene in the limit of very small deformation, shown as a function of the square of the concentration[93]. Values of $10^{-3}\ M_n$ ranged from 5.8 (lowest curve) to 800 (uppermost curve). Reprinted with permission from Li, Z., et al. (1989), *Macromolecules*, **22**. Copyright 1997 American Chemical Society.

The c^2 dependence of the modulus is consistent with the importance of pairwise encounters between chain segments in the gelation process.

More detailed comparisons of the modulus in the limits a $\alpha \to 1$ and $c \to c^*$ have also been carried out[93]. If a weight-average molecular weight M_w or viscosity-average molecular weight M_η is used for this purpose, then there is no correlation between the results for unfractionated polymers and fractionated ones, although the trends seem to be the same for the two types of samples. However, a common dependence does result when the same values of $[f^*]$ are plotted against the *number*-average molecular weight M_n, as shown in figure 7.7. Specifically, when the data are analyzed on this basis, the same value of the modulus is obtained, irrespective of the sample being a whole polymer or a fraction. There are two line segments making up the relationship shown in this figure. Below approximately $M_n = 10^4$ the modulus is independent of molecular weight. Above this molecular weight, the modulus increases linearly as the logarithm of the chain length. This demarcation corresponds to a similar change in the critical concentration with molecular weight. The major structural change at this point is from a extended-chain form to a "folded" lamellar structure[93].

The commonality observed demonstrates the importance of the large number of end groups associated with the short chains that contribute overwhelmingly to M_n but hardly at all to M_w and M_η. Since the end groups are predominantly located in the interlamellar, disordered structural regions, the major contribution of these structures to the initial modulus is emphasized by these results. The end-group concentration is also of considerable importance with regard to the drawing of highly oriented, high-strength, high-modulus fibers from gels of this type[101].

It is also possible to obtain some additional molecular information from the magnitudes of these moduli[93]. This can be done by making the basic assumption

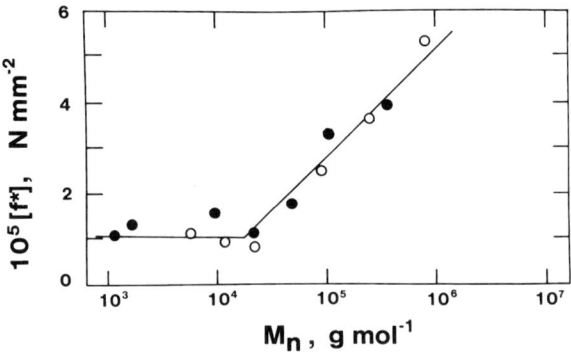

Figure 7.7 Moduli in the limit of very small deformation and at concentrations approaching the critical concentration c^*, shown as a function of the number-average molecular weight[93]. Results on the unfractionated polymers are shown by the open circles, and results on the fractionated polymers, by the filled circles. Reprinted with permission from Li, Z., et al. (1989), *Macromolecules*, **22**. Copyright 1997 American Chemical Society.

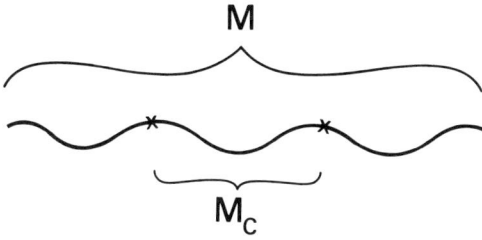

Figure 7.8 Sketch suggesting the structure of a polyethylene chain in a gel prepared in the vicinity of the critical concentration c^*, with the crosses representing cross-links (crystallites)[93]. A very large portion of such a chain would be in the form of dangling ends, which are elastically ineffective. Reprinted with permission from Li, Z., et al. (1989), *Macromolecules*, **22**. Copyright 1997 American Chemical Society.

that the lamellar crystallites act as cross-links of very high functionality ϕ. They are presumably much larger than the usual covalent cross-links, but this may not be important at the very high dilutions involved. The resulting calculated values of the molecular weight M_c between cross-links make up a considerable fraction of the total molecular weight M of the chains used to form the gel. In fact, there may be only two or three cross-links per chain, which is, of course, consistent with their preparation at $c \approx c^*$. This is shown schematically in figure 7.8. Thus, a considerable portion of each chain is in the form of dangling ends, which are elastically ineffective. This indicates that the gels are extremely fragile, not only because of their very low polymer content, but also because a great deal of the polymer present is in the form of elastically ineffective material.

The very low degrees of cross-linking in such materials gives them enormous swelling capacity, and this has been exploited in the case of "superabsorbent polymers"[102]. This capacity can give them a variety of important applications, ranging from diapers to transport agents for cleaning up oil spills.

References

(1) Flory, P. J. *Discuss. Faraday Soc.* 1974, 57, 1.
(2) Almdal, K.; Dyre, J.; Hvidt, S.; Kramer, O. *Makromol. Chem., Macromol. Symp.* 1993, 76, 49.
(3) Flory, P. J. *Principles of Polymer Chemistry.* Cornell University Press: Ithaca, NY. 1953.
(4) Horkay, F.; McKenna, G. B. In *Physical Properties of Polymers Handbook.* J. E. Mark, Ed. American Institute of Physics Press: Woodbury, NY. 1996.
(5) Dusek, K.; Patterson, D. *J. Polym. Sci.* 1968, A26, 1209.
(6) Dusek, K.; Prins, W. *Adv. Polym. Sci.* 1969, 6, 1.
(7) Tanaka, T. *Phys. Rev. Lett.* 1978, 40, 820.
(8) Erman, B.; Flory, P. J. *Macromolecules* 1986, 19, 2342.
(9) Janas, V. F.; Rodriguez, F.; Cohen, C. *Macromolecules* 1980, 13, 978.
(10) Tanaka, T.; Fillmore, D.; Sun, S. J.; Nishia, I.; Swislow, G.; Shah, A. *Phys. Rev. Lett.* 1980, 45, 1636.

(11) Tanaka, T. *Sci. Am.* 1981, 244, 124.
(12) Ilavsky, M. *Polymer* 1981, 22, 1687.
(13) Hrouz, J.; Ilavsky, M.; Ulbrich, K.; Kopecek, J. *Eur. Polym. J.* 1981, 17, 361.
(14) Ilavsky, M. *Polymer* 1982, 15, 782.
(15) Ilavsky, M.; Hrouz, J.; Ulbrich, K. *Polym. Bull.* 1982, 7, 107.
(16) Ohmine, I.; Tanaka, T. *J. Chem. Phys.* 1982, 77, 5725.
(17) Nicoli, D.; Young, C.; Tanaka, T.; Pollack, A.; Whitesides, G. *Macromolecules* 1983, 16, 887.
(18) Hirokawa, Y.; Tanaka, T.; Sato, E. *Macromolecules* 1985, 18, 2782.
(19) Amiya, T.; Tanaka, T. *Macromolecules* 1987, 20, 1162.
(20) Hirokawa, Y.; Tanaka, T. *J. Chem. Phys.* 1984, 81, 6379.
(21) Ilavsky, M.; Hrouz, J.; Havlicek, I. *Polymer* 1985, 26, 1514.
(22) Hoffman, A. S.; Afrassiabi, A.; Dong, L. C. *J. Controlled Release* 1986, 4, 213.
(23) Hirotsu, S.; Hirokawa, Y.; Tanaka, T. *J. Chem. Phys.* 1987, 87, 1392.
(24) Hirose, Y.; Amiya, T.; Hirokawa, T.; Tanaka, T. *Macromolecules* 1987, 20, 1342.
(25) Freitas, R. F. S.; Cussler, E. L. *Chem. Eng. Sci.* 1987, 42, 97.
(26) Hirotsu, S. *J. Chem. Phys.* 1988, 88, 427.
(27) Li, Y.; Tanaka, T. *J. Chem. Phys.* 1989, 90, 5161.
(28) Park, T. G.; Hoffman, A. S. *Appl. Biochem. Biotech.* 1988, 19, 1.
(29) Marchetti, M.; Prager, S.; Cussler, E. L. *Macromolecules* 1990, 23, 1760.
(30) Dusek, K., Ed. *Responsive Gels: Volume Transitions I and II*; Springer-Verlag: Berlin. 1993; Vols. 109 and 110.
(31) Nishio, I.; Sun, S. T.; Swislow, G.; Tanaka, T. *Nature* 1979, 281, 208.
(32) Swislow, G.; Sun, S. T.; Nishio, I.; Tanaka, T. *Phys. Rev. Lett.* 1980, 44, 796.
(33) Sun, S. T.; Nishio, I.; Swislow, G.; Tanaka, T. *J. Chem. Phys.* 1980, 73, 5971.
(34) Nishio, I.; Swislow, G.; Sun, S. T.; Tanaka, T. *Nature* 1982, 300, 243.
(35) Prange, M.; Hooper, H. H.; Prausnitz, J. M. *AIChE J.* 1989, 35, 803.
(36) Hooper, H. H.; Baker, J. P.; Blanch, H. W.; Prausnitz, J. M. *Macromolecules* 1990, 23, 1096.
(37) Onuki, A. *J. Phys. Soc. Japan* 1988, 57, 699.
(38) Li, Y.; Tanaka, T. *Ann. Rev. Mater. Sci.* 1992, 22, 243.
(39) Otake, K.; Inomata, H.; Konno, M.; Saito, S. *J. Chem. Phys.* 1989, 91, 1345.
(40) Katayama, S. *J. Phys. Chem.* 1992, 96, 5209.
(41) Osada, Y. *Adv. Mater.* 1991, 3, 107.
(42) Hirai, T.; Nemoto, H.; Hirai, M.; Hayashi, S. *J. App. Polym. Sci.* 1994, 53, 79.
(43) Yoshida, R.; Ichijo, H.; Hakuta, T.; Yamaguchi, T. *Macromol. Rapid Commun.* 1995, 16, 305.
(44) Hu, Z.; Zhang, X.; Li, Y. *Science* 1995, 269, 525.
(45) Horie, K.; Asano, M.; Hu, Y. *Macromol. Symp.* 1995, 93, 125.
(46) Osada, Y.; Okuzaki, H.; Gong, J. P. *TRIP* 1994, 2, 61.
(47) Kuroki, Y.; Sekimoto, K. *J. Chem. Phys.* 1995, 102, 8626.
(48) Zhang, X.; Li, Y.; Hu, Z.; Littler, C. L. *J. Chem. Phys.* 1995, 102, 551.
(49) Osada, Y.; Matsuda, A. *Nature* 1995, 376, 219.
(50) Yoshida, R.; Uchida, K.; Kaneko, Y.; Sakai, K.; Kikuchi, A.; Sakurai, Y.; Okano, T. *Nature* 1995, 374, 240.
(51) Asano, M.; Horie, K.; Yamashita, T. *Polym. Gels Netw.* 1995, 3, 281.
(52) Budtova, T.; Suleimenov, I.; Frenkel, S. *Polym. Gels Netw.* 1995, 3, 387.
(53) Nagasaki, Y.; Honzawa, E.; Kato, M.; Kataoka, K. *Macromolecules* 1994, 27, 4848.
(54) Tanaka, Y.; Kagami, Y.; Matsuda, A.; Osada, Y. *Macromolecules* 1995, 28, 2574.

(55) Tanaka, T. *Makromol. Chem., Macromol. Symp.* 1983, 70/71, 13.
(56) Matsuo, E. S.; Tanaka, T. *Nature* 1992, 358, 482.
(57) Zhang, Y. Q.; Tanaka, T.; Shibayama, M. *Nature* 1992, 360, 142.
(58) Annaka, M.; Tanaka, T. *Nature* 1992, 355, 430.
(59) Suzuki, A.; Tanaka, T. *Nature* 1990, 346, 345.
(60) Kokufata, E.; Zhang, Y. Q.; Tanaka, T. *Nature* 1991, 351, 302.
(61) Osada, Y.; Ross-Murphy, S. B. *Sci. Am.* 1993, 268(5), 82.
(62) Li, Y.; Hu, Z.; Zhang, X.; Littler, C. L. *Mater. Sci. Eng. C* 1995, 2, 221.
(63) Iwatsubo, T.; Ogasawara, K.; Yamasaki, A.; Masuoka, T.; Mizoguchi, K. *Macromolecules* 1995, 28, 6579.
(64) Ilavsky, M.; Sedlakova, Z.; Bouchal, K.; Plestil, J. *Macromolecules* 1995, 28, 6835.
(65) Brazel, C. S.; Peppas, N. A. *Macromolecules* 1995, 28, 8016.
(66) Tomari, T.; Doi, M. *Macromolecules* 1995, 28, 8334.
(67) Suzuki, A.; Suzuki, H. *J. Chem. Phys.* 1995, 103, 4706.
(68) Kawasaki, H.; Nakamura, T.; Miyamoto, K.; Tokita, M. *J. Chem. Phys.* 1995, 103, 6241.
(69) Suzuki, A.; Yamazaki, M.; Kobiki, Y. *J. Chem. Phys.* 1996, 104, 1751.
(70) Aalberts, D. *J. Chem. Phys.* 1996, 104, 4309.
(71) Ikehara, T.; Nishi, T.; Hayashi, T. *Polym. J.* 1996, 28, 169.
(72) Shibayama, M.; Mizutani, S.; Nomura, S. *Macromolecules* 1996, 29, 2019.
(73) Kurosu, H.; Shibuya, T.; Yasunaga, H.; Ando, I. *Polym. J.* 1996, 28, 80.
(74) Katakai, R.; Yoshida, M.; Hasegawa, S.; Iijima, Y.; Yonezawa, N. *Macromolecules* 1996, 29, 1065.
(75) Flory, P. J. *Discuss. Faraday Soc.* 1970, 49, 7.
(76) Eichinger, B. E.; Flory, P. J. *Trans. Faraday Soc.* 1968, 64, 2035.
(77) Flory, P. J.; Daoust, H. *J. Polym. Sci.* 1957, 25, 429.
(78) Vasilevskaya, V.; Khokhlov, A. R. *Macromolecules* 1992, 25, 384.
(79) Dormidontova, E.; Erukhimovitc, I. Y.; Khokhlov, A. R. *Macromol. Theory Simul.* 1994, 3, 661.
(80) Smith, P.; Lemstra, P. J.; Pijpers, J. P. L.; Kiel, A. M. *Coll. Polym. Sci.* 1981, 251, 1070.
(81) Edwards, C. O.; Mandelkern, L. *J. Polym. Sci., Lett. Ed.* 1982, 20, 355.
(82) Okabe, M.; Isayama, M.; Matsuda, H. *J. Appl. Polym. Sci.* 1985, 30, 4735.
(83) Domszy, R. C.; Alamo, R.; Edwards, C. O.; Mandelkern, L. *Macromolecules* 1986, 19, 310.
(84) Sawatari, E.; Okumura, T.; Matsuo, M. *Polym. J.* 1986, 18, 741.
(85) Koltisko, B.; Keller, A.; Litt, M.; Baer, E.; Hiltner, A. *Macromolecules* 1986, 19, 1207.
(86) Russo, P. S., Ed. *Reversible Polymeric Gels and Related Systems*. American Chemical Society: Washington, DC. 1987.
(87) Hiltner, A. In *Order in the Amorphous "State" of Polymers*. S. E. Keinath, R. L. Miller and J. K. Rieke, Ed. Plenum Press: New York. 1987; p. 119.
(88) McKenna, G. B.; Guenet, J.-M. *Polym. Commun.* 1988, 29, 58.
(89) Skukla, P.; Muthukumar, M. *Polym. Eng. Sci.* 1988, 28, 1304.
(90) Matsuda, H.; Kashiwagi, R.; Okabe, M. *Polym. J.* 1988, 20, 189.
(91) Hoffman, H.; Ebert, G. *Angew. Chem. Int. Ed.* 1988, 27, 902.
(92) McKenna, G. B.; Guenet, J.-M. *J. Polym. Sci., Polym. Phys. Ed.* 1988, 26, 267.
(93) Li, Z.; Mark, J. E.; Chan, E. K. M.; Mandelkern, L. *Macromolecules* 1989, 22, 4273.

(94) Klein, M.; Brulet, A.; Guenet, J.-M. *Macromolecules* 1990, 23, 540.
(95) Guenet, J.-M.; Klein, M. *Macromol. Symp.* 1990, 39, 85.
(96) Callister, S.; Keller, A.; Hikmet, R. M. *Macromol. Symp.* 1990, 39, 19.
(97) Broecke, P. V. D.; Berghmans, H. *Macromol. Symp.* 1990, 39, 59.
(98) Horton, J. C.; Donald, A. M. *Macromol. Symp.* 1990, 39, 131.
(99) Ichise, N.; Yang, Y.; Li, Z.; Yuan, Q.; Mark, J. E.; Chan, E. K. M.; Alamo, R. G.; Mandelkern, L. In *Synthesis, Characterization, and Theory of Polymeric Networks and Gels*, S. M. Aharoni, Ed. Plenum Press: New York. 1992; p. 217.
(100) Yang, Y.; Ichise, N.; Li, Z.; Yuan, Q.; Mark, J. E.; Chan, E. K. M.; Alamo, R. G.; Mandelkern, L. In *Complex Fluids*, E. B. Sirota, D. Weitz, T. Witten, and J. Israelachvili, Eds. Materials Research Society: Pittsburgh, PA. 1992; p. 325.
(101) Smith, P.; Lemstra, P. J.; Pijper, P. L. *J. Polym. Sci., Polym. Phys. Ed.* 1982, 20, 2229.
(102) Buchholz, F. L. *Trends Polym. Sci.* 1994, 2, 277.

8

Calculations and Simulations

The classical theories of rubber elasticity are based on the Gaussian chain model. The only molecular parameter that enters these theories is the mean-square end-to-end separation of the chains constituting the network. However, there are various areas of interest that require characterization of molecular quantities beyond the Gaussian description. Examples are segmental orientation, birefringence, rotational isomerization, and finite extensibility, and we will address these properties in the following chapters. One often needs a more realistic distribution function for the end-to-end vector, as well as for averages of the products of several vectorial quantities, as will be evident in these chapters. The foundations for such characterizations, and several examples of their applications, are given in this chapter.

8.1 Spatial Configurations of an Isolated Chain

Several aspects of rubber elasticity (such as the dependence of the elastic free energy on network topology, number of effective junctions, and contributions from entanglements) are successfully explained by theories based on the freely jointed chain and the Gaussian approximation. Details of the real chemical structure are not required at the length scales describing these phenomena. On the other hand, studies of birefringence, thermoelasticity, rotational isomerization upon stretching, strain dichroism, local segmental orientation and mobility, and characterization of networks with short chains require the use of more realistic network chain models. In this section, properties of rotational isomeric state models for the chains are discussed. The notation is based largely on the Flory book, *Statistical Mechanics of Chain Molecules*[1]. More recent information is readily found in the literature[2–5].

Due to the simplicity of its structure, a polyethylene-like chain serves as a convenient model for discussing the statistical properties of real chains. This simplicity can be seen in figure 8.1, which shows the planar form of a small portion of a polyethylene chain. Bond lengths and bond angles may be regarded

88 STRUCTURES AND PROPERTIES OF RUBBERLIKE NETWORKS

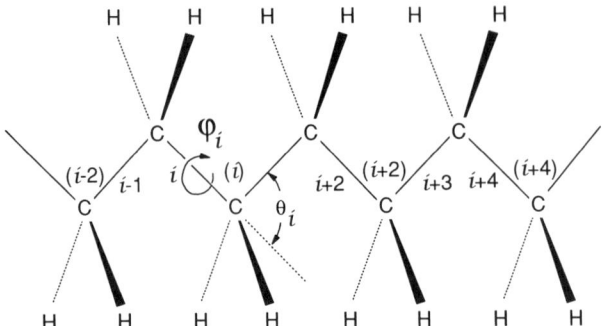

Figure 8.1 The planar form of a small portion of a polyethylene chain. The carbon–carbon backbone bonds are labeled from $i-1$ to $i+4$. The bond angles are labeled in parentheses from $i-2$ to $i+4$. The supplementary ith bond angle is shown by θ_i. Torsional rotations about the backbone bonds are shown by the angle ϕ_i for the ith bond; the value shown corresponds to $0°$.

as fixed in the study of rubber elasticity because their rapid fluctuations are usually in the range of only ± 0.05 Å and $\pm 5°$, respectively. The chain changes its configuration only through torsional rotations about the backbone bonds, shown, for example, by the angle ϕ_i for the ith bond in figure 8.1. The zero state for the angle ϕ_i is obtained when the $(i-1)$st, ith and $(i+1)$st bonds are planar. Keeping all the bonds to the left of the ith bond in figure 8.1 stationary, and changing ϕ_i, displaces the carbon and the hydrogen atoms on the right of the ith carbon, and hence changes the distances between the various atoms in the chain. Interactions between atoms resulting from changes of interatomic distances determine the energy of the chain as a function of conformation.

Rotations about each skeletal bond are subject to a potential. The potential for bond i, for example, is shown in figure 8.2 in units of kT as a function of the rotation angle ϕ_i. This potential or energy results primarily from two sources. (1) The first is the intrinsic rotational potential of each bond, which may be represented as being threefold. For the ith bond, for example,

$$E(\phi_i) = (E_0/2)[1 - \cos(3\phi_i)] \tag{8.1}$$

This function indicates three minima of equal energy at $0°$, $120°$, and $-120°$, referred to as *trans* (t), *gauche*$^+$ (g^+), and *gauche*$^-$ (g^-), respectively. (2) The other contribution involves the non-bonded interactions between various atom pairs separated by three or more bonds along the chain. The energy of interaction between two such pairs of atoms is usually approximated by the Lennard-Jones potential:

$$E = 4\varepsilon\left[\left(\frac{\sigma}{r}\right)^{12} - \left(\frac{\sigma}{r}\right)^{6}\right] \tag{8.2}$$

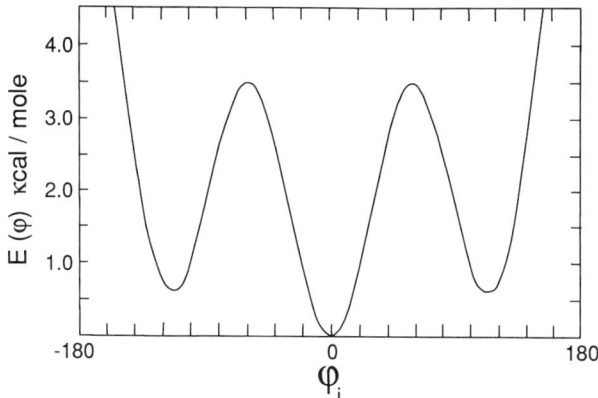

Figure 8.2 The potential for bond i, as a function of the rotation angle ϕ_i.

where r is the distance between the two interacting atoms, σ is the separation between the atoms when $E = 0$, and ε is the depth of the potential well at the minimum of E.

Superposition of the energies given by eqs. (8.1) and (8.2) results in a potential resembling that shown in figure 8.2. In general, the location and the energy values of the minima may be displaced from the values $0°$, $120°$, and $-120°$ indicated by eq. (8.1), and there may be fewer or more than three minima. This depends, of course, on the local chemical structure of the chain: for example, the occurrence of a double bond adjacent to the single bond under consideration. The walls around the energy wells such as those shown in figure 8.2 are generally sufficiently steep to confine the bond rotations to values close to the minima of the wells. Thus, the bond rotational angles, instead of taking a continuum of values between $-\pi$ to $+\pi$, may be allowed to reside only in the three minima represented by the t, g^+, and g^- states. Representation of chain configurations by confining the skeletal bond angles to isomeric minima alone is referred to as the rotational isomeric state (RIS) approximation. Thus, the chain configuration is represented as a sequence of skeletal bonds, in various discrete rotational states. This approximation was first proposed by Volkenstein[6] and was applied most extensively to the analysis of polymer chains by Flory and his research group[1,2,7].

Two atoms along the chain are called topologically distant if they are separated by several other atoms along the chain contour. Two atoms may be spatially close to each other even though they are topologically distant. Excluded-volume effects[8] are said to be present if the energy contribution of two topologically distant atoms, given by eq. (8.2), is included when evaluating the total energy of the chain. In the absence of excluded-volume effects, the chain exhibits its so-called "theta" dimensions or unperturbed dimensions. It is now well established that there are no excluded-volume effects in chains in the undeformed bulk state[8] and that the chains exhibit their theta or unperturbed dimensions.

A chain with n bonds, each having m isomeric states, has m^n possible rotational states or configurations. For a chain in dilute solution, as well as in the bulk state well above the glass temperature, these configurations are not frozen and the chain rapidly passes from one configuration to another as the result of transitions of bonds from one isomeric state to another. From computer molecular dynamics simulations and their correlation with various experimental results, it is now known that these transitions take place following angular oscillations or librations about the energy minima that are relatively large ($\sim \pm 40°$). At temperatures well above the glass transition temperature, transitions from one isomeric state to another take place at a rate corresponding to approximately 10^{-10}–10^{-9} s. Although the dynamics of chains is a subject beyond the scope of equilibrium molecular theories of elasticity, a dynamic picture of the network chain is implicitly adopted in these theories. In the absence of these transitions, the chains would be frozen and the network would be devoid of flexibility and deformability.

The two-dimensional projection of an instantaneous configuration of an undeformed 200-bond polyethylene chain is shown in figure 8.3[9]. It is an example of

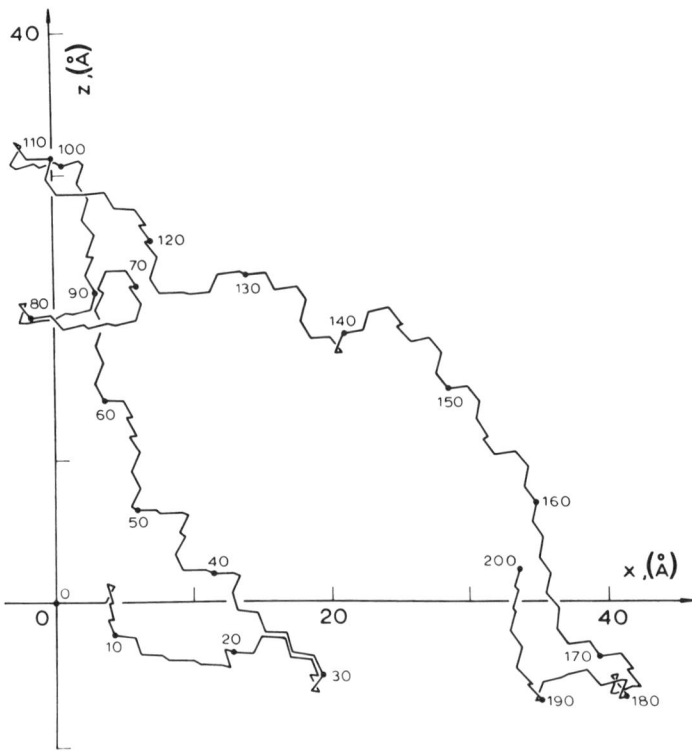

Figure 8.3 The two-dimensional projection of an instantaneous configuration of an undeformed 200-bond polyethylene chain[9]. This spatial configuration was computer generated using known values of the bond lengths, bond angles, rotational angles about skeletal bonds, preferences among the corresponding rotational isomeric states, and the cooperativity between these preferences.

CALCULATIONS AND SIMULATIONS   91

one instantaneous configuration of the chain out of a possible 3^{200} other isomeric configurations. This spatial configuration was computer generated using known values of the bond lengths, bond angles, rotational angles about skeletal bonds, preferences among the corresponding rotational isomeric states, and the cooperativity between these preferences. The method of generating such configurations forms the basis of the Monte Carlo (MC) technique, which is an efficient numerical technique for evaluating statistical averages for polymer chains, especially shorter ones.

A configuration of a chain is fixed if the positions of all backbone atoms are specified relative to a known coordinate system, such as the one labeled XYZ in figure 8.4(a). Such a coordinate system is referred to as a "laboratory-fixed system." The bonds of the chain are numbered from 1 to n and the backbone atoms from zero to n. The first skeletal bond is chosen along the X direction, as shown in figure 8.4(a), and the second bond is in the XY plane. Selecting the first two bonds in this way fixes the position of the chain in space. In figure 8.4(b), the local coordinate axes for internal bonds are shown, with such a local coordinate system defined for each bond. The x_i axis is in the direction of the ith bond; the y_i axis is

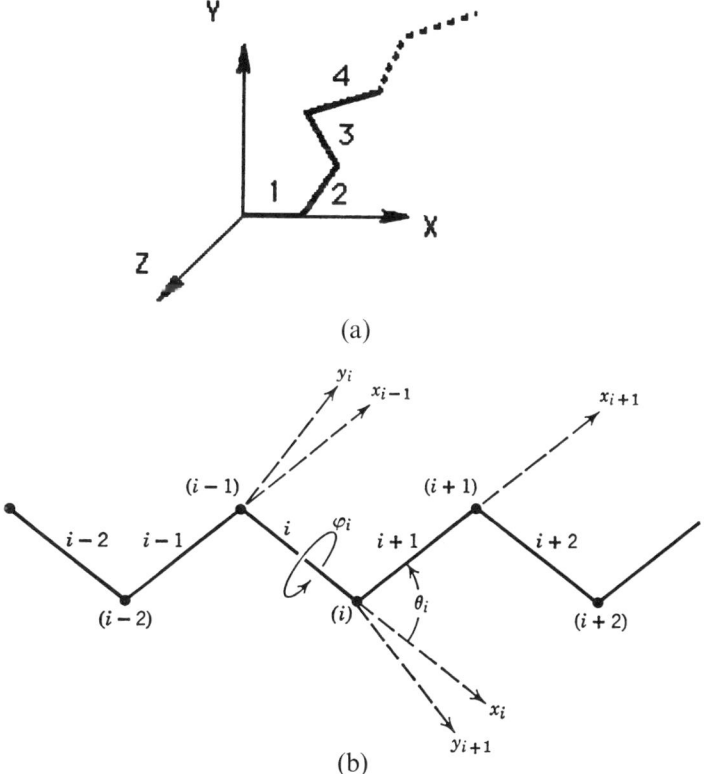

Figure 8.4 (a) Placement of a network chain in a right-handed Cartesian coordinate system. (b) Geometric features of a network chain.

in the plane of bond $i-1$ and i, and its direction is chosen to make an acute angle with y_{i-1}. The z axis completes a right-handed system. In figure 8.4(a), θ_i represents the supplementary bond angle and ϕ_i is the torsional rotation about bond i. The reference state for ϕ_i is chosen so that $\phi_i = 0$ when the bond is in the *trans* state.

The components $v_{x,i+1}, v_{y,i+1}, v_{z,i+1}$ of a vector **v** in the $(i+1)$st coordinate system are transformed into components $v_{x,i}, v_{y,i}, v_{z,i}$ in the ith coordinate system by

$$\mathbf{v}_i = \mathbf{T}_i(\theta, \phi)\mathbf{v}_{i+1} \tag{8.3}$$

where the transformation matrix $\mathbf{T}_i(\theta, \phi)$ is

$$\mathbf{T}_i(\theta, \phi) = \begin{bmatrix} \cos\theta_i & \sin\theta_i & 0 \\ \sin\theta_i\cos\phi_i & -\cos\theta_i\cos\phi_i & \sin\phi_i \\ \sin\theta_i\sin\phi_i & -\cos\theta_i\sin\phi_i & -\cos\phi_i \end{bmatrix} \tag{8.4}$$

and $\mathbf{v}_i$ and $\mathbf{v}_{i+1}$ represent the same vector in the ith and $i+1$st coordinate systems, respectively. Using this transformation operation, the position $\mathbf{r}_i$ of the ith atom relative to the laboratory-fixed coordinate system is written as

$$\mathbf{r}_i = \mathbf{l}_1 + \mathbf{T}_1\mathbf{l}_2 + \mathbf{T}_1\mathbf{T}_2\mathbf{l}_3 + \cdots + \mathbf{T}_1\mathbf{T}_2\ldots\mathbf{T}_{i-1}\mathbf{l}_i \tag{8.5}$$

Equation (8.5) may be written in concise form as[7]

$$\mathbf{r}_i = \sum_{k=1}^{i} \mathbf{T}_1^{(k-1)}\mathbf{l}_k \tag{8.6}$$

where $\mathbf{T}_1^{(k-1)}$ represents the serial product of transformation matrices, starting with the first and containing $(k-1)$ factors as identified by the superscript.

8.2 Statistical Averages of Configurational Variables

The set of position vectors $\{\mathbf{r}_i\}$ for all n backbone atoms uniquely prescribes a given instantaneous configuration of a chain. There is one-to-one correspondence between the set $\{\mathbf{r}_i\}$ and the set $\{\phi\}_i$ of all torsional angles of the chain. The probability of occurrence of each such configuration, as explained in more detail in appendix D, is given by

$$p(\{\phi\}_i) = \frac{\exp(-E(\{\phi\}_i/kT)}{\sum \exp(-E(\{\phi\}_i/kT)} \tag{8.7}$$

The average of any function of the set $\{\mathbf{r}_i\}$ may be calculated according to the relationship

$$\langle f(\{\mathbf{r}_i\})\rangle = \sum_{\{\phi\}_i} f(\{\mathbf{r}_i\})p(\{\phi\}_i) = \frac{\sum_{\{\phi\}_i} f(\{\mathbf{r}_i\})\exp(-E(\{\phi\}_i/kT)}{\sum_{\{\phi\}_i}\exp(-E(\{\phi\}_i/kT)} \tag{8.8}$$

where the summations are carried out over all sets of configurations. This averaging may be carried out in various exact or approximate ways. In this chapter, its approximate evaluation by Monte Carlo simulations will be outlined. A Fortran program for performing this averaging is given in appendix G.

The MC scheme to be outlined here is for a chain with independent bond torsional potentials, as represented in figure 8.2. Formulations for chains with more realistic interdependent bond torsional energies are described elsewhere[1,7,10,11].

The MC scheme is based on generating a chain configuration whose probability is proportional to the numerator of eq. (8.8). The steps are as follows: (1) Fix the first bond along the X axis, and the second in the XY plane of the laboratory-fixed coordinate system. (2) Choose the value of ϕ_2. This will determine the direction of the third bond in space. According to the rotational isomeric state picture, the values at the energy minima are assigned to each bond torsional angle, ϕ_i. Assuming *trans* (t), *gauche*$^+$ (g^+), and *gauche*$^-$ (g^-), states for the bonds, one defines the three statistical weights u_t, u_{g^+}, and u_{g^-} as

$$u_t = \exp(-E_t/RT), \qquad u_{g^+} = \exp(-E_{g^+}/RT), \qquad u_{g^-} = \exp(-E_{g^-}/RT) \tag{8.9}$$

where E_t, E_{g^+}, and E_{g^-} are the energies of the three states of the bond. The three statistical weights given in eq. (8.9) define the probabilities p_t, p_{g^+}, and p_{g^-} of the three states as

$$\begin{aligned} p_t &= u_t/(u_t + u_{g^+} + u_{g^-}) \\ p_{g^+} &= u_{g^+}/(u_t + u_{g^+} + u_{g^-}) \\ p_{g^-} &= u_{g^-}/(u_t + u_{g^+} + u_{g^-}) \end{aligned} \tag{8.10}$$

The state of the bond is assigned according to the probabilities given in eq. (8.10). A uniformly distributed random number in the interval (0.0–0.1) is generated for this purpose and compared with the three range of values given by eq. (8.10) (specifically the first range is from 0.0 to p_t, the second from p_t to p_{g^+}, and the third from p_{g^+} to 1.0). (3) This process is repeated for all bonds of the chain in a sequential manner. At the end of this step, the sets $\{\phi\}_i$ and therefore $\{\mathbf{r}_i\}$ have been determined. (4) Steps (1)–(3) are repeated for a set of N chains, with N thus determining the size of the subpopulation over which the averages are to be evaluated. Each of the N configurations generated by this scheme obeys the probability of occurrence given by eq. (8.7). Thus, the average $\langle f(\mathbf{r}_i) \rangle$ given by eq. (8.8) may be written as

$$\langle f(\mathbf{r}_i) \rangle = \frac{1}{N} \sum_{k=1}^{N} f(\mathbf{r}_i)_k \tag{8.11}$$

Here, $f(\mathbf{r}_i)_k$ is the function associated with the kth generation of the chain. Usually, values of N the order of 10^4 are sufficient to determine averages with high accuracy.

The sample program given in appendix G calculates: (1) the persistence length of a chain in the coordinate system of figure 8.4, (2) its mean-square end-to-end vector, and (3) the orientation configuration function of its middle skeletal bond. (See chapter 11 for the definition of the orientation configuration function.) Applications of the method for generating non-Gaussian distributions for the end-to-end separations of network chains are discussed in the next section. This is followed by the use of such distributions in predicting stress–strain isotherms.

8.3 Distributions for End-to-End Separations for Specific Types of Network Chains

Because of its nature, this approach[12–17] takes direct account of the structural differences between chemically different elastomers through the different rotational isomeric state representations of the chains[1–3]. Specifically, all the structural features that distinguish one type of elastomeric chain from another are taken into account, as was done in the generation of the spatial configuration shown in figure 8.3. This information is used in the Monte Carlo method described earlier to generate a large number of spatial configurations which are representative of the specified chain structure, at the specified chain length and temperature. The values of the end-to-end separation r for these various configurations are then calculated and, in effect, put into boxes corresponding to different ranges of r. Representation of the number of chains in a given range by the height of a bar, and displaying these bars as a function of r, then gives the usual type of bar graph. A smooth curve put through the levels of this bar graph then represents the distribution of r, which can be used to replace the approximate Gaussian distribution. Such distributions are particularly useful for chains that are known to be non-Gaussian, for example because of their shortness or because of their being stretched close to the limits of their extensibility. Analytical expressions for such distributions are given in appendix F.

Going from the usual "structureless" molecular theories of rubberlike elasticity[18] to the present detailed representations parallels going from the theory of ideal gases to the van der Waals theory of real gases. The advantage in both cases is a more realistic portrayal of the system, but at the loss of universality (in that additional information specific to the chosen system is required).

Some typical distributions calculated for poly(dimethylsiloxane) (PDMS) network chains having $n = 20$, 40, and 250 skeletal bonds are shown in figure 8.5[12]. The three solid curves show how the distributions change, moving to lower values of r as n increases. The Gaussian distribution is seen to be a relatively good approximation to the Monte Carlo distribution at the largest value of n, as expected. The effects of going to still smaller values of r are illustrated for PDMS chains in figure 8.6[13]. The values of r/nl for chains having $n = 10$ skeletal bonds are seen to be quite large, which is consistent with the very low extensibility of such short chains. The additional maximum and the shoulder on this distribution is apparently a real characteristic of short chains and not an artifact of the rotational isomeric state model. This is demonstrated by the multimodal distributions obtained for n-alkane chains, as is illustrated in figure 8.7[19]. The complexity

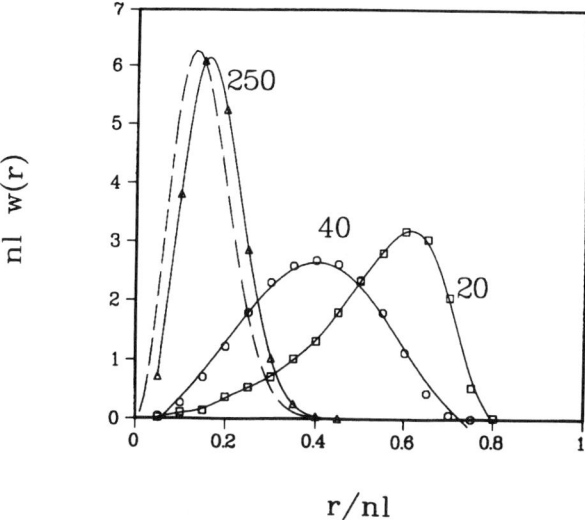

Figure 8.5 Simulated distributions of end-to-end separations for poly(dimethylsiloxane) chains, as obtained from Monte Carlo simulations[12] based on a rotational isomeric state model[1]. The PDMS chains have the specified number of skeletal bonds, each of length l, and the results pertain to 413.2 K. The Gaussian approximation for $n = 250$ is shown by the dashed line. The Monte Carlo curves represent cubic-spline fits to the discrete simulation data for 80,000 chains, and each curve is normalized to an area of unity.

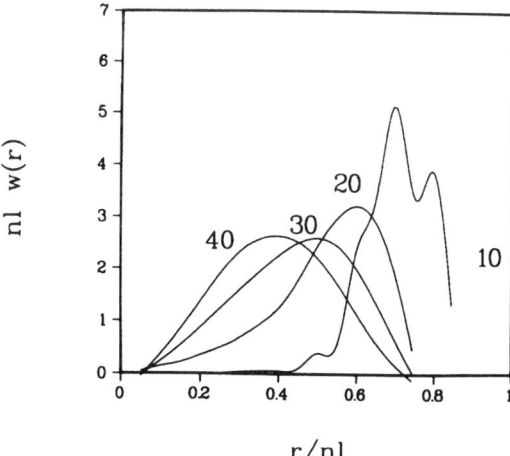

Figure 8.6 Monte Carlo distributions at 298 K for very short PDMS chains. The number of skeletal bonds is specified for each curve[13].

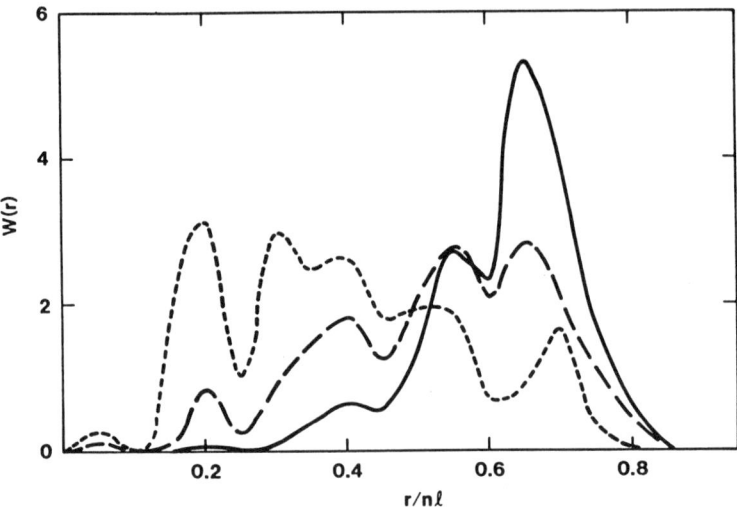

Figure 8.7 Monte Carlo distributions at 413.2 K for an *n*-alkane chain having 10 skeletal bonds[15]. The solid curve gives the results for interdependent bond rotations, the long-dashed curve for independent rotations, and the short-dashed curve for free rotations.

of the distribution is, in fact, increased in going from an interdependent-state RIS model to an independent-state RIS model to a freely rotating model (with no rotational states at all). This shift between models corresponds to an increase in chain flexibility, which causes a marked broadening of the distributions, through increased populations having very small values of r/nl.

Distributions obtained for PDMS and polyethylene (PE) chains having 20 skeletal bonds are shown in figure 8.8[12]. The distribution for the PDMS chains is seen to be shifted to values of r/nl somewhat larger than those for PE, as would be expected from its larger value of the characteristic ratio[1] associated with its unperturbed dimensions[8]. Also included in this figure is the Gaussian limit for this value of *n*. For such short chains, the Gaussian distribution is seen to give a very poor approximation to the Monte Carlo distributions. This is particularly true in the region of large *r*, which is a very important factor when interpreting limited chain extensibility and elastomer rupture. The discrepancy was found to become worse as *n* was decreased[13,15].

Chains of sulfur or selenium are known to have extremely high flexibility because of the lengths of their bonds and the lack of any chain substituents[1]. Distributions obtained for these chains are shown in figure 8.9, which also shows a corresponding distribution for polyethylene and the Gaussian limit[15,20]. The flexibility of the S and Se chains is demonstrated by the shifts in the most-probable values of *r* (as located by the maxima in the curves) to values significantly lower than that shown by polyethylene at the same chain length. Not surprisingly,

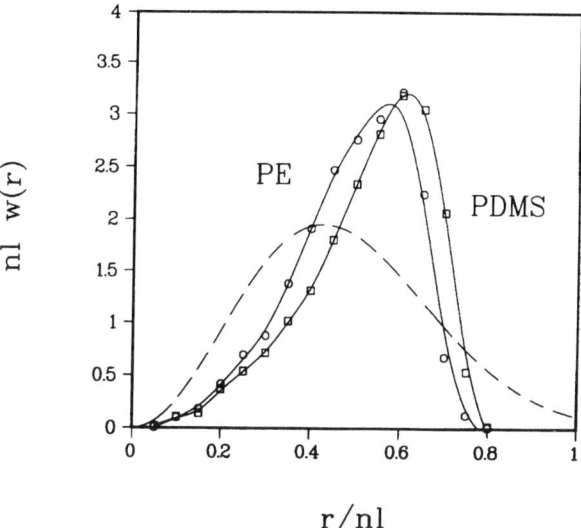

Figure 8.8 Comparisons among the Monte Carlo distributions at 413.2 K for polyethylene (O) and poly(dimethylsiloxane) (□) chains having $n = 20$ skeletal bonds, and the Gaussian approximation (- - -) to the PDMS distribution[12].

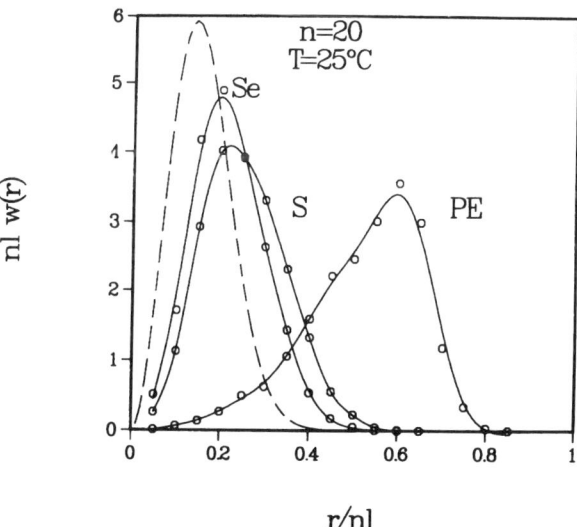

Figure 8.9 A comparison of Monte Carlo distributions at 298.2 K for sulfur and selenium with polyethylene, for chains of 20 skeletal bonds[14]. The dashed curve is the Gaussian approximation to a chain of the same root-mean-square end-to-end distance as the S chain.

the Gaussian distribution is a better approximation for these chains than it is for polyethylene.

Some chain molecules have a strong tendency to form helical regions within a conformation that is otherwise random-coil in nature. These chains would therefore be expected to have multimodal distributions of the end-to-end distance, particularly in the case of very short chains. One such chain molecule is poly(oxymethylene) (POM), with repeat unit $[-CH_2O-]_x$.[20,21] The bimodal character of its distribution, at $n = 40$ skeletal bonds and at 298.2 K, is shown in the uppermost part of figure 8.10[21]. The remaining parts of the figure show the contributions of specific conformations, specifically a perfect rod, a once-broken rod, a twice-broken rod, and so on. Such information can give valuable insights into the conformational makeup of helicogenic polymers and their elastomeric responses.

As expected, application of an external force to the ends of these chains was found to shift the coexistence equilibrium between helical and random-coil sequences[20,22]. As can be seen from figure 8.11, increase in force increases sig-

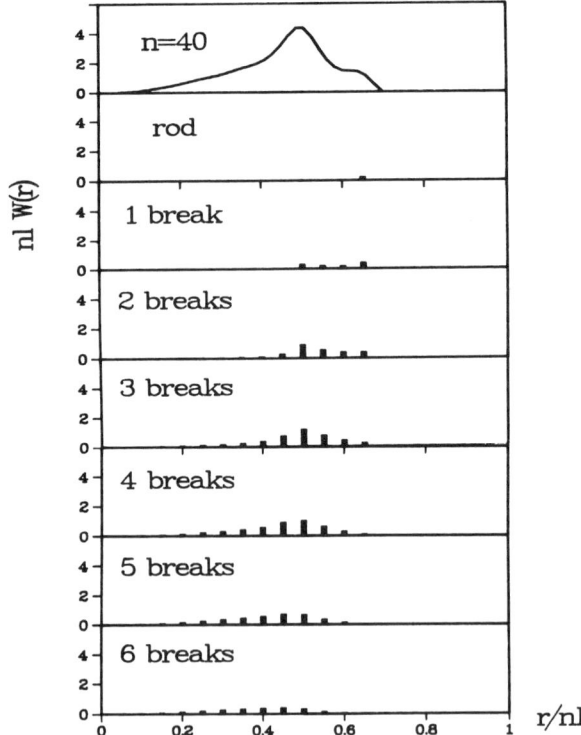

Figure 8.10 Resolution of the Monte Carlo distributions for poly(oxymethylene) POM $[-CH_2O-]$ at 298.2 K, for chains having 40 skeletal bonds[21]. The breaks specified for each contribution correspond to kinks in the otherwise-helical chains.

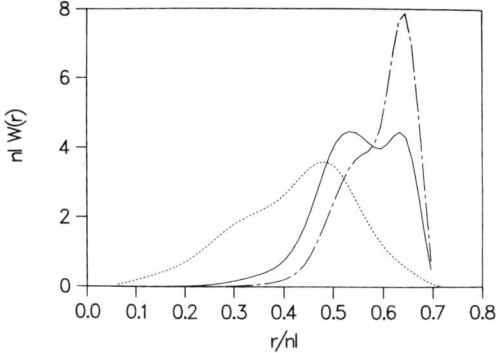

Figure 8.11 Effect of tension on the Monte Carlo distributions for the POM chains described in figure 8.10[22]. Null, moderate, and large tensions give the dotted, solid, and dashed curves, respectively. Reprinted with permission from Curro, J. G., et al. (1986), *Macromolecules*, **19**. Copyright 1997 American Chemical Society.

nificantly the contributions from the helical sequences, as identifiable by their larger values of the fractional extension r/nl. This has obvious implications in a number of areas, including the strain-induced crystallization of elastomeric materials, as described in chapter 12.

Monte Carlo computer simulations have been carried out on filled networks[23–26] in an attempt to obtain a better molecular interpretation of how such dispersed fillers reinforce elastomeric materials. The approach taken enables estimation of the effect of the excluded volume of the filler particles and the non-Gaussian characteristics of the chains on the elastic properties of the filled networks. In the earliest such study[23,24], it was assumed that (1) the filled polymer network consists of a cross-linked mixture of two types of chains, specifically those attached at one end to spherical filler particles and those that are unattached, and (2) the elastic modulus of the filled polymer is the sum of contributions from these two types of chains. In an extension of this approach, the filled polymer network was modeled as a composite of cross-linked polymer chains and spherical filler particles arranged in a regular array on a cubic lattice[25]. Composites of this type have actually been produced[27,28], and their mechanical properties are under investigation. The most realistic portrayal to date, however, randomly distributes the reinforcing particles within the elastomeric matrix[26]. In all cases, distribution functions for the end-to-end vectors of the chains were then obtained using the Monte Carlo rotational isomeric state technique. In the present application, however, conformations of attached chains which overlapped with a particle during the simulation were rejected.

Figure 8.12 shows some simulation results for PDMS distributions for chains as a function of filler content for the most realistic case[26]. One effect of the filler is obviously to increase the extensions of the chains, at least in the case of relatively small filler particles. The larger values of r are presumably due to a particle's impenetrable volume biasing the chain trajectory in the direction away from the

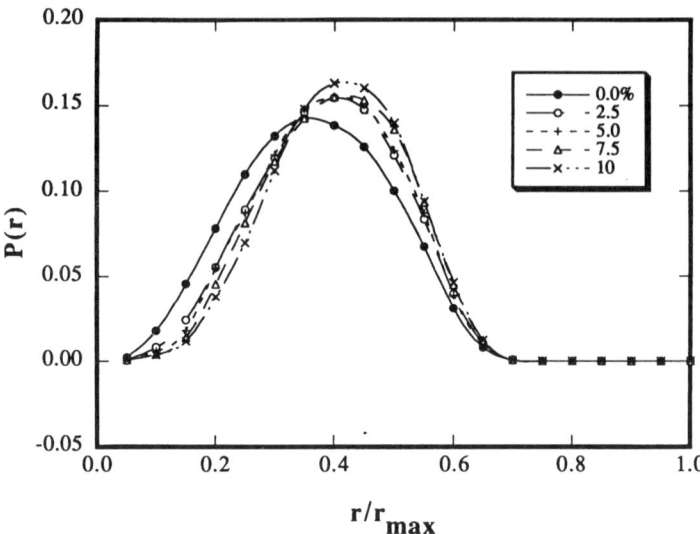

Figure 8.12 Radial distribution functions $P(r)$ for the end-to-end vector obtained by Monte Carlo simulations, shown as a function of the relative chain extension r/r_{max} for poly(dimethylsiloxane) networks[26]. In this figure, and in figures 8.16 and 8.17, $T = 500$ K, the chains have $n = 50$ skeletal bonds between cross-links, the weight percent of filler is as specified, and the radius of the filler particles is 50 nm.

particle. The effects of these changes in the distributions on the stress–strain isotherms are discussed in the following section.

8.4 Stress–Strain Isotherms Calculated from the Non-Gaussian Distributions

The Monte Carlo distributions thus obtained can be used in the standard "three-chain" model[29] for rubberlike elasticity. Some typical results, in terms of Mooney-Rivlin isotherms for networks made up of PDMS chains of various lengths, are presented in figure 8.13[12]. As expected, the network consisting of relatively long chains ($n = 250$) gives the Gaussian result $[f^*]/\nu kT = 1$. The upturns in $[f^*]$ obtained at smaller n are very similar to those found experimentally in bimodal networks containing large proportions of very short network chains[30]. Such networks having unusual network chain-length distributions are discussed in greater detail in chapter 13. The results in figure 8.13 also show that the shorter the network chains, the smaller the elongation at which the upturn occurs. The Monte Carlo simulations based on the rotational isomeric state model for the network chains have thus been very useful in interpreting these upturns in modulus. It is also possible to interpret the upturns in modulus in these isotherms using analytical expressions, for example the Fixman-Alben modification[31] of the Gaussian distribution function, combined with the constrained-junction theory and reasonable values of the constraint parameter κ.[32] Similarly useful distribu-

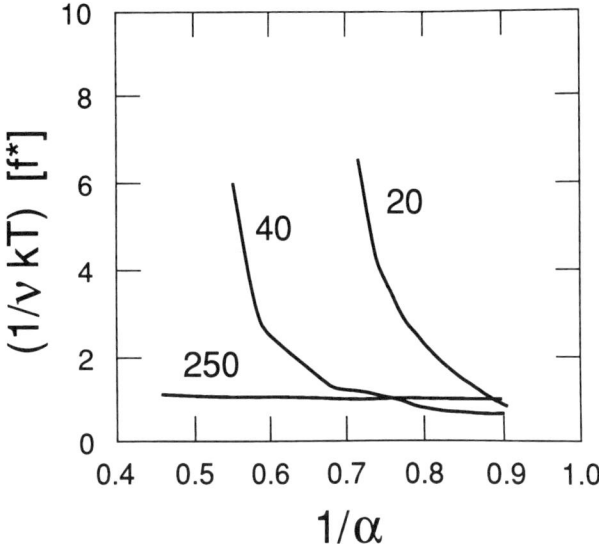

Figure 8.13 Moduli predicted for PDMS networks having n = 20, 40, and 250 skeletal bonds[12]. The values of $[f^*]$ are normalized by the Gaussian prediction for the modulus, νkT, where ν is the number of network chains and kT has the usual significance.

tion functions have been obtained by means of spherical harmonic representations of rotational operators[33,34].

Rather unusual stress–strain isotherms are obtained for networks consisting of the highly flexible S or Se chains, as is illustrated in figure 8.14[15]. The normalized modulus is seen to be approximately independent of elongation and close to the value of unity, as expected for Gaussian networks. This is quite different from the isotherm for polyethylene networks shown in the same figure, or for the short-chain PDMS networks described in figure 8.13.

Distributions obtained as described above have also been used to calculate stress–strain isotherms for networks of natural rubber[16]. The results are shown in figure 8.15. Of particular interest here were both the nature of the simulated upturns in modulus at high elongations, and the elongations at which they are first discernible. These seem to support the conclusion[30] that the experimental upturns reported in the literature for natural rubber were more likely due to strain-induced crystallization than to limited chain extensibility.

The isotherms for the PDMS chains in the vicinity of randomly placed filler particles are shown in figures 8.16 and 8.17[26]. Specifically, values of the stress (as reduced by the number of network chains, the gas constant, and absolute temperature) are shown as a function of elongation in figure 8.16. The alternative Mooney–Rivlin representations of the similarly reduced modulus vs. reciprocal elongation are shown in figure 8.17. The substantial increases in stress and modulus with increase in filler content and elongation are in at least qualitative agreement with experiment, as described in chapter 16.

102 STRUCTURES AND PROPERTIES OF RUBBERLIKE NETWORKS

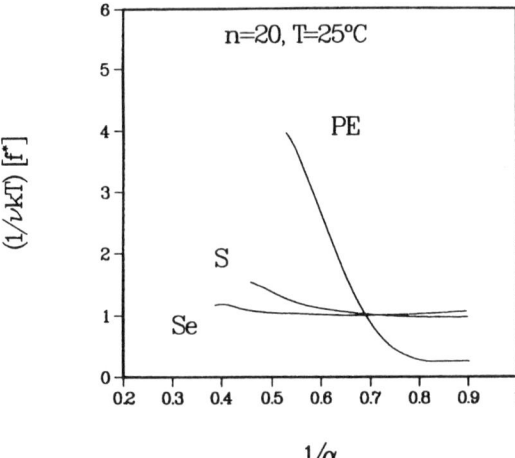

Figure 8.14 Comparisons among the stress–strain isotherms predicted for networks consisting of the unusually flexible sulfur and selenium chains, and one consisting of polyethylene chains[14]. The results pertain to 20 skeletal bonds and to 298.2 K.

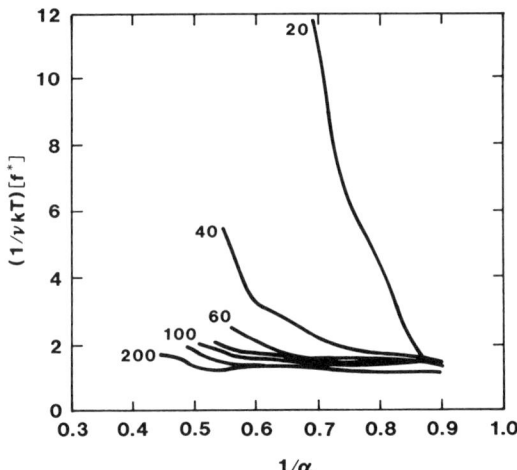

Figure 8.15 Normalized moduli predicted for networks consisting of natural rubber chains having the numbers of skeletal bonds specified for each isotherm[16].

There are a number of directions in which such filler simulations could be extended. For example, one should investigate different extents of particle aggregation, and different particle-size distributions. It would also be important to model physical adsorption of chains onto the filler surfaces using standard Lennard-Jones interaction potentials. Chemical adsorption, on the other hand, could be modeled by randomly distributing active particle sites, and then

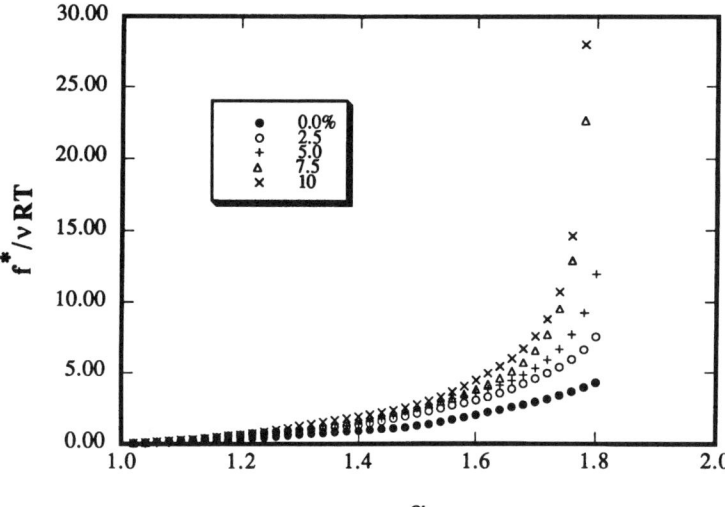

Figure 8.16 Dependence of the normalized nominal stress $f^*/\nu RT$ on the elongation α for the randomly filled PDMS networks[26].

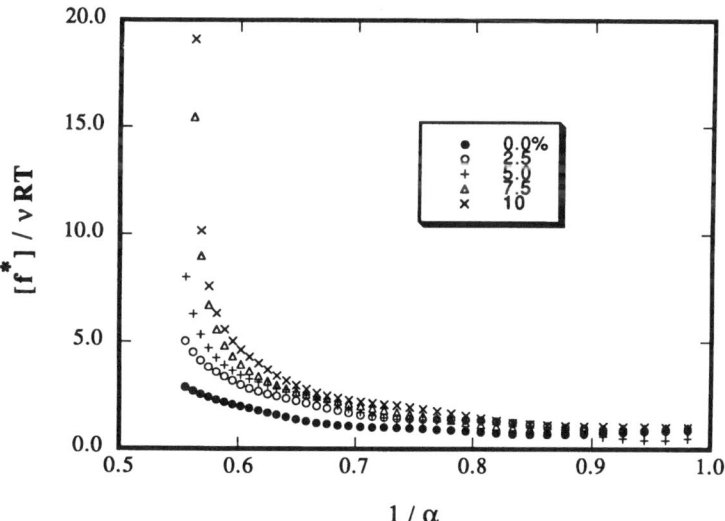

Figure 8.17 The normalized theoretical modulus $[f^*]/\nu RT$, shown as a function of the reciprocal elongation α^{-1}, for the randomly filled PDMS networks[26]

interacting chains with them through a Dirac δ-function type of potential (with chains at less than some short-range interaction distance becoming chemisorbed). Of particular interest would be simulations for chains sufficiently long to partially adsorb onto several filler particles, and to model chain-contour distributions between the bulk polymer and the filler particles. This approach could well give important and much-needed insight into molecular aspects of elastomer reinforcement.

For other MC simulations of non-Gaussian chain configurations, the reader is referred to the work of Haliloglu and collaborators[35,36]. In these studies, configurations of polyethylene, polyoxyethylene, and poly(dimethylsiloxane) chains were generated, and segmental orientation in highly stretched, non-Gaussian chains was evaluated.

8.5 Molecular Dynamics Calculations

The Monte Carlo calculations described in the preceding sections are single-chain calculations where the detailed behavior of the single chain is studied and the behavior of the network is predicted based on the additivity of the elastic free energies of the chains. A direct way of simulating a network without invoking assumptions on the additivity of the single-chain properties is obviously the molecular dynamics (MD) method, and this method is now being applied widely to networks. Although computational difficulties currently prevent simulations of realistic networks, noteworthy rapid progress is being made.

The earlier studies of network elasticity by MD were made by Gao and Weiner[37-40] who investigated the elasticity of networks with fixed junctions. Studies of networks with mobile junctions came later[41-45]. The direct study of the entanglement problem in networks is especially important inasmuch as the MD technique is capable of incorporating the full intermolecular potentials into the formulations without assumptions. This work by Kremer and collaborators[46] indicates contributions to the elastic modulus from intermolecular correlations, in addition to those from the network connectivity. The extent of these contributions is ambiguous at present, however, since only small systems have been studied. A thorough review of the application of MD to networks and a summary of the findings has been given by Kremer and Grest[46].

References

(1) Flory, P. J. *Statistical Mechanics of Chain Molecules*. Interscience: New York. 1969.
(2) Mattice, W. L.; Suter, U. W. *Conformational Theory of Large Molecules. The Rotational Isomeric State Model in Macromolecular Systems*. Wiley: New York. 1994.
(3) Honeycutt, J. D. In *Physical Properties of Polymers Handbook*, J. E. Mark, Ed. American Institute of Physics Press: Woodbury, NY. 1996; p. 39.
(4) Roe, R.-J. In *Physical Properties of Polymers Handbook*, J. E. Mark, Ed. American Institute of Physics Press: Woodbury, NY. 1996; p. 55.

(5) Kloczkowski, A. In *Physical Properties of Polymers Handbook*, J. E. Mark, Ed. American Institute of Physics Press: Woodbury, NY. 1996; p. 61.
(6) Volkenstein, M. *Configurational Statistics of Polymer Chains*. Interscience: New York. 1963.
(7) Flory, P. J. *Macromolecules* 1974, 7, 381.
(8) Flory, P. J. *Principles of Polymer Chemistry*. Cornell University Press: Ithaca NY. 1953.
(9) Mark, J. E. In *Physical Properties of Polymers*, J. E. Mark, A. Eisenberg, W. W. Graessley, L. Mandelkern, E. T. Samulski, J. L. Koenig, and G. D. Wignall, Eds. American Chemical Society: Washington, DC. 1993; p. 3.
(10) Flory, P. J.; Yoon, D. Y. *J. Chem. Phys.* 1974, 61, 5358.
(11) Yoon, D. Y.; Flory, P. J. *J. Chem. Phys.* 1974, 61, 5366.
(12) Mark, J. E.; Curro, J. G. *J. Chem. Phys.* 1983, 79, 5705.
(13) Curro, J. G.; Mark, J. E. *J. Chem. Phys.* 1984, 80, 4521.
(14) Mark, J. E.; Curro, J. G. *J. Chem. Phys.* 1984, 80, 5262.
(15) Mark, J. E.; Curro, J. G. In *Characterization of Highly Cross-Linked Polymers*, S. S. Labana and R. A. Dickie, Eds. American Chemical Society: Washington, DC. 1984; p. 47.
(16) Mark, J. E.; Curro, J. G. *J. Polym. Sci., Polym. Phys. Ed.* 1985, 23, 2629.
(17) Mark, J. E.; DeBolt, L. C.; Curro, J. G. *Macromolecules* 1986, 19, 491.
(18) Erman, B.; Mark, J. E. *Ann. Rev. Phys. Chem.* 1989, 40, 351.
(19) Mark, J. E.; Curro, J. G. *J. Chem. Phys.* 1984, 81, 6408.
(20) Mark, J. E. *Comp. Polym. Sci.* 1992, 2, 135.
(21) Curro, J. G.; Mark, J. E. *J. Chem. Phys.* 1985, 82, 3820.
(22) Curro, J. G.; Schweizer, K. S.; Adolf, D.; Mark, J. E. *Macromolecules* 1986, 19, 1739.
(23) Kloczkowski, A.; Sharaf, M. A.; Mark, J. E. *Comp. Polym. Sci.* 1993, 3, 39.
(24) Kloczkowski, A.; Sharaf, M. A.; Mark, J. E. *Chem. Eng. Sci.* 1994, 49, 2889.
(25) Sharaf, M. A.; Kloczkowski, A.; Mark, J. E. *Comp. Polym. Sci.* 1994, 4, 29.
(26) Yuan, Q. W.; Kloczkowski, A.; Mark, J. E.; Sharaf, M. A. *J. Polym. Sci., Polym. Phys. Ed.* 1996, 34, 1674.
(27) Sunkara, H. B.; Jethmalani, J. M.; Ford, W. T. *Chem. Mater.* 1994, 6, 362.
(28) Sunkara, H. B.; Jethmalani, J. M.; Ford, W. T. In *Hybrid Organic-Inorganic Composites*, J. E. Mark, C. Y-C Lee, and P. A. Bianconi, Eds. American Chemical Society: Washington, DC. 1995; p. 181.
(29) Treloar, L. R. G. *Trans. Faraday Soc.* 1946, 42, 77.
(30) Mark, J. E.; Erman, B. *Rubberlike Elasticity. A Molecular Primer*. Wiley-Interscience: New York. 1988.
(31) Fixman, M.; Alben, R. *J. Chem. Phys.* 1973, 58, 1553.
(32) Erman, B.; Mark, J. E. *J. Chem. Phys.* 1988, 89, 3314.
(33) Llorente, M. A.; Rubio, A. M.; Freire, J. J. *Macromolecules* 1984, 17, 2307.
(34) Menduina, C.; Freire, J. J.; Llorente, M. A.; Vilgis, T. *Macromolecules* 1986, 19, 1212.
(35) Erman, B.; Haliloglu, T.; Bahar, I.; Mark, J. E. *Macromolecules* 1991, 24, 901.
(36) Haliloglu, T.; Bahar, I.; Erman, B. *Comp. Polym. Sci.* 1991, 1, 151.
(37) Gao, J.; Weiner, J. H. *Macromolecules* 1991, 24, 1519.
(38) Gao, J.; Weiner, J. H. *Macromolecules* 1991, 24, 5179.
(39) Gao, J.; Weiner, J. H. *Macromolecules* 1992, 25, 1348.
(40) Gao, J.; Weiner, J. H. *J. Chem. Phys.* 1992, 97, 8698.
(41) Duering, E. R.; Kremer, K.; Grest, G. S. *Phys. Rev. Lett.* 1991, 67, 3531.

(42) Grest, G. S.; Kremer, K.; Duering, E. R. *Europhys. Lett.* 1992, 19, 195.
(43) Grest, G. S.; Kremer, K.; Duering, E. R. *Physica A* 1993, 330, 1993.
(44) Duering, E. R.; Kremer, K.; Grest, G. S. *Macromolecules* 1993, 26, 3241.
(45) Duering, E. R.; Kremer, K.; Grest, G. S. *J. Chem. Phys.* 1994, 101, 8169.
(46) Kremer, K.; Grest, G. S. In *Monte Carlo and Molecular Dynamics Simulations in Polymer Science*, K. Binder, Ed. Oxford University Press: Oxford. 1995; p. 194.

9

Thermoelasticity

The important postulate that intermolecular interactions are independent of extent of deformation[1-6] leads directly to the conclusion that such interactions cannot contribute to an energy of elastic deformation ΔE_{el} at constant volume. In the earliest theories of rubberlike elasticity[1,5,7], it was additionally assumed that *intra*molecular contributions to ΔE_{el} were likewise nil. In this idealization that the total ΔE_{el} is zero, the elastic retractive force exhibited by a deformed polymer network would be entirely entropic in origin. At the molecular level, this would correspond, of course, to assuming all configurations of a network chain to be of exactly the same conformational energy and thus the average configuration to be independent of temperature. Under these circumstances, the dependence of stress on temperature is strikingly simple, as shown, for example, by the equation[2]

$$f^* = \frac{\nu k T}{V}\left(\frac{\langle r^2\rangle_i}{\langle r^2\rangle_0}\right)(\alpha - \alpha^{-2}) \tag{9.1}$$

that characterizes a polymer network in elongation where, it should be recalled, $\langle r^2\rangle_i^{3/2}$ is proportional to the volume of the network. This additional assumption that $\langle r^2\rangle_0$ is independent of temperature would lead to the prediction that the elastic stress determined at constant volume and elongation α is directly proportional to the absolute temperature. Such network chains would be akin to the particles of an ideal gas, which would obey the equation of state $p = nRT(1/V)$ and thus exhibit a pressure at constant deformation $(1/V)$ likewise directly proportional to the temperature.

For a more careful consideration of possible intramolecular contributions to ΔE_{el}, it should be recalled that deformation of a polymer network results in an increase in the number of those chain configurations of relatively large end-to-end separation that, in conformational terms, requires a redistribution of rotational states about the bonds of the chain backbone. Since invariably some, or all, of the rotational states accessible to the skeletal bonds of a chain molecule differ in energy, different chain configurations generally do differ in energy[8,9] and, correspondingly, $\langle r^2 \rangle_0$ generally depends quite significantly on temperature. As a consequence, experimental values of the elastic force or stress are generally found *not* to be proportional to the temperature. Most important, as shown later in this chapter, the extent to which the experimentally observed relationship between stress and temperature differs from simple proportionality may be used to determine the importance of energetic contributions in the thermodynamics of network deformation[5,10–15]. Such resolution of the total retractive force into its energetic and entropic components is fundamental to an understanding of the thermodynamics of the elastic deformation process.

This chapter is thus concerned with network "thermoelasticity," by which we mean the dependence of stress on temperature[1–6,15–24]. Of primary interest is the interpretation of such results; thermodynamically, in terms of the fraction of the total elastic force that is of energetic origin, and molecularly, in terms of transitions between conformational states of the molecules, as described by the rotational isomeric state theory of chain configurations[8,9]. Also of consequence is the fact that the thermoelastic properties of a number of polymers have now been studied under a wide variety of experimental conditions. These and related results may be used to test the validity of the basic postulate underlying the molecular theories of rubberlike elasticity recounted earlier, namely that the retractive force is essentially entirely intramolecular in origin.

9.1 Theory

The primary goal of thermoelasticity studies is the quantitative resolution of the elastic force into its entropic and energetic components[1,5,7]

$$f = f_s + f_e \qquad (9.2)$$

As discussed in preceding chapters, the entropic contribution f_s is generally expected to be large and positive, arising from the decreased entropy of the stretched chains. The energetic contribution f_e may be presumed to be of either sign, depending on the conformational characteristics of the network chains, and thus may either augment or diminish f_s.

For concreteness, the equations characterizing the stress–temperature relationships for a polymer network are presented most clearly by consideration of a particular type of strain. Uniaxial deformation (elongation or compression) is chosen for this purpose here, since this type of deformation is particularly simple to visualize and characterize, and it is by far the most widely studied experimentally[5,13,14]. (The thermodynamic arguments leading to the equations presented below are given for this illustrative case, and are generalized to any type of deformation in the appendixes.) In the case of uniaxial deformation, the energetic component of the force is defined by[1]

$$f_e \equiv (\partial E / \partial L)_{T,V} \qquad (9.3)$$

with the symbols having their usual meanings. Its fractional contribution to the total force is given, without approximation and without reliance on any particular model of a polymer network, by the thermodynamically exact equation[2]

$$f_e/f = -T[\partial \ln(f/T)/\partial T]_{L,V} \qquad (9.4)$$

This important ratio is thus seen to be directly attainable from force–temperature measurements at constant length and volume.† The volume of the network may be held constant, as specified, in spite of the required changes in temperature, in either of two ways. The sample may be placed under a hydrostatic pressure, the magnitude of which is adjusted at each temperature so as to offset exactly the effect of thermal expansion[25-28]. Alternatively, the network may be studied swollen in equilibrium with a diluent or diluent mixture[29] in which the change in polymer–solvent interactions with temperature exactly nullify the same thermal effects. In order to minimize experimental difficulties, however, thermoelastic measurements are generally carried out with the networks at constant pressure rather than at constant volume[7,13,14]. Interpretation of such isobaric thermoelastic results is most conveniently accomplished through adoption of a specific elastic equation of state. The use of the theoretical equation of state derived in chapter 2 for networks in uniaxial deformation gives the result[2,10,11]

$$f_e/f = -T[\partial \ln(f/T)/\partial T]_{L,p} - \beta T/(\alpha^3 - 1) \qquad (9.5)$$

for data obtained at constant length and pressure, where $\beta = (\partial \ln V/\partial T)_p$ is the thermal expansion coefficient of the network.‡ Similarly, for conditions of constant elongation (or compression) and pressure:

$$f_e/f = -T[\partial \ln(f/T)/\partial T]_{\alpha,p} + \beta T/3 \qquad (9.6)$$

In the case of studies carried out at constant length, measurements in the vicinity of $\alpha = 1$ incur large uncertainties due to the form of the denominator in the difference term $\beta T/(\alpha^3 - 1)$, and the difficulty of obtaining sufficiently precise values of α under these conditions. The effect of this term can obviously be minimized by carrying out the force–temperature measurements at relatively high deformations. Then, $\beta T/(\alpha^3 - 1)$ approaches a minimum value of zero in the case of elongation ($\alpha > 1$) and $-\beta T$ in the case of compression ($\alpha < 1$). Under the alternative conditions, the elongation or compression α is itself maintained at a constant value by appropriate adjustment of the length of the deformed sample to offset exactly the changes in the undeformed length that accompany the changes in temperature. In such experiments at constant pressure, the required term is seen to be simply $\beta T/3$, independent of α.§ The above alternatives by no means exhaust the variety of techniques that may be employed to obtain thermoelastic data. Other approaches would include the extraction of stress–temperature information from a series of stress–strain isotherms obtained at different temperatures[34], and the measurement of sample length as a function

---

†As shown in eq. (9.4), only the fractional change in the force is relevant here, and, thus, the temperature derivative of $\ln(f^*/T)$ or of $\ln([f^*]/T)$ serves equally well in this equation.
‡In the case of networks containing diluent, the thermal expansion coefficient required is, of course, that of the *swollen* network.
§Some other types of deformation, for example torsion or shear, have the advantage of avoiding entirely such difference terms dependent on the deformation α.[5,30–33]

of temperature at constant force[35]. In any case, determination of reliable values of f_e/f from thermoelastic data acquired at constant pressure does require inclusion of such terms effecting the conversion to the thermodynamically required conditions of constant volume, since the factor βT is typically of the order of 0.2–0.3.

Implicit in the basic postulate of rubberlike elasticity is the assumption that in the deformation of a polymer network, any energy changes are intramolecular in origin. In this case, the theoretical equation of state derived in chapter 2 may also be used to obtain a molecular interpretation of thermoelastic data and the quantity f_e/f derived therefrom. The result is given by the equation[10–12]

$$f_e/f = T d\ln\langle r^2\rangle_0/dT \qquad (9.7)$$

where $\langle r^2\rangle_0$ represents the unperturbed dimensions[1] of the network chains. This equation is of considerable importance since it permits the comparison[13,36] of results of thermoelastic measurements on polymer chains in the bulk, in network structures, with results of viscosity measurements on chains of the same polymer, essentially isolated, in dilute solution. It also establishes the relationship between the purely thermodynamic quantity f_e/f and its molecular counterpart $d\ln\langle r^2\rangle_0/dT$, which can be interpreted in terms of the rotational isomeric state theory[8,9] of chain configurations.

9.2 Typical Stress-Temperature Data

Detailed discussions of the experimental techniques employed in thermoelastic studies may readily be found in the original literature, which is extensively cited in the following section. It is important to emphasize here, however, that use of the thermodynamic relationships presented in the previous section, and derived in the appendices, requires that the polymer network be brought as closely as possible to elastic equilibrium at each temperature of investigation. Thus, only those force–temperature (or length–temperature) curves which are found to be reproducible upon subsequent changes in temperature are suitable. A typical series of such force–temperature curves, which have been obtained on amorphous unswollen polyethylene networks at constant pressure and at constant length[12], is shown in figure 9.1. As was done in this study, the elastic force is typically divided by the cross-sectional area of the undistorted sample at a convenient temperature in order to facilitate comparisons among different samples. The value of the deformation α typically cited is that calculated from the rest length of the sample at the highest temperature investigated. Although α itself varies slightly with temperature at constant length, no significant error is introduced by use of a constant value of α in the calculation of the difference term $\beta(\alpha^3 - 1)^{-1}$. As is generally the case, the stress is found to vary linearly with temperature within experimental error, thus indicating that f_e/f itself does not vary significantly with temperature.

Stress–temperature relationships have also been determined for a number of bioelastomers, as described in chapter 15. Sufficient chain mobility for elastomeric behavior in network structures of this type generally requires the presence of relatively large amounts of low-molecular-weight diluent. They are therefore cus-

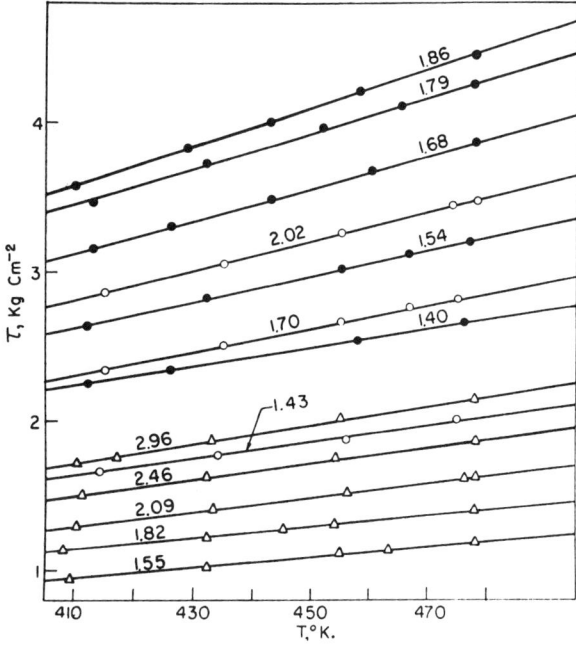

Figure 9.1 Some thermoelastic data[12] for amorphous polyethylene networks in the unswollen state, at constant length. The stress f^* is expressed relative to the undeformed cross-section at the highest temperature of measurement, 205°C, and each curve is characterized by the value of the elongation α at the same temperature. Reprinted with permission from Ciferri, A., et al. (1961), *J. Am. Chem. Soc.*, **83**. Copyright 1997 American Chemical Society.

tomarily studied in the highly swollen state. Some thermoelastic data obtained on elastin, a cross-linked protein described in chapter 15, are presented in figure 9.2[37]. This material was studied in the totally amorphous state, in swelling equilibrium, at constant pressure. Each line shown gives a good representation of data taken by either decreasing or increasing the temperature, thus illustrating clearly the high degree of reproducibility attainable in thermoelastic investigations.

The use of a compressive deformation in force–temperature measurements is illustrated in figure 9.3, which pertains to an amorphous polyoxyethylene network, swollen with a constant amount of nonvolatile diluent and studied at constant pressure[38]. The data shown were obtained indirectly from stress–strain isotherms determined for a compressed cylindrical sample at a number of temperatures. The isotherms had been represented in such a manner as to yield the equivalent of stress–temperature data at constant length. As a final example, some typical thermoelastic data in torsion are shown in figure 9.4[31]. They pertain to a natural rubber network studied in the amorphous unswollen state at constant length and pressure. In this type of deformation, a network is characterized by the torsion couple M and angle of torsion δ (which replace, respectively, the quantities f and L used to characterize a network in uniaxial deforma-

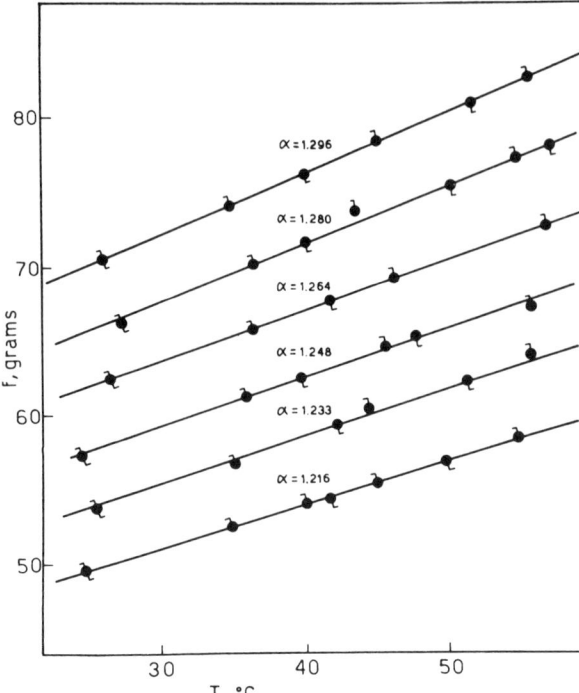

Figure 9.2 Thermoelastic data[37] for the tissue protein elastin in the amorphous state, in swelling equilibrium with dimethylsulfoxide. Measurements were carried out at constant length, and the specified values of the elongation pertain to 40°C. ⚲, results obtained upon decrease in temperature; ⚭, results obtained upon increase in temperature.
Reprinted with permission from Mistrali, F., et al. (1971), *J. Phys. Chem.*, **75**. Copyright 1997 American Chemical Society.

tion)[5,30,33,39–41]. As expected, the experimental stress–temperature relationships obtained using torsion as the deformation are very similar to those obtained in elongation or compression, as is illustrated in figures 9.1, 9.2, and 9.3.

9.3 Illustrative Thermoelastic Results

As was pointed out above, experimental studies of network thermoelasticity may be performed in a variety of ways. Those carried out at constant volume assume considerable importance since they yield values of f_e/f directly from eq. (9.4), a thermodynamically exact equation. Values thus obtained may therefore be used to check results obtained at constant pressure and interpreted through the use of equations such as eqs. (9.5) and (9.6) which are based on the statistical, molecular theory of rubberlike elasticity.

Because of experimental difficulties associated with maintaining a polymer network at constant volume in spite of changes in temperature, relatively few studies meeting this thermodynamic constraint have been performed successfully.

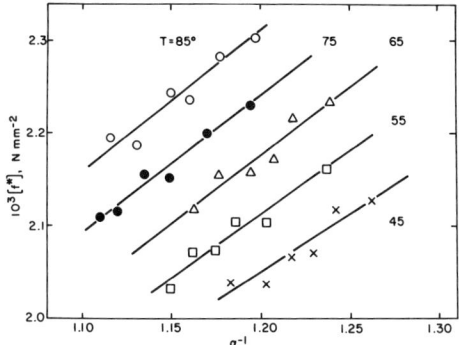

Figure 9.3 Thermoelastic data derived from a series of stress–strain isotherms obtained for a compressed amorphous network of polyoxyethylene, swollen with a constant amount of nonvolatile diluent[38].

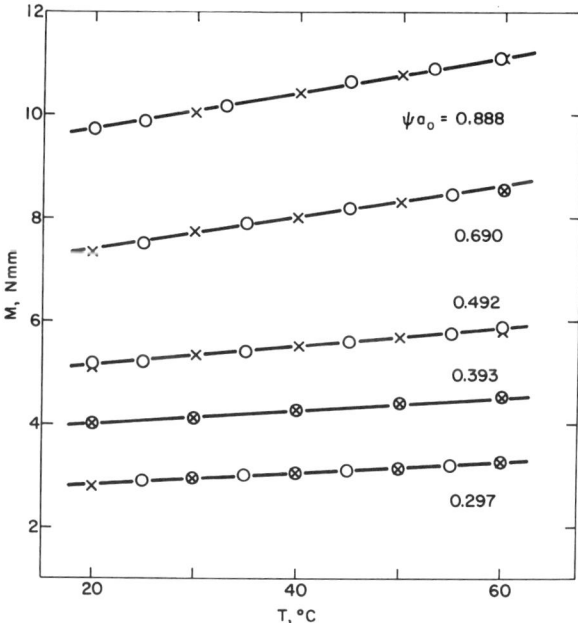

Figure 9.4 Thermoelastic data obtained on an amorphous network of natural rubber in torsion[31]. The stress is given by the torsional couple M, and the strain by the product of the twist ψ (in radians per unit length) and the unstrained radius a_0. Results obtained upon decrease in temperature are shown by open circles, and those upon increase in temperature, by crosses.

The most extensive of these are the studies[25-28] of Allen and coworkers, who have employed the method of variable hydrostatic pressure to nullify changes in the volume of the sample due to the usual thermal expansion. Unfortunately, only a small range in temperature could be covered. In these studies, carried out thus far on networks of natural rubber[25,28], poly(dimethylsiloxane)[27], and polyisobutylene[26], the required coefficient $(\partial f/\partial T)_{L,V}$ was generally obtained in two independent ways. The first was the direct measurement of the force as a function of temperature, with appropriate control of the network volume, as already described. The second involved measurement of values of the coefficients $(\partial f/\partial T)_{L,P}$, $(\partial p/\partial T)_{L,V}$, and $(\partial f/\partial p)_{L,V}$ and use of the thermodynamic identity:

$$(\partial f/\partial T)_{L,V} = (\partial f/\partial T)_{L,p} + (\partial p/\partial T)_{L,V}(\partial f/\partial p)_{L,T} \qquad (9.8)$$

These two approaches gave values of f_e/f in satisfactory agreement. The networks which have been most extensively studied in this regard are those of natural rubber and poly(dimethylsiloxane), and these results are therefore summarized in table 9.1[14]. Included for purposes of comparison are the most reliable of the corresponding results obtained under conditions of constant pressure (and either constant length, elongation, or force). All networks were studied in the unswollen state, unless specified otherwise. The agreement between the two types of results obtained should be considered satisfactory in view of the experimental difficulties involved, particularly in the studies at constant volume, over a narrow range in temperature.† Less complete data are available for polyisobutylene networks, but the results at constant volume and at constant pressure are also in good agreement, giving the average value $f_e/f = -0.06(\pm 0.04)$[12,14,26].

Using the less direct approach for obtaining thermoelastic data at constant volume, Sakurada[50] and Smith[51], and their coworkers, studied networks of atactic poly(vinyl alcohol) in swelling equilibrium in a water and ethylene glycol mixture of composition such that the thermal expansion coefficient of the swollen network was essentially zero. Thermoelastic measurements were carried out at this unique composition, and also at compositions at which volume changes were significant and therefore had to be corrected for by use of eq. (9.5), of the statistical theory. The two types of results were again found to be in satisfactory agreement, giving $f_e/f = -0.43(\pm 0.10)$[14,50,51].

The above comparisons strongly indicate that reliable values of f_e/f may be obtained under the relatively simple condition of constant pressure, rather than constant volume. As is also shown in table 9.1, and in more extensive compilations[13,14], the results obtained at constant pressure do not depend on the choice of the other constraint (constant L, α, or f), or on whether the system is thermo-

---

†In the case of natural rubber, the value of f_e/f obtained at constant volume is approximately 0.05 below the average value obtained at constant pressure. The difference, although small, could be due to the volume dependence of the Mooney-Rivlin $2C_2$ constant[5,7,19,42-47]. An approximate calculation[17,48] to account for such a contribution to f_e/f does improve the agreement between the two types of results in the case of natural rubber, but has the opposite effect[49] when applied to the equally extensive data obtained on networks of poly(dimethylsiloxane).

Table 9.1 Effect of Thermodynamic Constraints on $f_e/f = Td\ln\langle r^2\rangle_0/dT$

Polymer	Invariants	f_e/f
Natural rubber	V, L[a]	$0.12\,(\pm 0.02)$[d]
	p, L	$0.17\,(\pm 0.02)$[e]
	p, L[b]	$0.19\,(\pm 0.01)$[f]
	p, α	$0.13\,(\pm 0.02)$[g,h]
	p, f	$0.18\,(\pm 0.05)$[i]
Poly(dimethylsiloxane)	V, L[a]	$0.25\,(\pm 0.02)$[j]
	p, L	$0.19\,(\pm 0.02)$[e]
	p, α	$0.13\,(\pm 0.03)$[k]
	p, f	$0.18\,(\pm 0.06)$[h]
Polyisobutylene	V, L[a]	$-0.09\,(\pm 0.03)$[l]
	p, L	$-0.03\,(\pm 0.02)$[g]
Poly(vinyl alcohol) atactic	V, L[c]	$-0.49\,(\pm 0.07)$[m,n]
	p, L[b]	$-0.36\,(\pm 0.02)$[m,n]

[a] Constant V achieved by imposed hydrostatic pressure.
[b] Swelling equilibrium, with $\beta \neq 0$.
[c] Constant volume achieved by swelling equilibrium, with $\beta \approx 0$.
[d] Taken from references 25 and 28.
[e] Taken from reference 13.
[f] Taken from reference 51.
[g] Taken from reference 12.
[h] Taken from reference 91.
[i] Taken from reference 35.
[j] Taken from reference 27.
[k] Taken from reference 34.
[l] Taken from reference 26.
[m] Taken from reference 50.
[n] Taken from reference 51.

dynamically closed (at fixed composition) or open (at swelling equilibrium). This verification of the reliability of thermoelastic analyses carried out through recourse to the statistically derived equations of rubberlike elasticity now permits further, more detailed inquiry into the possible effects of other experimental variables on f_e/f or its molecular counterpart $f_e/fT = d\ln\langle r^2\rangle_0/dT$.

One variable which could possibly have an effect on f_e/f is the state of the sample when the required network structure was imposed. Although most polymer networks used in thermoelastic studies have been prepared in the undiluted, undeformed, amorphous state, a number of studies have employed networks prepared under more unusual conditions[13,14]. Some of the relevant results are summarized in table 9.2[14].† As is evident from these illustrative results, f_e/f is independent of the degree of deformation or orientation of the chains prior to cross-linking, of the degree of crystallinity of the sample, and of the presence of a diluent partially disentangling the network chains prior to their cross-linking. Similar but less extensive results have also been obtained for networks of cis-

---

†The average value of f_e/f and the average deviation from this value cited for a given polymer network will not necessarily be the same in each table since different sets of experiments must generally be cited when different experimental variables are under consideration.

Table 9.2 Effect of Cross-Linking Conditions on f_e/f

Polymer	State of sample during cross-linking	f_e/f
Polyethylene	Unoriented, amorphous	-0.30^a
	Unoriented, crystalline	$-0.47\,(\pm0.06)^a$
	Oriented, crystalline	$-0.42\,(\pm0.04)^a$
Natural rubber	Unoriented, amorphous	$0.17\,(\pm0.03)^b$
	Oriented, amorphous	$0.18\,(\pm0.03)^c$
	Oriented, crystalline	$0.17\,(\pm0.03)^d$
	In solution	$0.14\,(\pm0.04)^{e,f}$
Cis-1,4-polybutadiene	Unoriented, amorphous	$0.12\,(\pm0.04)^{b,g}$
	In solution	$0.13\,(\pm0.02)^{h,i}$
Poly(dimethylsiloxane)	Unoriented, amorphous	$0.20\,(\pm0.04)^b$
	In solution	$0.15\,(\pm0.02)^{j,k}$

[a] Taken from reference 12.
[b] Taken from reference 13.
[c] Taken from reference 92.
[d] Taken from reference 93.
[e] Taken from reference 94.
[f] Taken from reference 69.
[g] Taken from reference 39.
[h] Taken from reference 52.
[i] Taken from reference 53.
[j] Taken from reference 34.
[k] Taken from reference 74.

1,4-polybutadiene[13,14,39,40,52-54] and poly(dimethylsiloxane)[13,14,34,55]. Also, essentially identical values of f_e/f seem to be obtained for systems cross-linked by chemical means and those cross-linked by high-energy radiation[12]. This insensitivity of f_e/f to cross-linking conditions is of considerable interest, since changes in such conditions are thought to have a marked effect on the topology of the resulting network. Cross-linking in the presence of diluent should yield a network of relatively few permanent entanglements, as should also the prior, temporary orientation or lining up of the polymer chains, either through chain crystallization or by means of a mechanical deformation[4,56,57].

The possible dependence of f_e/f on degree of cross-linking is also of interest, but data relevant to this point are somewhat limited by the fact that, in general, only relatively narrow ranges of degree of cross-linking are studied in thermoelastic experiments[13]. Networks of relatively low degree of cross-linking are very slow in reaching equilibrium, and are usually insufficiently stable for measurements over the required range of temperature and time intervals required in such experiments. High degrees of cross-linking, on the other hand, prevent attainment of high deformations, which, of course, have the advantage of minimizing uncertainties from the difference term $\beta(\alpha^3 - 1)^{-1}$. Table 9.3[14] summarizes some results from those typical thermoelastic studies felt to be most pertinent with regard to the possible dependence of f_e/f on degree of cross-linking. Values of the effective degree of cross-linking, $\nu/2V$, cited in this table are in units of moles of cross-links per cubic centimeter of network. They were calculated from values of $[f^*]$ for the unswollen networks, extrapolated to $\alpha^{-1} = 0$, that is, from the Mooney-Rivlin

Table 9.3 Effect of Degree of Cross-Linking on f_e/f

Polymer	$(10^5 \nu/2V)^a$	f_e/f
Polyethylene[b]	0.528	$-0.41\,(\pm 0.04)$
	1.87	$-0.43\,(\pm 0.02)$
	2.80	$-0.44\,(\pm 0.04)$
Natural rubber[c]	1.85	$0.11\,(\pm 0.02)$
	2.43	$0.12\,(\pm 0.01)$
	2.78	$0.15\,(\pm 0.01)$
	6.42	$0.14\,(\pm 0.02)$
	7.15	$0.12\,(\pm 0.01)$
Natural rubber[d]	5.21	$0.16\,(\pm 0.02)$
	5.83	$0.17\,(\pm 0.01)$
	8.80	$0.17\,(\pm 0.02)$
	12.0	$0.17\,(\pm 0.02)$
	14.8	$0.14\,(\pm 0.01)$
Cis-1,4-polybutadiene[d]	6.10	$0.07\,(\pm 0.02)$
	10.9	$0.09\,(\pm 0.01)$
	13.6	$0.05\,(\pm 0.02)$
	16.8	$0.05\,(\pm 0.02)$
Trans-1,4-polyisoprene[d]	5.41	$-0.04\,(\pm 0.05)$
	9.47	$-0.09\,(\pm 0.02)$
	10.5	$-0.10\,(\pm 0.03)$
	13.3	$-0.08\,(\pm 0.03)$
	16.5	$-0.12\,(\pm 0.04)$

[a] Mols of cross-links per cm³ of unswollen network.
[b] Taken from reference 12.
[c] Taken from reference 28.
[d] Taken from references 39 and 40.

constant[5,7,42–47] $2C_1 = \nu kT/V$. Although a severalfold range of $\nu/2V$ is covered in each of these studies, there is no discernible dependence of f_e/f on degree of cross-linking. Results of similar studies[14,39,40] on cis-1,4-polybutadiene and trans-1,4-polyisoprene having $10^5\nu/2V = 6.1$–16.8 and 5.4–16.5, respectively, have been analyzed elsewhere and found to be confirmatory of the above conclusion.†

Relatively little work has been done on the possible dependence of f_e/f on network chain-length distribution. It does appear, however, that poly(dimethylsiloxane) elastomers having a bimodal distribution of chain lengths have values of f_e/f that are only about half the value exhibited by poly(dimethylsiloxane) elastomers having the unusual unimodal distributions. This has been demonstrated for both elongation[60] and for torsion[61], and may be due to the non-Gaussian nature of the very short chains present in such networks.

---

†It should be acknowledged that there are studies in which f_e/f has been found to depend on the degree of cross-linking. Two examples of such studies involve polyethylene and ethylene–propylene copolymer[58], and poly(methyl methacrylate) and natural rubber[59].

118 STRUCTURES AND PROPERTIES OF RUBBERLIKE NETWORKS

Although most thermoelasticity studies have been carried out on networks in uniaxial elongation[13], several studies in which the deformation was uniaxial compression, torsion, shear, or isotropic swelling have now been reported. Representative results obtained on polymers which have been studied in more than one type of deformation are presented in table 9.4[14]. These results indicate that f_e/f is independent of the type of deformation employed in its thermoelastic determination. Other relevant studies, carried out on atactic polystyrene[62,63] and on polyoxyethylene[37,64] in elongation and in compression, and on trans-1,4-polyisoprene[39,40,53] in elongation and in torsion, show equally good agreement[14] between values of f_e/f obtained using different types of deformation. Two more-recent investigations on a variety of elastomers in torsion and in elongation, however, have found good agreement between values of f_e/f, in general, but poor agreement in the case of some polybutadiene elastomers[33,41].

Considerable circumspection is required in the analysis of results relevant to the possible dependence of f_e/f on extent of deformation. There is now an abun-

Table 9.4 Effect of Type of Deformation on f_e/f

Polymer	Type of Deformation	f_e/f
Natural rubber	Elongation	0.17 (±0.02)[a]
	Compression	0.18 (±0.02)[b]
	Torsion	0.17 (±0.03)[c,d,e]
Cis-1,4-polybutadiene	Elongation	0.12 (±0.02)[a,f]
	Compression	0.14 (±0.02)[f]
	Torsion	0.10 (±0.02)[e,g]
Trans-1,4-polyisoprene	Elongation	−0.11 (±0.04)[f]
	Torsion	−0.14 (±0.04)[e]
Polystyrene	Elongation	0.16 (±0.03)[h]
atactic	Compression	0.17 (±0.05)[i]
Polyoxyethylene	Elongation	0.07 (±0.01)[j]
	Compression	0.06 (±0.01)[k]
Poly(dimethylsiloxane)	Elongation	0.20 (±0.04)[a]
	Compression	0.13 (±0.03)[l]
	Isotropic swelling	0.1 (±0.1)[m]
Poly(n-butyl acrylate), atactic	Elongation	−0.35[n]
	Shear	−0.36[o]

[a] Taken from reference 13.
[b] Taken from reference 14.
[c] Taken from reference 31.
[d] Taken from reference 70.
[e] Taken from references 39 and 40.
[f] Taken from reference 53.
[g] Taken from reference 54.
[h] Taken from reference 62.
[i] Taken from reference 63.
[j] Taken from reference 64.
[k] Taken from reference 38.
[l] Taken from reference 34.
[m] Taken from reference 95.
[n] Taken from reference 96.
[o] Taken from reference 32.

dance of data bearing on this point in the case of uniaxial elongation[7,13,14]. Such studies have been reviewed in this regard, with extensive tabulation of values of f_e/f as a function of the elongation α.[14] Table 9.5[14] contains a selection of representative results chosen from those which appear to have been most carefully obtained, and which employ the widest variation of α in the range where α is most reliably measured and thermoelastic results most readily interpreted. These results strongly suggest f_e/f to be independent of α.†

One of the most interesting types of thermoelasticity experiments involves the study of networks swollen with diluent, either at constant composition or in swelling equilibrium. The effects of such diluents on f_e/f are illustrated by the typical results summarized in table 9.6[14]. As is readily seen from the details given in the first three columns of this table, and in a more extensive survey reported elsewhere[14], a wide variety of polymers and diluents have been employed in such studies, and a wide range of degrees of swelling have been investigated. Nonetheless, no apparent effect of dilution on f_e/f has been observed.

The above evidence gathered from thermoelastic studies clearly demonstrates that the thermodynamic ratio f_e/f is independent of a variety of important experimental conditions: namely, the thermodynamic constraints chosen in the force–temperature measurements, the conditions during cross-linking, the degree of cross-linking, the type and extent of deformation, and the degree of dilution or swelling of the network. This evidence is in strong support of the basic postulate that network elasticity is an intramolecular effect, with intermolecular interactions being essentially independent of chain configuration and network deformation. Of particular importance is the absence of any effect of network swelling, since intermolecular interactions between chains are certainly greatly diminished by the relatively large amounts of diluent incorporated into the networks in these studies. The invariance of f_e/f with degree of swelling of the network thus constitutes very strong evidence against both the importance of intermolecular contributions to the force, and the presence of any intermolecular ordering in polymers in the amorphous state.

The insensitivity of f_e/f to the experimental variables cited is also of importance with regard to the deviations generally observed between experimental stress–strain relationships and the form of the relationship predicted by the statistical theories[5,42–47]. The semiempirical constant $2C_2$, used as a measure of this

---

† Evidence leading some workers to the contrary conclusion generally falls into two categories. Most studies showing a dependence of f_e/f on α are characterized by relatively constant values of this ratio at moderate elongations ($\alpha = 1.3–3.0$), with significant dependence on α at both higher and lower values of α. Such departures are almost certainly due to the unwarranted extension of measurements into a region of such low elongation that uncertainties in length measurements become very serious, or into a region of such high elongation that strain-induced crystallization or limited chain extensibility invalidates the desired thermodynamic analysis. Less comprehensible, however, are a number of studies, clearly a minority, in which f_e/f was found to depend on elongation, even over a presumably reliable range of α. Experimental difficulties may nonetheless be involved since, frequently, results obtained on networks of the same polymers by other investigators do exhibit the expected independence of f_e/f on α.

Table 9.5 Effect of Extent of Deformation on f_e/f

Polymer	α^a	f_e/f
Polyethylene[b]	1.55	−0.41
	1.82	−0.33
	2.09	−0.45
	2.46	−0.39
	2.96	−0.45
Natural rubber[c]	1.08	0.09
	1.48	0.14
	1.78	0.11
	1.95	0.09
Natural rubber[d]	1.21	0.16
	1.43	0.10
	1.62	0.15
	1.80	0.14
	2.04	0.12
Natural rubber[e]	1.30	0.18
	1.70	0.18
	2.08	0.18
	2.50	0.15
	3.02	0.14
Natural rubber[f]	1.15	0.12
	1.26	0.13
	1.32	0.14
	1.37	0.12
	1.41	0.14
Cis-1,4-polybutadiene[g]	1.20	0.17
	1.30	0.14
	1.41	0.11
	1.63	0.17
	1.84	0.11
Polyisobutylene[h]	1.73	−0.03
	2.07	0.01
	2.38	−0.02
	3.24	−0.02
	3.75	−0.02
Poly(dimethylsiloxane)[i]	1.48	0.30
	1.60	0.25
	1.71	0.27
	1.89	0.26
	2.10	0.25

[a] Elongation $\alpha = L/L_i$ for unswollen networks in uniaxial extension.
[b] Taken from reference 12.
[c] Taken from reference 28.
[d] Taken from reference 69.
[e] Taken from reference 97.
[f] Taken from reference 98.
[g] Taken from reference 52.
[h] Taken from reference 12.
[i] Taken from reference 75.

Table 9.6 Effect of Dilution on f_e/f

Polymer	Diluent	$v_2{}^a$	f_e/f
Polyethylene	None	1.00	$-0.42\,(\pm 0.05)^c$
	Diethylhexyl azelate	0.80–0.30	$-0.44\,(\pm 0.10)^c$
	n-$C_{30}H_{62}$	0.50	-0.64^c
	n-$C_{32}H_{66}{}^b$	~0.30	$-0.48\,(\pm 0.04)^c$
Natural rubber	None	1.00	$0.17\,(\pm 0.03)^{d,e}$
	n-$C_{16}H_{34}$	0.93–0.34	$0.18\,(\pm 0.04)^f$
	n-$C_{10}H_{22}$	0.65–0.36	$0.13\,(\pm 0.01)^g$
	Paraffin oil	0.40	$0.19\,(\pm 0.02)^e$
	Decalin^b	0.20	$0.14\,(\pm 0.02)^e$
Cis-1,4-polybutadiene	None	1.00	$0.12\,(\pm 0.04)^{d,e,h}$
	1-Chloronaphthalene	0.52	$0.14\,(\pm 0.01)^h$
	Paraffin oil	0.50	$0.10\,(\pm 0.02)^e$
	Decalin^b	0.24	$0.09\,(\pm 0.01)^e$
Trans-1,4-polyisoprene	None	1.00	$-0.10\,(\pm(0.05))^{e,h}$
	Paraffin oil	0.40	$-0.13\,(\pm 0.02)^e$
	Decalin^b	0.18	$-0.20\,(\pm 0.04)^e$
Polyisobutylene	None	1.00	$-0.03\,(\pm 0.02)^c$
	n-$C_{16}H_{34}$	0.54, 0.41	$-0.03\,(\pm 0.02)^c$
Poly(dimethylsiloxane)	None	1.00	$0.19\,(\pm 0.05)^d$
	n-$C_{16}H_{34}$	0.49	$0.15\,(\pm 0.03)^i$
Polystyrene	None	1.00	$0.16\,(\pm 0.03)^j$
	Toluene	0.29	$0.19\,(\pm 0.03)^k$

^{a} Volume fraction of polymer in the network.
^{b} Swelling equilibrium
^{c} Taken from reference 12.
^{d} Taken from reference 13.
^{e} Taken from references 39, 40.
^{f} Taken from references 92–94, 97, and 99.
^{g} Taken from reference 28.
^{h} Taken from reference 53.
^{i} Taken from reference 74.
^{j} Taken from reference 62.
^{k} Taken from reference 63.

deviation, has been found to be markedly dependent on each of the variables investigated in tables 9.2 through 9.6[5,7]. The results surveyed here thus suggest that the elastic force need not be resolved into its separate contributions from $2C_1$ and $2C_2$ in thermoelastic analyses.

9.4 Relevant Calorimetric Studies of Elastic Deformations

Direct calorimetric measurements carried out on a polymer network during the deformation process have also been used to determine values of f_e/f[15,54,65–70]. In this approach, the heat q of the deformation process is measured in a sensitive microcalorimeter, and the accompanying work w in the process is obtained from the differential quantity fdL integrated from the initial to the final length of the sample. Then, according to the first law of thermodynamics,

Table 9.7 Comparison of Calorimetric and Thermoelastic Values of f_e/f

Polymer	Type of deformation	Calorimetric values	Thermoelastic values
Natural rubber	Elongation	0.18 (±0.01)[a,b]	0.17 (±0.03)[d]
	Torsion	0.20 (±0.02)[a]	0.15 (±0.04)[e,f]
Cis-1,4-polybutadiene	Elongation	0.11 (±0.02)[c]	0.12 (±0.02)[d,g]
	Torsion	0.14 (±0.02)[c]	0.10 (±0.02)[f]

[a] Taken from reference 70.
[b] Taken from reference 69.
[c] Taken from reference 54.
[d] Taken from reference 13.
[e] Taken from reference 31.
[f] Taken from references 39 and 40.
[g] Taken from reference 53.

$$\Delta E = q + w \cong \Delta H \quad (9.9)$$

where the energy and enthalpy changes are essentially identical since $\Delta(PV) \cong 0$ for any deformation not carried out at extraordinarily high pressures. For elongation studies, the derivative $(\partial H/\partial L)_{T,p}$ is obtained from the experimental values of ΔH plotted as a function of L. Values of f_e/f are then calculated from[54,69,70]

$$f_e/f \equiv (\partial E/\partial L)_{T,V}/f \cong (\partial H/\partial L)_{T,p}/f - \beta T/(\alpha^3 - 1) \quad (9.10)$$

which is an obvious analogue to eq. (9.5). Similar equations pertain to calorimetric studies of networks in torsion[39,40,70].

The most reliable calorimetric values of f_e/f obtained to date are given in table 9.7[14], and are seen to be in good agreement with thermoelastic results reported for the same two polymers.

9.5 Relevant Viscosity–Temperature Results on Dilute Polymer Solutions

Conversion of the thermodynamic quantity f_e/f to $d\ln\langle r^2\rangle_0/dT = f_e/fT$ permits comparison of values of the temperature coefficient of the unperturbed dimensions of the network chains as obtained from thermoelasticity measurements, with values obtained from viscosity–temperature measurements on chains of the same polymer dispersed in a solvent at infinite dilution. The latter method is based on measurements of the temperature dependence of the intrinsic viscosity $[\eta]$ defined by[1]

$$[\eta] = \lim_{c \to 0} [(\eta_{\text{rel}} - 1)/c] \quad (9.11)$$

where η_{rel} is the relative viscosity (the ratio of the viscosity of a solution of the polymer relative to the viscosity of the pure solvent at the same temperature), and c is the concentration of the polymer in weight per unit volume, typically g dl^{-1}. According to theory[1,71,72],

$$[\eta] = \Phi \langle r^2 \rangle_0^{3/2} M^{-1} \alpha^3 \quad (9.12)$$

in which Φ is a constant and $\alpha = [\langle r^2 \rangle / \langle r^2 \rangle_0]^{1/2}$ is a chain expansion factor characterizing perturbations due to excluded volume effects. In an ideal or Θ-solvent, such perturbations are nullified by polymer–solvent interactions, α is unity, and the unperturbed dimensions $\langle r^2 \rangle_0$ are directly calculable from the intrinsic viscosity. Correspondingly, viscosity measurements in a series of solvents having Θ-points over a sufficiently wide range in temperature yield directly the coefficient $d \ln \langle r^2 \rangle_0 / dT$. It is, however, generally required that these Θ-solvents be of very similar chemical structure[73]. This is necessary in order to avoid changes in $\langle r^2 \rangle_0$ from specific solvent effects which, although generally small, may differ sufficiently from solvent to solvent, even at the respective Θ-points, to have a large effect on the relatively small temperature coefficient of $\langle r^2 \rangle_0$.

Viscosity–temperature measurements in a single solvent, usually one that is a thermodynamically good solvent for the polymer, may also be used to obtain an experimental value of $d \ln \langle r^2 \rangle_0 / dT$. In this case, the chain expansion factor is not unity, and it is imperative to take into account its variation with temperature[73]. The required value of the temperature dependence of α is generally obtained in an approximate way by recourse to current solution theories giving the relationship between α and the thermodynamic parameters characterizing the interactions between solvent molecules and polymer segments[1]. This approach is considerably simplified by the choice of a solvent closely similar in chemical structure to that of the polymer: for example, the choice of n-hexadecane in the study of polyethylene[55]. Such solvent–solute combinations mix essentially athermally, with an attendant simplification of the solution thermodynamics and the associated estimation of $d \ln \langle r^2 \rangle_0 / dT$[55,74-77].

Table 9.8[14] lists results obtained on the three polymers that have been studied most carefully with regard to their viscosity–temperature coefficients. Values of $d \ln \langle r^2 \rangle_0 / dT$ calculated therefrom are compared with the corresponding values obtained in thermoelastic investigations through use of eq. (9.7). There is excellent agreement between the two sets of values of $d \ln \langle r^2 \rangle_0 / dT$ thus obtained, as there is in the case of several other polymers studied using both of these techniques[14].

The excellent agreement obtained between values of $d \ln \langle r^2 \rangle_0 / dT$ for the same polymer chains under such vastly different conditions is extremely important in two respects. It clearly demonstrates that thermoelastic data analyzed with intentional disregard of $2C_2$ contributions give values of the temperature coefficient of the unperturbed dimensions that are in good agreement with values obtained in dilute solution. Most important, it is inconceivable that such agreement would be obtained if intermolecular interactions varied with configuration, or if significant molecular ordering did, in fact, occur in the amorphous state. This latter conclusion, drawn from the study of network thermoelasticity, is given compelling support from a variety of additional experimental evidence[36]. For example, the decrease in the reduced force $[f^*]$ upon incorporation of diluent into a polymer network is much less pronounced than would be expected from presumed dispersal of ordered arrangements in the unswollen network. Similarly, vapor pressure measurements and gas–liquid chromatographic analyses indicate that the free energy parameter χ, characterizing polymer–solvent interactions, does not show any abnormal dependence on composition that could be attributed to dissipation

Table 9.8 Comparison of Values of $d\ln\langle r^2\rangle_0/dT$ Deduced from Thermoelastic Measurements on Networks with Values from Viscometric Measurements on Isolated Chains

Polymer	$10^3 d\ln\langle r^2\rangle_0/dT$	
	$f - T$ [a]	$\eta - T$
Polyethylene	−1.05 (±0.10)	−1.2 (±0.2)[b,c]
		−1.19 (±0.04)[d,e,f]
		−0.8 (±0.1)[g,h]
Poly(n-pentene-1), isotactic	0.34 (±0.04)	0.52 (±0.05)[b,i]
Polyisobutylene	−0.19 (±0.11)	−0.28 (±0.05)[b,j]
		−0.10 (±0.03)[d,k]
		−0.4 (±)[g,h]
Polyoxyethylene	0.23 (±0.02)	0.2 (±0.2)[g,l]
Poly(dimethylsiloxane)	0.59 (±0.14)	0.52 (±0.20)[b,m,n]

[a] Taken from the thermoelastic results surveyed in reference 13.
[b] Measurements in an athermal solvent
[c] Taken from reference 55.
[d] Measurements in a series of structurally similar Θ solvents.
[e] Taken from reference 100.
[f] Taken from reference 101.
[g] Approximate results from measurements in a thermodynamically good solvent.
[h] Taken from reference 102.
[i] Taken from reference 76.
[j] Taken from reference 77.
[k] Taken from reference 103.
[l] Taken from reference 73.
[m] Taken from reference 74.
[n] Taken from reference 75.

of molecular order with dilution[36]. Finally, optical anisotropies determined from the depolarization of light scattered from undiluted amorphous polymers, or from the strain birefringence of both diluted and undiluted amorphous polymer networks, are comparable in magnitude to the corresponding quantities measured for low-molecular-weight liquids, and are very much smaller than would be expected from materials containing significant intermolecular ordering[36].

It should also be recalled that theoretical arguments presented in chapter 2 lead to the expectation that excluded volume interactions, although assuredly present in unswollen or swollen networks, have no effect on the end-to-end dimensions of the network chains at the relatively high polymer concentrations existing in such systems. That is, the chain molecules should be in their unperturbed states, characterized by the dimension $\langle r^2\rangle_0$. There are two important pieces of evidence that bear on this point. Ring-chain cyclization constants for undiluted dimethylsiloxane chains yield values of $\langle r^2\rangle_0$ in good agreement with values obtained from studies of these molecules in dilute solution[36]. Most directly, measurements of x-ray scattering from end-labeled polymer molecules in the undiluted state, and neutron scattering from mixtures of deuterated and undeuterated polymers, give values of the radii of gyration that are generally in good agreement with values obtained using dilute solutions of the same polymers[14,36]. The polymer chains are thus shown to be in their unperturbed configurations in the undiluted amorphous state, a finding which a fortiori directly contradicts the presence of order in such systems.

The basic postulates of rubber elasticity theory that intermolecular interactions are independent of chain configuration, and that the configurations of network chains are random in the undeformed polymer network, are thus seen to be verified by a great variety of experimental evidence.

An additional, indirect piece of evidence that bears on this point is the fact that experimental values of $f_e/fT = d\ln\langle r^2\rangle_0/dT$ can be interpreted very satisfactorily, using the rotational isomeric state theory of chain configurations[8,9,78], exclusively in terms of *intra*molecular interactions (the energies of which are obtained from the study of other configuration-dependent properties, such as dipole moments, light scattering intensities, optical anisotropies, and cyclization equilibrium constants, or by conformational energy calculations). This aspect of rotational isomeric state theory is covered briefly in the final section of the chapter.

9.6 Rotational Isomeric State Interpretation of Stress–Temperature Results

The energetic contribution f_e to the total force f arises from the fact that deformation of a polymer network requires a corresponding change in the configurations and dimensions of its constituent chains, and that different chain configurations have, in general, different conformational energies[8,9,78]. Such differences in conformational energy are also the origin of the dependence of the unperturbed dimensions on temperature, and, as already shown, $d\ln\langle r^2\rangle_0/dT$ has the same sign as f_e/f and is directly proportional to it. It is now appropriate to consider the interpretation of these quantities in terms of the characteristics of the skeletal bonds making up the network chains.

Values of $\langle r^2\rangle_0$ and its temperature coefficient may readily be calculated from a rotational isomeric state representation of these chains, in which each rotational skeletal bond is assumed to occur in one of a small number of discrete rotational states[8,9]. The states most commonly encountered, and the rotational angles at which they occur, are the *trans* ($\phi = 0°$), skew ($\pm 60°$), gauche ($\pm 120°$), and cis (180°) states. Rotational states about consecutive skeletal bonds are generally interdependent, and conformational sequences are therefore characterized in terms of statistical weights embodying the appropriate energies for conformations of groups of two or more such bonds[8,9]. The desired conformational energies are generally obtained by comparison of calculated and experimental values of configuration-dependent properties and their temperature coefficients. A brief outline of the use of rotational isomeric state theory to calculate $\langle r^2\rangle_0$ and its temperature coefficient[8,9,79] is given in appendix G.

Rotational isomeric state calculations thus serve to make quantitative the obvious qualitative connection that chain molecules in which conformations of high spatial extension are of high conformational energy have positive values of f_e/f, whereas a correspondence between high extension and low conformational energy gives negative values[13]. The former case corresponds, of course, to chains

Table 9.9 Experimental Thermoelastic Results which have been Interpreted by Rotational Isomeric State Theory[a]

Polymer	Repeat unit	f_e/f	$10^3 d\ln\langle r^2\rangle_0/dT$
Polyethylene	[CH$_2$CH$_2$]	-0.42 (± 0.05)	-1.05 (± 0.10)
Poly(dimethylsiloxane)	[Si(CH$_3$)$_2$O]	0.19 (± 0.05)	0.59 (± 0.14)
1,4-Polybutadiene	[CH$_2$CHCHCH$_2$]		
cis		0.12 (± 0.04)	0.39 (± 0.10)
trans		-0.25 (± 0.06)	-0.66 (± 0.17)
1,4-Polyisoprene	[CH$_2$C(CH$_3$)CHCH$_2$]		
cis[b]		0.17 (± 0.03)	0.53 (± 0.09)
trans[c]		-0.09 (± 0.05)	-0.27 (± 0.16)
Polyoxyethylene	[(CH$_2$)$_2$O]	0.08 (± 0.01)	0.23 (± 0.02)
Poly(butene-1)	[CH(C$_2$H$_5$)CH$_2$]		
isotactic[d]		0.04 (± 0.03)	0.09 (± 0.07)
atactic		0.22 (± 0.02)	0.50 (± 0.04)
Poly(n-pentene-1)	[CH(nC$_3$H$_7$)CH$_2$]		
isotactic		0.13 (± 0.02)	0.34 (± 0.04)
actactic		0.20 (± 0.02)	0.53 (± 0.05)
Poly(vinyl chloride)	[CHClCH$_2$]		
syndiotactic		-2.4 (± 0.1)	-7.0 (± 0.3)
Ethylene–Propylene			
copoolymer[e]	[(CH$_2$)$_2$, CHCH$_3$CH$_2$]	-0.43 (± 0.15)	-1.3 (± 0.5)
Polyisobutylene	[C(CH$_3$)$_2$CH$_2$]	-0.06 (± 0.03)	-0.19 (± 0.11)
Poly(dimethylsilmethylene)	[Si(CH$_3$)$_2$CH$_2$]	0.07 (± 0.07)	0.20 (± 0.20)
Poly(trimethylene oxide)	[(CH$_2$)$_3$O]	0.03 (± 0.03)	0.08 (± 0.08)
Poly(tetramethylene oxide)	[(CH$_2$)$_4$O]	-0.47 (± 0.05)	-1.33 (± 0.15)
Elastin[f]	[CHRCONH]	0.26 (± 0.09)	0.85 (± 0.29)
Poly(2-hydroxyethyl methacrylate) atactic[g]	[C(CH$_3$)-COOCHOHCH$_3$)CH$_2$]	0.78	2.4

[a] See references 8 and 9.
[b] "Natural rubber."
[c] "Gutta percha."
[d] Isotactic, atactic, and syndiotatic refer to meso, random, and racemic stereochemical placements, respectively.
[e] Approximately 50 mol% propylene.
[f] Taken from reference 104. See also chapter 15.
[g] Taken from reference 105. In principle, interpretable from the RIS model for poly(methyl methacrylate)[9], but may be a spurious result.

having unperturbed dimensions which increase with increasing temperature, while in the latter case they would decrease. Table 9.9[14] presents a survey of thermoelastic results which have been interpreted, to date, in terms of rotational isomeric state theory. It is thus now possible to provide a simple physical picture of the thermoelastic properties of these and related polymers. Some of this information is presented below in outline form.

In the case of polyethylene, a sequence of which is shown in figure 9.5, the negative values obtained for $d\ln\langle r^2\rangle_0/dT$ and f_e/f are primarily due to the fact that *trans* conformational states about skeletal bonds are of significantly lower energy than *gauche* states[8,9], in which relatively large CH$_2$ groups are brought into conflict. Since sequences of *trans* states in this chain are of very high extension,

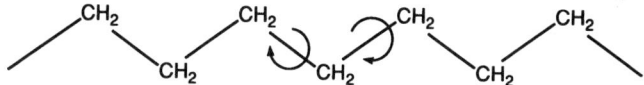

Figure 9.5 Schematic diagram of a section of a polyethylene chain. In this figure, and in the remaining ones in this chapter, the chains are shown in the planar, all-*trans* conformation with the skeletal bonds in the plane of the paper. Bonds extending toward the reader are shown by wedges and those away from the reader by dashed lines; the arrows specify the skeletal bonds of the repeat unit about which rotations may occur.

increase in temperature decreases $\langle r^2 \rangle_0$ due to the increase in the number of relatively compact *gauche* states. Similarly, elongation of these chains by deformation of the network decreases the energy because of the increase in the number of *trans* states of low conformational energy.

The chain molecule which has been most extensively studied with regard to its configurational characteristics is the semi-inorganic polymer poly(dimethylsiloxane) (PDMS)[8,9], which is illustrated in figure 9.6. *Trans* states about skeletal bonds in this molecule are also of relatively low energy, probably because of favorable interactions between pendant CH_3 groups. Although *trans* sequences are of high chain extension in the case of polyethylene, however, the opposite is the case for PDMS. Because of the inequality of bond angles about the Si and O atoms, the all-*trans* form of this polymer is essentially a closed figure, of very small end-to-end distance. This relatively low energy of the compact *trans* states in PDMS chains yields a positive value for f_e/f, the opposite of that found for polyethylene.

The *cis* and *trans* forms of 1,4-polybutadiene and 1,4-polyisoprene have also been investigated in detail[8,9,80,81]. Their structures are shown in figure 9.7, in which the choice of H for R yields polybutadiene and the choice of CH_3 yields polyisoprene. The *cis*-1,4 forms of polybutadiene and polyisoprene are characterized by positive values of f_e/f; they arise from the fact that the lowest energy conformations about $CH_2-CH=CH-CH_2$ and $CH_2-C(CH_3)=CH-CH_2$ pairs of rotatable bonds are $\pm 60°$, $\pm 60°$ and $\mp 60°$, $\pm 60°$ states, which are relatively compact[8,9,80,81]. In the *trans*-1,4 forms, extended $0°$ states about CH_2-CH_2 bonds are of lower energy than the very compact $\pm 120°$ states when the

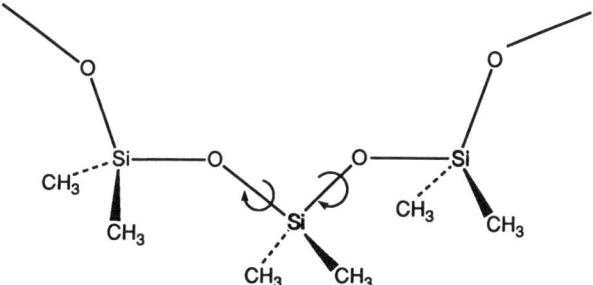

Figure 9.6 The poly(dimethylsiloxane) chain.

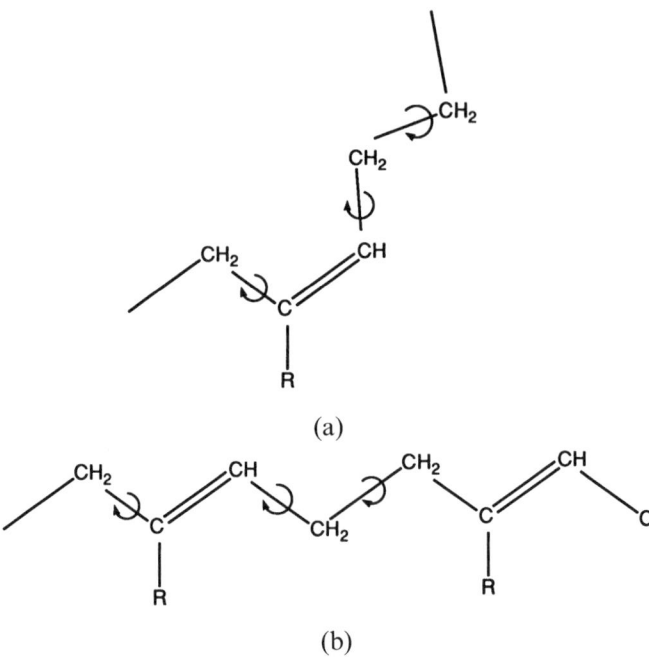

Figure 9.7 Diene polymer chains of (a) *cis*-1,4 structure and (b) *trans*-1,4 structure. Choice of H for R gives polybutadiene, and choice of CH_3 gives polyisoprene.

preceding bond is at 180°, thus providing a basis for the negative value obtained for f_e/f for these polymers[8,9,80,81].

A number of other, miscellaneous polymers have now been studied, both experimentally and theoretically. We consider first those having positive values of f_e/f. In these chains, the higher energy conformations are more extended than those of lower energy. The positive value of f_e/f reported for polyoxyethylene $[CH_2-CH_2-O-]_x$, for example, stands in particularly interesting contrast to the negative coefficient of polyethylene. It is due to the fact that in this molecule *trans* states about CH_2-CH_2 bonds are of higher energy than the *gauche* states which bring into proximity the relatively small O atoms[8,9]. Increase in temperature therefore increases the number of *trans* states of high spatial extension. The positive values reported for f_e/f for isotactic and atactic poly(butene-1) $[CH(C_2H_5)-CH_2-]_x$ and poly(n-pentene-1) $[CH(n-C_3H_7)-CH_2-]_x$ are thought to result from the fact that extended *trans, trans* states about $CH_2-CHR-CH_2$ pairs of bonds are of relatively high energy in these chains because of steric repulsions between the chain backbone and the articulated side groups R ($-CH_2CH_3$ and $-CH_2CH_2CH_3$, respectively)[8,9].

The following polymers, listed in table 9.9[14], all have negative values of f_e/f. In these chains, the higher energy conformations are less extended than those of lower energy. In the case of poly(tetramethylene oxide) $[CH_2-CH_2-CH_2-CH_2-O-]_x$, three of the five skeletal bonds exhibit a preference for *trans* states over the alternative *gauche* states (which bring into proximity the relatively large CH_2 groups). The increase in the population of compact *gauche* states about these bonds, upon increase in temperature, gives rise to the negative value of f_e/f obtained for this polymer[8,9]. Although the thermoelastic results reported for poly(vinyl chloride) $[CHCl-CH_2-]_x$, of presumably high syndiotacticity, are tentative because of possible complications from crystallization, a large negative value would be expected for f_e/f[82] on the basis of the configurational analysis of this molecule. The Cl atoms, being relatively small, interact favorably with the chain backbone in a manner that makes the highly extended *trans* states of relatively low energy in the syndiotactic form of this polymer. The variety of chemical and stereochemical sequences that occur in copolymeric chains such as the ethylene–propylene copolymer[83,84] makes it impossible to give a brief, qualitative description of the molecular origin of this copolymer's negative temperature coefficient of $\langle r^2 \rangle_0$ (and thus of f_e/f). One important contribution, of course, would come from the ethylene sequences, in which the low-energy conformations are the *trans* states, as already described. Another contribution would arise from isotactic polypropylene sequences, for which the low-energy conformations are *trans, gauche* pairs such as $(tg^+)(tg^+)(tg^+)\ldots$. This replication of conformations for pairs of skeletal bonds gives 3_1 helical sequences[8,9], obviously also of high spatial extension.

The ratio f_e/f may be used to define the concept of ideality in elastomeric materials. Thus, an *ideal elastomer* is one for which $(\partial E/\partial L)_{T,V} \equiv f_e$ is zero for all values of the length L, in parallel to the requirement that an ideal gas has $(\partial E/\partial V)_T = 0$ for all values of the pressure p. It must be emphasized, however, as shown above, that deviations from network ideality are, to an excellent approximation, entirely due to changes in *intra*molecular interactions along the network chains. For this reason, there are no limiting experimental conditions under which all elastomers exhibit ideal thermoelastic behavior. In this regard, ideality of elastomers is fundamentally different from that of gases, all of which behave ideally, of course, in the limit of infinitesimally small pressure. Thus, the only type of network that would be ideal in the sense of exhibiting purely entropic elasticity would be one made up of chains in which all conformations were of the same energy. This highly idealized construct is, of course, the "freely rotating chain[1]," which served as one of the simplest models in early studies of chain configurations. Several real polymer chains at least approach this idealized model in that they are conformationally highly random with rotational states, when accessible, of almost identical conformational energy.

Polyisobutylene[8,9], shown in figure 9.8, is a polymer of this type. In the simplest, most qualitative analysis, rotations about skeletal bonds would merely interchange skeletal CH_2 groups and pendant CH_3 groups, and these are known to be nearly identical in their intramolecular interactions[8,9]. As shown in table 9.9, f_e/f is found to be close to zero, as would be expected from the lack of strong conformational preferences in this molecule. The actual conformational analysis of this chain is much more complicated, however, partly because of strong steric interactions between methyl groups along the chains[85,86].

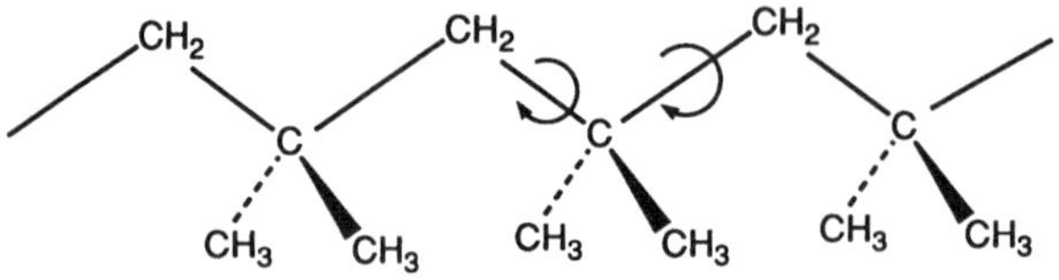

130 STRUCTURES AND PROPERTIES OF RUBBERLIKE NETWORKS

Figure 9.8 The polyisobutylene chain.

The semi-inorganic polymer poly(dimethylsilmethylene)[87-89], shown in figure 9.9, should be very similar to polyisobutylene in this regard, since its accessible conformations again differ in energy only to the extent that CH_2 and CH_3 groups differ in their intramolecular interactions. As shown in the table 9.9, its value of f_e/f is also very nearly zero. A final example of a real chain having some characteristics of free rotation is poly(trimethylene oxide), shown schematically in figure 9.10[90]. *Gauche* states about the two C—C skeletal bonds bring CH_2 groups and relatively small O atoms into proximity and are of lower energy than *trans* states; the preference is, however, reversed for the two C—O bonds in the repeat unit since *gauche* states about these bonds bring a pair of relatively large CH_2 groups into conflict. Thus, *gauche* states are preferred about the first two bonds in the repeat unit, and *trans* states are preferred about the remaining two, making the chain conformations highly irregular upon averaging over the entire repeat unit. Stretching a poly(trimethylene oxide) chain increases, of course, the total *trans* population, but *gauche*-to-*trans* transitions about C—C bonds increase the energy whereas such transitions about C—O bonds decrease the energy. The two effects would thus be expected to offset one another to a large extent. In agreement with this expectation, poly(trimethylene oxide) is found to be the most nearly ideal elastomer in that its value of f_e/f is one of the smallest of any reported to date and is, in fact, zero within experimental error.

As clearly demonstrated by the examples cited in this section, thermoelastic measurements and their interpretation by means of rotational isomeric state

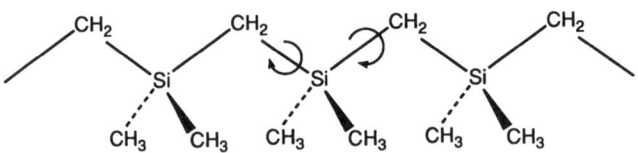

Figure 9.9 The poly(dimethylsilmethylene) chain.

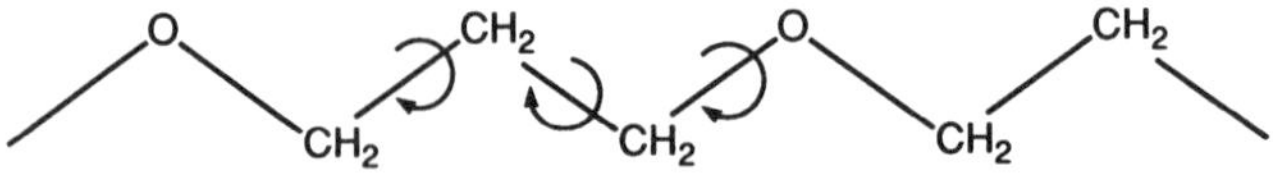

Figure 9.10 The poly(trimethylene oxide) chain.

theory provide a great deal of insight into both thermodynamic and molecular aspects of rubberlike elasticity.

References

(1) Flory, P. J. *Principles of Polymer Chemistry.* Cornell University Press: Ithaca, NY. 1953.
(2) Flory, P. J. *Trans. Faraday Soc.* 1961, 57, 829.
(3) Flory, P. J. *Rubber. Chem. Technol.* 1968, 41, G41.
(4) Mark, J. E. *J. Am. Chem. Soc.* 1970, 92, 7252.
(5) Treloar, L. R. G. *The Physics of Rubber Elasticity*, 3rd ed. Clarendon Press: Oxford. 1975.
(6) Price, C. *Proc. Roy. Soc. London, Ser. A* 1976, 351, 331.
(7) Mark, J. E.; Erman, B. *Rubberlike Elasticity. A Molecular Primer.* Wiley-Interscience: New York. 1988.
(8) Flory, P. J. *Statistical Mechanics of Chain Molecules.* Interscience: New York. 1969.
(9) Mattice, W. L.; Suter, U. W. *Conformational Theory of Large Molecules. The Rotational Isomeric State Model in Macromolecular Systems.* Wiley: New York. 1994.
(10) Flory, P. J.; Hoeve, C. A. J.; Ciferri, A. *J. Polym. Sci.* 1959, 34, 337.
(11) Flory, P. J.; Ciferri, A.; Hoeve, C. A. J. *J. Polym. Sci.* 1960, 45, 235.
(12) Ciferri, A.; Hoeve, C. A. J.; Flory, P. J. *J. Am. Chem. Soc.* 1961, 83, 1015.
(13) Mark, J. E. *Rubber Chem. Technol.* 1973, 46, 593.
(14) Mark, J. E. *Macromol. Rev.* 1976, 11, 135.
(15) Godovsky, Y. K. *Adv. Polym. Sci.* 1986, 76, 31.
(16) Vasko, M.; Bleha, T.; Romanov, A. *J. Macromol. Sci.—Chem.* 1976, C15, 1.
(17) Sullivan, J. L.; Smith, K. J., Jr. *J. Polym. Sci., Polym. Phys. Ed.* 1975, 13, 857.
(18) Staverman, A. J. *Polym. Eng. Sci.* 1979, 19, 260.
(19) Ogden, R. W. *Polymer* 1987, 28, 379.
(20) Gee, G. *Polymer* 1987, 28, 386.
(21) Boyer, R. F.; Miller, R. L. *Polymer* 1987, 28, 399.
(22) Vilgis, T. A. In *Elastomeric Polymer Networks*, J. E. Mark and B. Erman, Eds. Prentice Hall: Englewood Cliffs, NJ. 1992; p. 32.
(23) Smith, K. J., Jr. In *Elastomeric Polymer Networks*, J. E. Mark and B. Erman, Eds. Prentice Hall: Englewood Cliffs, NJ. 1992; p. 116.
(24) Ambacher, H.; Kilian, H. G. In *Elastomeric Polymer Networks*, J. E. Mark and B. Erman, Eds. Prentice Hall: Englewood Cliffs, NJ. 1992; p. 124.
(25) Allen, G.; Bianchi, U.; Price, C. *Trans. Faraday Soc.* 1963, 59, 2493.
(26) Allen, G.; Gee, G.; Kirkham, M. C.; Price, C.; Padget, J. *J. Polym. Sci., Part C* 1968, 23, 201.
(27) Price, C.; Padget, J. C.; Kirkham, M. C.; Allen, G. *Polymer* 1969, 10, 573.
(28) Allen, G.; Kirkham, M. J.; Padget, J.; Price, C. *Trans. Faraday Soc.* 1971, 67, 1278.
(29) Hoeve, C. A. J.; Flory, P. J. *J. Polym. Sci.* 1962, 60, 155.
(30) Treloar, L. R. G. *Polymer* 1969, 10, 291.
(31) Boyce, P. H.; Treloar, L. R. G. *Polymer* 1970, 11, 21.
(32) Chen, T. Y.; Ricica, P.; Shen, M. *J. Macromol. Sci.—Chem.* 1973, 7, 889.
(33) Mohsin, M. A.; Treloar, L. R. G. *Polymer* 1987, 28, 1893.

(34) Chen, R. Y. S.; Yu, C. U.; Mark, J. E. *Macromolecules* 1973, 6, 746.
(35) Barrie, J. A.; Standen, J. *Polymer* 1967, 8, 97.
(36) Flory, P. J. *Pure Appl. Chem., Macromol. Chem. 8* 1973, 33, 1.
(37) Mistrali, F.; Volpin, D.; Garibaldo, G. B.; Ciferri, A. *J. Phys. Chem.* 1971, 75, 142.
(38) Yu, C. U.; Mark, J. E. *Polymer* 1975, 16, 326.
(39) Gent, A. N.; Kuan, T. H. *J. Polym. Sci., Polym. Phys. Ed.* 1973, 11, 1723.
(40) Gent, A. N.; Kuan, T. H. *J. Polym. Sci., Polym. Phys. Ed.* 1974, 12, 633.
(41) Mohsin, M. A.; Berry, J. P.; Treloar, L. R. G. *Brit. Polym. J.* 1986, 18, 145.
(42) Mooney, M. *J. Appl. Phys.* 1940, 11, 582.
(43) Mooney, M. *J. Appl. Phys.* 1948, 19, 434.
(44) Rivlin, R. S. *Phil. Trans. Roy. Soc., Part A* 1948, 240, 459.
(45) Rivlin, R. S. *Phil. Trans. Roy. Soc., Part A* 1948, 240, 491.
(46) Rivlin, R. S. *Phil. Trans. Roy. Soc., Part A* 1948, 240, 509.
(47) Rivlin, R. S. *Phil. Trans. Roy. Soc., Part A* 1948, 241, 379.
(48) Smith, K. J., Jr. In *Polymer Science*, A. D. Jenkins, Ed. North-Holland: Amsterdam. 1972; p. 323.
(49) Mark, J. E. unpublished results.
(50) Sakurada, I.; Nakajima, A.; Shibatani, K. *Makromol. Chemie* 1965, 87, 103.
(51) Bashaw, J.; Smith, K. J., Jr. *J. Polym. Sci., Part A-2* 1968, 6, 1041.
(52) de Candia, F.; Amelino, L.; Price, C. *J. Polym. Sci., Part A-2* 1972, 10, 975.
(53) Becker, R. H.; Yu, C. U.; Mark, J. E. *Polym. J.* 1975, 7, 234.
(54) Price, C.; Allen, G.; Yoshimura, N. *Polymer* 1975, 16, 261.
(55) Flory, P. J.; Ciferri, A.; Chiang, R. *J. Am. Chem. Soc.* 1961, 83, 1015, 1023.
(56) Price, C.; Allen, G.; de Candia, F.; Kirkham, M. C.; Subramaniam, A. *Polymer* 1970, 11, 486.
(57) Mark, J. E. *J. Polym. Sci., Part C* 1970, 31, 97.
(58) Opshcoor, A.; Prins, W. *J. Polym. Sci., Part C* 1967, 16, 1095.
(59) Rehage, G.; Schafer, E. E.; Schwarz, J. *Angew. Makromol. Chemie* 1971, 16/17, 231.
(60) Zhang, Z.-M.; Mark, J. E. *J. Polym. Sci., Polym. Phys. Ed.* 1982, 20, 473.
(61) Wen, J.; Mark, J. E. *Polym. J.* 1994, 26, 151.
(62) Orofino, T. A.; Ciferri, A. *J. Phys. Chem.* 1964, 68, 3136.
(63) Dusek, K. *Coll. Czech. Chem. Commun.* 1967, 32, 2264.
(64) Mark, J. E.; Flory, P. J. *J. Am. Chem. Soc.* 1965, 87, 1415.
(65) Engelter, A.; Muller, F. H. *Kolloid-Z. u. Z. Polym.* 1958, 157, 89.
(66) Dick, W.; Muller, F. H. *Kolloid-Z., u. Z. Polym.* 1960, 172, 1.
(67) Goritz, D.; Muller, F. H. *Kolloid-Z. u. Z. Polym.* 1973, 251, 679.
(68) Goritz, D.; Muller, F. H. *Kolloid-Z. u. Z. Polym.* 1973, 251, 892.
(69) Price, C.; Evans, K. A.; de Candia, F. *Polymer* 1973, 14, 338.
(70) Allen, G.; Price, C.; Yoshimura, N. *J. Chem. Soc., Faraday Trans. I* 1975, 71, 548.
(71) Flory, P. J. *J. Chem. Phys.* 1949, 17, 303.
(72) Fox, T. G., Jr.; Flory, P. J. *J. Phys. Chem.* 1949, 53, 197.
(73) Bluestone, S.; Mark, J. E.; Flory, P. J. *Macromolecules* 1974, 7, 325.
(74) Ciferri, A. *Trans. Faraday Soc.* 1961, 57, 846.
(75) Mark, J. E.; Flory, P. J. *J. Am. Chem. Soc.* 1964, 86, 138.
(76) Mark, J. E.; Flory, P. J. *J. Am. Chem. Soc.* 1965, 87, 1423.
(77) Mark, J. E.; Thomas, G. B. *J. Phys. Chem.* 1966, 70, 3588.
(78) Flory, P. J. *Pure Appl. Chem.* 1971, 26, 309.
(79) Flory, P. J. *Macromolecules* 1974, 7, 381.

(80) Abe, Y.; Flory, P. J. *Macromolecules* 1971, 4, 219.
(81) Abe, Y.; Flory, P. J. *Macromolecules* 1971, 4, 230.
(82) Mark, J. E. *J. Chem. Phys.* 1972, 56, 451.
(83) Mark, J. E. *J. Chem. Phys.* 1972, 57, 2541.
(84) Mark, J. E. *J. Polym. Sci., Polym. Phys. Ed.* 1974, 12, 1207.
(85) Suter, U. W.; Saiz, E.; Flory, P. J. *Macromolecules* 1983, 16, 1317.
(86) DeBolt, L. C.; Suter, U. W. *Macromolecules* 1987, 20, 1424.
(87) Ko, J. H.; Mark, J. E. *Macromolecules* 1975, 8, 869.
(88) Mark, J. E.; Ko, J. H. *Macromolecules* 1975, 8, 874.
(89) Sundararajan, P. R. *Comput. Polym. Sci.* 1991, 1, 18.
(90) Takahashi, Y.; Mark, J. E. *J. Am. Chem. Soc.* 1976, 98, 3756.
(91) Wood, L. A.; Roth, F. L. *J. Appl. Phys.* 1944, 15, 781.
(92) Greene, A.; Smith, K. J., Jr.; Ciferri, A. *Trans. Faraday Soc.* 1965, 61, 2772.
(93) Greene, A.; Ciferri, A. *Kolloid-Z. u. Z. Polym.* 1962, 186, 1.
(94) Shen, M. *Macromolecules* 1969, 2, 358.
(95) Mark, J. E. *J. Phys. Chem.* 1964, 68, 1092.
(96) Cirlin, E. H.; Shen, M. *J. Macromol. Sci.—Chem.* 1971, 5, 1311.
(97) Smith, K. J., Jr.; Greene, A.; Ciferri, A. *Kolloid-Z. u. Z. Polym.* 1964, 194, 49.
(98) Shen, M.; Blatz, P. J. *J. Appl. Phys.* 1968, 39, 4937.
(99) Ciferri, A. *Makromol. Chemie* 1961, 43, 152.
(100) Chiang, R. *J. Phys. Chem.* 1966, 70, 2348.
(101) Nakajima, A.; Hamada, F.; Hayashi, S. *J. Polym. Sci., Part C* 1966, 15, 285.
(102) Bohdanecky, M. *Coll. Czech. Chem. Commun.* 1968, 33, 4397.
(103) Kotera, A.; Onda, N.; Saito, T. *Rep. Prog. Polym. Phys. Japan* 1974, 17, 25.
(104) DeBolt, L. C.; Mark, J. E. *Polymer* 1987, 28, 416.
(105) Dorrington, K. L.; McCrum, N. G.; Watson, W. R. *Polymer* 1977, 18, 712.

10

Model Elastomers

10.1 The Dependence of the Stress on Network Structure

10.1.1 General Approach

Until quite recently, there was relatively little reliable quantitative information on the relationship of stress to structure, primarily because of the uncontrolled manner in which elastomeric networks were generally prepared[1-5]. Segments close together in space were linked irrespective of their locations along the chain trajectories, thus resulting in a highly random network structure in which the number and locations of the cross-links were essentially unknown. Such a structure is shown in figure 10.1[3]. New synthetic techniques are now available, however, for the preparation of "model" polymer networks of known structure. More specifically, if networks are formed by end linking functionally terminated chains instead of haphazardly joining chain segments at random, then the nature of this very specific chemical reaction provides the desired structural information[3,4,6-61]. Thus, the functionality of the cross links is the same as that of the end-linking agent, and the molecular weight M_c between cross-links and the molecular weight distribution are the same as those of the starting chains prior to their being end-linked.

An example is the reaction shown in figure 10.2[8], in which hydroxyl-terminated chains of poly(dimethylsiloxane) (PDMS) are end-linked using tetraethyl orthosilicate. Characterizing the un-cross-linked chains with respect to molecular weight M_n and molecular weight distribution, and then carrying out the specified reaction to completion, gives elastomers in which the network chains have these characteristics; in particular, a molecular weight M_c between cross-links equal to M_n, a network chain-length distribution equal to that of the starting chains, and cross-links having the functionality of the end-linking agent. It is also possible to use chains having a known number of potential cross-linking sites placed as side chains along the polymer backbone, so long as their distribution is known as well[62-64].

do the data for the PDMS system. Thus, structurally different polymers can give networks with very different dependences of the modulus on degree of cross-linking.

The departure of the experimental points from the simple molecular theory when the network chains are long indicates the presence of additional contributions to the modulus. The reasons behind this departure are not yet clear and the problem clearly remains open. The understanding of entanglement phenomena is perhaps based on answering the question of whether the stated departures result from trapped entanglements (which scale with the number of entanglement loci N_E along the chain contour) or simply from the presence of other chains, sharing the volume of a given network chain (defined as the Flory number, N_F). This Flory number is defined by the expression[75]

$$N_F = \left(\frac{4\pi}{3} \langle r^2 \rangle_0^{3/2}\right) \frac{\nu}{V_0}$$
$$= \frac{4\pi}{3} \left[\frac{C_\infty \rho l^2}{M_u}\right]^{3/2} N_A \left(\frac{\nu}{V_0}\right)^{-1/2} \quad (10.1)$$

where $\langle r^2 \rangle_0$ is the mean-squared end-to-end distance of a network chain in the unperturbed state, ν/V_0 is the number of network chains per unit volume (obtained in the state of formation), $C_\infty = \langle r^2 \rangle_0 / nL^2$ is the characteristic ratio of a chain of n bonds of bond length L,[76] and l and M_u are the length and the molecular weight of the chain repeat unit, respectively. One important difference between the two types of constraints involves their dependence on network swelling. The localized, permanent entanglements should be independent of swelling, while the more diffuse interchain interactions should decrease with increase in swelling. The contributions from trapped entanglements should therefore persist, even in highly swollen networks, and therefore contribute to the phantom modulus. For these reasons, it is quite important to carry out measurements of the modulus at a series of degrees of swelling, with the results at the highest degrees of swelling presumably being least complicated by nonequilibrium effects.

It should be noted that if one makes a network in the extremely dilute state so that there is no chain interpenetration during its formation, N_F will equate to zero and the phantom-structure state will be obtained, by definition. Experimentally, one indeed sees that network formation in the dilute state decreases both the modulus and its dependence on elongation (the "C_2 effect")[4,77]. The Flory number, which varies as the inverse square root of the degree of cross-linking, is significantly different from the number of entanglement loci, N_E. The latter is defined as the number of chains of molecular weight M_e in a network chain, and varies linearly with the inverse of the degree of cross-linking as

$$N_E = \frac{M_c}{M_e} = \frac{\rho}{M_e}\left(\frac{\nu}{V_0}\right)^{-1} \quad (10.2)$$

Figure 10.7 compares values of N_F and N_E for PDMS chains, for which the molecular weight M_e between entanglements is 8100 g mol^{-1}.[72] It is very striking

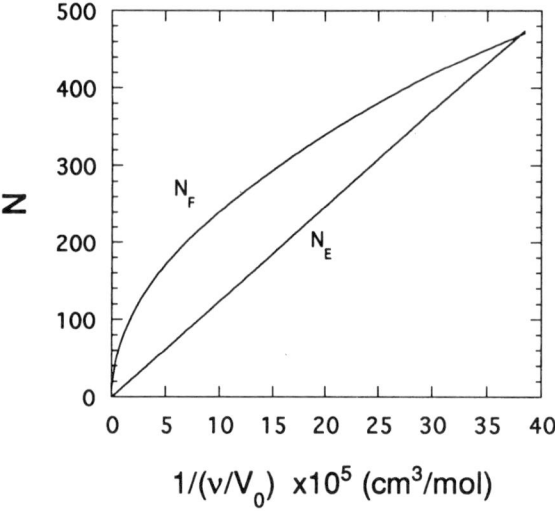

Figure 10.7 Comparisons between the Flory number N_F and the number of entanglement loci N_E for PDMS chains as a function of the degree of cross-linking (as represented by the number of network chains per unit volume).

to note that in the most commonly occurring range of ν/V_0, N_F is always larger than N_E. Thus, the number of entanglements resulting from the disperse interpenetration of chains in the cross-linked state is far greater than the N_E specific localized points along the chain defined by explicit reference to the plateau modulus.

As shown in chapters 2 and 3, the model emphasizing the generalized constraining effects of the neighboring chains gives, for the modulus,

$$G = \left(1 - \frac{2}{\phi}\right)\frac{\nu kT}{V_0}\left[1 + \frac{\phi}{\phi - 2} N \frac{\kappa^2(1+\kappa^2)}{(1+\kappa)^4}\right] \quad (10.3)$$

where the quantities have their usual significance. The corresponding expression for the localized-entanglement model is

$$G = G_{ch} + G_e T_e = A(n_{ch} + n_e T_e)RT \frac{\langle r^2 \rangle}{\langle r^2 \rangle_0} \quad (10.4)$$

where T_e is the entanglement "trapping factor." Utilization of eq. (10.3) requires comparisons between experimental and theoretical values of the modulus, so as to obtain values of the parameter N. These values would then give insight into the type of constraints that are operative. For example, the result $N = 1$ corresponds to the Flory-Erman constrained-junction theory, while $N = 0$ corresponds to the phantom limit. Definitive experiments of this type would do much to resolve this issue of the nature and importance of chain entanglements in network structures at elastic equilibrium.

10.3 Interpretation of Ultimate Properties

This section focuses on the discussion of unfilled elastomers at high elongations, with an emphasis on ultimate properties and how model networks can clarify issues in this area. Of particular interest is the upturn in modulus frequently exhibited by elastomers at very high elongations[1-3,78], and the commercially important increases in ultimate strength associated with this. Such an upturn in modulus is illustrated for natural rubber in the stress-strain isotherms shown in figures 10.8[3] and 10.9[79]. This increase is very important since it corresponds to a significant toughening of the elastomer. Its molecular origin, however, has been the source of some controversy[80]. It had been widely attributed to the "limited extensibility" of the network chains, that is, to an inadequacy in the Gaussian distribution function described in chapter 1 and in appendix F, specifically that it does not assign a zero probability to a configuration unless its end-to-end separation r is infinite. However, the increase in modulus had generally been observed only in networks that could undergo strain-induced crystallization, which could account for the increase in modulus, primarily because the crystallites thus formed would act as additional cross-links in the network. These effects of strain-induced crystallization are mentioned briefly below, and discussed further in chapter 12.

This type of reinforcement from strain-induced crystallization is typified by results reported on the ultimate properties of cis-1,4-polybutadiene networks[4,78,81]. The data indicated the higher the temperature, the lower the extent of crystallization, and this was found to diminish the ultimate properties. In fact, the upturns diminish and eventually disappear upon increase in temperature[81,82]. The effects of increase in swelling were found to parallel those for increase in temperature, as was expected, since diluent also suppresses network crystallization[1,83,84].

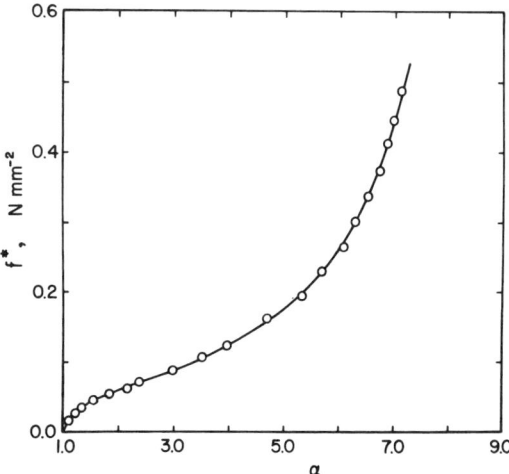

Figure 10.8 Stress–elongation curve showing the high-elongation upturn in modulus for natural rubber in the vicinity of room temperature[3].

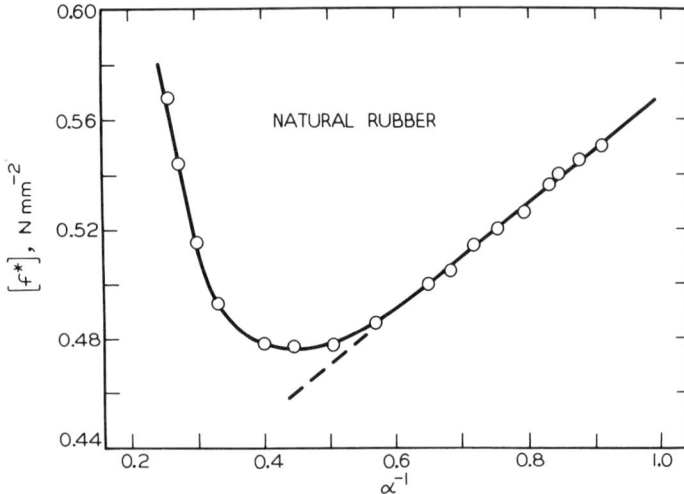

Figure 10.9 Representation of the upturn in modulus for natural rubber[79] in terms of the Mooney-Rivlin equation[4], $[f^*] = 2C_1 + 2C_2\alpha^{-1}$.

On the other hand, in those cases where the upturns are due to limited chain extensibility, increase in temperature has relatively little effect on the upturns[67,80]. Also, in these cases, swelling can even make the upturns more pronounced because of the already imposed stretching of the chains from the dilational effects of the swelling[85,86]. Thus, studying the effects of increase in temperature or introduction of a swelling solvent represents a way to determine whether the upturns in modulus are due to strain-induced crystallization or to a non-Gaussian contribution arising from limited-chain extensibility.

Attempts to observe upturns from non-Gaussian effects in noncrystallizable networks[79] were not successful, presumably because such networks were incapable of the large deformations required to bring about the effects of limited-chain extensibility. Such upturns have been observed[81,82,87], however, by the use of some of the end-linked, noncrystallizable model PDMS networks described above. These networks have high extensibilities, presumably because of their very low incidence of dangling-chain network irregularities. They have particularly high extensibilities when they are prepared from a mixture of very short chains (around a few hundred grams per mole) with relatively long chains (around 18,000 g mol^{-1}), giving a bimodal distribution of network chain lengths[4,55]. Apparently the very short chains in such networks are important because of their limited extensibilities, and the relatively long chains are important because of their ability to retard the rupture process. Such "bimodal" model networks are discussed further in chapter 13.

As will be documented in Chapter 13, stress–strain measurements on such bimodal PDMS networks exhibited upturns in modulus that were much less pronounced than those in crystallizable polymer networks such as natural rubber or cis-1,4-polybutadiene. Furthermore, they are independent of temperature and are not diminished by incorporation of solvent. These characteristics are what is to be expected in the case of limited-chain extensibility[80,88]. Thus, these results permit interpretation of properties such as the elongation at the upturn in the modulus and the elongation at rupture ("maximum extensibility").

10.4 Some Other Unusual Networks

10.4.1 Dangling-Chain Networks

These same very specific chemical reactions can also be used to prepare networks containing known numbers and lengths of dangling-chain irregularities. This is illustrated in figure 10.10[89]. If more chain ends are present than reactive groups on the end-linking molecules, then dangling ends will be produced, and their number is directly determined by the extent of the stoichiometric imbalance. However, their lengths are, of necessity, the same as those of the elastically effective chains, as shown in part (a) of the figure. This constraint can be removed by separately preparing monofunctionally terminated chains of the desired lengths and attaching them as shown in part (b).

Since dangling chains represent imperfections in a network structure, one would expect their presence to have a detrimental effect on the ultimate properties, α_r and $(f/A^*)_r$, of an elastomer. This expectation is confirmed by an exten-

(a) Excess difunctional chains

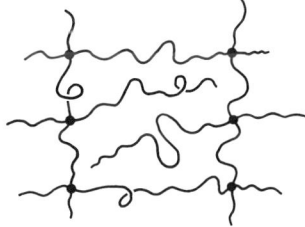

(b) Monofunctional chains

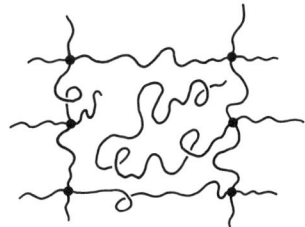

Figure 10.10 Two end-linking techniques for preparing networks with known numbers and lengths of dangling chains[16]. Reprinted with permission from Mark, J. E. (1985), *Acc. Chem. Res.*, **18**. Copyright 1997 American Chemical Society.

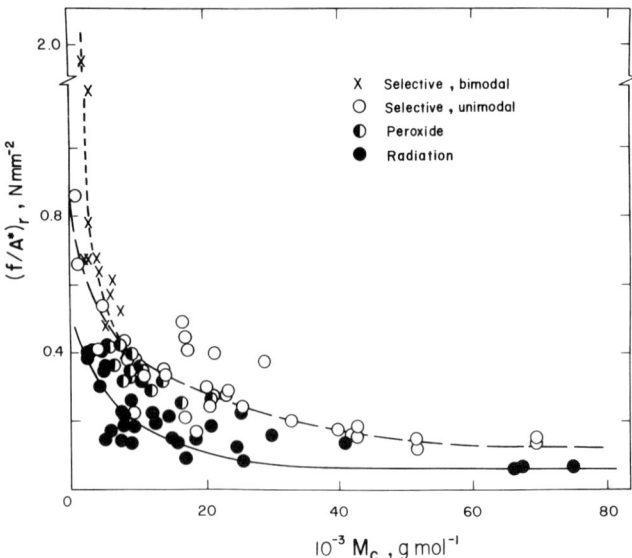

Figure 10.11 Values of the ultimate strength shown as a function of the molecular weight M_c between cross-links for (unfilled) tetrafunctional PDMS networks at 25°C[90].

sive series of results obtained on PDMS networks that had been tetrafunctionally cross-linked using a variety of techniques. Some pertinent results are shown, as a function of the molecular weight between cross-links, in figure 10.11[90]. The largest values of $(f/A^*)_r$ are obtained for the networks prepared by selectively joining functional groups occurring either as chain ends or as side groups along the chains. This is to be expected because of the relatively low incidence of dangling ends in such networks. (As already described, the effects are particularly pronounced when such model networks are prepared from a mixture of relatively long and very short chains.) Also, as expected, the lowest values of the ultimate properties generally occur for the networks cured by radiation (UV light, high-energy electrons, and γ-radiation)[90]. The peroxide-cured networks are generally intermediate to these two extremes, with the ultimate properties presumably depending on whether or not the free radicals generated by the peroxide are sufficiently reactive to cause some chain scission. Similar results were obtained for the maximum extensibility[90]. These observations are at least semi-quantitative and certainly interesting, but they are somewhat deficient in that information on the number of dangling ends in these networks is generally not available.

More definitive results have been obtained by investigation of a series of model networks prepared by end-linking vinyl-terminated PDMS chains[90]. The tetra-functional end-linking agent was used in varying amounts smaller than that corresponding to a stoichiometric balance between its active hydrogen atoms and the chains' terminal vinyl groups. The ultimate properties of these networks, with known numbers of dangling ends, were then compared with those obtained on networks previously prepared so as to have negligible numbers of these irregularities[90].

Values of the ultimate strength $(f/A^*)_r$ of the networks containing the dangling ends were found to be lower than those of the more nearly perfect networks, with the largest differences occurring at high proportions of dangling ends (low $2C_1$), as expected[90]. These results thus confirm the less definitive results shown in figure 10.11. The values of the maximum extensibility showed a similar dependence, as expected.

10.4.2 Networks Containing Reptating Chains

End-linking functionally terminated chains in the presence of chains whose ends are inert yields networks through which the unattached chains reptate[91]. Networks of this type have been used to determine the efficiency with which unattached chains can be extracted from an elastomer as a function of their lengths and the degree of cross-linking of the network[4,92]. This subject was covered in chapter 6 in the discussion on swelling.

10.4.3 Networks Prepared in Solution or in a State of Strain

Two techniques that may be used to prepare networks having simpler topologies[77,93] are illustrated in figure 10.12[67]. Basically, they involve separating the chains prior to their cross-linking, either by stretching or by dissolution. After the cross-linking, the stretching force or solvent is removed and the network is studied (unswollen) with regard to its stress–strain properties in elongation. Some results obtained on PDMS networks cross-linked in solution by means of γ-radiation[77,88] showed that there was a continual decrease in the time required to reach elastic equilibrium, and in the extent of stress relaxation, upon decrease in the volume fraction of polymer present during the cross-linking. Also, at higher dilutions there was a decrease in the Mooney-Rivlin $2C_2$ constant as well.

These observations are qualitatively explained in terms of network connectivity and topology. If a network is cross-linked in solution and the solvent is then removed, the chains collapse in such a way that there is reduced overlap in their configurational domains. It is primarily in this regard, namely decreased chain-junction entangling, that solution-cross-linked samples have simpler topologies, with correspondingly simpler elastomeric behavior.

It is appropriate to comment at this point on the opposite sort of experiment, that is, cross-linking a network in the undiluted state and then studying its stress–strain isotherms in the swollen state. Such a diluent might be introduced to suppress crystallization or to facilitate the approach to elastic equilibrium. As pointed out in chapter 5, the presence of such a swelling solvent can greatly decrease the $2C_2$ Mooney-Rivlin correction to the simplest elastic equation of state. The near-constancy of the modulus with deformation is an additional advantage in the interpretation of stress–strain isotherms.

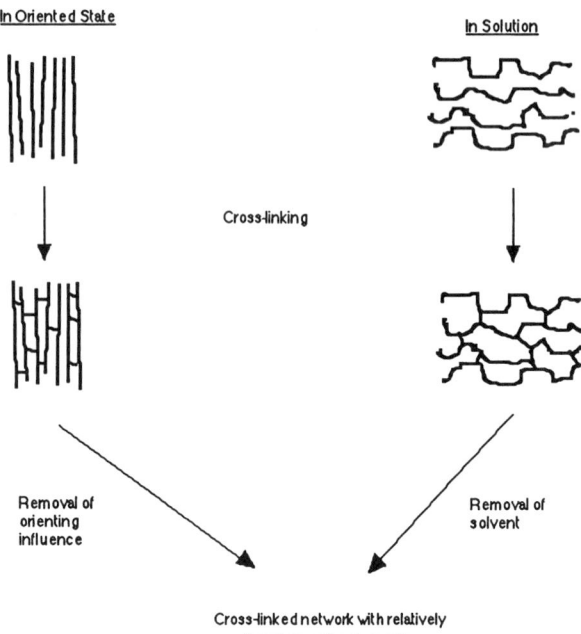

Figure 10.12 Two techniques that may be used to prepare networks of simpler topology[67]. Reprinted with permission from Mark, J. E. (1993), The Rubber Elastic State, in Mark, J. E., et al. (Eds.), *Physical Properties of Polymers*, 2nd ed., Washington, DC: American Chemical Society. Copyright 1997 American Chemical Society.

10.4.4 Networks Containing Unusual Diluents

There is a complication, however, which can occur in the case of networks of polar polymers at relatively high degrees of swelling[88,94] The observation is that different solvents, at the same degree of swelling, can have significantly different effects on the elastic force. This is apparently due to a "specific solvent effect" on the unperturbed dimensions, which appear in the various molecular forms on the elastic equations of state. Although frequently observed in studies of the solution properties of un-cross-linked polymers[95,96], the effect is not yet well understood. It is apparently partly due to the effect of the solvent's dielectric constant on the coulombic interactions between parts of a chain, but is probably also due to solvent–polymer segment interactions that change the conformational preferences of the chain backbone[94].

In the case of the optical properties of swollen elastomeric networks, the size and shape of the diluent molecule may also be of considerable importance[97–110]. For example, the birefringence may depend significantly on the degree of polymerization of a oligomeric diluent, and may even be increased by the presence of some diluents.

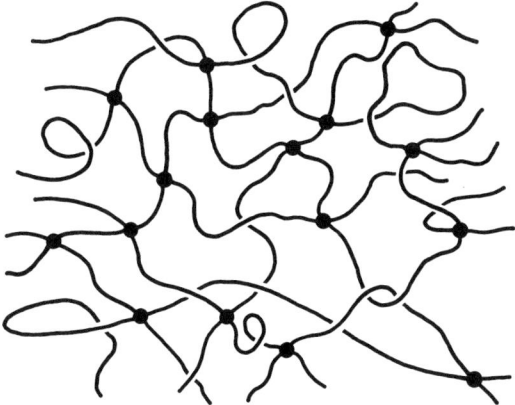

Figure 10.1 Schematic sketch of a typical elastomeric network[3].

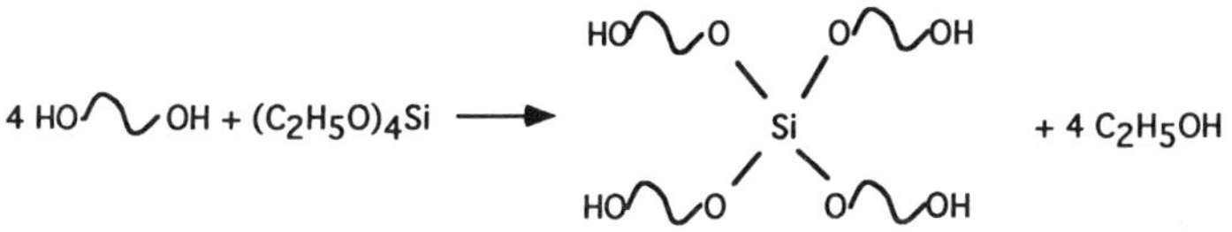

Figure 10.2 End-linking by a condensation reaction between hydroxyl groups at the ends of a polymer chain and the alkoxy groups on a tetrafunctional end-linking agent[8]. As is pointed out, the number-average molecular weight M_n of the precursor chains becomes the critically important molecular weight M_c between cross links, and the distribution of M_n also characterizes the distribution of M_c (i.e., network chain-lengths).

Because of their known structures, such model elastomers are now the preferred materials for the quantitative characterization of rubberlike elasticity. Such very specific cross-linking reactions have also been shown to be useful in the preparation of liquid-crystalline elastomers[65,66].

10.1.2 Effect of Junction Functionality

Trifunctional and tetrafunctional PDMS networks prepared in this way have been used to test the molecular theories of rubber elasticity with regard to the increase in non-affineness of the network deformation with increasing elongation. Some of these results are shown in figure 10.3[9]. The ratio $2C_2/2C_1$ decreases with increase in cross-link functionality from three to four because cross-links connecting four chains are more constrained than those connecting only three[67]. There is therefore less of a decrease in modulus brought about by the fluctuations that are enhanced at high deformation and give the deformation its non-affine character, as described in chapter 3. The decrease in $2C_2/2C_1$ with decrease in network chain molecular weight is due to the fact that there is less configurational interpenetration in the case of short network chains. This decreases the firmness with which

136 STRUCTURES AND PROPERTIES OF RUBBERLIKE NETWORKS

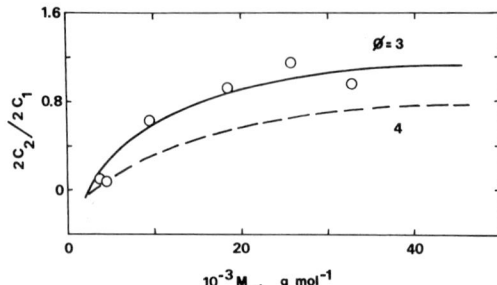

Figure 10.3 Experimental data showing values of the ratio $2C_2/2C_1$, which is a measure of the increase in nonaffineness of the deformation as the elongation increases[9]. The ratio decreases with increase in junction functionality and with decrease in network chain molecular weight, as predicted by theory[108–110].

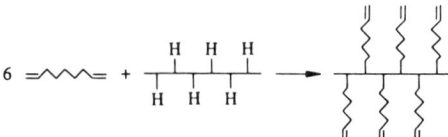

Figure 10.4 End-linking by an addition reaction between vinyl groups at the ends of a polymer chain and the active hydrogen atoms on silicon atoms in an oligomeric poly(methyl hydrogen siloxane)[67]. Reprinted with permission from Mark, J. E. (1993), The Rubber Elastic State, in Mark, J. E., et al. (Eds.), *Physical Properties of Polymers*, 2nd ed., Washington, DC: American Chemical Society. Copyright 1997 American Chemical Society.

the cross-links are embedded and thus the deformation is already highly nonaffine, even at relatively small deformations.

A more thorough investigation of the effects of cross-link functionality requires use of the more versatile chemical reaction illustrated in figure 10.4[67]. Specifically, vinyl-terminated PDMS chains are end-linked using a multifunctional silane[67]. In the study summarized in figure 10.5[12], this reaction was used to prepare PDMS model networks having functionalities in the range 3–11, with a relatively unsuccessful attempt to achieve a functionality of 37. As shown in the figure, the modulus $2C_1$ increases with increase in functionality, as expected from the increased constraints on the cross-links, and as predicted in chapter 3. Similarly, $2C_2$ and its value relative to $2C_1$ both decrease, for the reasons described in the discussion of figure 10.3.

10.2 The Issue of Entanglements

Such model networks may also be used to provide a direct test of molecular predictions of the modulus of a network of known degree of cross-linking. Some

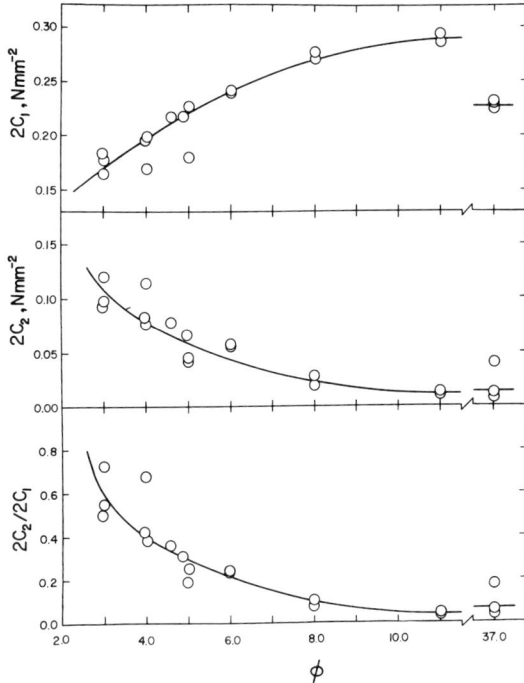

Figure 10.5 Experimental data showing the effect of cross-link functionality on $2C_1$ (a measure of the high-deformation modulus), $2C_2$, and $2C_2/2C_1$.[12] Reprinted with permission from Llorente, M. A. and Mark, J. E. (1980), *Macromolecules*, **13**. Copyright 1997 American Chemical Society.

experiments on model networks[8,9,12,40,57,68] have given values of the elastic modulus in good agreement with theory. Others[13,25,39,52,69] have given values significantly larger than predicted, and the increases in modulus have been attributed to contributions from "permanent" chain entanglements of the type shown in the lower right portion of figure 10.1. There are disagreements, and the issue has not yet been resolved. Since the relationship of modulus to structure is of such fundamental importance, there has been a great deal of research activity in this area.

Discussions and correspondence among the authors of this book and those groups[70] particularly involved with this issue have generated a number of helpful suggestions, several of which are explored in greater detail below. Aspects of greatest importance were cited to be: (1) studying the effects of cross-linking in solution, (2) studying the effects of swelling on networks cross linked in the bulk (dry) state, (3) building on the demonstration by Vilgis and Erman[71] that the constraint models and slip-link models have much in common, (4) studying the effects of cross-link functionality and degree of cross-linking, (5) studying a variety of elastomeric polymers, particularly those having very different values of the plateau modulus[72], and (6) generalizing rubber elasticity models to include viscoelastic effects as well.

Although a number of experimental studies have been important for providing a better understanding of these issues, it must be admitted at the outset that these problems have still not been unequivocally resolved or even unambiguously formulated. In any case, it is useful to consider the illustrative studies of Oppermann and Rennar[73,74] on end-linked PDMS networks of different functionalities, since some of these issues arose in the detailed and careful analysis of their experimental data. For example, Figure 10.6 shows their values of the small-strain shear modulus of end-linked PDMS networks with pentafunctional junctions as a function of network chain density. The solid curve is obtained by interpolating through the experimental points, and the dot–dashed and the dashed lines were obtained from the affine and phantom network models, respectively. The data agree with predictions of the molecular theories at larger values of the chain or junction densities. At lower junction densities, however, the measured moduli are significantly larger than the predicted ones. Part of the difference, but probably not all of it, is due to problems in bringing such lightly cross-linked networks to elastic equilibrium in the unswollen state. Finally, the moduli are seen to tend to zero as the junction density goes to zero. These data are representative of other PDMS networks, as described in the paper by Oppermann and Rennar[73]. Data for randomly cross-linked polybutadiene given by Rennar and Oppermann[74] also lie above the theoretically predicted values, but do not follow a pronounced sigmoidal shape as

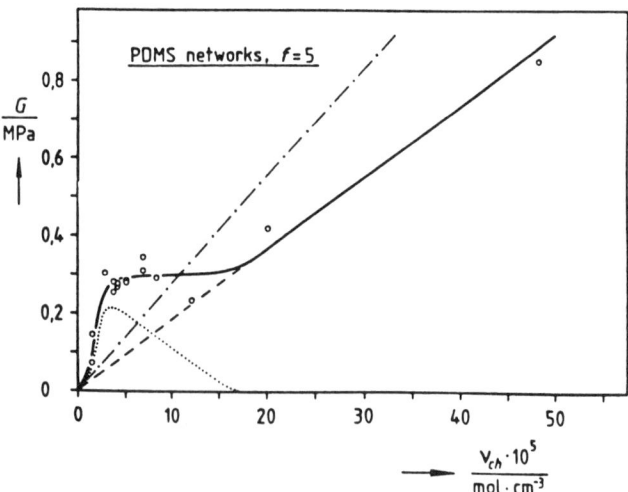

Figure 10.6 Shear modulus G as a function of the chemical network density ν_{ch} for end-linked PDMS networks having a junction functionality of five[74]. The temperature for the cross-linking and the mechanical measurements was 333 K. The lines correspond to the observed relationship (———), prediction of the affine network model (– · – · – · –), contributions from the chemical cross-links (– – – –), and contributions from additional restrictions (· · · · · ·).

do the data for the PDMS system. Thus, structurally different polymers can give networks with very different dependences of the modulus on degree of cross-linking.

The departure of the experimental points from the simple molecular theory when the network chains are long indicates the presence of additional contributions to the modulus. The reasons behind this departure are not yet clear and the problem clearly remains open. The understanding of entanglement phenomena is perhaps based on answering the question of whether the stated departures result from trapped entanglements (which scale with the number of entanglement loci N_E along the chain contour) or simply from the presence of other chains, sharing the volume of a given network chain (defined as the Flory number, N_F). This Flory number is defined by the expression[75]

$$N_F = \left(\frac{4\pi}{3} \langle r^2 \rangle_0^{3/2}\right) \frac{\nu}{V_0}$$
$$= \frac{4\pi}{3} \left[\frac{C_\infty \rho l^2}{M_u}\right]^{3/2} N_A \left(\frac{\nu}{V_0}\right)^{-1/2} \quad (10.1)$$

where $\langle r^2 \rangle_0$ is the mean-squared end-to-end distance of a network chain in the unperturbed state, ν/V_0 is the number of network chains per unit volume (obtained in the state of formation), $C_\infty = \langle r^2 \rangle_0 / nL^2$ is the characteristic ratio of a chain of n bonds of bond length L[76], and l and M_u are the length and the molecular weight of the chain repeat unit, respectively. One important difference between the two types of constraints involves their dependence on network swelling. The localized, permanent entanglements should be independent of swelling, while the more diffuse interchain interactions should decrease with increase in swelling. The contributions from trapped entanglements should therefore persist, even in highly swollen networks, and therefore contribute to the phantom modulus. For these reasons, it is quite important to carry out measurements of the modulus at a series of degrees of swelling, with the results at the highest degrees of swelling presumably being least complicated by nonequilibrium effects.

It should be noted that if one makes a network in the extremely dilute state so that there is no chain interpenetration during its formation, N_F will equate to zero and the phantom-structure state will be obtained, by definition. Experimentally, one indeed sees that network formation in the dilute state decreases both the modulus and its dependence on elongation (the "C_2 effect")[4,77]. The Flory number, which varies as the inverse square root of the degree of cross-linking, is significantly different from the number of entanglement loci, N_E. The latter is defined as the number of chains of molecular weight M_e in a network chain, and varies linearly with the inverse of the degree of cross-linking as

$$N_E = \frac{M_c}{M_e} = \frac{\rho}{M_e} \left(\frac{\nu}{V_0}\right)^{-1} \quad (10.2)$$

Figure 10.7 compares values of N_F and N_E for PDMS chains, for which the molecular weight M_e between entanglements is 8100 g mol^{-1}.[72] It is very striking

140 STRUCTURES AND PROPERTIES OF RUBBERLIKE NETWORKS

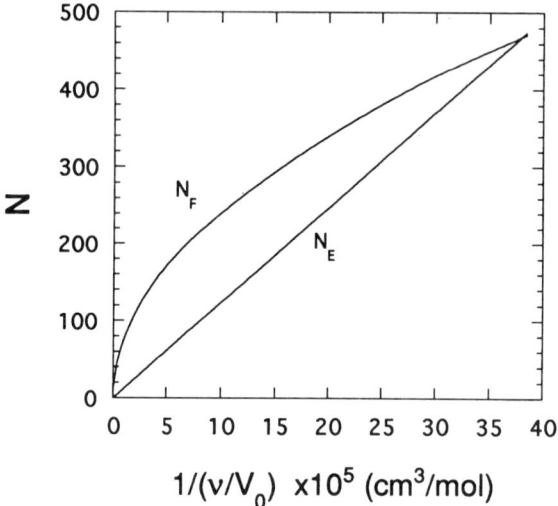

Figure 10.7 Comparisons between the Flory number N_F and the number of entanglement loci N_E for PDMS chains as a function of the degree of cross-linking (as represented by the number of network chains per unit volume).

to note that in the most commonly occurring range of ν/V_0, N_F is always larger than N_E. Thus, the number of entanglements resulting from the disperse interpenetration of chains in the cross-linked state is far greater than the N_E specific localized points along the chain defined by explicit reference to the plateau modulus.

As shown in chapters 2 and 3, the model emphasizing the generalized constraining effects of the neighboring chains gives, for the modulus,

$$G = \left(1 - \frac{2}{\phi}\right)\frac{\nu k T}{V_0}\left[1 + \frac{\phi}{\phi - 2} N \frac{\kappa^2(1 + \kappa^2)}{(1 + \kappa)^4}\right] \quad (10.3)$$

where the quantities have their usual significance. The corresponding expression for the localized-entanglement model is

$$G = G_{ch} + G_e T_e = A(n_{ch} + n_e T_e)RT \frac{\langle r^2 \rangle}{\langle r^2 \rangle_0} \quad (10.4)$$

where T_e is the entanglement "trapping factor." Utilization of eq. (10.3) requires comparisons between experimental and theoretical values of the modulus, so as to obtain values of the parameter N. These values would then give insight into the type of constraints that are operative. For example, the result $N = 1$ corresponds to the Flory-Erman constrained-junction theory, while $N = 0$ corresponds to the phantom limit. Definitive experiments of this type would do much to resolve this issue of the nature and importance of chain entanglements in network structures at elastic equilibrium.

10.3 Interpretation of Ultimate Properties

This section focuses on the discussion of unfilled elastomers at high elongations, with an emphasis on ultimate properties and how model networks can clarify issues in this area. Of particular interest is the upturn in modulus frequently exhibited by elastomers at very high elongations[1-3,78], and the commercially important increases in ultimate strength associated with this. Such an upturn in modulus is illustrated for natural rubber in the stress-strain isotherms shown in figures 10.8[3] and 10.9[79]. This increase is very important since it corresponds to a significant toughening of the elastomer. Its molecular origin, however, has been the source of some controversy[80]. It had been widely attributed to the "limited extensibility" of the network chains, that is, to an inadequacy in the Gaussian distribution function described in chapter 1 and in appendix F, specifically that it does not assign a zero probability to a configuration unless its end-to-end separation r is infinite. However, the increase in modulus had generally been observed only in networks that could undergo strain-induced crystallization, which could account for the increase in modulus, primarily because the crystallites thus formed would act as additional cross-links in the network. These effects of strain-induced crystallization are mentioned briefly below, and discussed further in chapter 12.

This type of reinforcement from strain-induced crystallization is typified by results reported on the ultimate properties of cis-1,4-polybutadiene networks[4,78,81]. The data indicated the higher the temperature, the lower the extent of crystallization, and this was found to diminish the ultimate properties. In fact, the upturns diminish and eventually disappear upon increase in temperature[81,82]. The effects of increase in swelling were found to parallel those for increase in temperature, as was expected, since diluent also suppresses network crystallization[1,83,84].

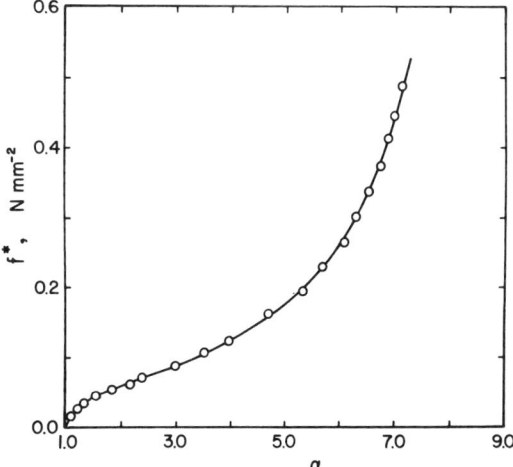

Figure 10.8 Stress–elongation curve showing the high-elongation upturn in modulus for natural rubber in the vicinity of room temperature[3].

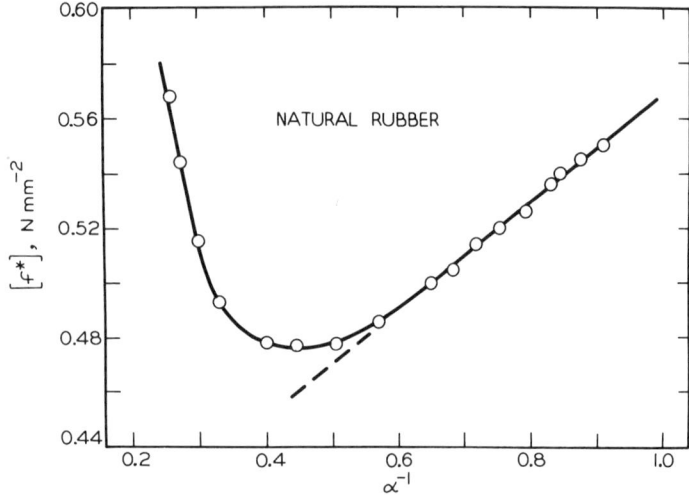

Figure 10.9 Representation of the upturn in modulus for natural rubber[79] in terms of the Mooney-Rivlin equation[4], $[f^*] = 2C_1 + 2C_2\alpha^{-1}$.

On the other hand, in those cases where the upturns are due to limited chain extensibility, increase in temperature has relatively little effect on the upturns[67,80]. Also, in these cases, swelling can even make the upturns more pronounced because of the already imposed stretching of the chains from the dilational effects of the swelling[85,86]. Thus, studying the effects of increase in temperature or introduction of a swelling solvent represents a way to determine whether the upturns in modulus are due to strain-induced crystallization or to a non-Gaussian contribution arising from limited-chain extensibility.

Attempts to observe upturns from non-Gaussian effects in noncrystallizable networks[79] were not successful, presumably because such networks were incapable of the large deformations required to bring about the effects of limited-chain extensibility. Such upturns have been observed[81,82,87], however, by the use of some of the end-linked, noncrystallizable model PDMS networks described above. These networks have high extensibilities, presumably because of their very low incidence of dangling-chain network irregularities. They have particularly high extensibilities when they are prepared from a mixture of very short chains (around a few hundred grams per mole) with relatively long chains (around 18,000 g mol^{-1}), giving a bimodal distribution of network chain lengths[4,55]. Apparently the very short chains in such networks are important because of their limited extensibilities, and the relatively long chains are important because of their ability to retard the rupture process. Such "bimodal" model networks are discussed further in chapter 13.

As will be documented in Chapter 13, stress–strain measurements on such bimodal PDMS networks exhibited upturns in modulus that were much less pronounced than those in crystallizable polymer networks such as natural rubber or *cis*-1,4-polybutadiene. Furthermore, they are independent of temperature and are not diminished by incorporation of solvent. These characteristics are what is to be expected in the case of limited-chain extensibility[80,88]. Thus, these results permit interpretation of properties such as the elongation at the upturn in the modulus and the elongation at rupture ("maximum extensibility").

10.4 Some Other Unusual Networks

10.4.1 Dangling-Chain Networks

These same very specific chemical reactions can also be used to prepare networks containing known numbers and lengths of dangling-chain irregularities. This is illustrated in figure 10.10[89]. If more chain ends are present than reactive groups on the end-linking molecules, then dangling ends will be produced, and their number is directly determined by the extent of the stoichiometric imbalance. However, their lengths are, of necessity, the same as those of the elastically effective chains, as shown in part (a) of the figure. This constraint can be removed by separately preparing monofunctionally terminated chains of the desired lengths and attaching them as shown in part (b).

Since dangling chains represent imperfections in a network structure, one would expect their presence to have a detrimental effect on the ultimate properties, α_r and $(f/A^*)_r$, of an elastomer. This expectation is confirmed by an exten-

(a) Excess difunctional chains

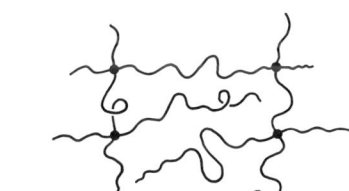

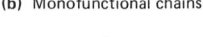

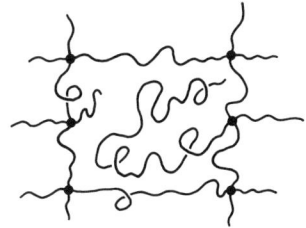

Figure 10.10 Two end-linking techniques for preparing networks with known numbers and lengths of dangling chains[16]. Reprinted with permission from Mark, J. E. (1985), *Acc. Chem. Res.*, **18**. Copyright 1997 American Chemical Society.

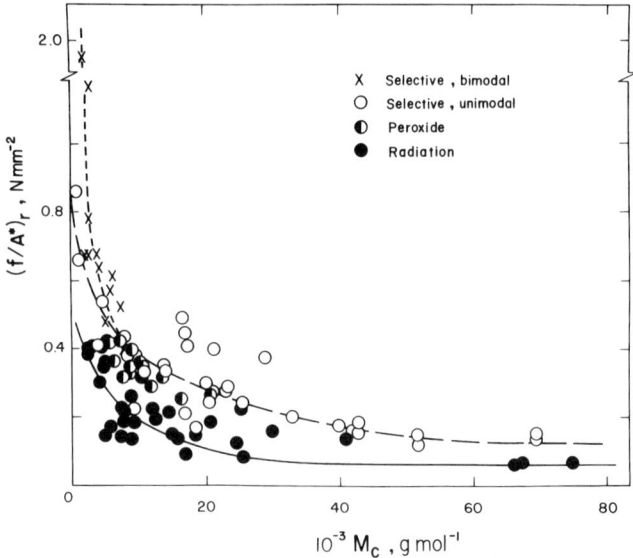

Figure 10.11 Values of the ultimate strength shown as a function of the molecular weight M_c between cross-links for (unfilled) tetrafunctional PDMS networks at 25°C[90].

sive series of results obtained on PDMS networks that had been tetrafunctionally cross-linked using a variety of techniques. Some pertinent results are shown, as a function of the molecular weight between cross-links, in figure 10.11[90]. The largest values of $(f/A^*)_r$ are obtained for the networks prepared by selectively joining functional groups occurring either as chain ends or as side groups along the chains. This is to be expected because of the relatively low incidence of dangling ends in such networks. (As already described, the effects are particularly pronounced when such model networks are prepared from a mixture of relatively long and very short chains.) Also, as expected, the lowest values of the ultimate properties generally occur for the networks cured by radiation (UV light, high-energy electrons, and γ-radiation)[90]. The peroxide-cured networks are generally intermediate to these two extremes, with the ultimate properties presumably depending on whether or not the free radicals generated by the peroxide are sufficiently reactive to cause some chain scission. Similar results were obtained for the maximum extensibility[90]. These observations are at least semi-quantitative and certainly interesting, but they are somewhat deficient in that information on the number of dangling ends in these networks is generally not available.

More definitive results have been obtained by investigation of a series of model networks prepared by end-linking vinyl-terminated PDMS chains[90]. The tetrafunctional end-linking agent was used in varying amounts smaller than that corresponding to a stoichiometric balance between its active hydrogen atoms and the chains' terminal vinyl groups. The ultimate properties of these networks, with known numbers of dangling ends, were then compared with those obtained on networks previously prepared so as to have negligible numbers of these irregularities[90].

Values of the ultimate strength $(f/A^*)_r$ of the networks containing the dangling ends were found to be lower than those of the more nearly perfect networks, with the largest differences occurring at high proportions of dangling ends (low $2C_1$), as expected[90]. These results thus confirm the less definitive results shown in figure 10.11. The values of the maximum extensibility showed a similar dependence, as expected.

10.4.2 Networks Containing Reptating Chains

End-linking functionally terminated chains in the presence of chains whose ends are inert yields networks through which the unattached chains reptate[91]. Networks of this type have been used to determine the efficiency with which unattached chains can be extracted from an elastomer as a function of their lengths and the degree of cross-linking of the network[4,92]. This subject was covered in chapter 6 in the discussion on swelling.

10.4.3 Networks Prepared in Solution or in a State of Strain

Two techniques that may be used to prepare networks having simpler topologies[77,93] are illustrated in figure 10.12[67]. Basically, they involve separating the chains prior to their cross-linking, either by stretching or by dissolution. After the cross-linking, the stretching force or solvent is removed and the network is studied (unswollen) with regard to its stress–strain properties in elongation. Some results obtained on PDMS networks cross-linked in solution by means of γ-radiation[77,88] showed that there was a continual decrease in the time required to reach elastic equilibrium, and in the extent of stress relaxation, upon decrease in the volume fraction of polymer present during the cross-linking. Also, at higher dilutions there was a decrease in the Mooney-Rivlin $2C_2$ constant as well.

These observations are qualitatively explained in terms of network connectivity and topology. If a network is cross-linked in solution and the solvent is then removed, the chains collapse in such a way that there is reduced overlap in their configurational domains. It is primarily in this regard, namely decreased chain–junction entangling, that solution-cross-linked samples have simpler topologies, with correspondingly simpler elastomeric behavior.

It is appropriate to comment at this point on the opposite sort of experiment, that is, cross-linking a network in the undiluted state and then studying its stress–strain isotherms in the swollen state. Such a diluent might be introduced to suppress crystallization or to facilitate the approach to elastic equilibrium. As pointed out in chapter 5, the presence of such a swelling solvent can greatly decrease the $2C_2$ Mooney-Rivlin correction to the simplest elastic equation of state. The near-constancy of the modulus with deformation is an additional advantage in the interpretation of stress–strain isotherms.

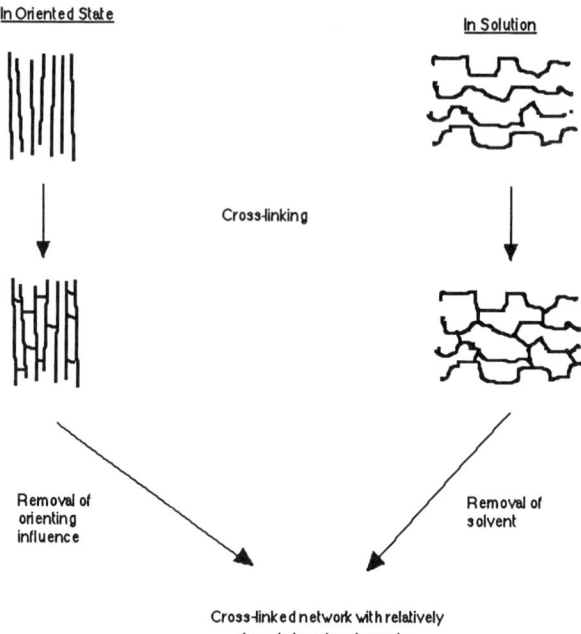

Figure 10.12 Two techniques that may be used to prepare networks of simpler topology[67]. Reprinted with permission from Mark, J. E. (1993), The Rubber Elastic State, in Mark, J. E., et al. (Eds.), *Physical Properties of Polymers*, 2nd ed., Washington, DC: American Chemical Society. Copyright 1997 American Chemical Society.

10.4.4 Networks Containing Unusual Diluents

There is a complication, however, which can occur in the case of networks of polar polymers at relatively high degrees of swelling[88,94] The observation is that different solvents, at the same degree of swelling, can have significantly different effects on the elastic force. This is apparently due to a "specific solvent effect" on the unperturbed dimensions, which appear in the various molecular forms on the elastic equations of state. Although frequently observed in studies of the solution properties of un-cross-linked polymers[95,96], the effect is not yet well understood. It is apparently partly due to the effect of the solvent's dielectric constant on the coulombic interactions between parts of a chain, but is probably also due to solvent–polymer segment interactions that change the conformational preferences of the chain backbone[94].

In the case of the optical properties of swollen elastomeric networks, the size and shape of the diluent molecule may also be of considerable importance[97–110]. For example, the birefringence may depend significantly on the degree of polymerization of a oligomeric diluent, and may even be increased by the presence of some diluents.

References

(1) Flory, P. J. *Principles of Polymer Chemistry*. Cornell University Press: Ithaca NY. 1953.
(2) Treloar, L. R. G. *The Physics of Rubber Elasticity*, 3rd ed. Clarendon Press: Oxford. 1975.
(3) Mark, J. E. *J. Chem. Ed.* 1981, 58, 898.
(4) Mark, J. E.; Erman, B. *Rubberlike Elasticity. A Molecular Primer*, Wiley-Interscience: New York. 1988.
(5) Mark, J. E.; Erman, B. In *Polymer Networks*, R. F. T. Stepto, Ed. Blackie Academic, Chapman & Hall: Glasgow. 1997; in press.
(6) Mark, J. E.; Sullivan, J. L. *J. Chem. Phys.* 1977, 66, 1006.
(7) Rempp, P.; Herz, J. E. *Angew. Makromol. Chemie* 1979, 76/77, 373.
(8) Mark, J. E. *Makromol. Chemie, Suppl.* 1979, 2, 87.
(9) Mark, J. E.; Rahalkar, R. R.; Sullivan, J. L. *J. Chem. Phys.* 1979, 70, 1794.
(10) Meyers, K. O.; Bye, M. L.; Merrill, E. W. *Macromolecules* 1980, 13, 1045.
(11) Kosfeld, R.; Hess, M.; Hansen, D. *Polym. Bull.* 1980, 3, 603.
(12) Llorente, M. A.; Mark, J. E. *Macromolecules* 1980, 13, 681.
(13) Gottlieb, M.; Macosko, C. W.; Benjamin, G. S.; Meyers, K. O.; Merrill, E. W. *Macromolecules* 1981, 14, 1039.
(14) Mark, J. E. *Rubber Chem. Technol.* 1982, 55, 762.
(15) Meyers, K. O.; Merrill, E. W. In *Elastomers and Rubber Elasticity*, J. E. Mark and J. Lal, Eds. American Chemical Society: Washington, DC. 1982; Vol. 193; p. 329.
(16) Mark, J. E. *Acc. Chem. Res.* 1985, 18, 202.
(17) Stadler, R.; Jacobi, M. M.; Gronski, W. *Makromol. Chem., Rapid Commun.* 1983, 4, 129.
(18) Sung, P.-H.; Pan, S.-J.; Mark, J. E.; Chang, V. S. C.; Lackey, J. E.; Kennedy, J. P. *Polym. Bull.* 1983, 9, 375.
(19) Lackey, J. E.; Chang, V. S. C.; Kennedy, J. P.; Zhang, Z.-M.; Sung, P.-H., Mark, J. E. *Polym. Bull.* 1984, 11, 19.
(20) Jiang, C.-Y.; Mark, J. E.; Chang, V. S. C.; Kennedy, J. P. *Polym. Bulletin* 1984, 11, 319.
(21) Kennedy, J. P. *J. Appl. Polym. Sci., Appl. Polym. Symp.* 1984, 39, 21.
(22) Entelis, S. G.; Evreinov, V. V.; Gorshkov, A. V. *Adv. Polym. Sci.* 1986, 76, 129.
(23) Lanyo, L. C.; Kelley, F. N. *Rubber Chem. Technol.* 1987, 60, 78.
(24) Kennedy, J. P.; Lackey, J. E. *J. Appl. Polym. Sci.* 1987, 33, 2449.
(25) Miller, D. R.; Macosko, C. W. *J. Polym. Sci., Polym. Phys. Ed.* 1987, 25, 2441.
(26) Sharaf, M. A.; Mark, J. E. *Makromol. Chemie* 1989, 190, 495.
(27) Smith, T. L.; Haidar, B.; Hedrick, J. L. *Rubber Chem. Technol.* 1990, 63, 256.
(28) Clarson, S. J.; Wang, Z.; Mark, J. E. *Eur. Polym. J.* 1990, 26, 621.
(29) Kornfield, J. A.; Spiess, H. W.; Nefzger, H.; Hayen, H.; Eisenbach, C. D. *Macromolecules* 1991, 24, 4787.
(30) Silva, L. K.; Mark, J. E.; Boerio, F. J. *Makromol. Chemie* 1991, 192, 499.
(31) Hanyu, A.; Stein, R. S. *Macromol. Symp.* 1991, 45, 189.
(32) Roland, C. M.; Buckley, G. S. *Rubber Chem. Technol.* 1991, 64, 74.
(33) Kennedy, J. P. *Makromol. Chemie, Makromol. Symp.* 1991, 51, 169.
(34) Schimmel, K.-H.; Heinrich, G. *Coll. Polym. Sci.* 1991, 269, 1003.
(35) Andrady, A. L.; Llorente, M. A.; Mark, J. E. *Polym. Bull.* 1992, 28, 103.
(36) Oikawa, H. *Polymer* 1992, 33, 1116.

(37) Sharaf, M. A. *J. Macromol. Sci., Macromol. Rep.* 1992, A30, 83.
(38) Hamurcu, E. E.; Baysal, B. M. *Polymer* 1993, 34, 5163.
(39) Patel, S. K.; Malone, S.; Cohen, C.; Gillmor, J. R.; Colby, R. H. *Macromolecules* 1992, 25, 5241.
(40) Sharaf, M. A. *Int. J. Polymeric Mater.* 1992, 18, 237.
(41) Fischer, M. *Adv. Polym. Sci.* 1992, 100, 313.
(42) Sharaf, M. A.; Mark, J. E.; Hosani, Z. Y. A. *Eur. Polym. J.* 1993, 29, 809.
(43) Subramanian, P. R.; Galiatsatos, V. *Macromol. Symp.* 1993, 76, 233.
(44) Mark, J. E. *New J. Chem.* 1993, 17, 703.
(45) Sharaf, M. A.; Mark, J. E. *Macromol. Symp.* 1993, 76, 13.
(46) Kennedy, J. P. *TRIP* 1993, 1, 381.
(47) Bontems, S. L.; Stein, J.; Zumbrum, M. A. *J. Polym. Sci., Polym. Chem. Ed.* 1993, 31, 2697.
(48) Sharaf, M. A.; Mark, J. E. *Polymer* 1994, 35, 740.
(49) Sharaf, M. A.; Mark, J. E.; Al-Ghazal, A. A.-R. *J. Appl. Polym. Sci. Symp.* 1994, 55, 139.
(50) Mark, J. E. *J. Inorg. Organomet. Polym.* 1994, 4, 31.
(51) Sharaf, M. A.; Mark, J. E.; Ahmad, E. *Coll. Polym. Sci.* 1994, 272, 504.
(52) Gent, A. N.; Liu, G. L.; Mazurek, M. *J. Polym. Sci., Polym. Phys. Ed.* 1994, 32, 271.
(53) Shibayama, M.; Takahashi, H.; Yamaguchi, H.; Sakurai, S.; Nomura, S. *Polymer* 1994, 35, 2944.
(54) Kennedy, J. P. *CHEMTECH* 1994, 24, 24.
(55) Mark, J. E. *Acc. Chem. Res.* 1994, 27, 271.
(56) Trautenberg, H. L.; Sommer, J.-U.; Goritz, D. *J. Chem. Soc. Faraday Trans.* 1995, 91, 2649.
(57) Sharaf, M. A.; Mark, J. E. *J. Polym. Sci., Polym. Phys. Ed.* 1995, 33, 1151.
(58) Out, G. J. J.; Turetskii, A. A.; Snijder, M.; Moller, M.; Papkov, V. S. *Polymer* 1995, 36, 3213.
(59) Takahashi, H.; Shibayama, M.; Fujisawa, H.; Nomura, S. *Macromolecules* 1995, 28, 8824.
(60) Besbes, S.; Bokobza, L.; Monnerie, L.; Bahar, I.; Erman, B. *Macromolecules* 1995, 28, 231.
(61) Sharaf, M. A.; Mark, J. E.; Alshamsi, A. S. *Polym. J.* 1996, 28, 375.
(62) Falender, J. R.; Yeh, G. S. Y.; Mark, J. E. *J. Chem. Phys.* 1979, 70, 5324.
(63) Falender, J. R.; Yeh, G. S. Y.; Mark, J. E. *J. Am. Chem. Soc.* 1979, 101, 7353.
(64) Falender, J. R.; Yeh, G. S. Y.; Mark, J. E. *Macromolecules* 1979, 12, 1207.
(65) Disch, S.; Finkelmann, H.; Ringsdorf, H.; Schuhmacher, P. *Macromolecules* 1995, 28, 2424.
(66) Zentel, R.; Brehmer, M. *CHEMTECH* 1995, 25(5), 41.
(67) Mark, J. E. In *Physical Properties of Polymers*, 2nd ed. J. E. Mark, A. Eisenberg, W. W. Graessley, L. Mandelkern, E. T. Samulski, J. L. Koenig, and G. D. Wignall, Eds. American Chemical Society: Washington, DC. 1993; p. 3.
(68) Eichinger, B. E.; Akgiray, O. In *Computer Simulation of Polymers*, E. A. Colbourn, Ed. Longman: White Plains, NY. 1994; p. 263.
(69) Venkatraman, S. *J. Appl. Polym. Sci.* 1993, 48, 1383.
(70) Dusek, K.; Edwards, S. F.; Ferry, J. D.; Gent, A. N.; Graessley, W. W.; Kramer, O.; Langley, N. R.; Oppermann, W.; Stein, R. S.; Vilgis, T. A. Private communications.
(71) Vilgis, T. A.; Erman, B. *Macromolecules* 1993, 26, 6657.

(72) Ferry, J. D. *Viscoelastic Properties of Polymers*, 3rd ed. Wiley: New York. 1980.
(73) Oppermann, W.; Rennar, N. *Prog. Coll. Polym. Sci.* 1987, 75, 49.
(74) Rennar, N.; Oppermann, W. *Coll. Polym. Sci.* 1992, 270, 527.
(75) Erman, B.; Flory, P. J. *Macromolecules* 1982, 15, 806.
(76) Flory, P. J. *Statistical Mechanics of Chain Molecules*, Interscience: New York. 1969.
(77) Johnson, R. M.; Mark, J. E. *Macromolecules* 1972, 5, 41.
(78) Mark, J. E.; Eisenberg, A.; Graessley, W. W.; Mandelkern, L.; Samulski, E. T.; Koenig, J. L.; Wignall, G. D. *Physical Properties of Polymers*, 2nd ed. American Chemical Society: Washington, DC. 1993.
(79) Mark, J. E.; Kato, M.; Ko, J. H. *J. Polym. Sci., Part C* 1976, 54, 217.
(80) Andrady, A. L.; Llorente, M. A.; Mark, J. E. *J. Chem. Phys.* 1980, 72, 2282.
(81) Su, T.-K.; Mark, J. E. *Macromolecules* 1977, 10, 120.
(82) Mark, J. E. *Polym. Eng. Sci.* 1979, 19, 409.
(83) Mandelkern, L. *Crystallization of Polymers*, McGraw Hill: New York. 1964.
(84) Mandelkern, L. In *Comprehensive Polymer Science*, G. Allen, Ed. Pergamon Press: Oxford. 1989; p. 363.
(85) Mark, J. E. *Macromolecules* 1984, 17, 2924.
(86) Clarson, S. J.; Galiatsatos, V. *Polym. Commun.* 1986, 27, 260.
(87) Chiu, D. S.; Su, T.-K.; Mark, J. E. *Macromolecules* 1977, 10, 1110.
(88) Mark, J. E.; Eisenberg, A.; Graessley, W. W.; Mandelkern, L.; Koenig, J. L. *Physical Properties of Polymers*, 1st ed. American Chemical Society: Washington, DC. 1984.
(89) Mark, J. E. In *Silicon-Based Polymer Science. A Comprehensive Resource*, J. M. Zeigler and F. W. G. Fearon, Ed. American Chemical Society: Washington, DC. 1990; p. 47.
(90) Andrady, A. L.; Llorente, M. A.; Sharaf, M. A.; Rahalkar, R. R.; Mark, J. E.; Sullivan, J. L.; Yu, C. U.; Falender, J. R. *J. Appl. Polym. Sci.* 1981, 26, 1829.
(91) de Gennes, P. G. *Scaling Concepts in Polymer Physics*, Cornell University Press: Ithaca, NY. 1979.
(92) Mark, J. E.; Zhang, Z.-M. *J. Polym. Sci., Polym. Phys. Ed.* 1983, 21, 1971.
(93) Langley, N. R.; Dickie, R. A.; Wong, C.; Ferry, J. D.; Chasset, R.; Thirion, P. *J. Polym. Sci., Part A-2* 1968, 6, 1371.
(94) Yu, C. U.; Mark, J. E. *Macromolecules* 1974, 7, 229.
(95) Hoeve, C. A. J.; O'Brien, M. K. *J. Polym. Sci. Part A* 1963, 1, 1947.
(96) Dondos, A.; Benoit, H. *Macromolecules* 1971, 4, 279.
(97) Frisman, E. V.; Dadivanyan, A. K. *Vysokomol. Soedin.* 1966, 8, 1359.
(98) Nagai, K. *J. Chem. Phys.* 1967, 47, 4690.
(99) Gent, A. N. *Macromolecules* 1969, 2, 262.
(100) Ishikawa, T.; Nagai, K. *J. Polym. Sci., Part A-2* 1969, 7, 1123.
(101) Ishikawa, T.; Nagai, K. *Polym. J.* 1970, 1, 116.
(102) Gent, A. N.; Kuan, T. H. *J. Polym. Sci., Part A-2* 1971, 9, 927.
(103) Stein, R. S.; Hong, S. D. *J. Macromol. Sci.—Phys.* 1976, B12, 125.
(104) Erman, B.; Flory, P. J. *Macromolecules* 1983, 16, 1601.
(105) Erman, B.; Flory, P. J. *Macromolecules* 1983, 16, 1607.
(106) Llorente, M. A.; Mark, J. E.; Saiz, E. *J. Polym. Sci., Polym. Phys. Ed.* 1983, 21, 1173.
(107) Galiatsatos, V.; Mark, J. E. *Polym. Bulletin* 1987, 17, 197.
(108) Flory, P. J. *Proc. Roy. Soc. London, A* 1976, 351, 351.
(109) Flory, P. J. *Polymer* 1979, 20, 1317.
(110) Flory, P. J.; Erman, B. *Macromolecules* 1982, 15, 800.

11

Segmental Orientation

Segmental or molecular orientation refers to the anisotropic distribution of chain-segment orientations in space, due to the orienting effect of some external agent. In the case of uniaxially stretched rubbery networks, which will be the focus of this chapter, segmental orientation results from the distortion of the configurations of network chains when the network is macroscopically deformed. In the undistorted state, the orientations of chain segments are random, and hence the network is isotropic because the chain may undertake all possible configurations, without any bias. In the other hypothetically extreme case of infinite degree of stretching of the network, segments align exclusively along the direction of stretch. The mathematical description of segmental orientation at all levels of macroscopic deformation is the focus of this chapter.

Segmental orientation in rubbery networks differs distinctly from that in crystalline or glassy polymers. Whereas the chains in glassy or crystalline solids are fully or partly frozen, those in an elastomeric network have the full freedom to go from one configuration to another, subject to the constraints imposed by the network connectivity. The orientation at the segmental level in glassy or crystalline networks is mostly induced by *inter*molecular coupling between closely packed neighboring molecules, while in the rubbery network *intra*molecular conformational distributions predominantly determine the degree of segmental orientation.

The first section of this chapter describes the state of molecular deformation. In section 11.2, the simple theory of segmental orientation is outlined, followed by the more detailed treatment of Nagai[1] and Flory[2]. The chapter concludes with a discussion of infrared spectroscopy and the birefringence technique for measuring segmental orientation.

11.1 Molecular Deformation

For uniaxial deformation, the deformation tensor λ takes the form $\lambda = \mathrm{diag}(\lambda, \lambda^{-1/2}, \lambda^{-1/2})$, where diag represents the diagonal of a square matrix, and

λ is the ratio of the stretched length of the rubbery sample to its undeformed reference length. The first element along the diagonal of the matrix represents the extension ratio along the direction of stretch, which may be conveniently identified as the X axis of a laboratory-fixed frame XYZ. The other two elements refer to the deformation along two lateral directions, Y and Z.

The homogeneous macroscopic deformation of a network results in the deformation of the network chains, described by the microscopic deformation tensor, Λ^2, as

$$\Lambda^2 \equiv \begin{bmatrix} \Lambda_x^2 & 0 & 0 \\ 0 & \Lambda_y^2 & 0 \\ 0 & 0 & \Lambda_z^2 \end{bmatrix} = \begin{bmatrix} \frac{\langle x^2 \rangle}{\langle x^2 \rangle_0} & 0 & 0 \\ 0 & \frac{\langle y^2 \rangle}{\langle y^2 \rangle_0} & 0 \\ 0 & 0 & \frac{\langle z^2 \rangle}{\langle z^2 \rangle_0} \end{bmatrix} \tag{11.1}$$

Here, Λ_x, Λ_y, and Λ_z denote the components of the microscopic deformation tensor along the X, Y, and Z directions, respectively, and the quantity $\langle x^2 \rangle$ is the mean-square x-component of chain end-to-end vectors in the deformed network. The angular brackets refer to the ensemble averages over all chains, either in the deformed state or in the undeformed reference state. The latter is indicated by the subscript zero appended to the angular brackets. Network chains in the reference state will be assumed to have the so-called *unperturbed* dimensions, characteristic of *theta* conditions[3], unless the network is isotropically swollen prior to deformation. Similar definitions apply to the other two directions.

The averages depicted in eq. (11.1) may be evaluated after the choice of a molecular model of the network. Results for the affine, phantom, and constrained-junction models will be described in this section.

In the affine network model, eq. (11.1) takes the form

$$\Lambda^2 \equiv \begin{bmatrix} \lambda^2 & 0 & 0 \\ 0 & \lambda^{-1} & 0 \\ 0 & 0 & \lambda^{-1} \end{bmatrix} \tag{11.2}$$

since the chain ends transform affinely.

For the phantom network model[4], the mean chain dimensions transform affinely but the fluctuations are independent of macroscopic deformation, and therefore the deformation at the molecular level is less than that given by eq. (11.2). Writing the average squared x-component of the end-to-end vector in the deformed network as

$$\langle x^2 \rangle = \langle \bar{x}^2 \rangle + \langle (\Delta x^2) \rangle \tag{11.3}$$

and replacing the average chain dimensions and fluctuations from the equations in appendix E leads to

$$\langle x^2 \rangle = \lambda_x^2 \langle \bar{x}^2 \rangle_0 + \frac{\langle (\Delta x)^2 \rangle}{\langle (\Delta x)^2 \rangle_0} \langle (\Delta x)^2 \rangle_0 \qquad (11.4)$$

$$= \left[\lambda_x^2 \left(1 - \frac{2}{\phi}\right) + \frac{2}{\phi}\right] \langle x^2 \rangle_0$$

The second term on the right-hand side of eq. (11.4) is an identity. The ratio depicts the ratio of the x components of chain dimensions in the deformed and undeformed states. Substituting eq. (11.5) into eq. (11.1), together with similar terms for the y and z components, leads to the molecular deformation tensor Λ^2 as

$$\Lambda^2 \equiv \begin{bmatrix} (1 - 2/\phi)\lambda^2 + 2/\phi & 0 & 0 \\ 0 & (1 - 2/\phi)\lambda^{-1} + 2/\phi & 0 \\ 0 & 0 & (1 - 2/\phi)\lambda^{-1} + 2/\phi \end{bmatrix} \qquad (11.5)$$

In real networks, the state of microscopic deformation is obtained by considering the fluctuations of junctions from their mean positions, as was first done by Erman and Flory[5]. According to the constrained-junction model, the ratio shown in eq. (11.4) can be approximated by the ratio of fluctuations of junctions in the deformed and undeformed states, that is,

$$\frac{\langle (\Delta x)^2 \rangle}{\langle (\Delta x)^2_{ph} \rangle} = \frac{\langle (\Delta X)^2 \rangle}{\langle (\Delta X)^2_{ph} \rangle} \qquad (11.6)$$

The ratio on the right-hand side of eq. (11.6) is given by $1 + B_x$, where B_x is given by eq. (3.36) for the constrained-junction model, and by eq. (4.29) for the constrained-chain model. Substituting into eq. (11.6) and then into eq. (11.4), together with similar terms for the y and z components, leads to the molecular deformation tensor Λ^2 for the constraint models as

$$\Lambda^2 \equiv \begin{bmatrix} (1 - 2/\phi)\lambda^2 \\ +(2/\phi)[1 + B(\lambda)] & 0 & 0 \\ 0 & (1 - 2/\phi)\lambda^{-1} \\ & +(2/\phi)[1 + B(\lambda^{-1})] & 0 \\ 0 & 0 & (1 - 2/\phi)\lambda^{-1} \\ & & +(2/\phi)[1 + B(\lambda^{-1})] \end{bmatrix} \qquad (11.7)$$

The constrained-junction model assumes that constraints acting on junctions exert springlike forces. These forces result from the entanglements of the junctions and their pendent chains with the surrounding chains. Thus, an additional strain field surrounding each junction has to be defined. Accordingly, the state of deformation around a junction, resulting from the distortion of the constraints, is taken to be

$$\Theta^2 \equiv \begin{bmatrix} \dfrac{\langle(\Delta s_{*x})^2\rangle}{\langle(\Delta s_x)^2\rangle} & 0 & 0 \\ 0 & \dfrac{\langle(\Delta s_{*y})^2\rangle}{\langle(\Delta s_y)^2\rangle} & 0 \\ 0 & 0 & \dfrac{\langle(\Delta s_{*z})^2\rangle}{\langle(\Delta s_z)^2\rangle} \end{bmatrix} \quad (11.8)$$

which may be written

$$\Theta^2 \equiv \begin{bmatrix} 1+D(\lambda) & 0 & 0 \\ 0 & 1+D(\lambda^{-1}) & 0 \\ 0 & 0 & 1+D(\lambda^{-1}) \end{bmatrix} \quad (11.9)$$

where the function $D(\lambda)$ is given by eq. (3.38).

11.2 Segmental Orientation in Network Chains: The Simple Picture

Segmental orientation in chains of a deformed network is characterized by the spatial orientation of a given vector **u**, rigidly affixed to one or more bonds along the network chain[6,7], as shown in figure 11.1. For example, transition moment vectors **u**, rigidly attached to each repeat unit along the chain, are considered in Fourier transform infrared dichroism measurements[8], provided that the dipole moment change implied by a given infrared-active normal vibration may be ascribed to the reorientation of **u**. The X axis makes an angle χ with **u**, and Θ denotes the angle between the chain vector **r** and the X axis. The angle Φ, not

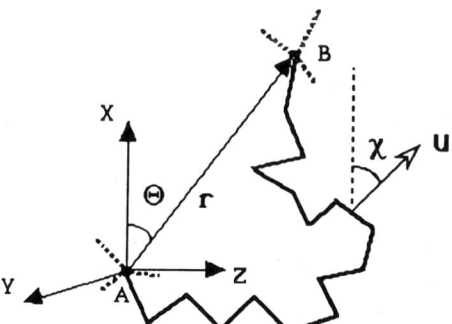

Figure 11.1 Spatial orientation of a vector **u** rigidly affixed to one or more bonds along the network chain. The broken solid line going from junction A to junction B represents a network chain. The dashed lines meeting at the two tetrafunctional junctions are bonds of other chains terminating at these junctions. The laboratory-fixed coordinate system is identified by the axes XYZ. The X axis makes an angle χ with **u**, and Θ denotes the angle between the chain vector **r** and the X axis.

shown in the figure, represents the angle between **u** and **r**, and is used in calculations of orientation as will be outlined below.

In uniaxial deformation, the X axis is conveniently identified with the direction of the applied load, which may be tensile or compressive. The orientation of **u** with respect to the X axis is expressed in terms of the orientation function $S(\chi)$, which is given by the second Legendre polynomial $P_2(\cos \chi)$ as

$$S(\chi) = \langle P_2(\cos \chi) \rangle = (1/2)(3\langle \overline{\cos^2 \chi} \rangle - 1) \tag{11.10}$$

The overbar refers to averaging over all configurations of the chain, subject to the conditions imposed on the positions of the junctions A and B. For the simplest case of the affine network model, for example, the junctions are assumed to be fixed in space. This model forms the starting point of the formulations in this chapter. The average $\langle \overline{\cos^2 \chi} \rangle$ results from two successive averages. First, the average over all configurations of a given chain with fixed ends is performed; this is designated by the overbar. Second, there is the averaging over all the chains of the network, subject to different molecular extension ratios compatible with the imposed macroscopic deformation; this is indicated by the angular brackets.

The average $\overline{\cos^2 \chi}$, appearing in the right-hand side of eq. (11.10), was first derived by Nagai[1], and subsequently in more complete form by Flory[2]. The first-order approximation for $\overline{\cos^2 \chi}$ is

$$\overline{\cos^2 \chi} = \frac{1}{3}\left\{1 + 2D_0\left[\frac{x^2}{\langle x^2 \rangle_0} - \frac{1}{2}\left(\frac{y^2}{\langle y^2 \rangle_0} + \frac{z^2}{\langle z^2 \rangle_0}\right)\right]\right\} \tag{11.11}$$

where D_0 is the configurational factor for segmental orientation given by

$$D_0 = \frac{3\langle r^2 \cos^2 \Phi \rangle_0 / \langle r^2 \rangle_0 - 1}{10} \tag{11.12}$$

For a given chain structure, the second-order moments $\langle r^2 \rangle_0$ and $\langle r^2 \cos^2 \Phi \rangle_0$ appearing in the front factor D_0 may be calculated by various theoretical schemes, among which is the rotational isomeric state formalism[2,9–12], and by Monte Carlo simulations[13,14]. The configurational factor has been shown to be inversely proportional to chain length, and is conveniently expressed as $D_0 \sim 1/n$, where n is the number of bonds in the network chain. For a freely jointed chain with N bonds, or for a real chain approximated by a freely jointed chain of N equivalent bonds ($N \leq n$), the front factor is $1/(5N)$, as was originally shown by Kuhn and Grün[15], and later elaborated by Roe and Krigbaum[16] and Jarry and Monnerie[17]. Accordingly, the front factor of $1/(5N)$ may be interpreted as the orientation modulus of a Kuhn chain. Departures of real network behavior from the simple Kuhn formulation have been discussed by Erman and Bahar[12].

The ensemble average of eq. (11.11) over all network chains in the deformed state leads to

$$\langle \overline{\cos^2 \chi} \rangle = \tfrac{1}{3}\{1 + 2D_0[\Lambda_x^2 - \tfrac{1}{2}(\Lambda_y^2 + \Lambda_z^2)]\} \tag{11.13}$$

Substitution of eq. (11.13) into eq. (11.10) leads to the segmental orientation function:

$$S = D_0[\Lambda_x^2 - \tfrac{1}{2}(\Lambda_y^2 + \Lambda_z^2)] \qquad (11.14)$$

Equation (11.14) may be expressed in terms of the macroscopic deformation tensor by substituting from eq. (11.2) for the affine network model, and from eq. (11.5) for the phantom network model. The results are

$$S(\chi) = \begin{cases} D_0(\lambda^2 - \lambda^{-1}) & \text{affine} \\ D_0(1 - 2/\phi)(\lambda^2 - \lambda^{-1}) & \text{phantom} \end{cases} \qquad (11.15)$$

Segmental orientation is thus conveniently separated into two factors: a front factor D_0 which is purely a function of the chain structure, and a deformation-dependent term which is a function of junction functionality and macroscopic deformation. The segmental orientation function for the constrained-junction model in uniaxial deformation follows as

$$S(\chi) = D_0[\Lambda_x^2 - \Lambda_y^2 + e(\Theta_x^2 - \Theta_y^2)] \qquad (11.16)$$

where e is a coefficient measuring the extent of coupling of the segment to the environment. This formulation is based on an earlier treatment of birefringence in constrained-junction networks[5,18,19].

In the case of networks swollen prior to deformation and/or cross-linked in solution, eq. (11.15) is replaced by

$$S(\chi) = \begin{cases} D_0(v_{2c}/v_2)^{2/3}(\alpha^2 - \alpha^{-1}) & \text{affine} \\ D_0(1 - 2/\phi)(v_{2c}/v_2)^{2/3}(\alpha^2 - \alpha^{-1}) & \text{phantom} \end{cases} \qquad (11.17)$$

where v_{2c} is the volume fraction of polymer during cross-linking, v_2 is its volume fraction during the stretching experiment, and α is the extension ratio relative to the swollen state.

11.3 Higher-Order Approximation for Segmental Orientation

Equation (11.11) represents a suitable first approximation for segmental orientation for relatively long chains. In a strict sense, this expression and the resulting relationship of $S(\chi)$ to deformation, given by eq. (11.17), are valid for networks with relatively long chains obeying Gaussian statistics and subjected to only small strains. Equation (11.11) consists of the first term of a series expansion in powers of $1/n$ (where n here is the number of repeat units in the chain), whose higher order terms are to be included for representing segmental orientation in shorter chains and at higher deformations.

The higher order approximation was first given by Nagai[1]. An improved version of the expression containing up to third-order terms in $1/n$ was recently obtained by Erman et al.[13]. The latter follows essentially from the work of Nagai with minor corrections, and has been adopted in the analysis of segmental orientation in short chains[14]. Here, we give the series expansion up to the second-order terms:†

$$S(\chi) = \tfrac{1}{2}[D_0(\lambda^2 - \lambda^{-1}) + D_1(\lambda^4 + \tfrac{1}{3}\lambda - \tfrac{4}{3}\lambda^{-2}) + D_2(\lambda^6 + \tfrac{3}{5}\lambda^3 - \tfrac{8}{5}\lambda^{-3})] \quad (11.18)$$

where

$$D_0 \equiv 2\eta_2 + 14\eta_4 + 126\eta_6 - 30\eta_2 g_4 \quad (11.19)$$

$$D_1 \equiv -\frac{18}{5}(\eta_4 + 18\eta_6 - 10\eta_2 g_4)\frac{\langle r^4 \rangle_0}{\langle r^2 \rangle_0^2} \quad (11.20)$$

$$D_3 \equiv \frac{54}{7}(\eta_6 - \eta_2 g_4)\frac{\langle r^6 \rangle_0}{\langle r^2 \rangle_0^3} \quad (11.21)$$

The various coefficients in eqs. (11.19)–(11.21) are defined in terms of the moments $\langle r^{2k} \rangle_0$ and $\langle r^{2k} \cos^2 \Phi \rangle_0$ of the unperturbed chains as

$$\eta_2 \equiv \frac{1}{10}\left(\frac{3\langle r^2 \cos^2 \Phi \rangle_0}{\langle r^2 \rangle_0} - 1\right) \quad (11.22)$$

$$\eta_4 \equiv \frac{1}{20}\left[\left(\frac{3\langle r^2 \cos^2 \Phi \rangle_0}{\langle r^2 \rangle_0} - 1\right) - \frac{3}{7}\left(\frac{3\langle r^4 \cos^2 \Phi \rangle_0}{\langle r^2 \rangle_0^2} - \frac{\langle r^4 \rangle_0}{\langle r^2 \rangle_0^2}\right)\right] \quad (11.23)$$

$$\eta_6 \equiv \frac{1}{80}\left[\left(\frac{3\langle r^2 \cos^2 \Phi \rangle_0}{\langle r^2 \rangle_0} - 1\right) - \frac{6}{7}\left(\frac{3\langle r^4 \cos^2 \Phi \rangle_0}{\langle r^2 \rangle_0^2} - \frac{\langle r^4 \rangle_0}{\langle r^2 \rangle_0^2}\right)\right.$$
$$\left. + \frac{1}{7}\left(\frac{3\langle r^6 \cos^2 \Phi \rangle_0}{\langle r^2 \rangle_0^3} - \frac{\langle r^6 \rangle_0}{\langle r^2 \rangle_0^3}\right)\right] \quad (11.24)$$

$$g_4 \equiv -\left(\frac{1}{2!2^2}\right)\left[1 - \frac{3\langle r^4 \rangle_0}{5\langle r^2 \rangle_0^2}\right] \quad (11.25)$$

Equation (11.18) is derived for a network whose junction points transform affinely with macroscopic deformation. The higher order expression corresponding to the phantom network is obtained by multiplying eq. (11.18) by $(1 - 2/\phi)$. The coefficient D_0 in eq. (11.18) is of the order n^{-1} while D_1 and D_2 are of the order n^{-2}. Retaining only the terms in η_2 in eq. (11.19) leads to the first-order approximation. Determination of segmental orientation up to any desired accuracy rests upon the evaluation of the statistical moments of the free chains. These moments may be evaluated, in principle, by the rotational isomeric state formalism[2,20,21], or, more easily, by the Monte Carlo technique[13,14]. For chains with N freely jointed segments, an expression for $S(\chi)$ similar to eq. (11.18) has been obtained by Roe and Krigbaum[16]:

---

†The coefficients D_0, D_1, and D_2 here were given, respectively, as D_1, D_2, and D_3 in the previous treatment.

$$S(\chi) = (1/5N)(\lambda^2 - 1/\lambda) + (1/15N^2)(\lambda^4 + \lambda/3 - 4/3\lambda^2)$$
$$+ (1/21N^3)(\lambda^6 + 3\lambda^2/5 - 8/5\lambda^3) \qquad (11.26)$$

Comparison with eq. (11.18) shows that the coefficients D_0, D_1, and D_2 for the freely jointed chain are given by $D_0 = 1/(5N)$, $D_1 = 1/(15N^2)$, and $D_2 = 1/(21N^3)$.

11.4 Experimental Determination of Segmental Orientation in Rubbery Networks

Strain birefringence experiments have been the most commonly adopted technique for determining segmental orientation in networks[5,18,19,22]. However, due to uncertainties in bond polarizabilities and the presence of large contributions from intermolecular interactions, this technique cannot reliably be used for quantitative determination of orientation of specific vectorial quantities in a network chain. For this reason, determination of segmental orientation by birefringence will not be treated here in detail. The interested reader may refer to the papers by Erman and Flory[5,18], and to the references cited therein, for either experimental details or theoretical interpretations. Polarized Fourier transform infrared (FTIR)[23-26] and deuterium nuclear magnetic resonance (^{2}H-NMR) spectroscopy[27-30], on the other hand, are two very specific and precise spectroscopic techniques[10]. In contrast to the situation with birefringence experiments, the analysis of segmental orientation with these techniques leads to information that is commensurate, in precision and in detail, with results from theoretical approaches based on the rotational isomeric state model of chain statistics[2]. Recent comparison of FTIR measurements on well-defined model poly(dimethylsiloxane) (PDMS) networks with rotational isomeric state calculations applied to segmental orientation[25], for example, gave satisfactory agreement between experiment and theory. The reader is referred to the work of Deloche et al.[31] for a comprehensive description of the application of the NMR technique to segmental orientation.

Both the FTIR and the ^{2}H-NMR techniques directly measure the orientation of specific labels on a chain relative to a laboratory-fixed axis. The orientation is conveniently induced by stretching the specimen uniaxially. Inasmuch as the applied deformation may be sustained indefinitely in a network, segmental orientation at equilibrium may be obtained by performing the measurements after allowing sufficient time for equilibration. Furthermore, the system may be swollen with a suitable diluent to eliminate local intermolecular orientational correlations[32].

The FTIR dichroism measurements permit the study of absorption bands associated with transition moments having a definite orientation with respect to the chain backbone. The vector **u** of figure 11.1 may, for instance, be assumed to be the transition moment observed in infrared measurements. As a common practice, the orientation of a *local chain axis* is considered in data interpretation, rather than the orientation of the specific transition moments. The chain axis is essentially a fictitious and somewhat ambiguous entity, defined as an axis of cylindrical symmetry with respect to **u**. If one denotes the angle between **u** and the axis of symmetry by α, and the angle between the axis of symmetry and the direction of stretch by ϑ, then the segmental orientation $S(\chi)$ may be related to the orientation function $S(\vartheta)$ of the symmetry axis by

$$S(\chi) = S(\vartheta)S(\alpha) \quad (11.27)$$

where $S(\alpha) = (1/2)(3\cos^2\alpha - 1)$, and $S(\vartheta)$ is given by an expression similar to eq. (11.10). Equation (11.27) was introduced by Fraser[33] and applied to orientation in deformed polymers by Read and Stein[34]. In the latter work, the angle ϑ has been referred to as the angle between the stretch direction and the "segments" of the chains.

For incident radiation polarized along the direction of stretching (identified above with the X axis), the absorbance of **u** may be resolved into two components: $a_\parallel$ parallel to this axis, and $a_\perp$ perpendicular to it. In terms of the angle χ, $a_\parallel$ and $a_\perp$ can be expressed as

$$a_\parallel = |\mathbf{u}|^2 \cos^2\chi$$
$$a_\perp = (1/2)|\mathbf{u}|^2 \sin^2\chi \quad (11.28)$$

The factor $1/2$ in eq. (11.28) results from averaging of all rotations of **u** about the stretch direction, which constitutes an axis of cylindrical symmetry. The closely related dichroic ratio R, measured in infrared studies, is defined as

$$R = \int_0^\pi a_\parallel h(\chi) \sin\chi\, d\chi \Big/ \int_0^\pi a_\perp h(\chi) \sin\chi\, d\chi \quad (11.29)$$

where $h(\chi)$ is the distribution function for the χ angles. Substituting eq. (11.28) into eq. (11.29), and using eq. (11.10) then leads to

$$S(\chi) = \frac{R-1}{R+2} \quad (11.30)$$

The components $a_\parallel$ and $a_\perp$ may alternatively be expressed in terms of the angles α and ϑ. In this case, the dichroic ratio takes the form[35,36]

$$R = \frac{1 + (2\cot^2\alpha - 1)\langle\cos^2\vartheta\rangle}{1 + (\cot^2\alpha - 1/2)(1 - \langle\cos^2\vartheta\rangle)} \quad (11.31)$$

which, using the expression for $S(\vartheta)$, leads to

$$S(\vartheta) = \frac{2}{3\cos^2\alpha - 1} \frac{R-1}{R+2} \quad (11.32)$$

Equation (11.32) is implicitly based on the presence of a chain-embedded local axis, with respect to which **u** is assumed to undergo cylindrically symmetric rotations.

11.5 Theoretical Interpretation of Infrared Dichroism Measurements of Segmental Orientation in Rubbery Networks

Strain-orientation data may suitably be interpreted in terms of the reduced orientation, $[S]$, defined as[6–8]

$$[S] \equiv \frac{S}{(\lambda^2 - \lambda^{-1})} = \frac{S}{(v_{2c}/v_2)^{2/3}(\alpha^2 - \alpha^{-1})} = \begin{cases} (1 - 2/\phi)D_0 & \text{Phantom} \\ D_0 & \text{Affine} \end{cases} \quad (11.33)$$

Defined in this manner, in analogy with the definition of the reduced force, $[S]$ may be seen as the orientational modulus of the network. In the first-order approximation, given by eq. (11.33), the reduced orientation is independent of deformation for both the affine and the phantom network models.

Results of polarized FTIR experiments on well-characterized PDMS networks from the work of Besbes et al.[25] are presented in figure 11.2, where the reduced orientation is plotted as a function of inverse extension ratio. Although agreement between experiment and theory appears reasonable, the data points are always

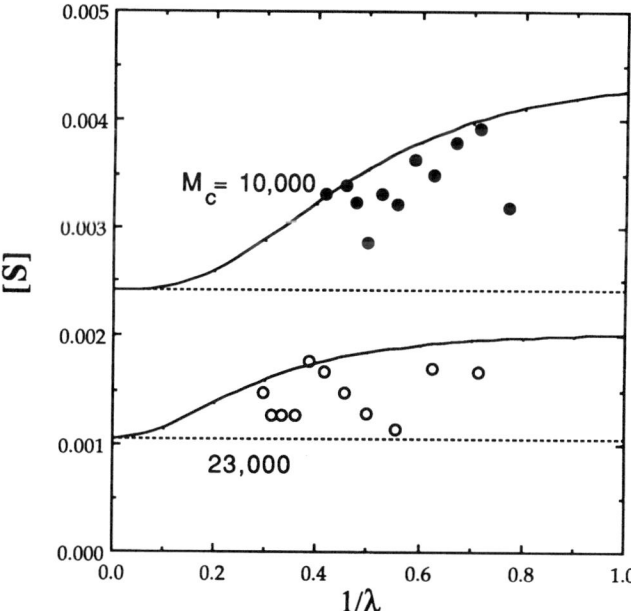

Figure 11.2 Results of polarized FTIR experiments on poly(dimethylsiloxane) PDMS networks[25]. The reduced orientation is shown along the ordinate as a function of inverse extension ratio. The circles represent experimental data for two networks with the indicated molecular weight M_c between cross-links. The horizontal dashed lines represent predictions for the phantom network model from the second of part eq. (11.33), with $\phi = 4$ and D_0 obtained theoretically from eq. (11.12) ($D_0 = 1.045$ for $M_c = 23,000$, and 2.405 for $M_c = 10,000 \, \text{g mol}^{-1}$).

above the phantom network prediction. Experiments on other systems have exhibited the same characteristic[18,23]. The departure from phantom network behavior is generally attributed to the effects of entanglements that constrain fluctuations at the molecular level[5,11]. This explanation has been recently verified for segmental orientation in well-defined PDMS networks, measured by polarized FTIR[38]. The curves in figure 11.2 illustrate the predictions of the constrained-junction model of rubber elasticity, applied to the segmental orientation of the PDMS networks with specified molecular weight and junction functionality[37]. It should be noted that the prediction for the constrained-junction model converges to that of the phantom network in the limit of infinite extension.

The dependence of segmental orientation on chain length is shown in Figure 11.3. The circles are from polarized FTIR experiments of Besbes et al.[25] for four different networks. The affine and phantom predictions, shown by the straight lines, are obtained from eq. (11.33). The Kuhn model prediction is obtained by simply taking D_0 as $1/(5N)$, where the number of bonds in the Kuhn segments is taken to be equal to 17 (i.e., $n/N = 17$), following the work of Flory and Chang[38]. This comparison shows that experimental data yield results closer to the phantom network model, and that the Kuhn model predicts much higher orientations. A critique of segmental orientation based on the Kuhn model has been presented elsewhere[12].

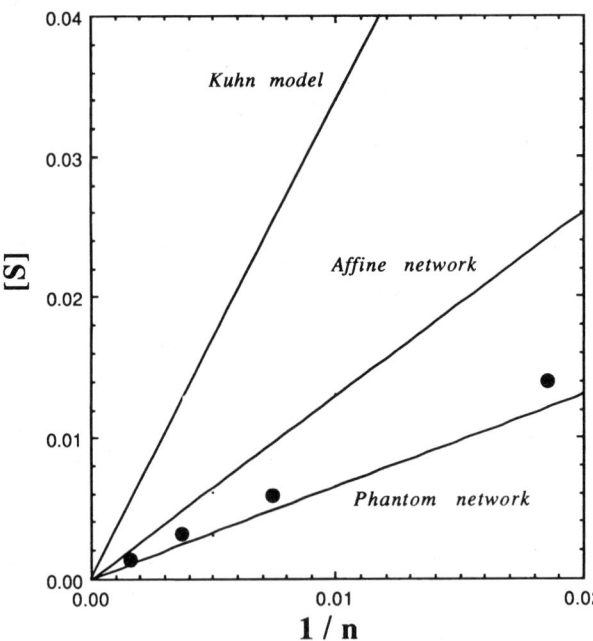

Figure 11.3 Dependence of segmental orientation on chain length. The circles are from polarized FTIR experiments[25] for four different networks. The Kuhn, affine, and phantom predictions are shown by the straight lines.

The results of the FTIR measurements on PDMS networks presented in the preceding paragraph are also in reasonable agreement with the ^{2}H-NMR data of Deloche et al.[27,28,31].

Although predictions of the theory[11] based on the constrained-junction model of rubber elasticity[39] agree reasonably well with experiment[23,37], a complete understanding of orientation in non-phantomlike networks is still lacking. In figure 11.4, for example, the reduced orientation obtained from eq. (11.18) with higher order terms is shown by the curve. The horizontal line is obtained by retaining only the first term and corresponds to the first-order theory. The strong upturn of the reduced orientation with higher order terms is representative of non-Gaussian contributions at large extensions. Such an upturn is not observed, however, in experimental data obtained from noncrystallizing networks.

A further contribution to orientation associated with trapped entanglements has been suggested by Herz and Deloche and their collaborators[40–42]. These contributions are asserted to persist, even in the swollen network, but their effects should vanish if the networks had been formed in the highly diluted state. One might therefore separate the contributions from intermolecular sources into two parts—a local and an entanglement component—and express the effective configurational factor as

$$D_0^{\text{eff}} = D_0 + D_1^{\text{int}} v_2 + D_2^{\text{int}} v_{2c} \qquad (11.34)$$

where D_1^{int} reflects the contributions from local intermolecular correlations and D_2^{int} is due to trapped entanglements. Strong contributions from D_1^{int} have been

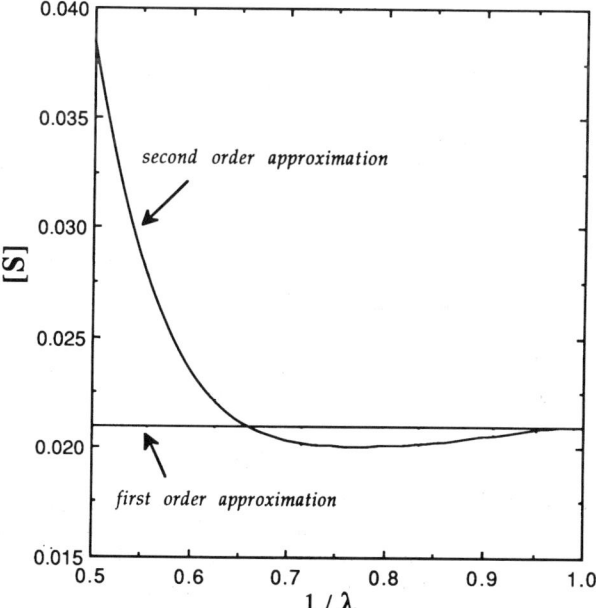

Figure 11.4 Reduced orientation obtained from eq. (11.18) with higher order terms.

reported for the orientation of polyisoprene (PIP) networks by polarized fluorescence measurements[23]. Results of FTIR experiments on swollen PDMS networks show, on the other hand, that D_1^{int} is not significant for this system[32]. The effects of swelling on segmental orientation are compared in figure 11.5 for PIP and PDMS networks. A very strong decrease in [S] is observed for small amounts of swelling as is shown in the figure; this could not be theoretically reproduced without a modification in the configurational front factor. This decrease would then be attributed to the disappearance of intermolecular effects upon swelling. The lowest theoretical curve and the indicated experimental data points were obtained for the PDMS networks with $v_2 = 1$ and 0.56^{37}. The differences between the two data sets are not discernible in the figure, in agreement with the predictions of the constrained-junction model. This indicates that intermolecular contributions to the configurational factor D are negligible in PDMS networks, in contrast to those in PIP. Calculations performed for PDMS networks having

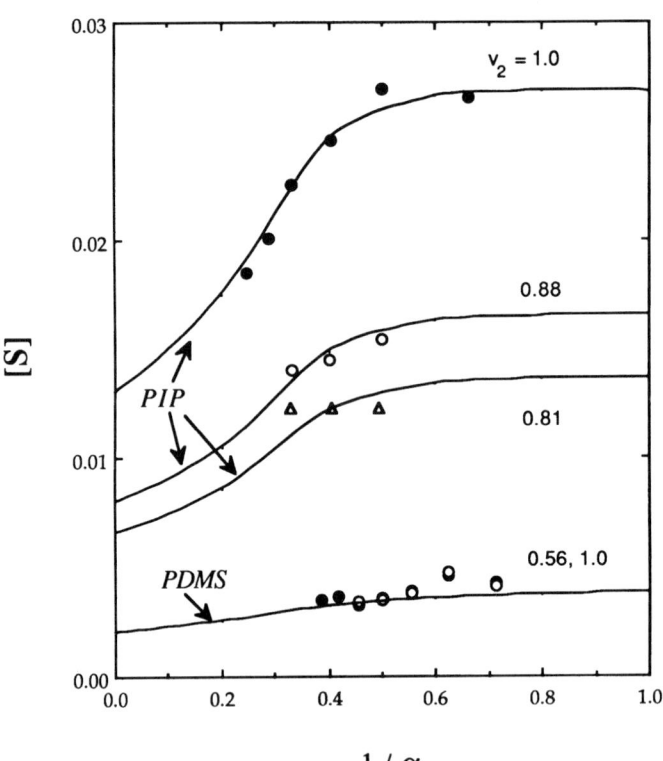

Figure 11.5 Effects of swelling on segmental orientation for polyisoprene (PIP) and PDMS networks. The upper set of filled points represent experimental measurements on a dry PIP network. The curve through the points was obtained using the constrained-junction model[23]. Results for PIP networks in the swollen state, with $v_2 = 0.88$ and 0.81, are shown by the open circles and triangles, respectively. The lower set of points is for the PDMS chains.

various solvent contents during cross-linking, that is, with different v_{2c} values, also yielded good agreement between theory and experiments[37], confirming further the negligibly weak effect of intermolecular interactions on segmental orientation in this polymer.

References

(1) Nagai, K. *J. Chem. Phys.* 1964, 40, 2818.
(2) Flory, P. J. *Statistical Mechanics of Chain Molecules.* Interscience: New York. 1969.
(3) Flory, P. J. *Principles of Polymer Chemistry.* Cornell University Press: Ithaca, NY. 1953; p. 351.
(4) Flory, P. J. *Proc. Roy. Soc. London, A* 1976, 351, 351.
(5) Erman, B.; Flory, P. J. *Macromolecules* 1983, 16, 1601.
(6) Bahar, I.; Erman, B.; Bokobza, L.; Monnerie, L. *Macromolecules* 1995, 28, 225.
(7) Sarac, Z.; Erman, B.; Bahar, I. *Macromolecules* 1995, 28, 582.
(8) Besbes, S.; Bokobza, L.; Monnerie, L.; Bahar, I.; Erman, B. *Macromolecules* 1995, 28, 231.
(9) Flory, P. J. *Macromolecules* 1974, 7, 381.
(10) Mark, J. E.; Erman, B. *Rubberlike Elasticity. A Molecular Primer.* Wiley-Interscience: New York. 1988.
(11) Erman, B.; Monnerie, L. *Macromolecules* 1985, 18, 1985.
(12) Erman, B.; Bahar, I. *Macromolecules* 1988, 21, 452.
(13) Erman, B.; Haliloglu, T.; Bahar, I.; Mark, J. E. *Macromolecules* 1991, 24, 901.
(14) Haliloglu, T.; Bahar, I.; Erman, B. *Comp. Polym. Sci.* 1991, 1, 151.
(15) Kuhn, W.; Grün, F. *Kolloid-Z.* 1942, 101, 248.
(16) Roe, R.-J.; Krigbaum, W. R. *J. Appl. Phys.* 1964, 35, 2215.
(17) Jarry, J. P.; Monnerie, L. *Macromolecules* 1979, 12, 316.
(18) Erman, B.; Flory, P. J. *Macromolecules* 1983, 16, 1607.
(19) Erman, B.; Queslel, J. P. In *Encyclopedia of Materials Science and Engineering.* John Wiley & Sons, New York. 1989; Supplement vol.; 2nd ed.; p. 18.
(20) Flory, P. J.; Yoon, D. Y. *J. Chem. Phys.* 1974, 61, 5358.
(21) Yoon, D. Y.; Flory, P. J. *J. Chem. Phys.* 1974, 61, 5366.
(22) Treloar, L. R. G. *The Physics of Rubber Elasticity*, 3rd ed. Clarendon Press: Oxford. 1975.
(23) Queslel, J. P.; Erman, B.; Monnerie, L. *Macromolecules* 1985, 18, 1991.
(24) Amram, B.; Bokobza, L.; Queslel, J. P.; Monnerie, L. *Polymer* 1986, 27, 877.
(25) Besbes, S.; Cermelli, I.; Bokobza, L.; Monnerie, L.; Bahar, I.; Erman, B.; Herz, J. *Macromolecules* 1992, 25, 1949.
(26) Noda, I.; Dowrey, A. E.; Marcott, C. In *Physical Properties of Polymers Handbook*, J. E. Mark, Ed. American Institute of Physics: Woodbury, NY. 1996; p. 291.
(27) Dubault, A.; Deloche, B.; Herz, J. *Polymer* 1984, 25, 1405.
(28) Dubault, A.; Deloche, B.; Herz, J. *Macromolecules* 1987, 20, 2096.
(29) Sotta, P.; Deloche, B. *Macromolecules* 1990, 23, 1999.
(30) Tonelli, A. E. In *Physical Properties of Polymers Handbook*, J. E. Mark, Ed. American Institute of Physics: Woodbury, NY. 1996; p. 271.
(31) Deloche, B.; Dubault, A.; Herz, J.; Lapp, A. *Europhys. Lett.* 1986, 1, 629.

(32) Besbes, S.; Bokobza, L.; Monnerie, L.; Bahar, I.; Erman, B. *Polymer* 1993, 34, 1179.
(33) Fraser, R. D. B. *J. Chem. Phys.* 1953, 21, 1511.
(34) Read, B. E.; Stein, R. S. *Macromolecules* 1968, 1, 116.
(35) Jasse, B.; Koenig, J. L. *J. Macromol. Sci. Macromol. Chem.* 1979, C17, 61.
(36) Jasse, B.; Tassin, J. F.; Monnerie, L. *Prog. Colloid Polym. Sci.* 1993, 92, 8.
(37) Erman, B.; Bahar, I.; Besbes, S.; Bokobza, L.; Monnerie, L. *Polymer* 1993, 34, 1858.
(38) Flory, P. J.; Chang, W. C. *Macromolecules* 1976, 9, 33.
(39) Erman, B.; Flory, P. J. *Macromolecules* 1982, 15, 800.
(40) Herz, J.; Munch, J. P.; Candau, S. *J. Macromol. Sci.—Phys.* 1980, B18(2), 267.
(41) Deloche, B.; Samulski, E. T. *Macromolecules* 1981, 14, 575.
(42) Dubault, A.; Deloche, B.; Herz, J. *Macromolecules* 1987, 20, 2096.

12

Networks with Semiflexible Chains and Networks Exhibiting Strain-Induced Crystallization

Classical theories of rubber elasticity are based on models of flexible polymer chains that are sufficiently long to exhibit Gaussian behavior as described in chapter 1 and in appendix F. Additionally, the chains are phantomlike in the sense that they do not interact with one other along their contours. The theories described in chapter 2 were based on this picture of the individual chain. In this chapter, we describe the elasticity of networks that depart substantially from those addressed in the classical theories. The departures may result from two sources: (1) the chains may be only semiflexible, as a result of which the segments of neighboring chains compete for space in the deformed network, and choose preferentially oriented configurations, and (2) the chains may form crystallites, upon deformation, as a result of which the homogeneous structure of the classical network model may be transformed into a nonhomogeneous one having microphases of crystalline and amorphous regions.

The subject of crystallization under deformation, for networks in general, is relatively old, and has been treated in some detail in previous studies[1-4]. For this reason, crystallization and some of its effects will be reviewed only briefly at the end of this chapter. The main emphasis will be given to networks with semiflexible chains[5].

12.1 Networks with Semiflexible Chains

Examples of networks with semiflexible chains are those in which the chains have rodlike segments separated by flexible spacers, or those where the chains have bond angles appreciably larger than tetrahedral. Incorporation of these chains into a network structure results in materials that exhibit segmental orientations significantly larger than those shown by classical networks. Specific examples would include networks prepared from aromatic polyamide chains[5-7] or from chains containing liquid-crystalline sequences along the direction of the backbone[8-10]. Because of their unique chain structures, these networks are easily orientable, at the molecular level, by macroscopic deformations. The orientational transitions may easily be controlled by application and removal of anisotropic strains, and are therefore of great technological interest for use in optical devices. Other examples of networks with easily orientable chains are those with rigid sequences in the side groups. These materials, referred to as side-chain liquid-crystalline networks, also display features that may be described by the theory to be presented here. Such liquid-crystalline networks have aroused wider interest than those of the main-chain type, perhaps due to the ease of their preparation, characterization, and handling[11-20]. More recent work in this area covers a variety of structures, with a number of novel physical properties[21-41].

The idea that mechanical stress applied to a cross-linked nematic polymer may induce an isotropic–nematic phase transition was first formulated by de Gennes[42], and was followed by several additional theoretical treatments[43-53]. The basic molecular effect in deformation of networks with semiflexible chains involves the fact that the chains are embedded in a lattice which deforms with macroscopic strain, a feature that is absent in the phantom network model described in chapter 2. This liquidlike picture of the deforming network was first treated with a lattice model by DiMarzio[54], followed by Tanaka and Allen[55], and Jarry and Monnerie[56]. These theories predict increases in segmental orientation due to local intermolecular correlations. However, the results do not show a dependence on the stiffness of the chains, in spite of its expected importance as a parameter. The theory of Warner and collaborators[49,57], and the theory of Erman and collaborators[47,48], incorporate the stiffness of the segments into the network free energy and therefore provide more realistic descriptions of the phenomenon. The theory of Warner and collaborators introduces the concept of a semiflexible wormlike chain into the calculation of the elasticity of a network, while the theory of Erman and collaborators is based on the semiflexible chain lattice model of Flory[58,59]. The predictions of the two theories are quite similar. The lattice model is chosen to illustrate work in this area, and is described in detail below.

12.1.1 The Lattice Model for the Semiflexible Chain

In this section, we describe the network as a lattice into which n_2 freely jointed chains are embedded, together with n_1 solvent molecules. The characteristic size of a lattice site is taken as the size of a solvent molecule. A chain refers to a network chain linked to the network at its two ends by active junctions. Each network chain contains m freely jointed segments, referred to as "Kuhn segments," of length l. Depending on the degree of stiffness of the chain, the length l may be significantly larger than that obtained for ordinary flexible chains. The important quantity for theis theory is the axial ratio x, denoting the ratio of the length of a segment to its lateral dimension. The lateral dimensions of the rods are assumed to be of the size of solvent molecules or lattice sites.

Figure 12.1(a) displays an illustrative network chain consisting of 20 Kuhn segments (i. e., freely jointed rods) between two tetrafunctional junctions A and B. For simplicity, the network chains are assumed to be monodisperse (i. e., composed of the same number m of rods, with all rods having identical values of the axial ratio x). Following an early treatment by Flory[60], the model may

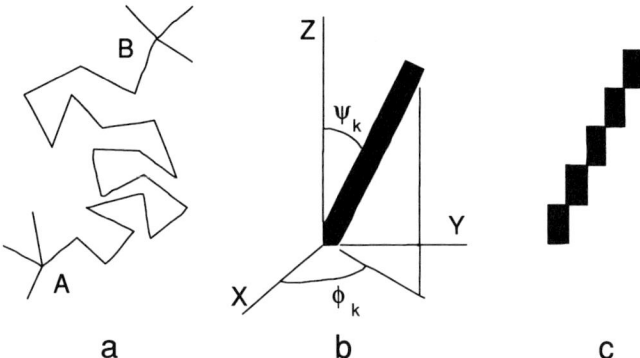

Figure 12.1 (a) An illustrative network chain consisting of 20 Kuhn segments between the two tetrafunctional junctions A and B[47]. (b) The orientation of a given segment as defined by the two Euler angles, ψ_k and ϕ_k. The Z direction is the preferred direction, that is, the direction of stretch. (c) A sequence of $y_k = x \sin \psi_k$ submolecules, occupying each x/y_k site, and oriented along the preferred direction. Reprinted with permission from Erman, B., et al. (1990), *Macromolecules*, **23**. Copyright 1997 American Chemical Society.

easily be adapted to the case of more than one lattice site occupied by either solvent molecules or the rodlike segments in the lateral direction. The orientation of a given segment shown in figure 12.1(b) is defined by the two Euler angles, ψ_k and ϕ_k. The Z direction in the figure indicates the preferred direction, also referred to as the domain axis. In simple extension, this direction coincides with the direction of stretch. The subscript k indicates that the orientation of the segment lies within the kth solid angle. In conformity with the lattice treatment of Flory[58,59,61], the accommodation of the rod within the lattice is achieved through its representation by a sequence of $y_k = x \sin \psi_k$ submolecules, occupying x/y_k sites and oriented along the preferred direction as shown in figure 12.1(c). Thus, y_k characterizes the orientation of the given rod. It is expressed in terms of ψ_k and ϕ_k as[59]

$$y_k = x \sin \psi_k (|\cos \phi_k| + |\sin \phi_k|) \tag{12.1}$$

for the rod exhibiting that particular orientation. According to this definition, the value of y_k increases as the rod becomes disoriented, and is therefore referred to as the "disorientation index."

The junctions of the network at the extremities of each chain are assumed to displace affinely with the macroscopic deformation. This assumption, which is also employed in the classic paper by Kuhn and Grün[62] and the affine network model of Wall and Flory[63], is not exactly correct but simplifies the problem considerably. Improvements in the theory to include non-affine displacements of the junctions may be considered, however, in a mean-field approach, as has been done previously. Additionally, the volume of the network is assumed to be constant throughout the deformation.

12.1.2 The Partition Function and the Free Energy of Mixing of Network Chains

This discussion adapts the Flory lattice treatment of semiflexible chains to networks. The reader is referred to the literature[58] for further details concerning the lattice model. For a system of n_2 network chains and n_1 solvent molecules in a lattice consisting of n_0 sites, the respective volume fractions are given by

$$v_2 = mxn_2/n_0$$
$$v_1 = n_1/n_0 \qquad (12.2)$$

The expected number ν_{j+1} of ways for adding the $(j+1)$st chain molecule to n_0 lattice sites with mxj of them already occupied by j chain molecules can be obtained from the lattice theory of rodlike particles[58]. The result is

$$\nu_{j+1} = (z-1)^{m-1} \frac{(n_0 - mxj)![n_0 - m(x-\bar{y})(j+1)]!}{[n_0 - mx(j+1)]![n_0 - m(x-\bar{y})j]!} \frac{1}{n_0^{m\bar{y}-1}} \qquad (12.3)$$

where z is the coordination number of the lattice, and the mean disorientation index $\bar{y}$ is defined by

$$\bar{y} = \frac{1}{n_2 m} \sum_{l=1}^{n_2} \sum_k n_{l,k} y_k \qquad (12.4)$$

Here, $n_{l,k}$ indicates the number of segments of the lth chain whose orientation lies within the kth solid angle, and y_k is the corresponding disorientation index. The index k in the last summation denotes all sets of solid angles available to the rods.

The computation of ν_{j+1} enables us to calculate the combinatorial part Z_{comb} of the partition function for the lattice from

$$Z_{\text{comb}} = \left(\prod_{j=1}^{n_2} \nu_j \right) / n_2! \qquad (12.5)$$

The orientational part of the partition function is

$$Z_{\text{orient}} = \frac{1}{n_2!} \prod_{j=1}^{n_2} m! \prod_k \frac{\omega_k^{n_{j,k}}}{n_{j,k}!} \qquad (12.6)$$

where ω_k is the kth fractional range of the solid angle. The total configuration partition function of the mixture is found from

$$Z_m = Z_{\text{comb}} Z_{\text{orient}} \qquad (12.7)$$

In the original treatment of the network problem[47], the additional contribution to the partition function resulting from the juxtaposition of the ends of chains at the sites of junctions was not considered. Inasmuch as this term is independent of orientation, its omission does not affect the calculations of segmental orientation and nematic–isotropic transitions presented below. The term $-\ln Z_{\text{comb}}$, which will be used in the calculations of the averages given below, may be written in simplified form by substituting eq. (12.3) into eq. (12.5) and using Stirling's approximation:

$$-\ln Z_{\text{comb}} = n_1 \ln v_1 + n_2 \ln(v_2/mx) - (n_1 + n_2 m\bar{y}) \ln[1 - v_2(1 - \bar{y}/x)] \\ + n_2(m\bar{y} - 1) - n_2(m-1)\ln(z-1) \quad (12.8)$$

Similarly, $-\ln Z_{\text{orient}}$ is obtained from eq. (12.6) as

$$-\ln Z_{\text{orient}} = n_2 \ln n_2 - n_2 - n_2 m \ln m + \sum_{j=1}^{n_2} \sum_k n_{j,k} \ln \frac{n_{j,k}}{\omega_k} \quad (12.9)$$

Equations (12.8) and (12.9) may be substituted into the expression $\Delta A_m = -k_B T \ln Z_m$ to obtain the free energy of mixing.

12.1.3 A Set of Nonlinear Equations for Evaluating the Orientational Distribution under Deformation

Characterization of the orientational distribution function $n_{j,k}$ is the first step for solving the problem outlined in the preceding section. In this section, we will show that the solution requires the determination of four Lagrange multipliers to be evaluated from a nonlinear equation. The formal solution is outlined here and the results for the approximate closed-form solution to the linearized equation are presented in section 12.1.4.

If there are no constraints imposed on the system, then an equilibrium distribution of rods among different orientations is obtained by minimizing the free energy of the system with respect to $n_{j,k}$. The imposition of external constraints requires the use of Lagrange multipliers while minimizing the free energy. In the treatment here, the end-to-end vectors for each of the n_2 chain molecules are assumed to be fixed. This condition may be expressed by the three equations

$$\frac{r_{Z,j}}{l} = \sum_k n_{j,k} \cos \psi_k \quad (12.10)$$

$$\frac{r_{X,j}}{l} = \sum_k n_{j,k} \sin \psi_k \cos \phi_k \quad (12.11)$$

$$\frac{r_{Y,j}}{l} = \sum_k n_{j,k} \sin \psi_k \sin \phi_k \quad (12.12)$$

where, $r_{X,j}$, $r_{Y,j}$, and $r_{Z,j}$ are the X, Y, and Z components of the end-to-end vector $\mathbf{r}$ for the jth molecule, and l is the length of a rod. It should be noted that the constraints given by eqs. (12.10)–(12.12), where each cartesian component of the end-to-end vector is kept fixed, are different from those of the Kuhn and Grün treatment[62]. In the latter, only the component along the direction of stretch is fixed. The distribution of $n_{j,k}$ is additionally subject to the constraint

$$m = \sum_k n_{j,k} \quad (12.13)$$

The equilibrium distribution of rods between different orientations with constraints given by eqs. (12.10)–(12.13) is determined by differentiating the Helmholtz free energy expression and equating to zero:

$$\frac{\partial}{\partial n_{j,k}}\left[\ln Z_{\text{comb}} + \ln Z_{\text{orient}} + \alpha \sum_{k'} n_{j,k'} + \beta \sum_{k'} n_{j,k'} \cos\psi_{k'}\right.$$
$$\left. + \gamma \sum_{k'} n_{j,k'} \sin\psi_{k'} \cos\phi_{k'} + \delta \sum_{k'} n_{j,k'} \sin\psi_{k'} \sin\phi_{k'}\right]_{n_1, n_2, n_{j,k'}} = 0 \quad (12.14)$$

Here, α, β, γ, and δ are Lagrange multipliers incorporating the conditions imposed by eqs. (12.10)–(12.14), and the summations are performed over all orientations k' different from k. Substituting $-\ln Z_{\text{comb}}$ and $-\ln Z_{\text{orient}}$ given by eqs. (12.8) and (12.9), respectively, and performing the above differentiation, leads to the following expression for the distribution $n_{j,k}$ of rods:

$$n_{j,k} = \omega_k \exp[-ay_k + \alpha - 1 + \beta\cos\psi_k + \gamma\sin\psi_k \cos\phi_k + \delta\sin\psi_k \sin\phi_k] \quad (12.15)$$

where

$$a = -\ln\left[1 - v_2\left(1 - \frac{\bar{y}}{x}\right)\right] \quad (12.16)$$

It should be noted that Z_{comb} depends implicitly on $n_{j,k}$ through $\bar{y}$, and the partial derivative $\partial \bar{y}/\partial n_{j,k} = y_k/n_2 m$ has been used in deriving eq. (12.15). This relies on the reasonable approximation that the distribution of orientations for a given rod can be equated to that of the whole ensemble. This approximation becomes rigorous when the network chain is sufficiently long. Equation (12.15) reflects the probability distribution of different orientations among the rods composing the jth chain. It results from two effects: (1) orientation due to stretching of the chain from its two ends, that is, orientation of the network chain in the gaslike medium, and (2) orientation resulting from intermolecular correlations or from local exclusion of the volume of one segment to its neighbors. The latter effect enters eq. (12.15) through the ay_k term in the exponent. In the absence of intermolecular correlations, the difference between $\bar{y}$ and x introduced by the lattice model vanishes. Equating $\bar{y}/x$ to unity in eq. (12.16) leads to $a = 0$. The distribution given by eq. (12.15) then reduces to

$$n_{j,k;\text{isotropic}} = \frac{\sin\psi_k d\psi_k d\phi_k}{4\pi} \exp(\alpha')\exp[\beta\cos\psi_k + \gamma\sin\psi_k \cos\phi_k + \delta\sin\psi_k \sin\phi_k] \quad (12.17)$$

In eq. (12.17), the solid angle ω_k of eq. (12.15) has been replaced by $(1/4\pi)\sin\psi_k d\psi_k d\phi_k$; $\alpha - 1$ in eq. (12.15) has been replaced by α'. Correspondence to the Kuhn and Grün result[62] is obtained by equating both γ and δ in eq. (12.17) to zero.

The Lagrange multipliers α, β, γ, and δ are to be determined from eqs. (12.10)–(12.13). Using eq. (12.15) and replacing summations over directions by integrals, the conditions imposed by the constrains may be written as

$$m = \frac{1}{4\pi} \int_0^{2\pi} d\phi_k \int_0^{\pi} d\psi_k \sin \psi_k \qquad (12.18)$$
$$\times \exp\{-ay_k - 1 + \alpha + \beta \cos \psi_k + \gamma \sin \psi_k \cos \phi_k + \delta \sin \psi_k \sin \phi_k\}$$

$$\frac{r_{Z,j}}{l} = \frac{1}{4\pi} \int_0^{2\pi} d\phi_k \int_0^{\pi} d\psi_k \sin \psi_k \cos \psi_k$$
$$\times \exp\{-ay_k - 1 + \alpha + \beta \cos \psi_k + \gamma \sin \psi_k \cos \phi_k + \delta \sin \psi_k \sin \phi_k\}$$
$$(12.19)$$

$$\frac{r_{X,i}}{l} = \frac{1}{4\pi} \int_0^{2\pi} d\phi_k \int_0^{\pi} d\psi_k \sin^2 \psi_k \cos \phi_k$$
$$\times \exp\{-ay_k - 1 + \alpha + \beta \cos \psi_k + \gamma \sin \psi_k \cos \phi_k + \delta \sin \psi_k \sin \phi_k\}$$
$$(12.20)$$

$$\frac{r_{Y,i}}{l} = \frac{1}{4\pi} \int_0^{2\pi} d\phi_k \int_0^{\pi} d\psi_k \sin^2 \psi_k \sin \phi_k$$
$$\times \exp\{-ay_k - 1 + \alpha + \beta \cos \psi_k + \gamma \sin \psi_k \cos \phi_k + \delta \sin \psi_k \sin \phi_k\}$$
$$(12.21)$$

It is noted that the Lagrange multiplier α may be eliminated from eqs. (12.19)–(12.21) by dividing each of them by m, which is given by eq. (12.18). A further simplification is made in the theory by equating $r_{X,j}$, $r_{Y,j}$, and $r_{Z,j}$ to their root mean-square values, that is,

$$r_{Z,j} = \langle r_Z^2 \rangle^{1/2}$$
$$r_{X,j} = \langle r_X^2 \rangle^{1/2} \qquad (12.22)$$
$$r_{Y,j} = \langle r_Y^2 \rangle^{1/2}$$

The subscript j in eqs. (12.19)–(12.21) may be omitted in the case of these approximations.

For freely jointed chains under affine deformation, the mean-square components of **r** are related to those in the free state by

$$\langle r_Z^2 \rangle = \lambda_Z^2 \langle r_Z^2 \rangle_0 = \lambda_Z^2 ml^2/3$$
$$\langle r_X^2 \rangle = \lambda_X^2 \langle r_X^2 \rangle_0 = \lambda_X^2 ml^2/3 \qquad (12.23)$$
$$\langle r_Y^2 \rangle = \lambda_Y^2 \langle r_Y^2 \rangle_0 = \lambda_Y^2 ml^2/3$$

where $\langle r_i^2 \rangle_0$ is the mean-square ith component of **r** prior to deformation, and λ_i is the extension ratio along the direction i, defined as the ratio of the final length to the reference length. For uniaxial deformation,

$$\lambda_Z = \lambda$$
$$\lambda_X = \lambda_Y = 1/(\lambda v_2)^{1/2} \qquad (12.24)$$

172 STRUCTURES AND PROPERTIES OF RUBBERLIKE NETWORKS

Substitution from eq. (12.24) into eq. (12.23) leads to

$$\langle r_Z^2 \rangle^{1/2}/l = \lambda(m/3)^{1/2} \tag{12.25}$$

$$\langle r_X^2 \rangle^{1/2}/l = \langle r_Y^2 \rangle^{1/2}/l = (m/3\lambda v_2)^{1/2} \tag{12.26}$$

The Lagrange multipliers γ and δ are equal to each other for uniaxial deformation. Combining eqs. (12.25) and (12.26) with eqs. (12.18)–(12.22) in this particular case leads to the expression

$$\frac{\lambda}{(3m)^{1/2}} = \frac{\int_0^{2\pi} d\phi \int_0^{\pi} d\psi \cos\psi f(\phi, \psi)}{\int_0^{2\pi} d\phi \int_0^{\pi} d\psi f(\phi, \psi)} \tag{12.27}$$

and

$$\frac{1}{(3m\lambda v_2)^{1/2}} = \frac{\int_0^{2\pi} d\phi \int_0^{\pi} d\psi \sin\psi \cos\phi f(\phi, \psi)}{\int_0^{2\pi} d\phi \int_0^{\pi} d\psi f(\phi, \psi)} \tag{12.28}$$

where $f(\phi, \psi)$ is the orientational distribution function given by

$$f(\phi, \psi) = \sin\psi \exp[-ax \sin\psi[|\cos\phi| + |\sin\phi|] \\ + \beta\cos\psi + \gamma\sin\psi(\cos\phi + \sin\phi)] \tag{12.29}$$

The dummy subscript k has been omitted in writing eqs. (12.27) and (12.28). These two equations additionally require knowledge of $\bar{y}$ since the latter is contained through a, as defined by eq. (12.16). Equations (12.15) and (12.4) lead to

$$\frac{\bar{y}}{x} = \frac{\int_0^{2\pi} d\phi \int_0^{\pi} d\psi \sin\psi[|\cos\phi| + |\sin\phi|] f(\phi, \psi)}{\int_0^{2\pi} d\phi \int_0^{\pi} d\psi f(\phi, \psi)} \tag{12.30}$$

Equations (12.27), (12.28), and (12.30) form a system of three nonlinear equations to be solved simultaneously for $\bar{y}$, β, and γ, which are all functions of λ. The evaluation of $\bar{y}$, β, and γ enables one to calculate other quantities of interest, such as

$$\langle \cos\psi \rangle = \frac{1}{m}\sum_{i=1}^{m} \cos\psi_i = \frac{1}{m}\sum_k n_{j,k} \cos\psi_k = \frac{\lambda}{(3m)^{1/2}} \tag{12.31}$$

and

$$\langle \cos^2\psi \rangle = \frac{\int_0^{2\pi} d\phi \int_0^{\pi} d\psi \cos^2\psi f(\phi, \psi)}{\int_0^{2\pi} d\phi \int_0^{\pi} d\psi f(\phi, \psi)} \tag{12.32}$$

12.1.4 The Linearized Closed-Form Solution

Solution of eqs. (12.27), (12.28), and (12.30) requires somewhat extensive numerical calculations. The equations may, however, be solved analytically for small β and γ (and also small a), that is, small ($\bar{y}/x - 1$), by expanding the trigonometric functions in the exponential part of eq. (12.29) in a Taylor series and keeping only the terms that are linear in α, β, and γ. The details of the calculations and the results obtained are described elsewhere[47] and will not be described in detail here. The resulting simple expressions for the Lagrange multipliers β, γ, and a are:

$$\beta \approx (3/m)^{1/2}\lambda \tag{12.33}$$

$$\gamma \approx (3/m\lambda v_2)^{1/2} \tag{12.34}$$

$$a = \left[\frac{5v_2/8}{1 - (xv_2/x_a)}\right]\frac{1}{5m}\left(\lambda^2 - \frac{1}{\lambda v_2}\right) \tag{12.35}$$

with

$$x_a = \frac{3}{(4/\pi) - 1} = 10.98 \tag{12.36}$$

In eq. (12.36), x_a is the length-to-width ratio of segments above which the system is totally anisotropic. A value of 12.38 is obtained for it in the Flory-Ronca model for solutions of rodlike particles. Using these relations in eq. (12.29) leads to the average $\langle \cos^2 \psi \rangle$ which is used in the expression for the orientation function $S = (1/2)(3\langle \cos^2 \psi \rangle - 1)$. The latter quantity is

$$S = \frac{1}{5m}\left(\lambda^2 - \frac{1}{\lambda v_2}\right)\left[1 + \frac{5x_a}{64}\frac{xv_2/x_a}{1 - (xv_2/x_a)}\right] = S_0 + S_i \tag{12.37}$$

where

$$S_i \equiv S_0 \left[\frac{5x_a}{64}\frac{xv_2/x_a}{1 - (xv_2/x_a)}\right] \tag{12.38}$$

is the intermolecular contribution to segmental orientation arising from competition for space of segments in an oriented environment.

From eq. (12.37) it is seen that S_i vanishes for $x/x_a \ll 1$. In this case, the problem reduces to orientation in a network of phantom chains. The value of S_i also decreases with increasing solvent content and equates to zero in the limiting case of $v_2 = 0$, as is readily seen from eq. (12.37).

Results of calculations for the orientation function are in agreement with those obtained by Walasek[64] by similar arguments using constraints. It is to be noted that the result proposed in the phenomenological treatment of Jarry and Monnerie[56] and the analytical expression given by eq. (12.37) are of the same functional form. Approximating $5x_a/64$ by unity in the latter, and comparing the resulting expression for segmental orientation with that obtained by Jarry and Monnerie shows that their parameter V may be identified with xv_2/x_a in the present treatment. However, this identification may be more apparent than real.

The basis for the interactions in the Jarry-Monnerie theory is energetic in nature, whereas xv_2/x_a is obtained by considering the liquidlike nature of the Kuhn segments and hence is of entropic origin.

In the presence of thermotropic interactions between neighboring repeat units, the orientation function for the dry network is found to be[48]

$$S = \frac{1}{5m}\left(\lambda^2 - \frac{1}{\lambda}\right)\left[\frac{u}{1 - u/5\tilde{T}}\right] \tag{12.39}$$

where

$$u \equiv 1 + \frac{5x}{64}\left[1 - \frac{x}{x_a}\right]^{-1} \tag{12.40}$$

and $\tilde{T}$ is defined as[59]

$$\tilde{T} = \frac{k_B T}{x}\left[\frac{Cz_c(\Delta\alpha)^2}{r_*^6}\right]^{-1} \tag{12.41}$$

In eq. (12.41), k_B is the Boltzmann constant, C is a constant, $\Delta\alpha$ is the mean optical anisotropy of all segments, r_* is the distance between subsegments for dense packing, and z_c is the number of first neighbors surrounding a given segment. The derivation of eq. (12.41) is given by Bahar et al.[48].

12.1.5 The Relationship of Stress to Deformation and Orientation

The true stress in uniaxial deformation is obtained from the elastic free energy by the thermodynamic relation

$$\sigma = \frac{1}{V}\lambda\left(\frac{\partial \Delta A_m}{\partial \lambda}\right)_{T,V,n_1,n_2,(n_{j,k})_{eq}} \tag{12.42}$$

where V is the total volume of the system, and the elastic free energy ΔA_m is related to the partition function Z_m by eq. (12.7). The explicit expression for Z_m is obtained by substituting the expressions for $n_{j,k}$ from eq. (12.15) into eqs. (12.8) and (12.9). The final form is

$$\begin{aligned}-\ln Z_m|_{eq} = &\; n_1 \ln v_1 + n_2 \ln(v_2/mx) - n_1 \ln[1 - v_2(1 - \bar{y}/x)] \\ &+ n_2(m\bar{y} - 1) - n_2(m - 1)\ln(z - 1) - mn_2 \ln(Z/4\pi) \\ &+ \beta n_2 \lambda \sqrt{m/3} + 2\gamma n_2 \sqrt{m/3\lambda v_2}\end{aligned} \tag{12.43}$$

where

$$Z = \int_0^{2\pi} d\phi \int_0^{\pi} d\psi\, f(\phi, \psi) \tag{12.44}$$

and $f(\phi, \psi)$ is given by eq. (12.29). Performing the differentiation indicated by eq. (12.42), with use of the expression for Z_m given by eq. (12.43), gives for the stress:

$$\sigma = \frac{k_B T}{V} \left\{ mn_2 \left[1 - \frac{v_1}{1 - v_2(1 - \bar{y}/x)} \right] \frac{\partial \bar{y}}{\partial \lambda} \right.$$
$$\left. + n_2 \sqrt{m/3} \left(\beta - \gamma/\sqrt{v_2 \lambda^3} + \lambda \frac{\partial \beta}{\partial \lambda} + 2 \frac{\partial \gamma}{\partial \lambda} / \sqrt{\lambda v_2} \right) - mn_2 \frac{\partial \ln Z}{\partial \lambda} \right\} \quad (12.45)$$

Using the identities

$$\frac{\partial \ln Z}{\partial \lambda} = \frac{\partial \ln Z}{\partial a} \frac{\partial a}{\partial \lambda} + \frac{\partial \ln Z}{\partial \beta} \frac{\partial \beta}{\partial \lambda} + \frac{\partial \ln Z}{\partial \gamma} \frac{\partial \gamma}{\partial \lambda} \quad (12.46)$$

in eq. (12.45) with

$$\frac{\partial \ln Z}{\partial a} = -\bar{y}, \qquad \frac{\partial \ln Z}{\partial \beta} = \lambda/\sqrt{3m}, \qquad \frac{\partial \ln Z}{\partial \gamma} = 2/\sqrt{3m\lambda v_2} \quad (12.47)$$

leads to

$$\sigma = \frac{k_B T}{V} n_2 \sqrt{m/3} \lambda (\beta - \gamma/\sqrt{v_2 \lambda^3}) \quad (12.48)$$

Substituting the expressions for β and γ given by eqs. (12.33) and (12.34), we obtain the simple expression for the stress in the first approximation as

$$\sigma = \frac{k_B T}{V} n_2 (\lambda^2 - 1/\lambda v_2) \quad (12.49)$$

Equation (12.49) shows that, in this approximation, the relationship between stress and deformation is independent of the degree of rigidity of the chains. This observation is in distinct contrast with the relationship between orientation and deformation indicated by the term S_i in eq. (12.38). The latter relationship becomes discontinuous at a specific amount of solvent below which the network becomes perfectly oriented, indicating occurrence of a phase transition. The possibility of a first-order phase transition is discussed in the following section.

Eliminating the extension ratio between eqs. (12.37) and (12.49), one obtains

$$S = \frac{1}{5mG} \left[1 + \frac{5x_a}{64} \frac{xv_2/x_a}{1 - (xv_2/x_a)} \right] \sigma \quad (12.50)$$

where

$$G = k_B T n_2 / V \quad (12.51)$$

is the shear modulus of the network.

12.1.6 Isotropic–Nematic Phase Transitions in Deformed Polymer Networks

This section describes the effects of isotropic–nematic phase transitions in stretched networks and is based on the work of Kloczkowski et al.[52], which is the continuation of the formulation described in sections 12.1.3 and 12.1.5.

There are two thermodynamic parameters which may easily be controlled during the phase transition: the length of the sample (measured by the extension ratio λ) and the applied force f. The isotropic–nematic transition at constant length occurs when the free energies of the nematic and the isotropic phase are equal:

$$\Delta A_m(\text{isotropic}) = \Delta A_m(\text{nematic}) \tag{12.52}$$

The phase with the lowest free energy always prevails. For athermal systems, the isotropic phase is stable at low axial ratios x and at low concentrations v_2 of the semi-rigid chains. For sufficiently high axial ratios and high values of v_2, the nematic phase has a lower free energy than the isotropic one. There is a critical value of the axial ratio x_{crit} below which the nematic solution never exists regardless of the deformation λ[48].

The proper free energy ΔG_m for studying phase transitions at constant force f is a Legendre transform of ΔA_m

$$\Delta G_m = \Delta A_m - fL \tag{12.53}$$

where ΔA_m is the Helmholtz free energy. The phase transition at constant force occurs when the free energies ΔG_m of the two phases are equal:

$$\Delta G_m(\text{isotropic}) = \Delta G_m(\text{nematic}) \tag{12.54}$$

One may also consider the phase transition at constant stress σ, as was done by Warner and Wang[49,50,57], although it is experimentally difficult to keep σ constant.

Figure 12.2 shows the effect of the extension ratio λ on the force f (measured in $n_0 V_0^{-1/3} k_B T$ units)[47]. The horizontal line joins two solutions which have the same

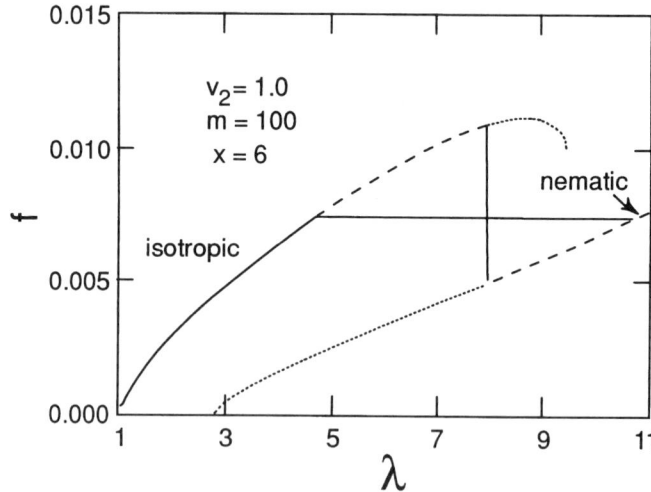

Figure 12.2 Relationship between the force f and the extension ratio λ.[52] The upper part of the figure represents the nematic solution, and the lower part, the isotropic solution. The calculations were performed for dry networks, with chains composed of $m = 100$ Kuhn segments with axial ratios $x = 6$. The horizontal line joins two solutions which have the same free energy ΔG at constant force f.

free energy ΔG at constant force f, and is described by eq. (12.38). The straight vertical line satisfies eq. (12.54) and joins nematic and isotropic solutions which have equal free energies ΔA at constant length. The broken part of both curves is for the solution, which always has a larger free energy and therefore is metastable irrespective of the process of deformation. Starting with the polymer strip in the isotropic state, fixing one of its ends, and increasing the load on the second end leads to the horizontal line in figure 12.2. The system will spontaneously transform to the nematic state, which has lower free energy at constant force. The length (extension) of the system will increase dramatically because of this transition to the more ordered phase.

The transition occurs at lower λ in the presence of anisotropic thermotropic effects. For the constant-length experiment, when both ends of the sample are mounted in clamps at a certain extension, the free energy of the nematic phase will become lower than that of the isotropic phase. There will then be a phase transition at constant length, accompanied by a large drop of the force (represented by a vertical line on the plot). In some cases, the force may even drop to negative values, that is, the sample after the phase transition will be compressed.

Figure 12.3 shows the change of the orientation function (or order parameter) $S = \langle P_2(\cos\theta) \rangle$ for the mechanically induced phase transformation at constant length[52]. The orientation function S increases smoothly with deformation and then abruptly jumps upward at the transition point. Similar curves were obtained in previous calculations[48] for network chains of $m = 20$ rigid segments with various axial ratios.

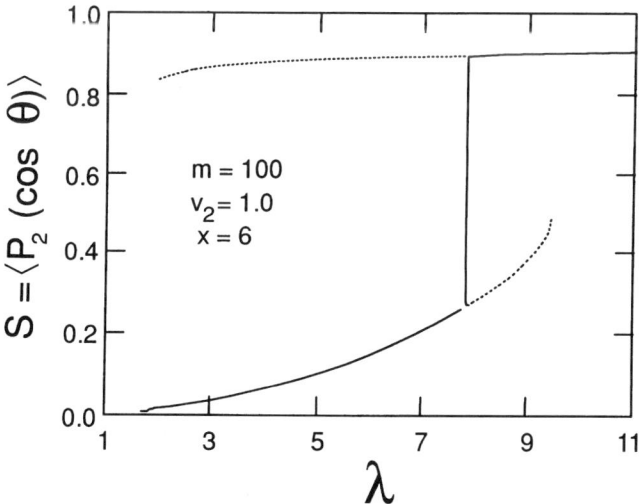

Figure 12.3 Change of the orientation function (or order parameter) $S = \langle P_2(\cos\theta) \rangle$ for the mechanically induced phase transformation at constant length[52]. All values of the parameters of the model are the same as in figure 12.2, and the dotted lines show metastable solutions for both phases.

178 STRUCTURES AND PROPERTIES OF RUBBERLIKE NETWORKS

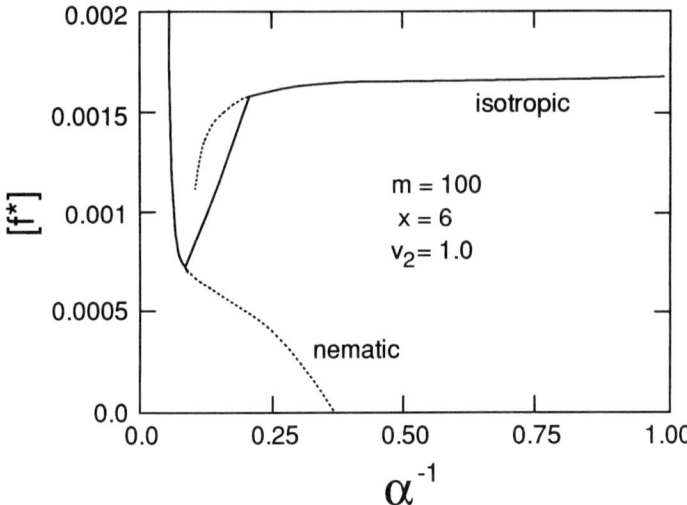

Figure 12.4 Modulus $[f^*]$ of the network as a function of inverse elongation α^{-1} for the polymer sample undergoing a mechanically induced nematic–isotropic phase transition at constant force[52]. Dotted lines indicate metastable states.

Figure 12.4 represents the behavior of the modulus $[f^*]$ of the system as a function of the inverse elongation α^{-1} for the polymer sample undergoing a mechanically induced nematic–isotropic phase transition at constant force[52].

12.2 Strain-Induced Crystallization

12.2.1 General Features

The previous sections described the effects of chain flexibility on segmental orientation in deformed networks. By the use of a lattice model representing the liquid-like nature of the polymer, it was found that when the stiffness of the chains exceeds a critical value, spontaneous orientational ordering takes place upon deforming the network. This transition from the less-oriented to the highly oriented state takes place irrespective of temperature, and is essentially an isotropic to nematic like transition. During such a transition, the segments assume an orientationally ordered state not associated with long-range spatial arrangements. In other words, the system is spatially disordered. The assumption of only nematic like effects in the model in the previous section therefore did not result in the prediction of three-dimensional crystallites distinctly separated from the remaining amorphous regions. A wide class of networks[65–68], on the other hand, do exhibit strain-induced crystallization, where well-defined crystallites form and grow during an anisotropic deformation, as may be detected by x-ray diffraction measurements. In the remaining part of this chapter, the molecular theories of strain-induced crystallization will be briefly reviewed.

SEMIFLEXIBLE CHAINS AND STRAIN-INDUCED CRYSTALLIZATION

The first quantitative description of strain-induced crystallization was by Alfrey and Mark[69], in which the behavior of a single chain (but not a network) was studied. This analysis was improved and extended by Flory[70] into a theory of strain-induced crystallization of networks, and this treatment has served as the basis for all later theoretical models. In the following section, we present the basic assumptions and results of this theory. Also mentioned are several more recent theoretical models that were essentially constructed to improve the Flory model.

12.2.2 Models for Strain-Induced Crystallization in Stretched Networks

The assumptions of the Flory theory of strain-induced crystallization in networks are as follows:

1. The displacements of the junctions of the network transform affinely with macroscopic strain.
2. The network chains consist of freely jointed segments, and crystallization is assumed to begin when a small number σ of neighboring segments align parallel to each other to form a crystallite nucleus. As crystallization progresses, the number of segments of a chain participating in the crystallite increases.
3. The network is subject to simple tension. The direction of stretch is identified by the z axis shown in figure 12.5, and the angle between the crystallite direction c and the direction of stretch is designated ζ.
4. Crystallization is assumed to evolve as an equilibrium process, which may be realized, for example, by stretching the network to a fixed length in the uncrystallized state and then decreasing the temperature to a point at which crystallization takes place.
5. Only incipient crystallization is treated.
6. There is no significant change in entropy associated with the formation of the nuclei.

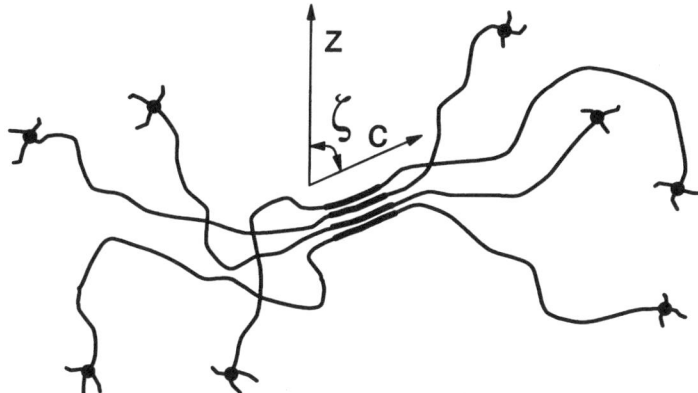

Figure 12.5 A small number of neighboring segments aligning parallel to one other to form a crystallite nucleus.

7. The directions of the crystallites coincide with the direction of stretch, that is, $\zeta = 0$. (This simplifying assumption was later removed by Allegra and coworkers[71,72] in a more comprehensive treatment.)
8. A chain passes through a crystallite only once, and along the direction going from one end of the chain to the other. (Permitting the chains to enter the crystallite along the reverse direction as well was considered subsequently by Smith[73].)
9. The formation of crystallites by chain folding is ignored. (The possibility of chain folding was later treated in terms of a non-Gaussian model by Gaylord[74].)

The possible effects of fluctuations of junctions from their mean positions were first introduced into the model by Allegra[71], along the lines of the Ronca-Allegra model[75], and later by Kloczkowski and collaborators, using the constrained-junction model[76] and the constrained-chain model[77]. The latter two treatments generalize the Flory theory[70] to nonaffine displacement of junctions.

12.2.3 Predictions of the Molecular Theories

The Flory theory of strain-induced crystallization explicitly expresses the effect of strain on: (1) the elevation of the melting point T_m, (2) the degree of crystallinity w, and (3) the elastic force f exhibited by the network. These relations are

$$w = 1 - \{(3/2 - \phi(\lambda))/(3/2 - \theta)\}^{1/2} \tag{12.55}$$

$$\frac{1}{T_m^0} - \frac{1}{T_m} = \frac{k}{h_f} \phi(\lambda) \tag{12.56}$$

and

$$f = \nu RT[(\lambda - \lambda^{-2}) - (6n/\pi)^{1/2} w]/(1 - w) \tag{12.57}$$

The temperature function θ in eq. (12.55) is

$$\theta = (h_f/k)\left(\frac{1}{T_m^0} - \frac{1}{T}\right) \tag{12.58}$$

and the strain function $\phi(\lambda)$ in eq. (12.56) is

$$\phi(\lambda) = (6/\pi n)^{1/2}\lambda - (\lambda^2/2 + \lambda^{-1})/n \tag{12.59}$$

In these equations, h_f is the heat of fusion per segment, k is the Boltzmann constant, and $T_m^0 = h_f/s_f$ is the incipient crystallization temperature of the undeformed polymer (with s_f being the entropy of fusion per segment). As usual, λ is the elongation in simple tension (defined as the ratio of the deformed length to the length at the state of formation of the network), n is the number of segments (statistical links) per chain, and ν is the number of network chains per unit volume.

Some results of the Flory affine network model and the Flory constrained-junction model are presented in figures 12.6, 12.7, and 12.8. The curves in these figures are obtained for $n = 150$, h_f (per mole of repeat units) = 9.195 kJ mol^{-1}, and $T_m^0 = 272$ K. Figure 12.6 shows the dependence of the incipient crystallization temperature T_m on elongation for the Flory affine model, and the constrained-

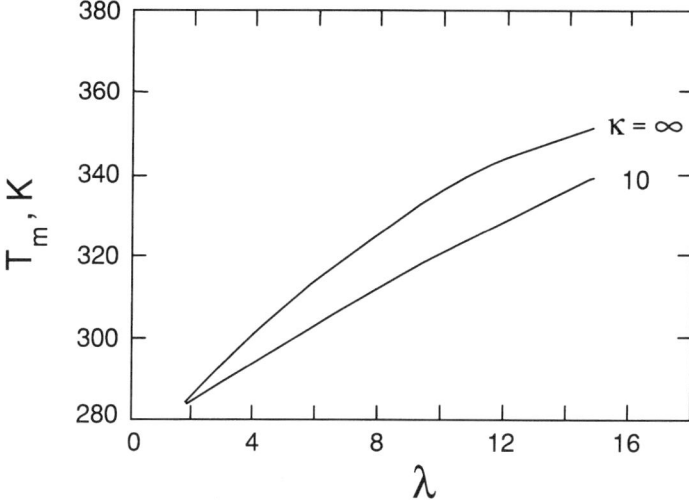

Figure 12.6 The dependence of incipient crystallization temperature T_m on elongation for $\kappa = 10$ for the constrained-junction model, and for the Flory affine model $(\kappa = \infty)$[76].

junction model with $\kappa = 10$.[76] Correspondingly, figure 12.7 gives the degree of crystallinity as a function of extension[77]. Figure 12.8 shows the behavior of the reduced nominal stress (modulus) $[f^*]/\nu RT$ as a function of the inverse elongation λ^{-1}.[77] For the affine limit ($\kappa = \infty$), the modulus is a monotonically increas-

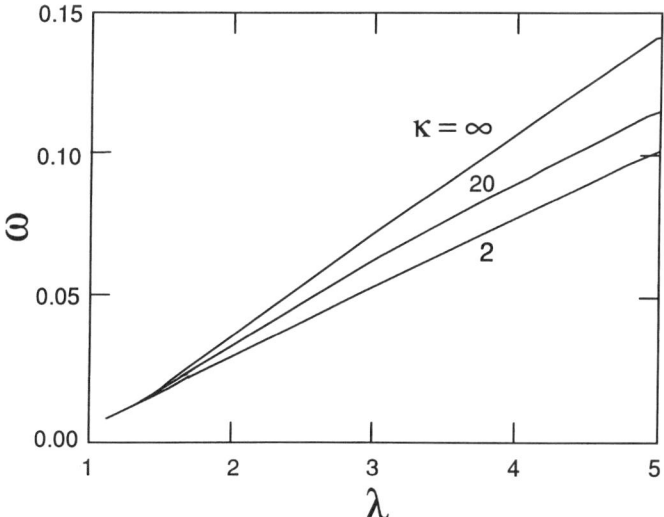

Figure 12.7 The degree of crystallinity as a function of elongation λ for different values of κ at $n = 150$, $h_f = 9.195\,\mathrm{kJ\,mol^{-1}}$, and $T_m^0 = 272\,\mathrm{K}$[77].

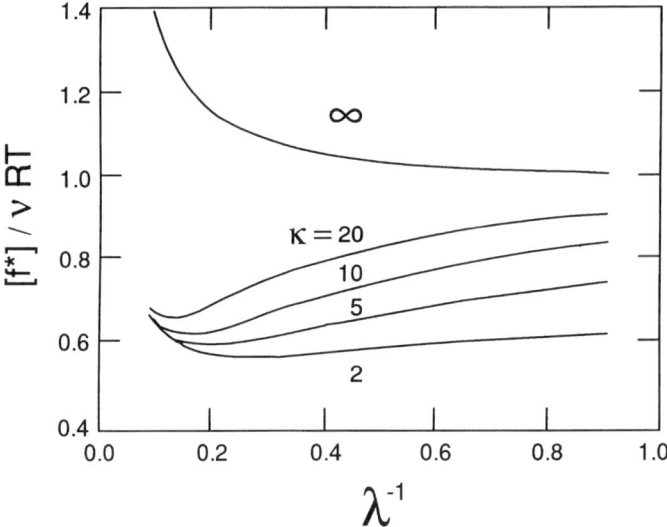

Figure 12.8 Reduced nominal stress (modulus) $[f^*]/\nu RT$ as a function of the inverse elongation λ^{-1}, for the affine limit $(\kappa = \infty)$[77].

ing function of the elongation. It also becomes a monotonically increasing function of the elongation in the phantom limit ($\kappa = 0$), which is not shown in the figure. For intermediate values of κ, there is a downturn in the modulus just prior to its upturn. As described below, these features have been observed in several experimental studies[68,78–82], and the stress–strain isotherms shown in figure 12.8 are in satisfactory qualitative agreement with experiment.

12.2.4 The Effects of Strain-Induced Crystallization on Mechanical Properties

Of direct relevance to the predictions of the previous section, crystallizable polymer networks such as those of natural rubber or *cis*-1,4-polybutadiene exhibit stress–strain isotherms in which there are pronounced upturns in modulus at high elongations. Such upturns have already been illustrated in figures 10.8 and 10.9. They are primarily due to the reinforcing effects of the crystallites thus generated, as described in chapter 10. They are much more pronounced than upturns due to limited chain extensibility, as exhibited, for example, by networks having bimodal distributions of network chain lengths, a subject described in the following chapter. Also, for a crystallizable network, these upturns disappear upon sufficient swelling, as illustrated in figure 12.9[83,84]. As expected, they also diminish and eventually disappear upon increase in temperature[84,85]. On the other hand, in those cases where the upturns are due to limited chain extensibility, increase in temperature has relatively little effect on the upturns[68,86]. Also, in these cases,

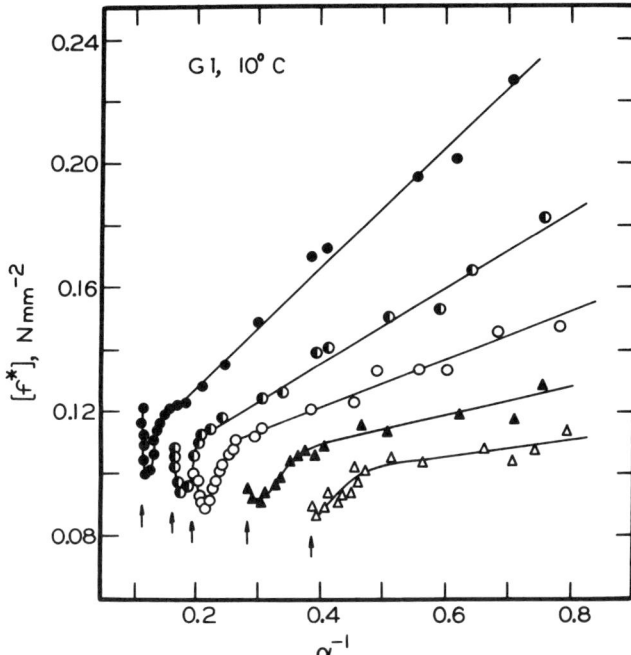

Figure 12.9 Stress–strain isotherms for a highly crystallizable cis-1,4-polybutadiene network, swollen with 1,2-dichlorobenzene to values of the volume fraction v_2 of polymer of 1.00 (●), 0.80 (◐), 0.60 (O), 0.40 (▲), and 0.20 (△)[83]. Reprinted with permission from Chiu, D. S., et al. (1977), *Macromolecules*, **10**. Copyright 1997 American Chemical Society.

swelling can even make the upturns more pronounced because of the already imposed stretching of the chains from the dilational effects of the swelling[68,86].

Two additional features in figure 12.9, however, merit additional comments[68]. First, the initiation of the strain-induced crystallization (as evidenced by the departure of the isotherm from linearity) is facilitated by the presence of the low-molecular-weight diluent. Thus, in a sense this kinetic effect acts in opposition to the thermodynamic effect (which is primarily the depression of the polymer melting point by the diluent). The second interesting point has to do with the downturn in the modulus prior to its increase. As shown schematically in figure 12.10[79,80], this is probably due to the fact that the crystallites are oriented along the direction of stretching (figure 12.10(a), and the chain sequences within a crystallite are in regular, highly extended conformations. The straightening and aligning of portions of the network chains thus decreases the deformation in the remaining amorphous regions, with an accompanying decrease in the stress (figure 12.10(b)). Both the downturn and subsequent upturn in the modulus are in at least qualitative agreement with the predictions of section 12.2.3.

In summary, the anomalous upturn in modulus observed for crystallizable polymers such as natural rubber and cis-1,4-polybutadiene is largely, if not

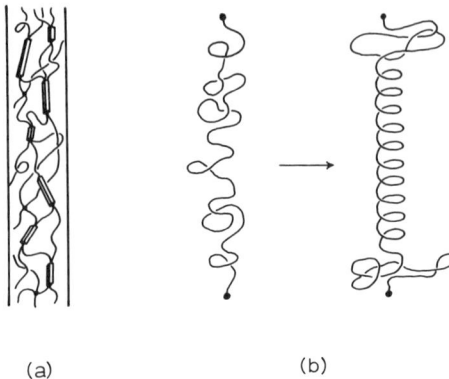

Figure 12.10 Strain-induced crystallization in a polymer network that has been elongated by a force along the vertical direction[79]. Reprinted with permission from Mark, J. E. (1979), *Acc. Chem. Res.*, **12**. Copyright 1997 American Chemical Society.

entirely, due to strain-induced crystallization. In the case of noncrystallizable model networks such as those of PDMS, it is clearly due to limited chain extensibility, and thus the results on this system will be extremely useful for reliable evaluation of the various non-Gaussian theories of rubberlike elasticity. Such distributions are described in appendix F.

Some illustrative results on the effects of strain-induced crystallization on ultimate properties are given for *cis*-1,4-polybutadiene networks in table 12.1[68,85]. The higher the temperature, the lower the extent of crystallization and, correspondingly, the lower the ultimate properties. The effects of increase in swelling parallel those for increase in temperature, since diluent also suppresses network crystallization. For noncrystallizable networks, however, neither change is very important, as is illustrated by the results shown for PDMS networks in Table 12.2[68,83].

Table 12.1 *Cis*-1,4-Polybutadiene Networks at High Elongation

Temperature (°C)	α at upturn	Ultimate properties	
		Max. upturn in $[f^*]$ (%)	α at rupture
5	3.27	54.2	6.64
10	3.48	30.1	6.22
25	4.03	4.3	5.85
40	—	0.0	5.68

Reprinted with permission from Mark, J. E. (1993), The Rubber Elastic State, in Mark, J. E., et al. (Eds.), *Physical Properties of Polymers*, 2nd ed., Washington, DC: American Chemical Society. Copyright 1997 American Chemical Society.

Table 12.2 Ultimate Properties, PDMS Networks

v_2	λ_r[a]	$[f^*]_r$
1.00	4.90	0.0362
0.80	4.42	0.0342
0.60	4.12	0.0338
0.40	4.16	0.0336

[a] The parameter λ_r is the value of the total elongation at the rupture point, using L_i (unswollen).
Reprinted with permission from Mark, J. E. (1993), The Rubber Elastic State, in Mark, J. E., et al. (Eds.), *Physical Properties of Polymers*, 2nd ed., Washington, DC: American Chemical Society. Copyright 1997 American Chemical Society.

References

(1) Mandelkern, L. *Crystallization of Polymers*. McGraw-Hill: New York. 1964.
(2) Mandelkern, L. In *Comprehensive Polymer Science*, G. Allen, Ed. Pergamon Press: Oxford. 1989.
(3) Mandelkern, L. *Rubber Chem. Technol.* 1993, 66, G61.
(4) Magill, J. H. *Rubber Chem. Technol.* 1995, 68, 507.
(5) Aharoni, S. M., Ed. *Synthesis, Characterization, and Theory of Polymeric Networks and Gels*, Plenum Press: New York. 1992.
(6) Aharoni, S. M.; Edwards, S. F. *Macromolecules* 1989, 22, 3361.
(7) Aharoni, S. M.; Hatfield, G. R.; O'Brien, K. P. *Macromolecules* 1990, 23, 1330.
(8) Zentel, R.; Reckert, G. *Makromol. Chem.* 1986, 187, 1915.
(9) Davis, F. J.; Gilbert, A.; Mann, J.; Mitchell, G. R. *J. Polym. Sci., Part A: Polym. Chem.* 1990, 28, 1455.
(10) Pakula, T.; Zentel, R. *Makromol. Chem.* 1991, 192, 2401.
(11) Finkelmann, H.; Kock, H. J.; Gleim, W.; Rehage, G. *Makromol. Chem. Rapid Commun.* 1984, 5, 287.
(12) Canessa, G.; Reck, B.; Reckert, G.; Zentel, R. *Makromol. Chem., Macromol. Symp.* 1986, 4, 91.
(13) Mitchell, G. R.; Davis, F. J.; Ashman, A. *Polymer* 1987, 28, 639.
(14) Zentel, R.; Schmidt, G. F.; Meyer, J.; Benalia, M. *Liq. Cryst.* 1987, 2, 651.
(15) Schatzle, J.; Finkelmann, H. *Mol. Cryst. Liq. Cryst.* 1987, 142, 85.
(16) Schatzle, J.; Kaufhold, W.; Finkelmann, H. *Makromol. Chemie* 1989, 190, 3269.
(17) Hammerschmidt, K.; Finkelmann, H. *Makromol. Chemie* 1989, 190, 1089.
(18) Hanus, K. H.; Pechhold, W.; Soergel, F.; Stoll, B.; Zentel, R. *Coll. Polym. Sci.* 1990, 268, 222.
(19) Loffler, R.; Finkelmann, H. *Makromol. Chem., Rapid Commun.* 1990, 11, 321.
(20) Kaufhold, W.; Finkelmann, H. *Makromol. Chem.* 1991, 192, 2555.
(21) Barclay, G. G.; Ober, C. K. *Prog. Polym. Sci.* 1993, 18, 899.
(22) Warner, M. *Phil. Trans. Roy. Soc. Lond. A* 1993, 344, 403.
(23) Finkelmann, H.; Brand, H. R. *TRIP* 1994, 2, 222.
(24) Bosch, A. T.; Varichon, L. *Macromol. Theory Simul.* 1994, 3, 533.
(25) Brehmer, M.; Zentel, R. *Macromol. Chem. Phys.* 1994, 195, 1891.
(26) Poths, H.; Zentel, R. *Macromol. Rapid Commun.* 1994, 15, 433.
(27) Zentel, R.; Brehmer, M. *CHEMTECH* 1995, 25(5), 41.

(28) Shilov, S. S.; Okretic, S.; Siesler, H. W.; Zentel, R.; Oge, T. *Macromol. Rapid Commun.* 1995, 16, 125.
(29) Brehmer, M.; Zentel, R. *Macromol. Rapid Commun.* 1995, 16, 659.
(30) Fisher, P.; Schmidt, C.; Finkelmann, H. *Macromol. Rapid Commun.* 1995, 16, 435.
(31) Kundler, I.; Finkelmann, H. *Macromol. Rapid Commun.* 1995, 16, 769.
(32) Verwey, G. C.; Warner, M. *Macromolecules* 1995, 28, 4296.
(33) Verwey, G. C.; Warner, M. *Macromolecules* 1995, 28, 4299.
(34) Verwey, G. C.; Warner, M. *Macromolecules* 1995, 28, 4303.
(35) Semmler, K.; Finkelmann, H. *Macromol. Chem. Phys.* 1995, 196, 3197.
(36) Benne, I.; Semmler, K.; Finkelmann, H. *Macromolecules* 1995, 28, 1854.
(37) Disch, S.; Finkelmann, H.; Ringsdorf, H.; Schuhmacher, P. *Macromolecules* 1995, 28, 2424.
(38) Hsu, C.-S.; Hsiue, G.-H. *Pure Appl. Chem.* 1995, 67, 2005.
(39) Sautter, E.; Belusov, S. I.; Pechold, W.; Makarova, N. N.; Godovsky, Y. K. *Polym. Sci., USSR* 1996, 38, 39.
(40) Molenberg, A.; Siffrin, S.; Moller, M. *Macromol. Symp.* 1996, 102, 199.
(41) Zubarev, E. R.; Talroze, R. V.; Yuranova, T. I.; Vasilets, V. N.; Plate, N. A. *Macromol. Rapid Commun.* 1996, 17, 43.
(42) de Gennes, P. G. *C. R. Hebd. Seances Acad. Sci. B.* 1975, 28, 101.
(43) Schwarz, J. *Makromol. Chem., Rapid Commun.* 1986, 7, 21.
(44) Warner, M.; Gelling, K.; Vilgis, T. *J. Chem. Phys.* 1988, 88, 4408.
(45) Warner, M. In *Side Chain Liquid Crystal Polymers*, B. McArdle, Ed. Blackie & Sons, Glasgow. 1989; Vol. 88, p. 7.
(46) Brand, H. *Makromol. Chem., Rapid Commun.* 1989, 10, 57.
(47) Erman, B.; Bahar, I.; Kloczkowski, A.; Mark, J. E. *Macromolecules* 1990, 23, 5335.
(48) Bahar, I.; Erman, B.; Kloczkowski, A.; Mark, J. E. *Macromolecules* 1990, 23, 5341.
(49) Warner, M.; Wang, X. J. *Macromolecules* 1991, 24, 4932.
(50) Warner, M.; Wang, X. J. *Macromolecules* 1992, 25, 445.
(51) Erman, B.; Bahar, I.; Kloczkowski, A.; Mark, J. E. In *Elastomeric Polymer Networks*, J. E. Mark and B. Erman, Eds. Prentice Hall: NJ. 1992; p. 142.
(52) Kloczkowski, A.; Mark, J. E.; Erman, B.; Bahar, I. In *Polymer Solutions, Blends and Interfaces*, I. Noda and D. N. Rubingh, Eds. Elsevier: Amsterdam. 1992; p. 221.
(53) Bladon, P.; Warner, M. *Macromolecules* 1993, 26, 1078.
(54) DiMarzio, E. A. *J. Chem. Phys.* 1962, 36, 1563.
(55) Tanaka, T.; Allen, G. *Macromolecules* 1977, 10, 426.
(56) Jarry, J. P.; Monnerie, L. *Macromolecules* 1979, 12, 316.
(57) Warner, M.; Wang, X. J. In *Elastomeric Polymer Networks*, J. E. Mark and B. Erman, Eds. Prentice Hall: Englewood Cliffs, NJ. 1992; p. 63.
(58) Flory, P. J. *Macromolecules* 1978, 11, 1141.
(59) Flory, P. J.; Ronca, G. *Mol. Cryst. Liq. Cryst.* 1979, 54, 289, 311.
(60) Flory, P. J. *Proc. Roy. Soc., London, Ser. A* 1956, 234, 73.
(61) Flory, P. J. *Proc. Roy. Soc., London, Ser. A* 1956, 234, 60.
(62) Kuhn, W.; Grün, F. *Kolloid-Z.* 1942, 101, 248.
(63) Wall, F. T.; Flory, P. J. *J. Chem. Phys.* 1951, 19, 1435.
(64) Walasek, J. *J. Polym. Sci., Part B: Polym. Phys.* 1988, 26, 1907.
(65) Gent, A. N. *J. Polym. Sci., Part A* 1965, 3, 3787.
(66) Gent, A. N. *J. Polym. Sci., Part A* 1966, 4, 447.
(67) Mark, J. E.; Erman, B. *Rubberlike Elasticity. A Molecular Primer.* Wiley-Interscience: New York. 1988.

(68) Mark, J. E. In *Physical Properties of Polymers*, 2nd ed., J. E. Mark, A. Eisenberg, W. W. Graessley, L. Mandelkern, E. T. Samulski, J. L. Koenig, and G. D. Wignall, Eds. American Chemical Society: Washington, DC. 1993; p. 3.
(69) Alfrey, T.; Mark, H. F. *J. Chem. Phys.* 1942, 46, 112.
(70) Flory, P. J. *J. Chem. Phys.* 1947, 15, 397.
(71) Allegra, G. *Makromol. Chem.* 1980, 181, 1127.
(72) Allegra, G.; Bruzzone, M. *Macromolecules* 1983, 16, 1167.
(73) Smith, K. J., Jr. *Polym. Eng. Sci.* 1976, 16, 168.
(74) Gaylord, R. J. *J. Polym. Sci: Polym. Phys. Ed.* 1976, 14, 1827.
(75) Ronca, G.; Allegra, G. *J. Chem. Phys.* 1975, 63, 4990.
(76) Sharaf, M. A.; Kloczkowski, A.; Mark, J. E.; Erman, B. *Comp. Polym. Sci.* 1992, 2, 84.
(77) Kloczkowski, A.; Mark, J. E.; Sharaf, M. A.; Erman, B. In *Synthesis, Characterization, and Theory of Polymeric Networks and Gels*, S. M. Aharoni, Ed. Plenum Press: New York. 1992; p. 227.
(78) Smith, K. J., Jr.; Greene, A.; Ciferri, A. *Kolloid Z—Z. Polym.* 1964, 194, 49.
(79) Mark, J. E. *Acc. Chem. Res.* 1979, 12, 49.
(80) Mark, J. E. *Polym. Eng. Sci.* 1979, 19, 409.
(81) Mark, J. E.; Erman, B. *Rubberlike Elasticity. A Molecular Primer*. Wiley-Interscience: New York. 1988.
(82) Chou, D.-G.; Smith, K. J., Jr. *Polym. Eng. Sci.* 1994, 34, 290.
(83) Chiu, D. S.; Su, T.-K.; Mark, J. E. *Macromolecules* 1977, 10, 1110.
(84) Mark, J. E. *Polym. Eng. Sci.* 1979, 19, 254.
(85) Su, T.-K.; Mark, J. E. *Macromolecules* 1977, 10, 120.
(86) Andrady, A. L.; Llorente, M. A.; Mark, J. E. *J. Chem. Phys.* 1980, 72, 2282.

13

Networks Having Multimodal Chain-Length Distributions

13.1 Ultimate Properties and Non-Gaussian Effects

As was mentioned in chapter 10, end-linking reactions can be used to make networks of known structures, including those having unusual chain-length distributions[1-5]. One of the uses of networks having a bimodal distribution is to clarify the dependence of ultimate properties on non-Gaussian effects arising from limited-chain extensibility, as was already pointed out. The following chapter provides more detail on this application, and others.

In fact, the effect of network chain-length distribution[6,7], is one aspect of rubberlike elasticity that has not been studied very much until recently, because of two primary reasons. On the experimental side, the cross-linking techniques traditionally used to prepare the network structures required for rubberlike elasticity have been random, uncontrolled processes, as was mentioned in chapter 10[6-8]. Examples are vulcanization (addition of sulfur), peroxide thermolysis (free-radical couplings), and high-energy radiation (free-radical and ionic reactions). All of these techniques are random in the sense that the number of cross-links thus introduced is not known directly, and two units close together in space are joined irrespective of their locations along the chain trajectories. The resulting network chain-length distribution is unimodal and probably very broad. On the theoretical side, it has turned out to be convenient, and even necessary, to assume a distribution of chain lengths that is not only unimodal, but *monodisperse*![7,8]

There are a number of reasons for developing techniques to determine or, even better, control network chain-length distributions. One is to check the "weakest-link" theory[9] for elastomer rupture, which states that a typical elastomeric network consists of chains with a broad distribution of lengths, and that the shortest of these chains are the "culprits" in causing rupture. This is attributed to the very limited extensibility associated with their shortness that is thought to cause them to break at relatively small deformations and then act as rupture nuclei. Another reason is to determine whether control of chain-length distribution can be used to maximize the ultimate properties of an elastomer.

As was described in chapter 10, a variety of model networks can be prepared using the new synthetic techniques that closely control the placements of cross-links in a network structure[5,8,10–27]. Thus, end-linking functionally terminated chains, instead of haphazardly joining chain segments at random, controls the structure of the resulting network. As already mentioned, the functionality ϕ of a cross-link is the same as the number of functional groups on the end-linking agent. More important in the present context, the molecular weight M_c between cross-links and the molecular weight distribution are the same as those of the starting chains prior to their being end-linked.

13.2 Bimodal Networks

Bimodal networks prepared by these end-linking techniques have very good ultimate properties, and there is currently much interest in preparing and characterizing such networks[8,19,21,27–36], and developing theoretical interpretations for their properties[37–41].

13.2.1 Materials and Synthetic Techniques

Most bimodal networks synthesized to date have been prepared from poly(dimethylsiloxane) (PDMS). One reason for this choice is the fact that the polymer is readily available with either hydroxyl or vinyl end groups, and the reactions that these groups participate in are relatively free of complicating side reactions[42]. Also, the polymer is noncrystallizable under most conditions because of its very low melting point ($\sim -40°C$)[43]. These bimodal networks are, in fact, generally studied at sufficiently high temperatures ($\sim 25°C$) to easily avoid contributions from strain-induced crystallization, when desired. The end-linking reactions have generally involved hydroxyl-terminated chains, which are readily obtained from the usual ring-opening polymerizations of the cyclic trimer or tetramer[44]. The ends of the chains are reacted with the alkoxy groups in a multifunctional organosilicate, such as tetraethylorthosilicate[42], as described in chapter 10. In the application of interest here, a mixture of two polymers is being cross-linked, one component consisting of very short chains and the other of much longer chains. The resulting network is shown schematically in figure 13.1[28]. An alternative approach involves the addition reaction between vinyl groups at the ends of a polymer chain and the active hydrogen atoms on silicon atoms in the [$Si(CH_3)HO-$] repeat units in an oligomeric poly(methyl hydrogen siloxane). As a final example, a network may be cross-linked using a trifunctional reactant carrying side chains terminated in some reactive group such as NH_2.[45] Complexing two of these NH_2 groups with a metal ion provides a set of additional, rather short, chains that could be elastically effective. This approach is illustrated in figure 13.2. These ideas can obviously be extended to higher modalities (trimodal, etc., eventually approaching an extremely broad, effectively unimodal distribution)[46–49].

The distribution of network chain lengths in a bimodal elastomer can be extremely unusual, and much different from the usual unimodal distribution obtained

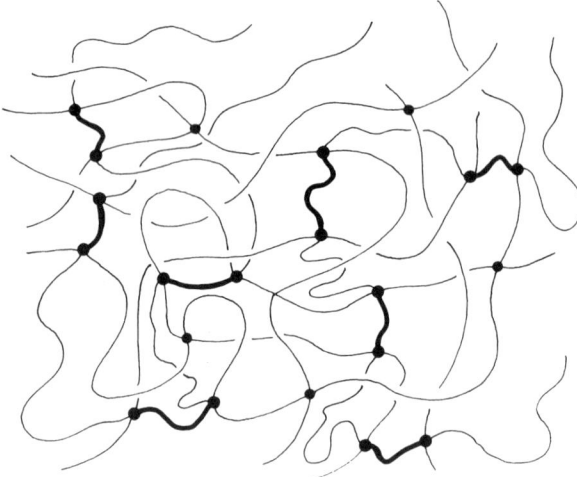

Figure 13.1 Sketch of a bimodal network in which the short chains are arbitrarily drawn in heavier lines than the long chains, which are more typically associated with rubberlike elasticity[28]. The dots represent the cross-links. Reproduced with permission; Copyright 1979, Huthig & Wepf Verlag, Basel.

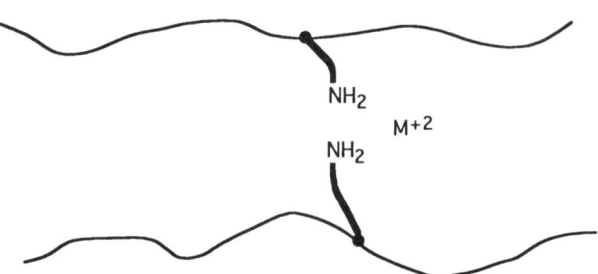

Figure 13.2 End-linking of side chains by a complexation reaction[26]. Since the number of bonds formed between the chains being joined is more than twice the number in each side chain, the cross-linking chains could be elastically effective and act as the short chains in a bimodal network[26,27]. See legend to figure 13.1.

in less controlled methods of cross-linking. Figure 13.3 shows a schematic distribution for the important example in which there is simultaneously a large number percent of short chains and a large weight percent of long chains. The major difference is the *large* amounts of both very short chains and very long chains in the bimodal network, contrasting sharply with the *small* amounts of such chains in a typical unimodal distribution. The case shown here, where the short chains predominate numerically, is of particular interest and can be used to illustrate the important difference between number distributions and weight

NETWORKS HAVING MULTIMODAL CHAIN-LENGTH DISTRIBUTIONS

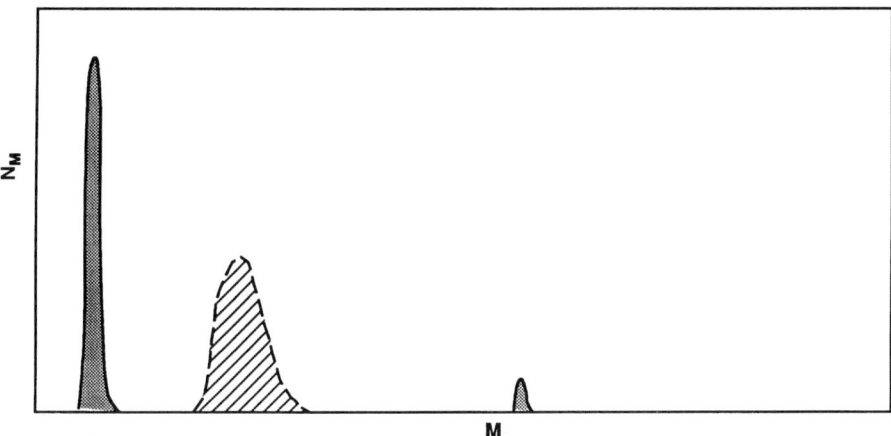

Figure 13.3 Network chain-length distributions, in which N_M is the number of chains in an infinitesimal interval around the specified value of the molecular weight M.[27] A broad unimodal distribution is shown by the hatched area, for reference purposes. Reprinted with permission from Mark, J. E. (1994), *Acc. Chem. Res.*, **27**. Copyright 1997, American Chemical Society.

distributions. For example, a bimodal network consisting of 95 *mol*% chains of molecular weight 200 g mol^{-1} and 5 mol% of chains of 20,000 g mol^{-1} would have only 16 *wt*% of the short chains! Obviously, such peculiar distributions can be obtained only by carefully controlled techniques such as the end-linking reactions being described. One of their uses is to test a molecular description of elastomer rupture, as described in the following section.

13.2.2 Testing of the Weakest-Link Theory

The weakest-link theory[9] was tested by preparing end-linked networks containing increasing amounts of short chains, of the order of 10–20 mol%[8,28]. Remarkably, there were no significant decreases in ultimate properties with these increases in the numbers of short chains, in striking disagreement with the suggested mode of elastomer failure. Networks are apparently much more resourceful than given credit for in this theory. As described in several other chapters, the strain is continually being reapportioned during the deformation, in such a way that the much more easily deformed long chains bear most of the burden of the deformation. Thus, the short chains do not contribute significantly until just prior to rupture. This is consistent with Nature preferring the low-energy route, unless overruled by entropy increases. The flaw in the weakest-link theory is thus the implicit assumption that all parts of the network deform in exactly the same way, that is, affinely, whereas the deformation is markedly nonaffine[8]. These two limiting behaviors are illustrated schematically in figure 13.4.

The above conclusion is supported by the common observation that stress–strain isotherms are generally reversible right up to the rupture points[8]. This

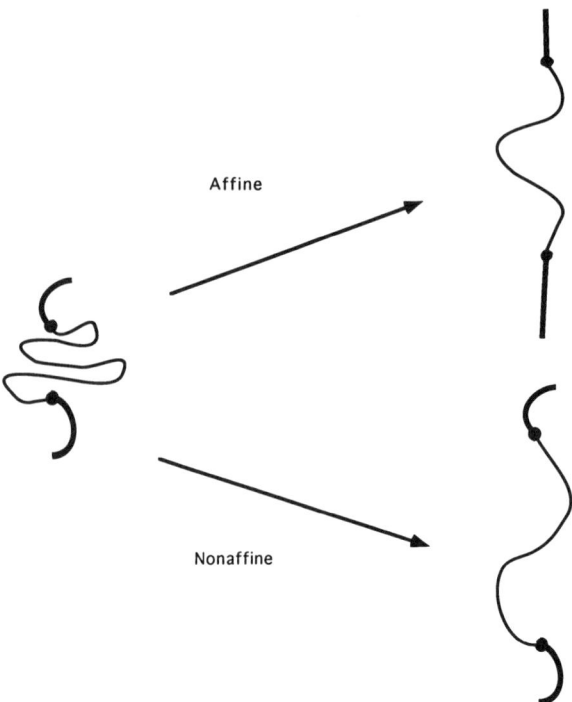

Figure 13.4 Sketch showing the vertical deformation of a network segment consisting of a long chain bracketed by two short chains[26]. In the limit of an affine deformation, all chains deform by the same fractional amount, whereas in the nonaffine deformation, the more easily deformed long chains deform more than the average and the short chains deform less than the average.

contradicts the weakest-link-theory assumption that short-chain breakage occurs first, well before the rupture point. Also, in an affine deformation, a decrease in the number of short chains in a bimodal network should decrease the magnitude of the upturn in modulus but not the elongation at which it begins. Experiments, however, indicate that the fewer the short chains in the bimodal network, the higher the elongation required to bring about the upturn in modulus[50]. Thus, a decreased number of short chains apparently facilitates the reapportioning of the stress.

The weakest-link issue having been resolved, it became of interest to see what would happen in the case of bimodal networks having such overwhelming numbers of short chains that they cannot be ignored in the network's response. In fact, there turns out to be an exciting bonus if one does put a very large number of short chains into the bimodal network. Specifically, there is a synergistic effect leading to mechanical properties that are better than those obtainable from the usual unimodal distribution! These results are the focus of most of the remaining sections of this chapter.

13.2.3 Elongation Results

13.2.3.1 Characterization of Upturns in the Modulus Stress–strain isotherms in elongation for bimodal PDMS networks were generally obtained in the vicinity of 25°C, a temperature sufficiently high to suppress strain-induced crystallization. The results thus determined were of considerable interest since they indicated that the bimodal nature of the distribution greatly *improved* the ultimate properties of the elastomer[5,8,13,23]. This is illustrated in figure 13.5[51], in which data on PDMS

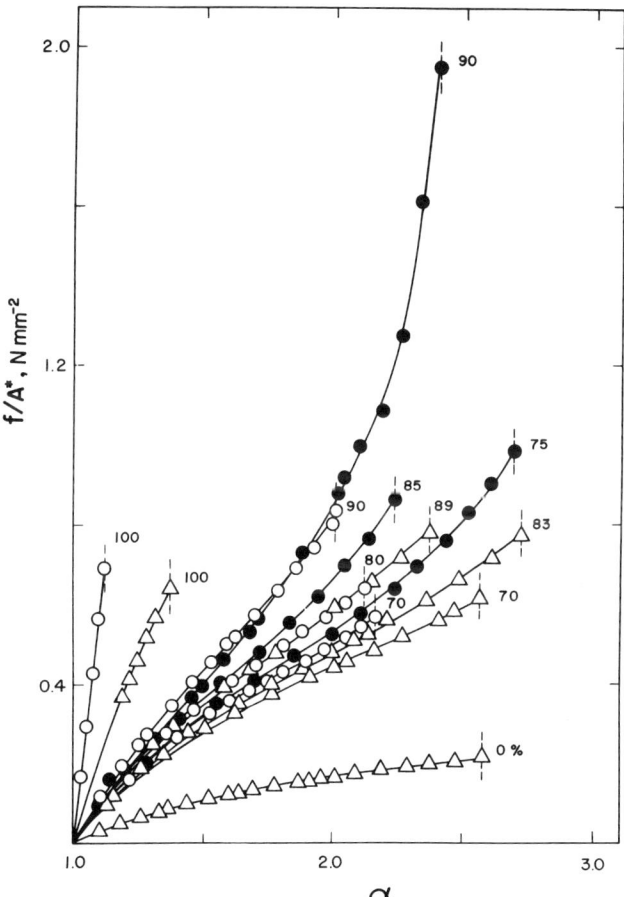

Figure 13.5 Typical plots of nominal stress against elongation for (unswollen) bimodal PDMS networks consisting of relatively long chains ($M_c = 18,500 \, \text{g mol}^{-1}$) and very short chains [1100 ($\triangle$), 660 ($\bigcirc$), and 220 ($\bullet$) in g mol^{-1}]. Each curve is labeled with the mol% of short chains that the network contains, and the area under each curve represents the rupture energy (a measure of the "toughness" of the elastomer). Reproduced with permission from Llorente, M. A., et al. (1981), *J. Polym. Sci., Polym. Phys. Ed.*, **19**. Copyright 1981, John Wiley & Sons, Inc.

networks are plotted in such a way that the area under a stress–strain isotherm corresponds to the energy required to rupture the network. If the network consists entirely of the short-chain component, then it is brittle (which means that the maximum extensibility is very small). If the network consists of only the long-chain component, the ultimate strength is very low. In neither case is the material a tough elastomer. As can readily be seen from the figure, the bimodal networks are much improved elastomers in that they can have a high ultimate strength without the usual diminished maximum extensibility. This corresponds to high values of the energy required for rupture, which makes them unusually tough elastomers, even in the unfilled state. The Mooney-Rivlin representation of some similar results is shown in figure 13.6[3]. Related improvements in mechanical properties have also been reported for polyurethane elastomers[52].

Since this internally generated improvement in properties is of considerable practical and fundamental interest, a number of studies on PDMS elastomers were carried out to determine its dependence on the molecular weights of the short chains[53], the proportions of short and long chains[53,54], and the cross-link

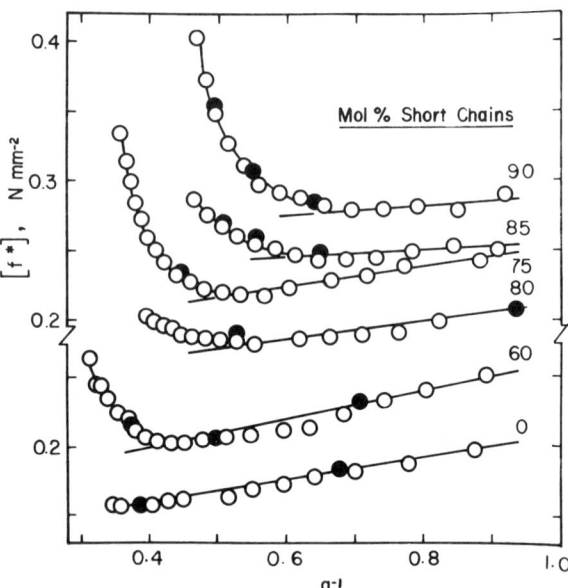

Figure 13.6 Representative stress–strain isotherms for bimodal (unswollen) PDMS networks in elongation at $-45°C$[3]. Each curve is labeled with the mole percent of the much shorter chains ($10^{-3} M_n = 0.220$ vs. $18.5\,\mathrm{g\,mol^{-1}}$). The open circles represent the results obtained using a series of increasing elongations, and the filled circles the results obtained out of sequence to test for reversibility. There is seen to be reversibility in the isotherms, even at these low temperatures. The short extensions of the linear portions of the isotherms help locate the values of the elongation at which the upturn in the modulus becomes discernible.

functionality[2,54,55]. These results can be used to optimize the improvements in the properties obtained.

The syntheses and properties of several additional types of bimodal networks were also investigated. The first of these involves prereacting the short chains prior to incorporating the long ones into the bimodal structure[1,56]. This makes the resulting network spatially, as well as compositionally heterogeneous in that many of the short chains will be segregated into densely cross-linked domains which are only lightly cross-linked to other such domains. This is shown schematically in figure 13.7[1]. These networks could serve as models for inhomogeneously cross-linked elastomers, for example those cured by thermolysis of a partially immiscible peroxide. Another type is based on the fact that the condensation and addition types of end-linking reactions can be carried out simultaneously and independently. This can give rise to the interpenetrating network structure[15,57] illustrated schematically in figure 13.8[15]. The distribution of network chains can obviously be made bimodal, as is shown, even though the short chains and long chains communicate only through their entanglements. Finally, it is also possible to reinforce any type of bimodal network with filler particles, thereby further improving its mechanical properties[58-61].

Measurements of the mechanical and optical properties of these networks as a function of temperature and degree of swelling were used to test further the conclusion cited above that crystallization or other intermolecular organization was not the origin of the improved properties[3,5,8,23,59,62,63]. As already mentioned, stress–strain measurements on such bimodal PDMS networks exhibited upturns in modulus that were much less pronounced than those in crystallizable polymer networks such as natural rubber or *cis*-1,4-polybutadiene. In another experiment, illustrated in figure 13.9, temperature was found to have little effect on the Mooney-Rivlin isotherms for bimodal PDMS networks, as would be expected in the case of limited chain extensibility[3]. Stress–temperature ("thermoelastic") and birefringence–temperature measurements showed no discontinuities or

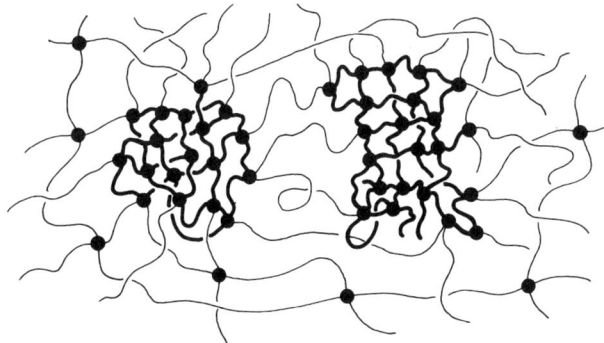

Figure 13.7 Sketch of a bimodal network that is spatially as well as compositionally heterogeneous with regard to chain length[1]. (See legend to figure 13.1.) Such a network can be prepared by prereacting the short chains in the first step of a two-step process.

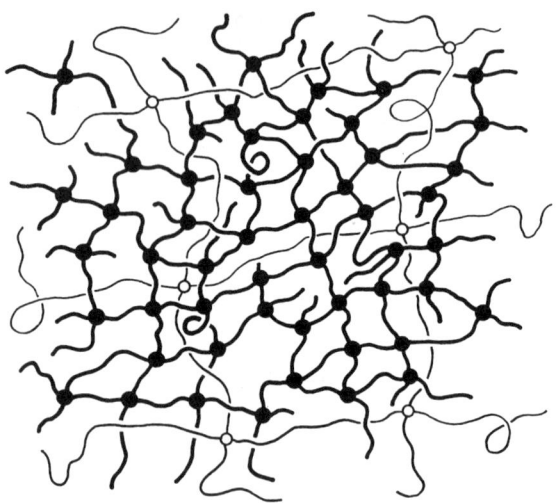

Figure 13.8 Sketch of a simultaneous interpenetrating network having a bimodal distribution of chain lengths, even though none of the short chains are attached to any of the long chains[15]. (See legend to figure 13.1.) Reprinted with permission from Mark, J. E. (1985), *Acc. Chem. Res.*, **18**. Copyright 1997, American Chemical Society.

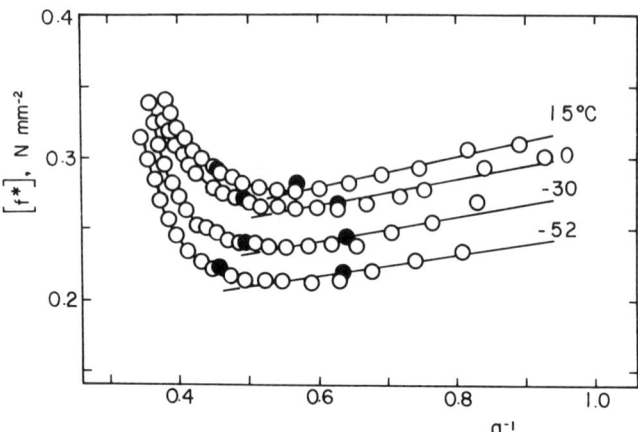

Figure 13.9 Typical results obtained to determine the effect of temperature on stress–strain isotherms of a bimodal PDMS network containing 75 mol% of short chains (220 g mol^{-1}) combined with much longer chains (18.5 × 10^3 g mol^{-1})[3].

discernible changes of slope, as is demonstrated in figures 13.10 and 13.11, respectively[3]. Similarly, swelling can even make the upturns in modulus *more pronounced*[50,64], as is illustrated in figure 13.12. Apparently, the enhanced upturns are due to the chains being stretched out in the solvent dilation process, prior to their being stretched further in the elongation experiments. In contrast, as has

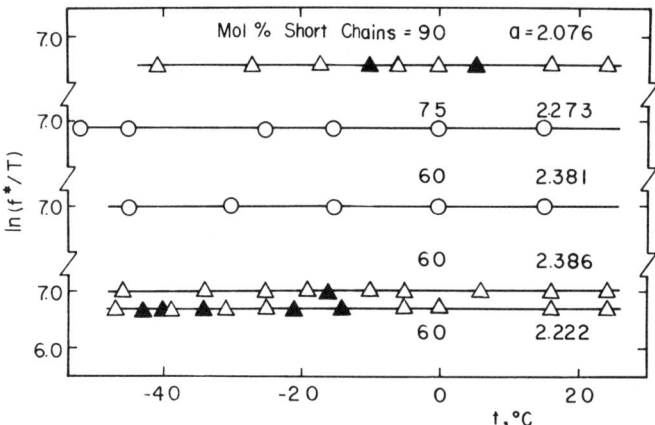

Figure 13.10 Typical thermoelastic data on bimodal PDMS networks, with circles referring to measurements at constant elongation and triangles to measurements at constant length[3]. Filled symbols represent data obtained to check for reversibility.

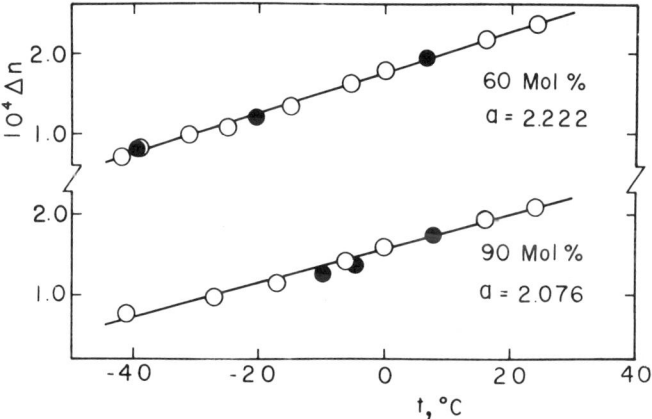

Figure 13.11 Representative birefringence–temperature relations on bimodal PDMS networks, with filled circles representing results obtained to check for reversibility[3].

already been mentioned in chapter 12, the upturns in crystallizable polymer networks disappear upon sufficient swelling[65,66]. A final experiment of relevance concerns the spatially-heterogeneous PDMS networks mentioned earlier[1,64]. Any ordering that might have been present would presumably have been effected by this change in spatial heterogeneity, but there was no discernible effect on the measured elastomeric properties.

The above findings argue against the presence of any crystallization or other type of intermolecular ordering[50,64], and the upturns thus seem to be intramolecular in origin. In addition, infrared and Raman spectroscopy indicate that bond-

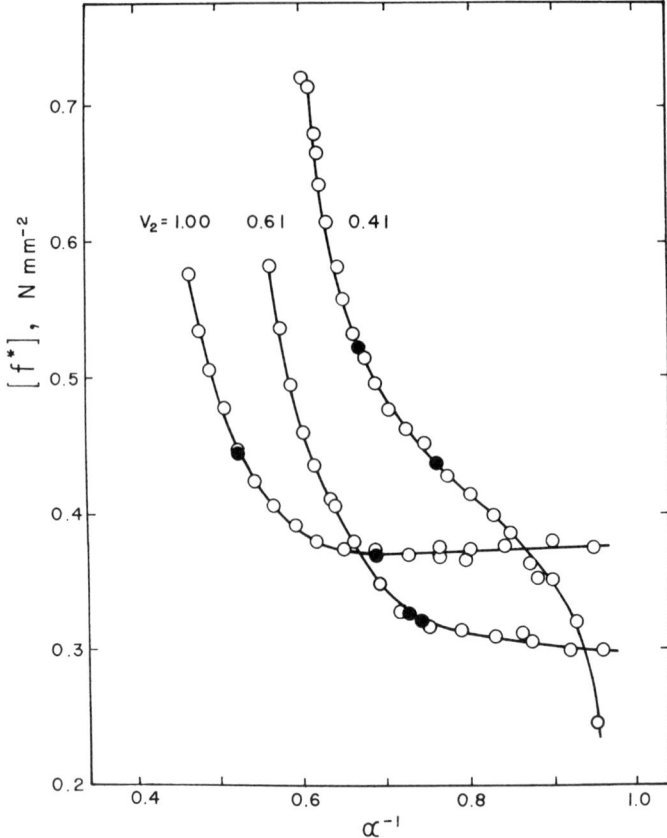

Figure 13.12 Stress–strain isotherms at 25°C for a PDMS bimodal network containing 90 mol% of short chains (220 g mol^{-1}) combined with much longer chains (18.5 × 10^3 g mol^{-1})[62]. Each curve is labeled with the volume fraction of polymer present in the network, as swollen with dimethylsiloxane oligomer. The filled circles represent results obtained out of sequence to check for reversibility. Reprinted with permission from Mark, J. E. (1984), *Macromolecules*, **17**. Copyright 1997, American Chemical Society.

angle distortion is of relatively minor importance, at least in PDMS networks[29]. The observed increases in modulus and ultimate strength therefore have to be due to the limited extensibility of the very short network chains. In qualitative terms, the chains soon exhaust their spatial configurations, their entropies plummet, and the elastic force rises correspondingly. It is possible to characterize this non-Gaussian limited extensibility more quantitatively in a number of ways, some of which are described below.

The first involves the interpretation of limited-chain extensibility in terms of the configurational characteristics of the PDMS chains making up the network structure[67]. The first important characteristic of limited-chain extensibility is the elongation α_u at which the increase in modulus first becomes discernible.

Although the deformation is nonaffine in the vicinity of the upturn, it is possible to provide at least a semiquantitative interpretation of such results in terms of the network chain dimensions[50]. At the beginning of the upturn, the average extension r of a network chain having its end-to-end vector along the direction of stretching is simply the product of the unperturbed dimension[67] $\langle r^2 \rangle_0^{1/2}$ and α_u. Similarly, the maximum extensibility r_m is the product of the number n of skeletal bonds and the factor (1.34 Å) which gives the axial component of a skeletal bond in the most extended helical form of PDMS, as obtained from the geometric analysis of the PDMS chain[50,64]. The ratio r/r_m at α_u thus represents the fraction of the maximum extensibility occurring at this point in the deformation. The values obtained indicate that the upturn in modulus generally begins at approximately 60–70% of maximum chain extensibility[5,50]. This is approximately twice the value that had been estimated previously[7] from stress–strain isotherms of elastomers which may have been undergoing strain-induced crystallization not accounted for in reaching this conclusion.

It is also of interest to compare the values of r/r_m at the beginning of the upturn with some theoretical results of Flory and Chang[4,68] on distribution functions for PDMS chains of finite length. Of relevance here are the calculated values of r/r_m at which the Gaussian distribution function starts to overestimate the probability of extended configurations, as judged by comparisons with the results of Monte Carlo simulations. The theoretical results[64,68] suggest, for example, that the network of PDMS chains having $n = 53$ skeletal bonds, which was experimentally studied, should show an upturn at a value of r/r_m a little less than 0.80. The observed value was 0.77,[50] which is thus in excellent agreement with theory.

More quantitative characterization of this limited chain extensibility requires, of course, a non-Gaussian distribution function[7] for the end-to-end separation r of the short network chains. Ideal for this case is the distribution obtained by Fixman and Alben[69], and recommended particularly for very short chains. It was found to give a much better approximation to the highly reliable Monte Carlo simulated results than does the Gaussian limit, particularly in the most important region of very high extension[37]. The parameters in the Fixman-Alben distribution function can be adjusted to give very good approximations to the Monte Carlo distributions for the short PDMS chains. These Fixman-Alben distributions can then be combined with the constrained-junction theory and reasonable values[37] of the constraint parameter κ to calculate stress–strain isotherms in elongation for bimodal PDMS networks. This is shown in figure 13.13. The observed upturns in the reduced stress $[f^*]$ at high values of the elongation α are seen to be well reproduced by the calculated results[37]. Other non-Gaussian distribution functions have also been successfully used for this purpose, as is shown in figures 13.14 and 13.15[70,71]. The experimental isotherms can also be interpreted using the van der Waals theory of rubberlike elasticity[72,73], as illustrated in figure 13.16.

Another approach, Monte Carlo simulations, utilizes the wealth of information that rotational isomeric state theory provides on the spatial configurations of chain molecules. Some typical results on bimodal networks are described in chapter 8. In brief, Monte Carlo calculations based on the rotational isomeric state approximation are used to simulate spatial configurations and thus distribution

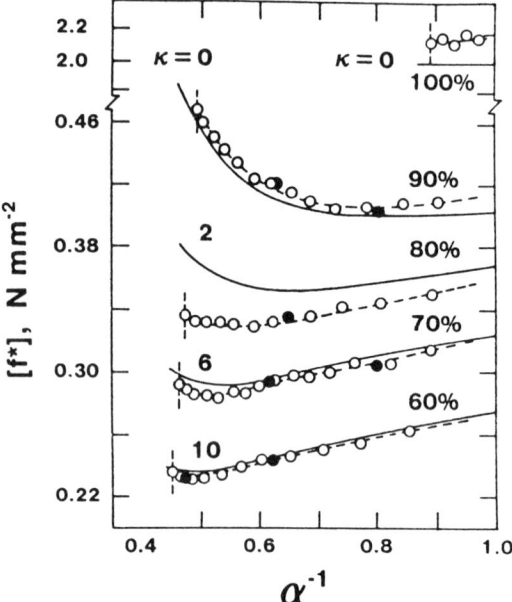

Figure 13.13 Mooney-Rivlin plots for some bimodal PDMS networks ($M_c = 660$ and $18,500 \, \text{g mol}^{-1}$)[37]. The circles and dashed lines represent experimental data, and the filled circles represent data taken out of sequence to test for reversibility. The solid curves give the results of theoretical calculations based on the Fixman-Alben distribution function[69]. Each set of experimental data and corresponding theoretical curve is labeled with the mole percent of short chains present in the network. The values of the parameter κ specified for the theoretical curves are those required to approximate the observed decreases in reduced stress prior to the upturns being modeled by the non-Gaussian distribution.

functions for the end-to-end separations of the chains[8,74]. These distribution functions are then used in place of the Gaussian function in the standard three-chain model[7] in the affine limit to give the desired non-Gaussian theory of rubberlike elasticity. Stress–strain isotherms calculated in this way are strikingly similar to the experimental isotherms obtained for the bimodal networks[8,74]. The overall theoretical interpretations are thus quite satisfactory and would encourage other applications of these distributions, for example, to segmental orientation in networks containing very short chains. Such segmental orientation is of critical importance, for example, with regard to strain-induced crystallization, as discussed in chapter 12.

13.2.3.2 Characterization of the Rupture Process A second important characteristic of an elastomeric network is the value α_r of the elongation at which rupture occurs. The corresponding values of r/r_m show that rupture generally occurred at approximately 80–90% of maximum chain extensibility[50]. These quantitative results on chain dimensions are very important but may not apply directly to

NETWORKS HAVING MULTIMODAL CHAIN-LENGTH DISTRIBUTIONS

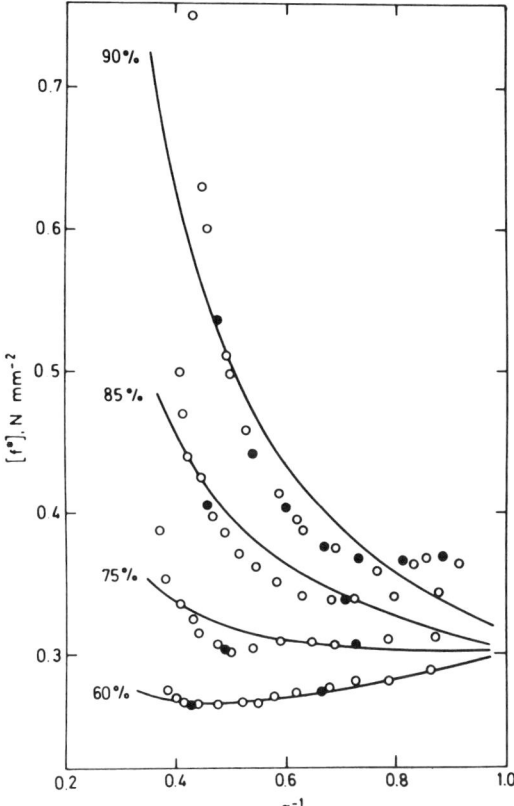

Figure 13.14 Comparison between calculated and experimental Mooney Rivlin isotherms for bimodal PDMS networks consisting of chains having values of $M_c = 220$ and 18,500 g mol^{-1}.[70] Each set of experimental data and the corresponding theoretical curve is labeled with the mole percent of short chains present in the network. Reprinted with permission from Llorente, M. A., et al. (1984), *Macromolecules*, **17**. Copyright 1997, American Chemical Society.

other networks, in which the chains could have very different configurational characteristics and in which the chain length distribution would presumably be quite different from the very unusual bimodal distribution intentionally produced in the present networks.

13.2.3.3 Strain-Induced Crystallization Poly(dimethylsiloxane) networks were found to be unsuitable for characterizing the effects of bimodality on strain-induced crystallization, because of their very low crystallization temperatures. The polymer chosen instead for these end-linked bimodal networks was poly(ethylene oxide), which has a relatively high melting point ($\sim 65°$)[43] and thus readily undergoes strain-induced crystallization[75]. The aspect of relevance here is the use of these networks to elucidate the dependence of strain-induced crystal-

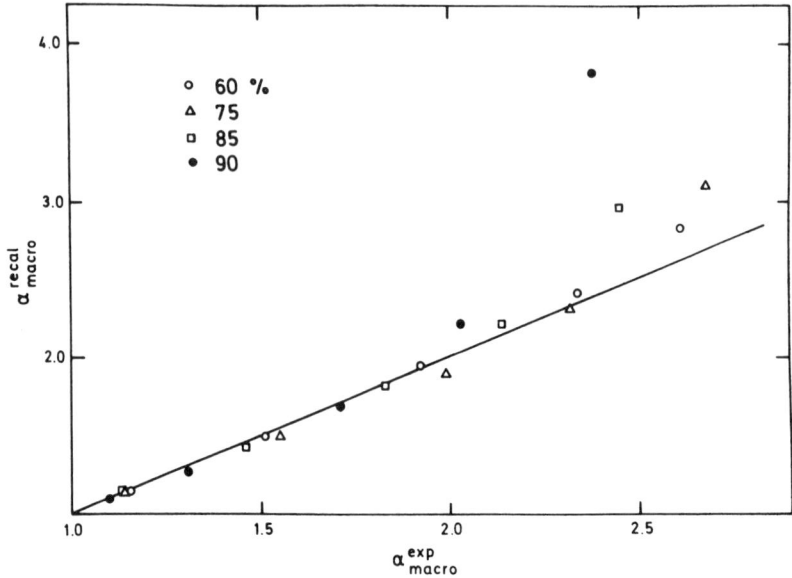

Figure 13.15 Comparisons between calculated and experimental values of the macroscopic ("macro") elongation α for bimodal PDMS networks consisting of chains having values of $M_c = 220$ and $18{,}500$ g mol^{-1}.[71] The designation "exp" refers to the experimental results, and "recal" to a revised calculation that avoids the preaveraging used in some earlier treatments. The values of the mole percent of short chains is indicated in the key.

lization on network chain length distribution. As shown in figure 13.17, decrease in temperature was found to increase the extent to which the values of the ultimate strength of the bimodal networks exceed those of the unimodal ones[75]. This suggests that bimodality facilitates strain-induced crystallization, possibly through increased orientation of the more easily crystallizable chains into crystallization nuclei. Similar conclusions have been reached in studies of elongated bimodal networks of poly(tetrahydrofuran)[76].

13.2.3.4 Thermosets In practical terms, the above results demonstrate that short chains of limited extensibility may be bonded into a long-chain network to improve its toughness. It is also possible to achieve the converse effect. Thus, bonding a small number of relatively long elastomeric chains into a short-chain PDMS thermoset greatly improves both its energy of rupture and impact resistance, as is illustrated in figure 13.18[60]. Approximately 95 mol% short chains gives the maximum effect for the molecular weights involved. Lower concentrations give smaller improvements than can otherwise be achieved, and higher concentrations will gradually convert the composite from a relatively hard material to one that is more rubberlike.

NETWORKS HAVING MULTIMODAL CHAIN-LENGTH DISTRIBUTIONS

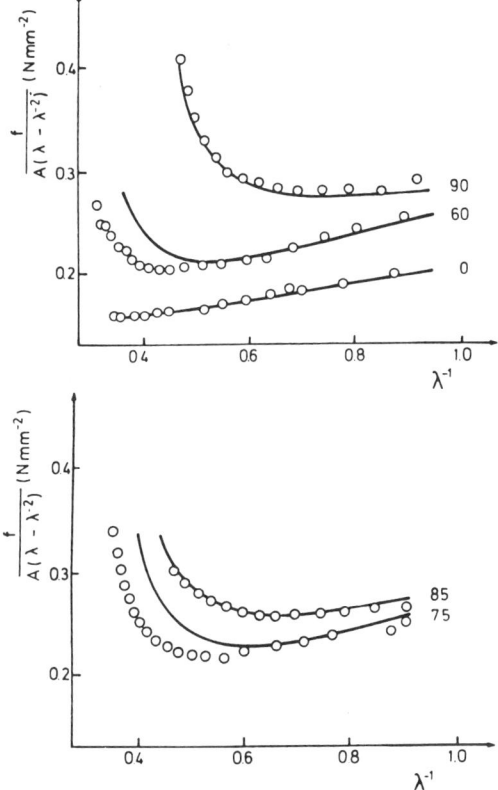

Figure 13.16 Stress–strain isotherms for unimodal and bimodal PDMS networks consisting of short chains (220 g mol^{-1}) combined with much longer chains (18.5 × 10^3 g mol^{-1})[72]. The experimental results are shown by the circles[3], and the van der Waals representation of them by the curves. Each set of experimental data and the corresponding theoretical curve is labeled with the mole percent of short chains present in the network.

13.2.4 Results in Other Mechanical Deformations

13.2.4.1 Biaxial Extension There are numerous other deformations of interest, including compression, biaxial extension, shear, and torsion. As described in appendix C, all of the associated equations of state may be derived from the general equations for the free energy of deformation by proper specification of the deformation ratios[6-8]. In the case of compression ($\alpha < 1$), the equation of state is the same as that for elongation ($\alpha > 1$). Some of these deformations are considerably more difficult to study than simple elongation and, unfortunately, have therefore not been as extensively investigated.

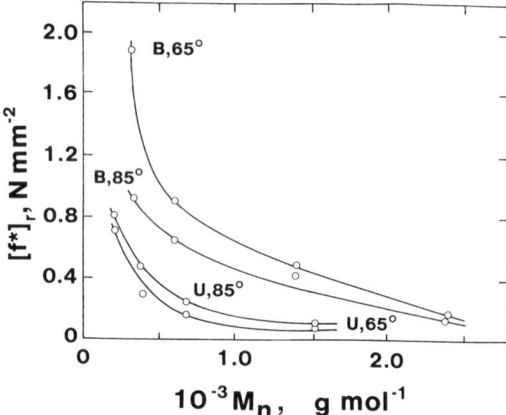

Figure 13.17 The ultimate strength shown as a function of the molecular weight $M_n = M_c$ between cross-links for unimodal (U) and bimodal (B) networks of crystallizable poly(ethylene oxide)[75].

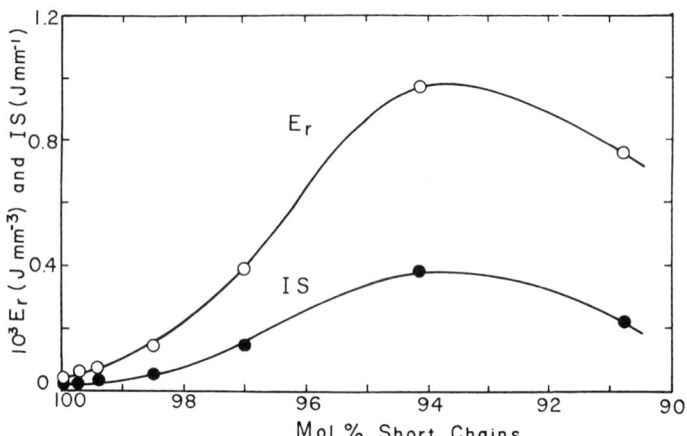

Figure 13.18 The energy required for rupture (E_r) and the impact strength (IS) (as measured by the falling-dart test) shown as a function of composition for bimodal PDMS networks in the vicinity of room temperature[60]. The molecular weights of the chains were 220 and 18,500 g mol^{-1}.

Some of the earliest studies of biaxial extension have involved the direct stretching of a sample sheet in two perpendicular directions within its plane, by two independently variable amounts. Equibiaxial extension, however, is equivalent to compression, and can more conveniently be imposed by inflation of sheets of the elastomer[7]. A good account of experimental biaxial extension results on typical unimodal networks[77] has been given by the simple molecular theory, with

NETWORKS HAVING MULTIMODAL CHAIN-LENGTH DISTRIBUTIONS 205

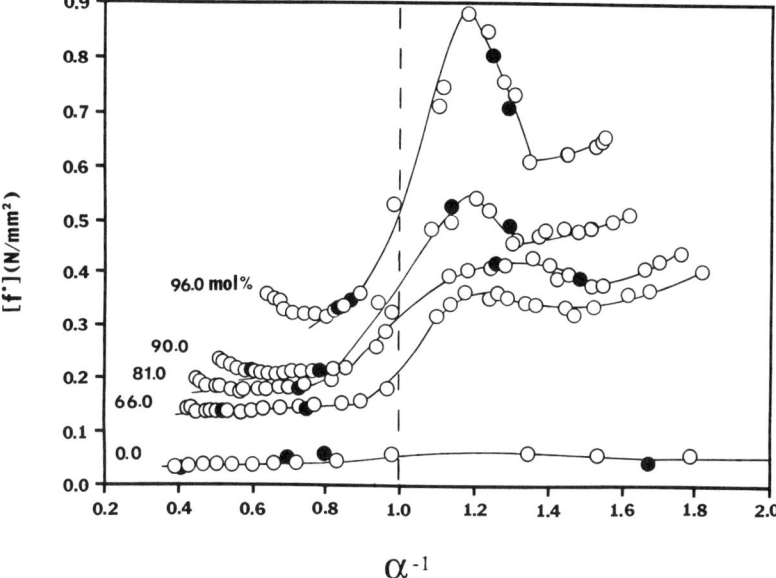

Figure 13.19 Representative stress–strain isotherms for unimodal and bimodal PDMS networks in uniaxial extension (left side of figure), and biaxial extension (right side)[78]. Each curve is labeled with the mole percent of the short chains present in the network. The open circles represent data points measured using increasing deformations, and the filled circles represent data obtained out of sequence to test for reversibility.

improvements being made at lower extensions upon use of the constrained-junction theory[8].

Equibiaxial extension results obtained by inflating sheets of unimodal and bimodal networks of PDMS are illustrated in figure 13.19[78,79]. Upturns in the modulus are seen to occur at high biaxial extensions, as expected. Also of interest, however, are the pronounced maxima preceding the upturns. This represent a challenging feature to be explained by molecular theories addressed to bimodal elastomeric networks, in general.

13.2.4.2 Shear Elastomeric networks can be studied in pure shear by stretching a sheet of the material that has a large ratio of width to length, in the direction perpendicular to the width. Experimental results on natural rubber networks in shear[80] are not well accounted for by the simple molecular theory of rubberlike elasticity. The constrained-junction theory, however, was found to give excellent agreement with experiment, as was described in chapter 5.

Results of such shear measurements on some unimodal and bimodal networks of PDMS are shown in figure 13.20[81]. The bimodal PDMS networks showed large upturns in the pure-shear modulus at high strains that were similar to those reported for elongation and biaxial extension. End-linking in solution gave smal-

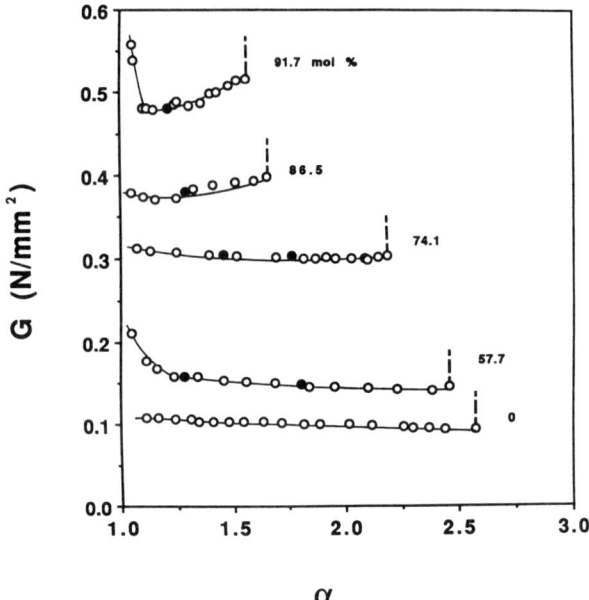

Figure 13.20 Stress–strain isotherms for unimodal and bimodal PDMS networks in shear[81]. Each curve is labeled with the mole percent of the short chains present in the network, and the vertical dashed lines locate the rupture points. The open circles represent data points measured using increasing deformations, and the filled circles represent data obtained out of sequence to test for reversibility.

ler upturns in modulus, presumably because of diminished chain-junction entangling.

13.2.4.3 Torsion Very little work has been done on elastomers in torsion (twisting a cylindrical sample around its long axis). There are, however, some results on stress–strain behavior and network thermoelasticity[7,82,83], as was described in chapter 9. More results are presumably forthcoming, particularly on bimodal networks and on networks containing some of the unusual fillers described in chapter 16.

The same types of bimodal PDMS networks showed rather different behavior in torsion[84]. Specifically, no unambigous upturns in modulus were observed at large deformations. It has not yet been established whether this is due to the inability, to date, of reaching sufficiently large torsions, or whether this is some inherent difference in this type of deformation.

13.2.4.4 Tear Tear tests have been carried out on bimodal PDMS elastomers[19,21] by using the standard "trouser-tear" method illustrated in figure 13.21. Tear energies were found to be considerably increased by the use of a bimodal distribution, with documentation of the effects of compositional changes and changes

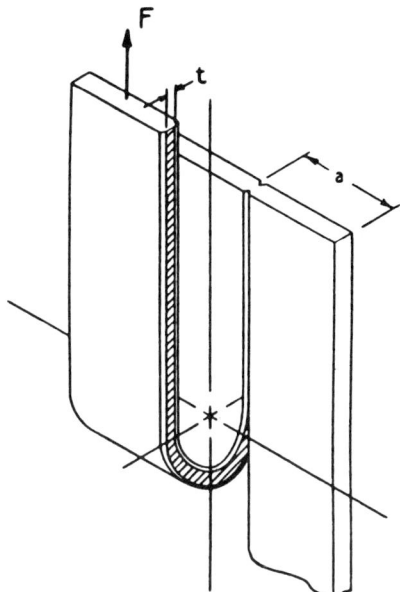

Figure 13.21 Sketch of the type of sample used in the trouser-tear test[19].

in the ratio of molecular weights of the short and long chains. As is shown in figure 13.22, the increase in tear energy did not seem to depend on tear rate[19], an important observation that seems to suggest that viscoelastic effects are not of paramount importance in explaining the observed improvements.

A subsequent series of shear tests[21] established the dependence of the tearing properties on the composition of the bimodal networks and the lengths of the chains used to prepare them. For example, figure 13.23 shows the dependence of the tearing energy on the amount of the short-chain component. The maxima in the curves obtained locates the composition giving the greatest increases in tear energy. Figure 13.24 shows how the tensile strength depends on the ratio of the molecular weights of the two components. The increase in strength with decease in the molecular weight of the short chains must eventually become a decrease when the chains become too short. It would obviously be important to carry out additional studies to establish the molecular weight at which this occurs.

13.2.4.5 Cyclic Deformations Some Rheovibron viscoelasticity results have been reported for bimodal PDMS networks[85]. To provide guidance for the desired interpretations, measurements were first carried out on unimodal networks consisting of the types of chains used in combination in the bimodal networks. Typical results are shown in figure 13.25. One of the important observations was the dependence of crystallinity on the network chain-length distribution.

Some measurements have been made on permanent set for PDMS networks in compressive cyclic deformations[86]. As can be seen from figure 13.26, there appears to be less permanent set or creep in the case of the bimodal elastomers.

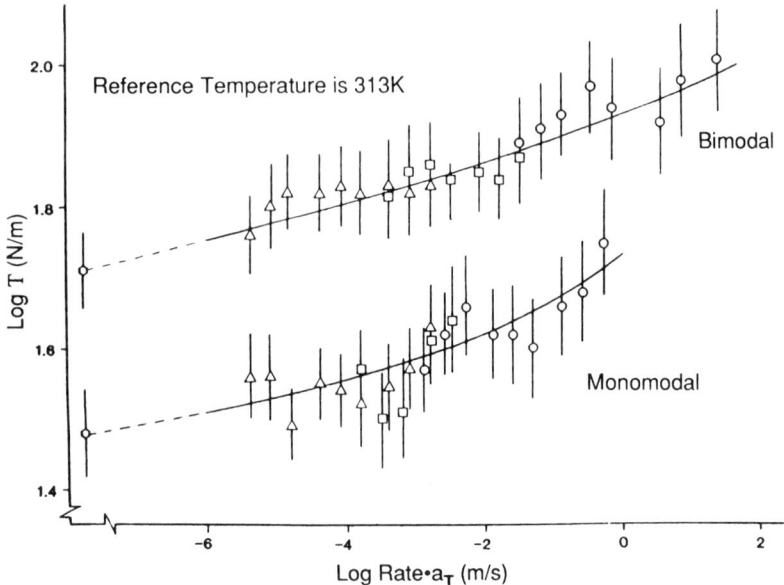

Figure 13.22 Tearing-energy master curves for monomodal (unimodal) and bimodal PDMS networks as a function of the tear rate[19]. Both types of network had an average M_c of 6,800 g mol^{-1}, and the bimodal one consisted of 70 mol% chains having $M_c = 930$ g mol^{-1} combined with chains having $M_c = 20.5 \times 10^3$ g mol^{-1}.

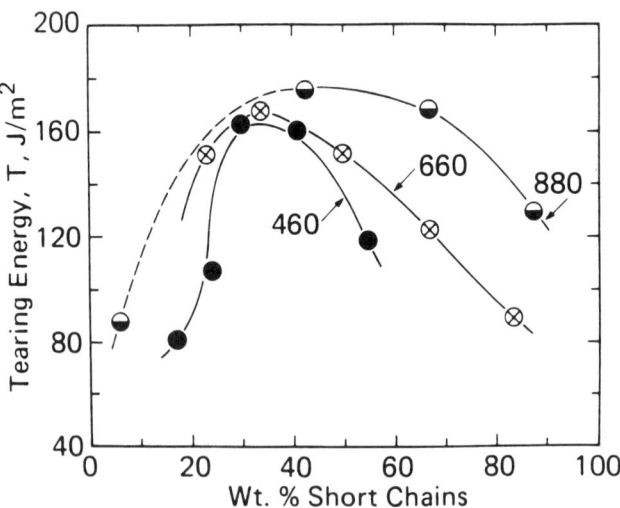

Figure 13.23 Dependence of the tearing energy on the concentration of short chains in bimodal PDMS networks[21]. The values of M_c for the short chains are given in the figure, and the value for the long chains was $M_c = 21.3 \times 10^3$ g mol^{-1}. The maxima occur at approximately 95 mol% short chains.

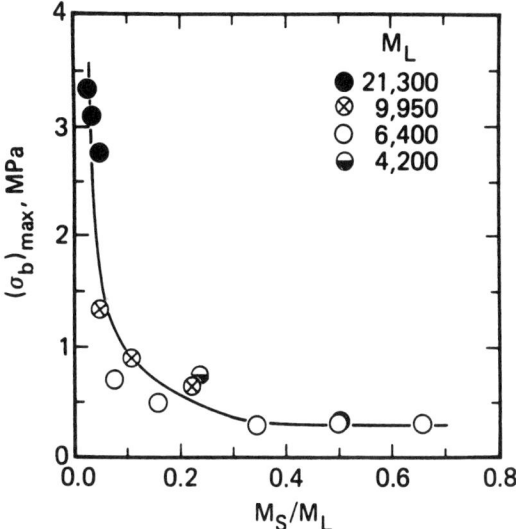

Figure 13.24 Maximum tensile strength of bimodal PDMS networks shown as a function of the ratio of the molecular weights of the short chains to the long chains[21]. The molecular weights of the short chains are listed in the key, and the molecular weights of the short chains can be obtained from the abscissa.

This is consistent with some early results for polyurethane elastomers, shown in figure 13.27[87]. Specifically, cyclic elongation measurements on unimodal and bimodal networks indicated that the bimodal ones survived as much as an order of magnitude more cycles before the occurrence of fatigue failure.

13.2.5 Results on Nonmechanical Properties

13.2.5.1 Birefringence Birefringence measurements have been shown to be very sensitive to bimodality, and have therefore also been used to characterize non-Gaussian effects resulting from it in PDMS bimodal elastomers[88–91]. This is illustrated in Figure 13.28, which shows the effect of strain on the stress–optical coefficient (defined as the ratio of the birefringence to the stress)[88]. There is seen to be a large decrease in this coefficient over a relatively small range in elongation, presumably due to the limited extensibility of the short chains in the bimodal structure. It is therefore a sensitive way of characterizing non-Gaussian effects in network deformation.

13.2.5.2 Freezing Points of Absorbed Solvents This type of experiment is based on the fact that solvent molecules constrained to small volumes form only relatively small crystallites upon crystallization, and therefore exhibit lower crystallization temperatures[92–96]. Some differential scanning calorimetry measurements on solvent molecules constrained in the pores of PDMS elastomers gave evidence

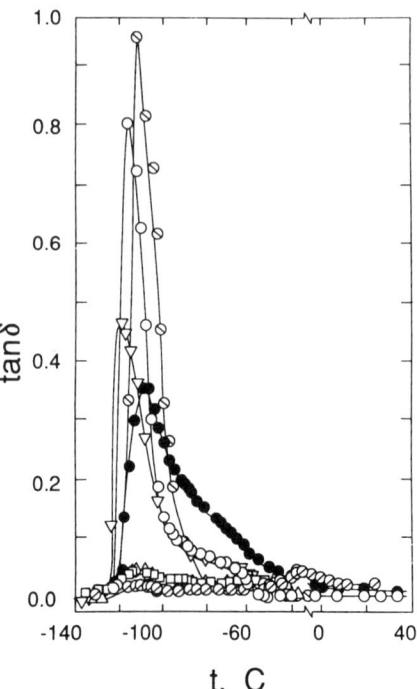

Figure 13.25 Temperature dependence of the dynamic-mechanical loss tangent for unimodal PDMS networks, measured at 11 Hz[85]. The values of $10^{-3} M_n$ are 0.220 (●), 0.660 (◐), 1.10 (○), 1.10 (trifunctional instead of tetrafunctional junctions) (▽), 4.00 (△), 11.3 (□), and 18.5 (⊘) in g mol^{-1}.

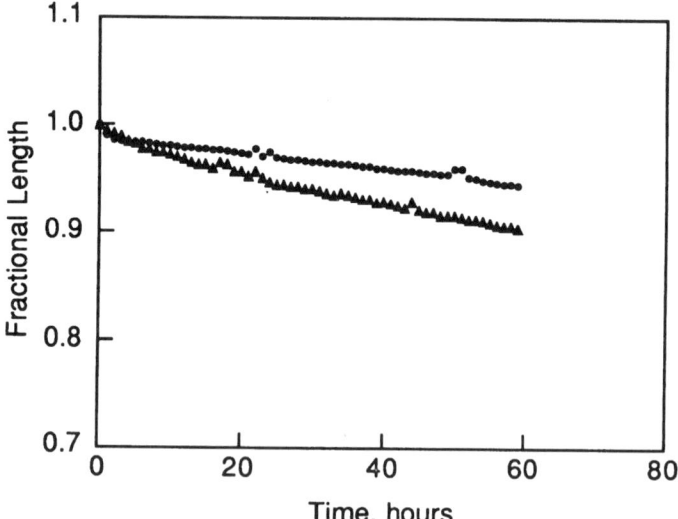

Figure 13.26 Effect of cyclic compressive stress on unimodal (▲) and bimodal (●) PDMS networks represented as the dependence of fractional length on time[86]. Reprinted from Wen, J., Mark, J. E., Fitzgerald, J. J. (1994), *J. Macromol. Sci., Macromol. Rep.*, **A31**, by courtesy of Marcel Dekker, Inc.

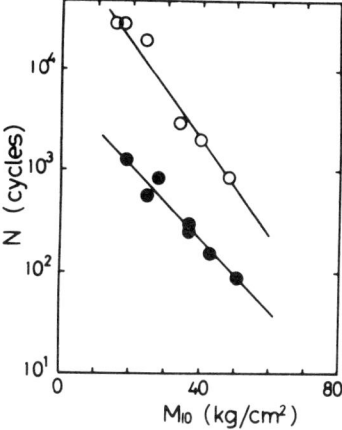

Figure 13.27 Relationship between the number of cycles to break and the modulus at 10% deformation for unimodal (filled circles) and bimodal (unfilled circles) polyurethane networks[87].

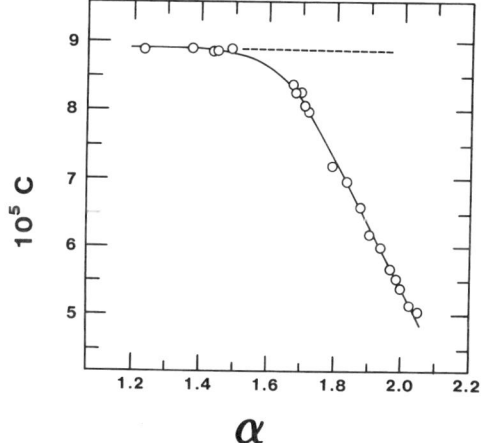

Figure 13.28 Strain dependence of the stress–optical coefficient for a bimodal network containing 93.4 mol% short chains (880 g mol^{-1}), with the rest being considerably longer (21.3 × 10^3 g mol^{-1})[88].

for several crystallization temperatures, which could be indicative of an unusual distribution of pore sizes. The effect seemed to be most pronounced for trimodal networks, which are discussed later in this chapter.

13.2.5.3 Related Investigations Calorimetric measurements on bimodal poly(ethylene oxide) networks indicated that the short chains seemed to decrease the amount of crystallinity in the unstretched state[97]. This is an intriguing result since, as mentioned above, they *increase* the extent of crystallization in the stretched state. A similar study on a different polymer, poly(tetrahydrofuran), did not show any decrease, however[31].

As described in chapter 6, when cyclic molecules are present during the end-linking reaction, the larger ones tend to get trapped by being threaded with chains that are subsequently bonded into the network structure. Such experiments have also been carried out using a bimodal distribution of end-linkable PDMS chains[98].

Dynamic light scattering experiments[32], neutron scattering calculations[39], and computer simulations[40] have also been used to obtain insight into the dynamics and structure of bimodal elastomers.

13.3 Trimodal Networks

As mentioned above, there have been differential scanning calorimetry measurements on solvent molecules constrained in the pores of a variety of PDMS elastomers. Some results on networks having a trimodal distribution of network chain lengths are presented in figure 13.29[48]. The several crystallization temperatures observed for the benzene in this network could possibly be used to obtain additional information on the pore sizes present.

Although there have been attempts to evaluate the mechanical properties of trimodal elastomers, this has not been done in any organized manner. The basic problem is the large number of variables involved, specifically three molecular weights and two independent composition variables (mole fractions); this makes it virtually impossible to do an exhaustive series of relevant experiments. For this reason, the only mechanical property experiments that have been carried out have

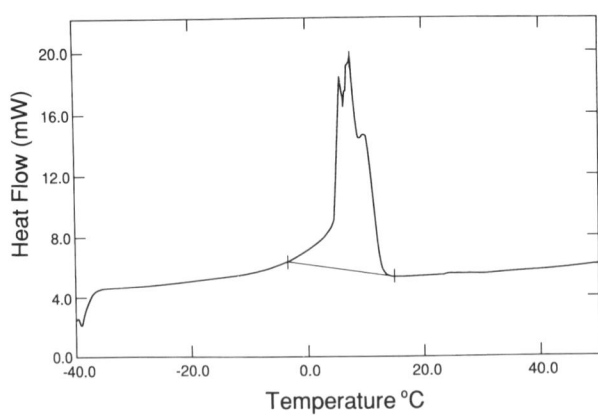

Figure 13.29 Typical DSC scan for a trimodal PDMS network consisting of 13 mol% chains having a value of M_n of 417 g mol^{-1}, 75 mol% chains having 2,760 g mol^{-1}, and 12 mol% chains having 18.9 × 10^3 g mol^{-1}.[48] The swelling solvent whose crystallization temperatures were monitored was benzene.

involved arbitrarily chosen molecular weights and compositions[47,48,99]. Not surprisingly, only modest improvements have been obtained over the bimodal materials.

Some recent computational studies[40], however, indicate that it is possible to do simulations to identify those molecular weights and compositions that should maximize further improvements in mechanical properties. Such simulations are being extended in a search for optimum properties of trimodal, trifunctional networks[100], specifically: (1) the elastic modulus, (2) maximum extensibility, (3) tensile strength, and (4) segmental orientability. Since this is an important problem, it is worth outlining the steps required for its solution. First, it is necessary to generate a three-dimensional network structure in the bulk state, with the Helmholtz free energy of the generated system being minimized, for the phantom-network limit. The most convenient statistical mechanical approach taken[100] can combine the basic principles that characterize phantom networks, in general, with more recent methods developed for generating phantom networks having multimodal chain-length distributions[39].

An important step in the simulations involves generation of a phantom network in which the mean junction positions are known[100]. A randomly generated n-functional network that has three types of chains will contain several combinations of the three chains at each n-functional junction. As a first approximation, each combination can be chosen randomly at each junction. Once the types of chains emanating from a junction are established, their mean positions can be determined according to the Gaussian distribution of mean chain vectors. This can be done by extension of previous work[39] to obtain the variance of the mean chain dimensions in the ensemble. This approach can yield a network whose mean chain vectors are known. Generation of the network should be terminated when the density of the chains in a given cube reaches the known density of the network being modeled.

The trimodal network models may now be deformed by imposing an affine translation on the junctions at the boundaries, and then calculating the parameters of interest[100]. The most important such parameter is the reduced force acting on the network as a function of macroscopic deformation. Since one can now calculate the average deformation of each chain, the macroscopic variables can be related to microscopic ones. Similarly, segmental orientation may be of interest, particularly with regard to strain-induced crystallization[66]. Finally, one can calculate how many of the chains will be subject to a strain that exceeds a given threshold value, which would cause them to break. Once a chain breaks, the result is a network of lower functionality but of higher average chain length (because some of the junctions will now be only bifunctional)[100]. This would increase the extensibility of the network, and such changes in this ultimate property would also be of interest.

The approach described should identify the molecular weights and compositions that should optimize the properties of trimodal networks. Elastomers could then be prepared, using these results for guidance, in an attempt to obtain materials having the best balance of mechanical properties.

214 STRUCTURES AND PROPERTIES OF RUBBERLIKE NETWORKS

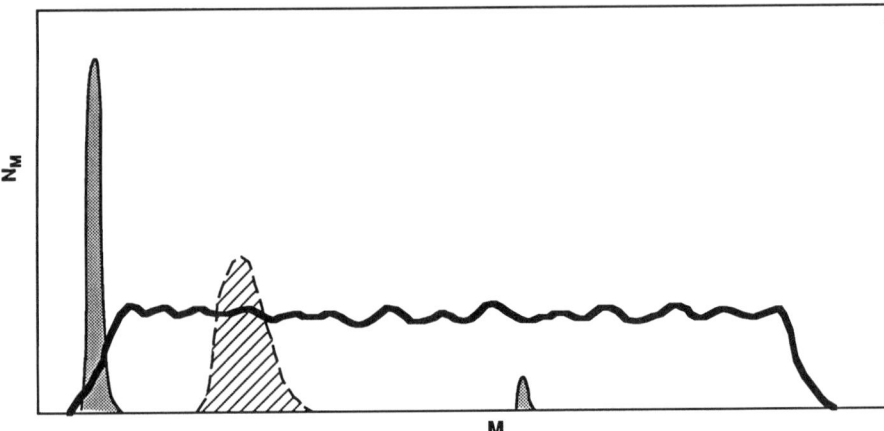

Figure 13.30 Network chain-length distributions, in which N_M is the number of chains in an infinitesimal interval around the specified value of the molecular weight M.[27] A typical unimodal distribution is shown by the area filled with diagonal slashes, a bimodal distribution by the dotted areas, and an extremely broad, pseudo-unimodal distribution (obtained by combining separately prepared polymers) by the unfilled area. Reprinted with permission from Mark, J. E. (1994), *Acc. Chem Res.*, **27**. Copyright 1997 American Chemical Society.

13.4 Networks of Very High Modality

The interpretation of the attractive mechanical properties of bimodal networks has been in terms of a "delegation of responsibilities," with the short chains serving in one role and the long chains in another. If this is picture is true, then it would be interesting to prepare and study networks having extraordinarily broad molecular weight distributions, in that there would be network chains of all conceivable lengths, available for any conceivable mechanism that would improve properties[49]. Polymer prepared from single polymerization would not have a broad enough distribution, but the combination of a series of samples of gradually differing average molecular weight could provide the desired broadness. It should be possible to achieve a polydispersity index of 50 or higher. This is illustrated in figure 13.30. An elastomer of this type has apparently never been synthesized, and might well have unusually attractive mechanical properties.

13.5 Elastomers that May Have Been Inadvertently Bimodal

Elastomers cured with sulfur frequently have improved mechanical properties when the curing conditions are chosen to give cross-links that consist of *chains of sulfur atoms*[101]. It has been suggested that if such polysulfidic cross-links can themselves act as elastomeric network chains, then a bimodal network is produced, albeit inadvertently[102]. This is shown schematically in figure 13.31.

Figure 13.31 Sketch showing the difference between a monosulfidic cross-link (left), and a polysulfidic one (right). In the latter case, the chains of sulfur atoms can possibly act as additional elastically effective chains in what is essentially a bimodal network. (See legend to figure 13.1.)

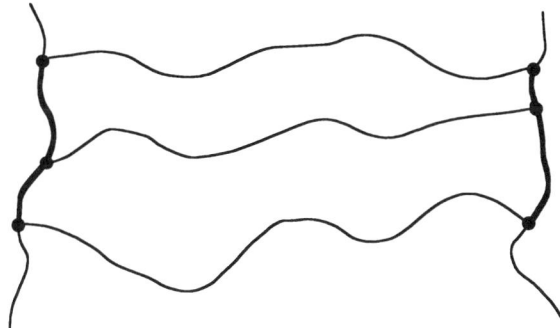

Figure 13.32 Sketch illustrating how incompleteness of reaction in the case of a vinyl hydrosilation end-linking can lead to additional network chains. The oligomeric poly(methyl hydrogen siloxane) chains are shown vertically, and the segments between reacted SiH sites on them possibly contribute elastically effective short chains in what would be an inadvertently bimodal distribution of chain lengths. (See legend to figure 13.1.

Calculations conducted to take into account this possible bimodality gave results in good agreement with experiment[102].

A similar situation may occur in the case of networks end-linked using the addition reaction involving vinyl chain ends and hydrogen atoms in an oligomeric poly(methyl hydrogen siloxane)[103–106]. In the case of incomplete reactions, the segments between the reacted silicon atoms on the oligomer may be long enough to act as elastically effective chains in a bimodal structure. A sketch of this possible situation is shown in figure 13.32. Finally, a bimodal chain-length distribution has also been proposed to explain some unusual properties of polysiloxane networks that have been postcured[107]

13.6 Other Materials in which Bimodality May Be Advantageous

There appear to be other cases where a bimodal distribution of chain length or some other physical property can be advantageous, again possibly through this idea of a delegation of responsibilities.

For example, in the area of thermosets, there seems to be an improvement in mechanical properties when the polymer being cured has a bimodal distribution of molecular weights[108]. In this case, the improvements may be due to different morphologies resulting from the fact that the long chains in a bimodal distribution could have considerably lower solubilities than the short chains. Also, it is well known that the flow characteristics of a polymer during processing[109] can frequently be adjusted by the addition of a small amount of polymer of either very low or very high molecular weight. Another example is in the area of rubber-toughened thermoplastics, in which an elastomer is dispersed as domains within the thermoplastic matrix to improve its mechanical properties[110,111]. It has been found that a bimodal distribution of particle sizes gives the largest improvements[112,113]. Perhaps the small particles are most efficient at stopping one type of failure mechanism, and the large particles, another type. In a related application, there is the possibility that a mixture of two chemically different particles, such as silica [SiO_2] and titania [TiO_2][114–116], could have significant advantages in elastomer reinforcement, as described in chapter 16.

References

(1) Mark, J. E.; Andrady, A. L. *Rubber Chem. Technol.* 1981, 54, 366.
(2) Llorente, M. A.; Andrady, A. L.; Mark, J. E. *Coll. Polym. Sci.* 1981, 259, 1056.
(3) Zhang, Z.-M.; Mark, J. E. *J. Polym. Sci., Polym. Phys. Ed.* 1982, 20, 473.
(4) Mark, J. E. In *Elastomers and Rubber Elasticity*, J. E. Mark and J. Lal, Eds. American Chemical Society: Washington, DC. 1982; p. 349.
(5) Mark, J. E.; Eisenberg, A.; Graessley, W. W.; Mandelkern, L.; Samulski, E. T.; Koenig, J. L.; Wignall, G. D. *Physical Properties of Polymers*, 2nd ed. American Chemical Society: Washington, DC. 1993.
(6) Flory, P. J. *Principles of Polymer Chemistry*. Cornell University Press: Ithaca, NY. 1953.
(7) Treloar, L. R. G. *The Physics of Rubber Elasticity;* 3rd ed. Clarendon Press: Oxford. 1975.
(8) Mark, J. E.; Erman, B. *Rubberlike Elasticity. A Molecular Primer*. Wiley-Interscience: New York. 1988.
(9) Bueche, F. *Physical Properties of Polymers*. Wiley-Interscience: New York. 1962.
(10) Mark, J. E. *J. Chem. Ed.* 1981, 58, 898.
(11) Gottlieb, M.; Macosko, C. W.; Benjamin, G. S.; Meyers, K. O.; Merrill, E. W. *Macromolecules* 1981, 14, 1039.
(12) Mark, J. E. *Adv. Polym. Sci.* 1982, 44, 1.
(13) Mark, J. E.; Lal, J., Eds. *Elastomers and Rubber Elasticity*. American Chemical Society: Washington, DC. 1982; Vol. 193.
(14) Queslel, J. P.; Mark, J. E. *Adv. Polym. Sci.* 1984, 65, 135.
(15) Mark, J. E. *Acc. Chem. Res.* 1985, 18, 202.
(16) Mark, J. E. *Polym. J.* 1985, 17, 265.
(17) Mark, J. E. *Brit. Polym. J.* 1985, 17, 144.
(18) Miller, D. R.; Macosko, C. W. *J. Polym. Sci., Polym. Phys. Ed.* 1987, 25, 2441.
(19) Lanyo, L. C.; Kelley, F. N. *Rubber Chem. Technol.* 1987, 60, 78.
(20) Mark, J. E. In *Frontiers of Macromolecular Science*, T. Saegusa, T. Higashimura and A. Abe, Eds. Blackwell Scientific Publishers: Oxford. 1989; p. 289.

(21) Smith, T. L.; Haidar, B.; Hedrick, J. L. *Rubber Chem. Technol.* 1990, 63, 256.
(22) Mark, J. E. *J. Inorg. Organomet. Polym.* 1991, 1, 431.
(23) Mark, J. E. *Angew. Makromol. Chemie* 1992, 202/203, 1.
(24) Sharaf, M. A.; Mark, J. E.; Hosani, Z. Y. A. *Eur. Polym. J.* 1993, 29, 809.
(25) Sharaf, M. A.; Mark, J. E. *Makromol. Chemie* 1994, 76, 13.
(26) Mark, J. E. *J. Inorg. Organomet. Polym.* 1994, 4, 31.
(27) Mark, J. E. *Acc. Chem. Res.* 1994, 27, 271.
(28) Mark, J. E. *Makromol. Chemie, Suppl.* 1979, 2, 87.
(29) Silva, L. K.; Mark, J. E.; Boerio, F. J. *Makromol. Chemie* 1991, 192, 499.
(30) Hanyu, A.; Stein, R. S. *Macromol. Symp.* 1991, 45, 189.
(31) Roland, C. M.; Buckley, G. S. *Rubber Chem. Technol.* 1991, 64, 74.
(32) Oikawa, H. *Polymer* 1992, 33, 1116.
(33) Hamurcu, E. E.; Baysal, B. M. *Polymer* 1993, 34, 5163.
(34) Subramanian, P. R.; Galiatsatos, V. *Macromol. Symp.* 1993, 76, 233.
(35) Sharaf, M. A.; Mark, J. E.; Al-Ghazal, A. A-R. *J. Appl. Polym. Sci. Symp.* 1994, 55, 139.
(36) Besbes, S.; Bokobza, L.; Monnerie, L.; Bahar, I.; Erman, B. *Macromolecules* 1995, 28, 231.
(37) Erman, B.; Mark, J. E. *J. Chem. Phys.* 1988, 89, 3314.
(38) Termonia, Y. *Macromolecules* 1990, 23, 1481.
(39) Kloczkowski, A.; Mark, J. E.; Erman, B. *Macromolecules* 1991, 24, 3266.
(40) Sakrak, G.; Bahar, I.; Erman, B. *Macromol. Theory Simul.* 1994, 3, 151.
(41) Bahar, I.; Erman, B.; Bokobza, L.; Monnerie, L. *Macromolecules* 1995, 28, 225.
(42) Mark, J. E.; Allcock, H. R.; West, R. *Inorganic Polymers*. Prentice Hall: Englewood Cliffs, NJ, 1992.
(43) Brandrup, J.; Immergut, E., Eds. *Handbook of Polymer Science*. Wiley: New York. 1975.
(44) Clarson, S. J.; Mark, J. E. In *Siloxane Polymers*, S. J. Clarson and J. A. Semlyen, Eds. Prentice Hall: Englewood Cliffs, NJ. 1993; p. 616.
(45) Sur, G. S.; Mark, J. E. *Polym. Bull.* 1985, 13, 505.
(46) Madkour, T. Ph. D. Thesis in Chemistry, University of Cincinnati, OH. 1993.
(47) Madkour, T.; Mark, J. E. *J. Macromol. Sci., Macromol. Rep.* 1994, A31, 153.
(48) Madkour, T.; Mark, J. E. *Polym. Bull.* 1993, 31, 621.
(49) Viers, B. D.; Mark, J. E. Research in progress.
(50) Andrady, A. L.; Llorente, M. A.; Mark, J. E. *J. Chem. Phys.* 1980, 72, 2282.
(51) Llorente, M. A.; Andrady, A. L.; Mark, J. E. *J. Polym. Sci., Polym. Phys. Ed.* 1981, 19, 621.
(52) Kim, C. S. Y.; Ahmad, J.; Bottaro, J.; Farzan, M. *J. Appl. Polym. Sci.* 1986, 32, 3027.
(53) Mark, J. E.; Tang, M.-Y. *J. Polym. Sci., Polym. Phys. Ed.* 1984, 22, 1849.
(54) Tang, M.-Y.; Mark, J. E. *Macromolecules* 1984, 17, 2616.
(55) Tang, M.-Y.; Garrido, L.; Mark, J. E. *Polymer* 1984, 25, 347.
(56) Pan, S.-J.; Mark, J. E. *Polym. Bull.* 1982, 7, 553.
(57) Mark, J. E.; Ning, Y.-P. *Polym. Eng. Sci.* 1985, 25, 824.
(58) Tang, M.-Y.; Mark, J. E. *Polym. Bull.* 1984, 11, 573.
(59) Jiang, C.-Y.; Mark, J. E.; Stebleton, L. *J. Appl. Polym. Sci.* 1984, 29, 4411.
(60) Tang, M.-Y.; Letton, A.; Mark, J. E. *Coll. Polym. Sci.* 1984, 262, 990.
(61) Tang, M.-Y.; Mark, J. E. *Polym. Eng. Sci.* 1985, 25, 29.
(62) Mark, J. E. *Macromolecules* 1984, 17, 2924.
(63) Clarson, S. J.; Galiatsatos, V.; Mark, J. E. *Macromolecules* 1990, 23, 1504.

(64) Mark, J. E.; Eisenberg, A.; Graessley, W. W.; Mandelkern, L.; Koenig, J. L. *Physical Properties of Polymers*, 1st ed. American Chemical Society: Washington, DC. 1984.
(65) Chiu, D. S.; Su, T.-K.; Mark, J. E. *Macromolecules* 1977, 10, 1110.
(66) Mark, J. E. *Polym. Eng. Sci.* 1979, 19, 409.
(67) Flory, P. J. *Statistical Mechanics of Chain Molecules*. Interscience: New York. 1969.
(68) Flory, P. J.; Chang, W. C. *Macromolecules* 1976, 9, 33.
(69) Fixman, M.; Alben, R. *J. Chem. Phys.* 1973, 58, 1553.
(70) Llorente, M. A.; Rubio, A. M.; Freire, J. J. *Macromolecules* 1984, 17, 2307.
(71) Menduina, C.; Freire, J. J.; Llorente, M. A.; Vilgis, T. *Macromolecules* 1986, 19, 1212.
(72) Kilian, H.-G. *Coll. Polym. Sci.* 1981, 259, 1151.
(73) Kilian, H.-G. *Proceedings, Network Group Meeting, Jerusalem* 1990.
(74) Curro, J. G.; Mark, J. E. *J. Chem. Phys.* 1984, 80, 4521.
(75) Sun, C.-C.; Mark, J. E. *J. Polym. Sci., Polym. Phys. Ed.* 1987, 25, 2073.
(76) Hanyu, A.; Stein, R. S. *Makromol. Chem., Macromol. Symp.* 1991, 45, 189.
(77) Obata, Y.; Kawabata, S.; Kawai, H. *J. Polym. Sci., Part A-2* 1970, 8, 903.
(78) Xu, P.; Mark, J. E. *J. Polym. Sci., Polym. Phys. Ed.* 1991, 29, 355.
(79) Xu, P.; Mark, J. E. *Polymer* 1992, 33, 1843.
(80) Rivlin, R. S.; Saunders, D. W. *Philos. Trans. Roy. Soc. London, A* 1951, 243, 251.
(81) Wang, S.; Mark, J. E. *J. Polym. Sci., Polym. Phys. Ed.* 1992, 30, 801.
(82) Gent, A. N.; Kuan, T. H. *J. Polym. Sci., Polym. Phys. Ed.* 1973, 11, 1723.
(83) Gent, A. N.; Kuan, T. H. *J. Polym. Sci., Polym. Phys. Ed.* 1974, 12, 633.
(84) Wen, J.; Mark, J. E. *Polym. J.* 1994, 26, 151.
(85) Andrady, A. L.; Llorente, M. A.; Mark, J. E. *Polym. Bull.* 1991, 26, 357.
(86) Wen, J.; Mark, J. E.; Fitzgerald, J. J. *J. Macromol. Sci., Macromol. Rep.* 1994, A31, 429.
(87) Kaneko, Y.; Watanabe, Y.; Okamoto, T.; Iseda, Y.; Matsunaga, T. *J. Appl. Polym. Sci.* 1980, 25, 2467.
(88) Galiatsatos, V.; Mark, J. E. *Macromolecules* 1987, 20, 2631.
(89) Galiatsatos, V.; Mark, J. E. In *Advances in Silicon-Based Polymer Science. A Comprehensive Resource*, J. M. Zeigler and F. W. G. Fearon, Eds. American Chemical Society: Washington, DC. 1990.
(90) Riande, E.; Saiz, E. *Dipole Moments and Birefringence of Polymers*. Prentice Hall: Englewood Cliffs, NJ. 1992.
(91) Galiatsatos, V.; Neaffer, R. O.; Sen, S.; Sherman, B. J. In *Physical Properties of Polymers Handbook*, J. E. Mark, Ed. American Institute of Physics Press: Woodbury, NY. 1996; p. 535.
(92) Kuhn, W.; Peterlin, E.; Majer, H. *Rubber Chem. Technol.* 1960, 33, 245.
(93) Phalippou, J.; Ayral, A.; Woignier, T.; Quinson, J. F. *Europhys. Lett.* 1991, 14, 249.
(94) Jackson, C. L.; McKenna, G. B. *J. Non-Cryst. Solids* 1991, 131–133, 221.
(95) Goldstein, A. N.; Esher, C. M.; Alivisatos, A. P. *Science* 1992, 256, 1425; and pertinent references cited therein.
(96) Grobler, J. H. A.; McGill, W. J. *J. Polym. Sci., Polym. Phys. Ed.* 1993, 31, 575.
(97) Clarson, S. J.; Mark, J. E.; Sun, C.-C.; Dodgson, K. *Eur. Polym. J.* 1992, 28, 823.
(98) Clarson, S. J.; Mark, J. E.; Semlyen, J. A. *Polym. Commun.* 1987, 28, 151.
(99) Burns, G. T. unpublished results, Dow Corning Corporation.
(100) Erman, B.; Mark, J. E. Work in progress.

(101) Nasir, M.; Teh, G. K. *Eur. Polym. J.* 1988, 24, 733.
(102) Sharaf, M. A.; Mark, J. E. *J. Macromol. Sci., Macromol. Rep.* 1991, 1, 67.
(103) Sharaf, M. A. *Int. J. Polymeric Mater.* 1992, 18, 237.
(104) Sharaf, M. A. *J. Macromol. Sci., Macromol. Rep.* 1992, A30, 83.
(105) Sharaf, M. A.; Mark, J. E. *Polym. Gels Netw.* 1993, 1, 33.
(106) Sharaf, M. A.; Mark, J. E. *J. Polym. Sci., Polym. Phys. Ed.* 1995, 33, 1151.
(107) Quan, X. *Polym. Eng. Sci.* 1989, 29, 1419.
(108) Holmes, G. A.; Letton, A. *Polym. Eng. Sci.* 1994, 34, 1635.
(109) Ferry, J. D. *Viscoelastic Properties of Polymers*, 3rd ed. Wiley: New York. 1980.
(110) Bucknall, C. B. *Toughened Plastics*; Applied Science: London. 1977.
(111) Donald, A. M.; Kramer, E. J. *J. Appl. Polym. Sci.* 1982, 27, 3729.
(112) Okamoto, Y.; Miyagi, H.; Kakugo, M.; Takahashi, K. *Macromolecules* 1991, 24, 5639.
(113) Chen, T. K.; Jan, Y. H. *J. Mater. Sci.* 1992, 27, 111.
(114) Wen, J.; Mark, J. E. *Rubber Chem. Technol.* 1994, 67, 806.
(115) Wen, J.; Mark, J. E. *Polym. J.* 1995, 27, 492.
(116) Wen, J.; Mark, J. E. *J. Appl. Polym. Sci.* 1995, 58, 1135.

14

Small-Angle Neutron Scattering

Small-angle neutron scattering (SANS) experiments from networks were initiated by Benoit and collaborators[1] in the mid-1970s. Currently, SANS is an important major technique used in studying network structure and behavior. Its importance lies in its being a direct method with which observations may be made at the molecular-length scale without the need for a theoretical model for interpreting the data. This feature makes neutron scattering a valuable tool for testing various molecular theories on which current understanding of elastomeric networks is based.

The general features of the technique are explained in section 14.1, followed in section 14.2 by a review of relevant experimental work. Section 14.3 then describes different theories of neutron scattering from networks, and compares them with experimental results.

14.1 General Features of SANS

The technique of neutron scattering and its application to polymers in the dilute and bulk states, to blends, and to networks are described in several review articles[2-9] and a book[10]. The reader is referred to this literature for a more comprehensive understanding of the technique and the underlying theory.

The neutrons incident on a sample during a typical experiment are from a nuclear reactor. Neutrons leaving the source are first collimated so that they arrive at the sample in the form of plane waves. Figure 14.1 shows such an incident neutron wave on two scattering centers i and j. After interacting with the scattering centers, the neutrons move in various directions. In a neutron scattering experiment, the intensity of the scattered neutron wave is measured as a function of the angle θ shown in the figure, in which the vectors $\mathbf{k}_0$ and $\mathbf{k}$ are the wave propagation vectors for incident and scattered neutron rays, respectively. In general, the magnitudes of $\mathbf{k}_0$ and $\mathbf{k}$ differ if there is energy change upon scattering, and in this case the scattering is called *inelastic*. Inelastic scattering

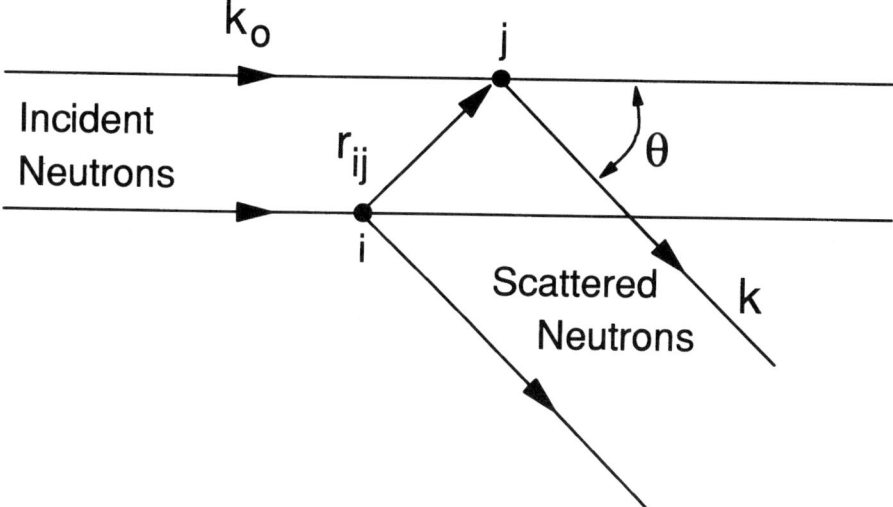

Figure 14.1 A neutron wave incident on two scattering centers i and j. The intensity of the scattered neutron wave is measured as a function of the angle θ shown in the figure. The vectors $\mathbf{k}_0$ and $\mathbf{k}$ are the wave propagation vectors for incident and scattered neutron rays, respectively, and $\mathbf{r}_{ij}$ is the vector from the ith to the jth center.

experiments are particularly useful in studying the dynamics of a system, such as relaxation or diffusion.

Scattering is called *elastic* if the energy change upon scattering is zero and the magnitudes of $\mathbf{k}_0$ and $\mathbf{k}$ are equal; more specifically, $|\mathbf{k}_0| = |\mathbf{k}| = 2\pi/\lambda$. Here, λ is the wavelength of the radiation, and the scattering vector $\mathbf{q}$ is defined by

$$\mathbf{q} = \mathbf{k} - \mathbf{k}_0 \tag{14.1}$$

For the elastic case, the magnitude of q is

$$|\mathbf{q}| = \frac{4\pi}{\lambda} \sin \frac{\theta}{2} \tag{14.2}$$

as can be obtained from the geometry in figure 14.1. The two scattering centers shown in the figure may be labeled points along a network chain or they may belong to different chains, depending on the type of experiment. There are several advantages of SANS over other kinds of scattering such as light, x-ray, or electron scattering[8-10]. First, the wavelengths of the neutrons are suitable for investigating the characteristic dimensions of polymer molecules. If the scattering angle θ is less than a degree, objects the size of a polymer molecule may be studied with readily available neutrons with the proper wavelengths. The length scale to be investigated is approximated by the inverse of $|\mathbf{q}|$ given in eq. (14.2). Thus, distances of 1–100 nm may be probed with neutrons having wavelengths between 0.2 and 0.8 nm if the angle θ is equal to or less than a degree. The adjective "small-angle" refers to angles below one degree. A second advantage of neutron scattering over x-ray scattering is that the energy of neutrons is much less than the energy of electrons of the same wavelength. Therefore, experiments may be performed with much less damage to the samples than with x-ray scattering.

Neutrons interact with the nuclei of atoms, and the strength of scattering is described by the terms "scattering length" or "scattering cross section." Some nuclei, such as that of deuterium, scatter much more strongly than others, for example hydrogen. Thus, some hydrogen atoms of a polymer molecule may be replaced by deuterium atoms to achieve "contrast." The process of replacing hydrogen with deuterium is referred to as deuteration or labeling. Chemical differences between deuterated and undeuterated polymer molecules are relatively small and the technique may conveniently be used with only a small number of deuterated chains.

There are two distinct forms of scattering by neutrons, called coherent and incoherent. Coherent scattering depends on the scattering vector **q** and therefore gives information on the structure of the scattering particles. Incoherent scattering is independent of **q** and can be regarded as background. Various parts of a network may be selectively labeled. Examples are: (1) networks labeled at junctions, (2) networks with labeled elementary chains, that is, chains between two cross-links, (3) networks with labeled chains extending over several cross-links, forming labeled paths, and (4) networks swollen with deuterated solvents.

In a scattering experiment, the intensity of scattered radiation $I(\mathbf{q})$, from a collection of n labeled particles, is obtained from the expression

$$I(\mathbf{q}) = \left\langle \sum_{i,j} b_i b_j \exp(i\mathbf{q} \cdot \mathbf{r}_{ij}) \right\rangle \tag{14.3}$$

Here, b_i is the scattering cross section of the ith scattering center and $\mathbf{r}_{ij}$ is the vector from the ith to the jth center, as shown in figure 14.1. The double summation is carried over all scattering centers and the angular brackets denote that the intensity is from all possible configurations of the scattering centers. The ratio

$$S(\mathbf{q}) = I(\mathbf{q})/I(0) \tag{14.4}$$

of $I(\mathbf{q})$) to the intensity $I(0)$ obtained at zero scattering angle is referred to as the scattering law and is the basic quantity reported in experimental work.

In the study of uniaxially deformed networks, it is convenient to resolve the scattering vector **q** into a component $q_\parallel$, parallel to the direction of stretch, and another, $q_\perp$, perpendicular to it. One can then write $S(\mathbf{q})$ in terms of the two experimentally measured components $S(q_\parallel)$ and $S(q_\perp)$.

The dynamics of chains can be studied by quasi-elastic scattering, for example, in the spin-echo technique, which probes the motions of chain segments of length scales of a few nanometers. This technique is particularly useful for the study of the dynamics of junction fluctuations, and the fluctuations of chain segments in the network.

14.2 Experimental Studies

The first studies of small-angle neutron scattering were made on amorphous un-cross-linked polymers by Kirste[11,12], Ballard et al.[13], and Cotton and collaborators[14,15]. These investigations led to the important conclusion that the dimensions of polymer chains in the bulk state are identical with their unperturbed dimensions[16] in theta solvents. Scattering measurements on networks containing deuterated polystyrene chains end-linked into a nondeuterated polystyrene network, performed by Benoit and collaborators[14], indicated further that the dimensions of chains in the network are identical to their unperturbed dimensions in the bulk un-cross-linked state. In a more extensive study, Beltzung et al.[17], showed that the radii of gyration of poly(dimethylsiloxane) chains in networks are identical to those in the un-cross-linked bulk state. Current studies of rubber elasticity are based on these important observations.

The effects of (1) swelling and (2) uniaxial extension of networks on chain dimensions have been studied extensively by various researchers since 1980[17–39]. The results of these studies may be summarized as follows[40]. (1) Transformations of chain dimensions generally, but not always, fall between the predictions of the affine and phantom models. In experiments performed on polystyrene networks, however, chain dimensions were observed to transform less than that of the phantom model. This phenomenon is described as network unfolding. (2) The state of dilution during cross-linking has a very pronounced effect on the transformations of chain dimensions.

Small-angle neutron scattering has also been applied to the analysis of networks while they were relaxing after a suddenly applied constant uniaxial deformation[32]. Plots of isointensity curves (defined in the following section) gave butterfly or lozenge shapes, although ellipses are predicted under normal conditions. This suggests the presence of inhomogeneities in networks, which could be attributed either to structural fluctuations built into the network during cross-linking or to nonuniformities arising from the deformation[28,30,32].

Earlier dynamic studies of polymer networks were carried out by Higgins and collaborators[38,41,42] and involved measurements of quasi-elastic incoherent scattering. Their results indicate that the network chain segments diffuse around in the network, with a temperature dependence of these motions corresponding to activation energies smaller than those for the center-of-mass motion of the entire chains. Later work in dynamic scattering from networks, by Ewen and collaborators[43,44], involved neutron spin-echo measurements on poly(dimethylsiloxane) networks with labeled junctions, and comparison of the results with those from free end-labeled chains. The neutron spin-echo technique is essentially a quasi-elastic neutron scattering technique that directly measures the time-dependent scattering function $S(\mathbf{q}, t)$ at sub-molecular length scales. The results of Ewen and collaborators showed that the range of fluctuations of junctions are substantial, but somewhat less than the fluctuations of a phantom network model. Their results also indicated that the motions of the junctions are diffusive and are similar to those expected from the Rouse model. Furthermore, the motions of the deuterated junctions were found to be much slower than those of deuterated free chain ends, as expected.

14.3 Theory of SANS from Networks

Theories of SANS from polymer networks require the adoption of a molecular model of the network. The first theoretical work in this field was by Benoit and collaborators[1] for a network chain whose segments transform affinely. Scattering from a chain in a phantom network was later formulated in a very concise and elegant way by Pearson[45], and this approach is presented in the following sections. The derivation is limited to unimodal phantom networks with tetrafunctional junctions, in the interest of emphasizing the basic features of scattering from networks. Derivations for other junction functionalities, bimodal networks, and polydisperse networks have been given by Kloczkowski et al.[46-48] and Higgs and Ball[49].

14.3.1 The Scattering Law

The basic quantity of interest is the scattering law $S(\mathbf{q})$, defined in eq. (14.4) as the ratio of scattering intensity at angle θ to that at $\theta = 0$. For a network chain containing n scattering centers, it is

$$S(\mathbf{q}) \equiv I(\mathbf{q})/I_0$$
$$= (n+1)^{-2} \sum_{i,j} \int \exp(i\mathbf{q} \cdot \mathbf{r}_{ij}) W(\mathbf{r}_{ij}) d\mathbf{r}_{ij} \qquad (14.5)$$

where $i = \sqrt{-1}$ and $W(\mathbf{r}_{ij})$ is the distribution function for the vector $\mathbf{r}_{ij}$ connecting the points i and j. This form of the scattering law follows from eq. (14.3) by assuming all scattering cross-sections to be identical, that is, $b_i = b_j = b$. The intensity of radiation at the scattering angle $\theta = 0$ follows from eq. (14.3) as

$$I(0) = b^2(n+1)^2 \qquad (14.6)$$

Dividing eq. (14.3) by eq. (14.6), and averaging over all configurations of the pair of scatterers i and j indicated by the integration after multiplying with the distribution $W(\mathbf{r}_{ij})$ gives eq. (14.5). The term $W(\mathbf{r}_{ij})$ represents the intramolecular pair correlation function if the scattering centers designated by i and j belong to the same chain. If labeled chains are well separated from each other, then there are no contributions from interchain interactions and the intensity given by eq. (14.5) represents scattering from a single chain. The scattering law given by eq. (14.5) was first derived by Zimm[50].

The scattering law $S(\mathbf{q})$ given by eq. (14.5) is a general one and further calculations are possible only after a molecular model of the network is chosen. In the following sections, scattering is derived from phantom and affine network chains, and then from a chain all of whose segments transform affinely. This is followed by scattering from a labeled path of chains.

14.3.2 Scattering from a Phantom Network Chain

As described earlier, a sufficiently long chain may be represented as a freely jointed chain. Some of the local configurational features of the chain (such as the orientation of a specific vector affixed to a bond), however, are lost in passing to the freely jointed model. This idealized chain nonetheless provides significant simplifications for analytical calculations, and has been extensively used, sometimes indiscriminately. All current theories of SANS from networks are based on the freely jointed long-chain model whose end-to-end distribution is satisfactorily represented as Gaussian. Further information on the problem of scattering from phantom network chains may be obtained from the original works by Pearson[45] and Ullman[51]. The vector $\mathbf{r}_{ij}$ connecting a labeled point i on a freely jointed network chain to another labeled point j on the same chain obeys the same Gaussian distribution function as that given for the end-to-end vector. Thus,

$$W(\mathbf{r}_{ij}) = \left(\frac{3}{2\pi \langle r_{ij}^2 \rangle_0}\right)^{3/2} \exp\left(-\frac{3 r_{ij}^2}{2 \langle r_{ij}^2 \rangle_0}\right) \quad (14.7)$$

where $\langle r_{ij}^2 \rangle_0$ is the mean-square distance between points i and j in the undeformed state, identified by the subscript zero.

The distribution function $W(\mathbf{r}_{ij})$ in the deformed state is

$$\begin{aligned} W(\mathbf{r}_{ij}) &= \left(\frac{3}{2\pi \langle r_{ij}^2 \rangle}\right)^{3/2} \exp\left(-\frac{3 r_{ij}^2}{2 \langle r_{ij}^2 \rangle}\right) \\ &= [(2\pi)^3 \langle x_{ij}^2 \rangle \langle y_{ij}^2 \rangle \langle z_{ij}^2 \rangle]^{-1/2} \exp(-x_{ij}^2/2\langle x_{ij}^2 \rangle - y_{ij}^2/2\langle y_{ij}^2 \rangle - z_{ij}^2/2\langle z_{ij}^2 \rangle) \end{aligned} \quad (14.8)$$

where $\langle x_{ij}^2 \rangle$, $\langle y_{ij}^2 \rangle$, and $\langle z_{ij}^2 \rangle$ are the mean-square components of the vector $\mathbf{r}_{ij}$ in the deformed state. Substituting the expression for $W(\mathbf{r}_{ij})$ given by eq. (14.8) into eq. (14.5) and performing the integration over $d\mathbf{r}_{ij} = dx_{ij}\,dy_{ij}\,dz_{ij}$ leads to

$$S(\mathbf{q}) = \frac{1}{n^2} \sum_{i,j=1} \exp(-q_x^2 \langle x_{ij}^2 \rangle / 2 - q_y^2 \langle y_{ij}^2 \rangle / 2 - q_z^2 \langle z_{ij}^2 \rangle / 2) \quad (14.9)$$

where $n+1$ is replaced by n for sufficiently long chains and q_x, q_y, and q_z are the components of the scattering vector $\mathbf{q}$. The vector $\mathbf{r}_{ij}$ between two scattering centers is related to its time-averaged component $\bar{\mathbf{r}}_{ij}$ and the instantaneous fluctuation $\Delta \mathbf{r}_{ij}$ from this by the relation

$$\mathbf{r}_{ij} = \bar{\mathbf{r}}_{ij} + \Delta \mathbf{r}_{ij} \quad (14.10)$$

Squaring both sides of the above equation and taking the ensemble average leads to

$$\langle r_{ij}^2 \rangle = \langle \bar{r}_{ij}^2 \rangle + \langle (\Delta \mathbf{r}_{ij})^2 \rangle \quad (14.11)$$

since the cross-product $\langle \bar{\mathbf{r}}_{ij} \cdot \Delta \mathbf{r}_{ij} \rangle$ vanishes, inasmuch as the mean values and instantaneous fluctuations are uncorrelated. This relation follows from the properties of the phantom network model presented in chapters 2 and 8. According to the model, mean chain vectors transform affinely with macroscopic deformation and their fluctuations are independent of deformation. These fluctuations are given by

$$\langle \bar{x}_{ij}^2 \rangle = \lambda_x^2 \langle \bar{x}_{ij}^2 \rangle_0$$
$$\langle (\Delta x_{ij})^2 \rangle = \langle (\Delta x_{ij})^2 \rangle_0 \qquad (14.12)$$

with similar expressions for the y and z components. Here, λ_x is the x component of the principal deformation gradient tensor λ:

$$\lambda = \mathrm{diag}(\lambda_x, \lambda_y, \lambda_z) \qquad (14.13)$$

and represents the ratio of the final length of the network in the x direction to the corresponding length in the state of reference. The x component of eq. (14.11) may be written as

$$\langle x_{ij}^2 \rangle = \left[\lambda_x^2 + (1 + \lambda_x^2) \frac{\langle (\Delta x_{ij})^2 \rangle_0}{\langle x_{ij}^2 \rangle_0} \right] \langle x_{ij}^2 \rangle_0 \qquad (14.14)$$

with similar expressions for the y and z components. For a freely jointed chain, the average quantity $\langle x_{ij}^2 \rangle_0$ for x_{ij} is related to the average $\langle r^2 \rangle_0$ of the whole chain by

$$\langle x_{ij}^2 \rangle_0 = \langle r_{ij}^2 \rangle_0 / 3 = \eta \langle r^2 \rangle_0 / 3 \qquad (14.15)$$

where η is the fractional distance:

$$\eta = |i - j|/n \qquad (14.16)$$

The relationship between $\langle (\Delta r_{ij})^2 \rangle$ and $\langle r^2 \rangle_0$ is derived in appendix E as

$$\langle (\Delta r_{ij})^2 \rangle = (\eta - \tfrac{1}{2}\eta^2) \langle r^2 \rangle_0 \qquad (14.17)$$

This relationship is valid for fluctuations of the distance r_{ij} between two points i and j belonging to the same chain. Substituting this relation into eq. (14.14) and using the resulting expression in eq. (14.9) leads to the scattering law:

$$S(\mathbf{q}) = \frac{1}{n^2} \sum_{i,j=1}^{n} \exp[-\nu \eta [1 - (1 - \lambda^{*2}) \eta / 2]] \qquad (14.18)$$

where

$$\nu = q^2 \langle r^2 \rangle_0 / 6 \qquad (14.19)$$

and

$$\lambda^{*2} = (q_x^2 \lambda_x^2 + q_y^2 \lambda_y^2 + q_z^2 \lambda_z^2)/q^2 \qquad (14.20)$$

For scattering parallel to the direction of extension, $\lambda^* = \lambda_\parallel$, while for scattering perpendicular to this direction, $\lambda^* = \lambda_\perp = 1/\sqrt{\lambda_\parallel}$. Equation (14.18), first obtained by Pearson[45], may be simplified by reducing the double summation to a single summation, using the identity

$$\sum_{i,j}^{n} f(|i-j|) = 2\sum_{j=0}^{n-1}(n-j)f(j) \tag{14.21}$$

and converting to an integral. The result is

$$S(\mathbf{q}) = 2\int_0^1 d\eta(1-\eta)\exp[-\nu\eta[1-\eta(1-\lambda^{*2})/2] \tag{14.22}$$

After integration, the scatterings parallel and perpendicular to the direction of stretch are obtained, respectively, as

$$S(q_\parallel) = 2\left(\frac{2}{\nu\alpha}\right)^{1/2}\left(\frac{\alpha+1}{\alpha}\right)\left\{F_1\left(\frac{\nu}{2\alpha}\right)^{1/2}-\exp[-\nu(\alpha+2)/2]F_1\left[\left(\frac{\nu}{2\alpha}\right)^{1/2}(\alpha+1)\right]\right\}$$
$$+\frac{2}{\nu\alpha}\{\exp[-\nu(\alpha+2)/2]-1\} \tag{14.23}$$

$$S(q_\perp) = 2\left(\frac{2}{\nu\beta}\right)^{1/2}\left(\frac{\beta-1}{\beta}\right)$$
$$\times\left\{-\exp[-\nu(2-\beta)/2]F_2\left[\left(\frac{\nu}{2\beta}\right)^{1/2}(1-\beta)\right]F_2\left(\frac{\nu}{2\beta}\right)^{1/2}\right\} \tag{14.24}$$
$$-\frac{2}{\nu\beta}\{\exp[-\nu(2-\beta)/2]-1\}$$

where

$$\alpha = \lambda_\parallel^2 - 1 \tag{14.25}$$

$$\beta = 1 - \lambda_\perp^2 \tag{14.26}$$

$$F_1(x) = e^{x^2}\int_x^\infty e^{-t^2}dt = \frac{\pi^{1/2}}{2}e^{x^2}\text{erfc }x \tag{14.27}$$

$$F_2(x) = e^{-x^2}\int_0^x e^{t^2}dt \tag{14.28}$$

and erfc denotes the complementary error function. By making use of the series and asymptotic expansions of F_1 and F_2, that is,

$$\lim_{x\to\infty} F_1(x) \approx \frac{1}{2x} - \frac{1}{4x^3} \tag{14.29}$$

and

$$\lim_{x\to\infty} F_2(x) \approx \frac{1}{2x} + \frac{1}{4x^3} \tag{14.30}$$

it can be shown that $S(\mathbf{q})$ may be simplified to the following limiting forms:
1. Small-deformation limit:

$$\lim_{\lambda \to 1} S(\mathbf{q}) = \frac{2}{\nu^2}(e^{-\nu} + \nu - 1) \tag{14.31}$$

which corresponds to the scattering from an unperturbed Gaussian coil[52].

2. Long-wavelength limit ($\nu \to 0$): eq. (14.23) gives[45]

$$\lim_{\nu \to 0} S(q_\parallel) = 1 - \frac{\nu}{3}[\lambda^2/4 + 3/4] \tag{14.32a}$$

and eq. (14.24) gives

$$\lim_{\nu \to 0} S(q_\perp) = 1 - \frac{\nu}{3}[1/4\lambda + 3/4] \tag{14.32b}$$

where λ is the extension ratio.

3. Short-wavelength limit ($\nu \to \infty$): eqs. (14.23) and (14.24) give, respectively[45],

$$\lim_{\nu \to \infty} S(q_\parallel) = \frac{2}{\nu}(1 - \lambda^2/4) \tag{14.33a}$$

and

$$\lim_{\nu \to \infty} S(q_\perp) = \frac{2}{\nu}(1 - 1/4\lambda) \tag{14.33b}$$

14.3.3 Scattering from an Affine Network Chain

An affine network chain has the property that its end points transform affinely with macroscopic deformation and the points along the chain are phantomlike, while no restrictions are imposed on them. As was stated earlier, the affine network chain is obtained from a phantom network chain in the limit of infinitely large junction functionality. To obtain the scattering from the affine network chain from the phantom network chain result, one therefore has to express $S(q)$ for the phantom network chain in terms of network functionality. These expressions have been obtained by Kloczkowski et al.[46-48], for the general ϕ-functional network. They are:

1. Long-wavelength limit ($\nu \to 0$):

$$\lim_{\nu \to 0} S(q_k) = 1 - \frac{\nu}{3}\left[\lambda_k^2\left(\frac{1}{2} - \frac{1}{\phi}\right) + \frac{1}{2} + \frac{1}{\phi}\right] \tag{14.34}$$

where k indicates either the parallel or the perpendicular direction, relative to the stretch direction. In the limit of infinitely large ϕ, eq. (14.34) reduces to

$$\lim_{\nu \to 0} S(q_\parallel) = 1 - \frac{\nu}{6}[\lambda^2 + 1] \tag{14.35a}$$

$$\lim_{\nu \to 0} S(q_\perp) = 1 - \frac{\nu}{6}[\lambda^{-1} + 1] \tag{14.35b}$$

2. Small-**q** limit ($\nu \to \infty$):

$$\lim_{\nu \to \infty} S(q_k) = \frac{2}{\nu}\left\{1 - \frac{1}{\nu}\left[\lambda_k^2 \frac{2(\phi-2)}{\phi} + \frac{4-\phi}{\phi}\right]\right\} \quad (14.36)$$

which leads in the limit of infinitely large ϕ to

$$\lim_{\nu \to \infty} S(q_\parallel) = \frac{2}{\nu}\left\{1 - \frac{1}{\nu}(2\lambda^2 - 1)\right\} \quad (14.37a)$$

$$\lim_{\nu \to \infty} S(q_\perp) = \frac{2}{\nu}\left\{1 - \frac{1}{\nu}(2\lambda^{-1} - 1)\right\} \quad (14.37b)$$

Equations (14.31)–(14.33), (14.35), and (14.37) provide expressions relating the changes in chain dimensions to the macroscopic state of deformation. The parameter $S(\mathbf{q})$ is proportional to the scattered intensity and is the parameter measured in SANS experiments. The expressions given in this section may therefore be tested directly by experiments. Results of such experiments on various polymeric networks are reported in the following section.

14.3.4 Scattering from a Chain whose Individual Segments Deform Affinely

This mode of deformation of a chain applies to a chain in a highly restrictive environment, perhaps when the system is very close to the glassy state. The scattering law for such a chain was obtained by Benoit and collaborators[1]. In terms of the discussion presented above, the scattering law may be obtained by equating the fluctuation $\langle \Delta \mathbf{r}_{ij} \rangle$ to zero for all i and j in eq. (14.12), substituting into eq. (14.9), and then performing operations identical to those used for the phantom network. The resulting expressions for the scattering law are

$$S(q_\parallel) = \frac{2}{\nu^2 \lambda^4}(e^{-\nu\lambda^2} + \nu\lambda^2 - 1) \quad (14.38a)$$

$$S(q_\perp) = \frac{2}{\nu^2 \lambda^{-2}}(e^{-\nu/\lambda} + \nu/\lambda - 1) \quad (14.38b)$$

14.3.5 Scattering from a Labeled Path in the Network

This mode of scattering is obtained when a long, labeled precursor chain is mixed with unlabeled chains, and the resulting system is randomly cross-linked.

The first theory of SANS from labeled paths in the polymer network was proposed in 1978 by Warner and Edwards[53], who used the replica formalism introduced earlier by Deam and Edwards[54]. They thus derived an expression for $S(\mathbf{q})$ for a randomly cross-linked, tetrafunctional phantom network. Ullman was the first to study the SANS from labeled cross-linked paths (with specified numbers of cross-links along the path) in a unimodal, Gaussian, phantom, ϕ-functional network with treelike topology[55]. He derived the general expression for the scattering form factor from a path including dangling ends, which do not deform under strain. However, his results were not exact because of an approximation he made on the correlations among vectors along the path.

The exact solution of the problem for cross-linked paths in a phantom network was given by Higgs and Ball[49], and independently by Kloczkowski et al.[46,47]. The latter group gives the following expression for $S(\mathbf{q})$ for a labeled path composed of n chains (i.e., $n-1$ cross-links along the path):

$$S(\mathbf{q}) = \frac{1}{n^2} \sum_{n_i=1}^{n} \sum_{n_j=1}^{n} \int_0^1 d\theta \int_0^1 d\zeta \exp\left\{-\nu\left[\lambda^2|n_j + \theta - n_i + \zeta|\right.\right.$$
$$+ (1-\lambda^2)\left\{\frac{2(\phi-1)}{\phi(\phi-2)}\left[1 - \frac{1}{(\phi-1)^{|n_j-n_i|}}\right]\right.$$
$$\left.\left.\left. + \frac{\phi-2}{\phi}\left[\zeta(1-\zeta) + \theta(1-\theta) - \frac{\zeta+\theta-2\zeta\theta}{(\phi-1)^{|n_j-n_i|}} + \frac{\eta-|n_j-n_i|}{(\phi-1)^{|n_j-n_i|}}\right]\right\}\right]\right\}$$

(14.39)

The expression for $S(\mathbf{q})$ from cross-linked paths given by eq. (14.39) differs substantially from eq. (14.22) for scattering from end-linked chains. However, in the limiting case when $n = 1$, eq. (14.39) reduces to eq. (14.22). The main reason for this difference is that for cross-linked paths, the double integration cannot be simplified to a single integral and can only be evaluated numerically[46,47]. This is because the variables θ and ζ in eq. (14.39) are not separable and cannot be replaced by a single variable $\eta = |\theta - \zeta|$, as for end-linked chains. Also, the double summation over different chains belonging to the path cannot be simplified, and a double integration has to be carried out for each member of a double sum. The expression given by eq. (14.39) is for scattering from paths without dangling chains at the ends. The extension to paths with dangling chains would have to follow the method introduced by Ullman[55].

14.4 Typical Results from Experiments and Comparison with Theory

The transformation of the radius of gyration of chains in uniaxially stretched or swollen networks can be calculated using the theoretical expressions given in the preceding section. The first experiments measuring these dimensions directly by SANS experiments were performed on polyisoprene networks by Yu et al.[39], and their results are shown in figure 14.2. The dashed lines represent results calculated from the affine and the phantom network models, using eqs. (14.32) and (14.35)[40]. Comparison of the curves with experimental data shows that the affine and phantom limits form upper and lower bounds to the microscopic deformation of chains in a real network. The solid curves were obtained by use of the constrained-junction theory[40]. Figure 14.3 presents the tranformation of the radius of gyration of chains in an isotropically swollen polyisoprene network. The circles

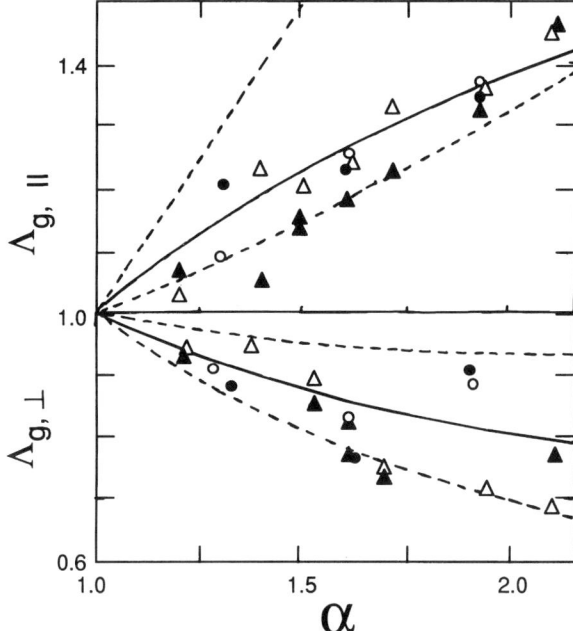

Figure 14.2 Results of measurements of chain dimensions by SANS on polyisoprene networks[39]. The circles and triangles represent experimental data for different molecular weights M_c between cross-links. The data in the upper half are for deformation parallel to the direction of stretch, and those in the lower half are for deformation in the transverse direction. The dashed lines represent results calculated from the affine and the phantom network models[40], and the solid curves were obtained from the constrained-junction theory[40].

represent data from the neutron scattering experiments of Bastide et al.[21], and the solid curve was obtained[40] from the constrained-junction theory.

The results of neutron scattering experiments are usually represented in the form of Kratky plots, as $q^2 S(q)$ versus q. Figure 14.4 shows such plots for uniaxially deformed and swollen networks[25]. The points represent experimental results for parallel and perpendicular directions to the extension direction, as indicated in the figure. The dot–dashed curves were calculated from the affine network chain model, and the dotted curves from the phantom network chain model. Comparisons with experimental data show a large discrepancy with theory that has not been resolved thus far. Several factors may be contributing to this discrepancy, such as contributions from intermolecular correlations, structural or deformational heterogeneities at the microscopic level, and the effects of pendant chains. These problems have been the subject of research since the late 1980s and the interested reader is referred to the original papers in this field[26–30,32,35,44,56–58]

The study of the dynamics of points along a chain is another line of research that has been developing in recent years. It was initiated by the theoretical work of

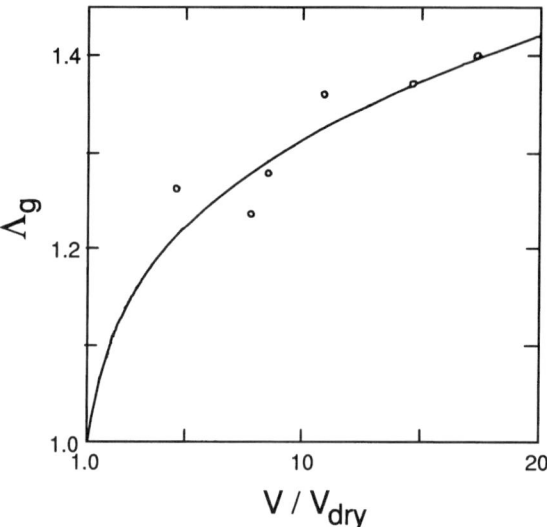

Figure 14.3 Transformation of the radius of gyration of chains in an isotropically swollen polyisoprene network. The circles are from neutron scattering experiments[21], and the solid curve was obtained[40] from the constrained-junction model.

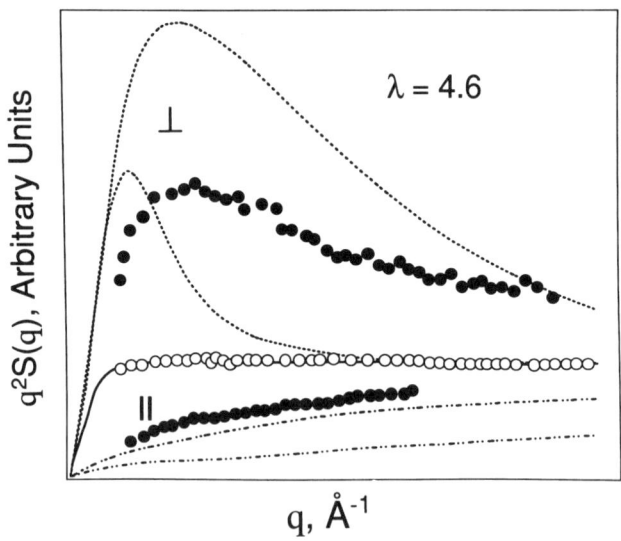

Figure 14.4 Kratky plots [$q^2 S(q)$ vs. q] for uniaxially deformed and swollen PDMS networks[25]. The filled points represent experimental results for directions parallel and perpendicular to the extension direction, as indicated. The open circles show results obtained from isotropic samples. The dot–dashed and dotted curves were calculated from the affine and the phantom network chain models, respectively.

Warner[59], and the experimental studies of Higgins and collaborators[60,61] using incoherent neutron scattering techniques. Neutron spin-echo measurements of Oeser et al.[35,44,56] clearly show that network junctions are not frozen but fluctuate in time, and the spatial range of these fluctuations is of the order of network chain dimensions, as predicted by the James-Guth theory.

References

(1) Benoit, H.; Duplessix, R.; Ober, R.; Cotton, J. P.; Farnoux, B.; Jannink, G. *Macromolecules* 1975, 8, 451.
(2) Ullman, R. *Ann. Rev. Mater. Sci.* 1980, 10, 261.
(3) Higgins, J. S. *J. Appl. Cryst.* 1978, 10, 261.
(4) Maconnachie, A.; Richards, R. W. *Polymer* 1978, 19, 739.
(5) Picot, C. In *Static and Dynamic Properties of the Polymeric Solid State*. Reidel, Dordrecht. 1982; p. 127.
(6) Sperling, L. H. *Polym. Eng. Sci.* 1984, 24, 1.
(7) Lohse, D. J. *Polym. News* 1986, 12, 8.
(8) Wignall, G. D. In *Physical Properties of Polymers*, J. E. Mark, A. Eisenberg, W. W. Graessley, L. Mandelkern, E. T. Samulski, J. L. Koenig and G. D. Wignall, Eds. American Chemical Society: Washington, DC. 1993; p. 313.
(9) Wignall, G. D. In *Physical Properties of Polymers Handbook*, J. E. Mark, Ed. American Institute of Physics Press: Woodbury, NY. 1996; p 299.
(10) Higgins, J. S.; Benoit, H. *Neutron Scattering from Polymers*. Clarendon Press: Oxford. 1994.
(11) Kirste, R. G.; Kruse, W. A.; Schelten, J. *Macromol. Chem.* 1973, 162, 299.
(12) Kirste, R. G.; Kruse, W. A.; Ibel, K. *Polymer* 1975, 16, 120.
(13) Ballard, D. G.; Schelten, J.; Wignall, G. D. *Eur. Polym. J.* 1973, 9, 965.
(14) Cotton, J. P.; Decker, D.; Benoit, H.; Farnoux, B.; Higgins, J. S.; Jannink, G.; Ober, R.; Picot, C.; des Cloizeaux, J. *Macromolecules* 1974, 7, 863.
(15) Daoud, M.; Cotton, J. P.; Farnoux, B.; Jannink, G.; Sarma, G.; Benoit, H.; Duplessix, R.; Picot, C.; DeGennes, P. G. *Macromolecules* 1975, 8, 804.
(16) Flory, P. J. *Principles of Polymer Chemistry*. Cornell University Press: Ithaca, NY. 1953.
(17) Beltzung, M.; Picot, C.; Rempp, P.; Herz, J. E. *Macromolecules* 1982, 15, 1594.
(18) Benoit, H.; Decker, D.; Duplessix, R.; Picot, C.; Rempp, P.; Cotton, J. P.; Farnoux, B.; Jannink, G.; Ober, R. *J. Polym. Sci., Polym. Phys. Ed.* 1976, 14, 2119.
(19) Hinckley, J. A.; Han, C. C.; Moser, B.; Yu, H. *Macromolecules* 1978, 11, 836.
(20) Bastide, J.; Picot, C.; Candau, S. *J. Macromol. Sci. Phys.* 1981, B19, 13.
(21) Bastide, J.; Duplessix, R.; Picot, C. *Macromolecules* 1984, 17, 83.
(22) Bastide, J.; Herz, J. E.; Boué, F. *J. Physique* 1985, 46, 1967.
(23) Bastide, J. In *Physics of Finely Divided Matter*, Springer (Proceedings in Physics): Heidelberg. 1985.
(24) Bastide, J.; Boué, F. *Physica* 1986, 27, 1154.
(25) Bastide, J.; Boué, F. *Physica* 1986, 104A, 251.
(26) Bastide, J.; Leibler, L. *Macromolecules* 1988, 21, 2647.
(27) Bastide, F.; Buzier, M.; Boué, F. In *Polymer Motion in Dense Systems*. Springer Verlag: Berlin. 1988; p. 112.
(28) Bastide, J.; Leibler, L. *Macromolecules* 1988, 21, 2647.

(29) Bastide, J.; Boué, F.; Buzier, M. In *Molecular Basis of Polymer Networks*, A. Baumgartner and C. Picot, Eds. Springer Verlag: Berlin. 1989; p. 48.
(30) Bastide, J.; Leibler, L.; Prost, J. *Macromolecules* 1990, 23, 1821.
(31) Boué, F.; Bastide, J.; Buzier, M.; Collette, C.; Lapp, A.; Herz, J. *Prog. Coll. Polym. Sci.* 1987, 75, 152.
(32) Boué, F.; Bastide, J.; Buzier, M.; Lapp, A.; Herz, J.; Vilgis, T. A. *Coll. Polym. Sci.* 1991, 269, 195.
(33) Boué, F.; Farnoux, B.; Bastide, J.; Lapp, A.; Herz, J.; Picot, C. *Europhys. Lett.* 1986, 1, 637.
(34) Boué, F.; Bastide, J.; Buzier, M.; Collette, C.; Lapp, A.; Herz, J. *Prog. Coll. Polym. Sci.* 1987, 75, 152.
(35) Oeser, R.; Bastide, J.; Farago, B.; Picot, C.; Richter, D. In *Polymer Motion in Dense Systems*. Springer Verlag: Berlin. 1988; p. 208.
(36) Beltzung, M.; Herz, J.; Picot, C. *Macromolecules* 1983, 16, 580.
(37) Beltzung, M.; Picot, C.; Herz, J. *Macromolecules* 1984, 17, 663.
(38) Clough, S. B.; Maconnachie, A.; Allen, G. *Macromolecules* 1980, 13, 774.
(39) Yu, H.; Kitano, T.; Kim, C. Y.; Amis, E. J.; Chang, T.; Landry, M. R.; Wesson, J. A.; Han, C. C.; Lodge, T. J.; Glinka, C. J. In *Advances in Elastomers and Rubber Elasticity*, J. Lal and J. E. Mark, Eds. Plenum Press: New York. 1986; p. 407.
(40) Erman, B. *Macromolecules* 1987, 20, 1917.
(41) Allen, G.; Brier, P. N.; Goodyear, G.; Higgins, J. S. *Faraday Symp. Chem. Soc.* 1972, 6, 169.
(42) Allen, G.; Higgins, J. S.; Wright, C. J. *Faraday Symp. Chem. Soc.* 1973, 7, 348.
(43) Ewen, B.; Richter, D. *Festkoerperprobleme* 1987, 27, 1.
(44) Oeser, R.; Ewen, B.; Richter, D.; Farago, B. *Phys. Rev. Lett.* 1988, 60, 1041.
(45) Pearson, D. S. *Macromolecules* 1977, 10, 696.
(46) Kloczkowski, A.; Mark, J. E.; Erman, B. *Macromolecules* 1989, 22, 1423.
(47) Kloczkowski, A.; Mark, J. E.; Erman, B. *Macromolecules* 1989, 22, 4502.
(48) Kloczkowski, A.; Mark, J. E.; Erman, B. *Macromolecules* 1991, 24, 3266.
(49) Higgs, P. G.; Ball, R. C. *J. Phys. France* 1988, 49, 1785.
(50) Zimm, B. H. *J. Chem. Phys.* 1948, 16, 1093.
(51) Ullman, R. *J. Chem. Phys.* 1979, 71, 436.
(52) Debye, P. *J. Phys. Coll. Chem.* 1947, 51, 18.
(53) Warner, M.; Edwards, S. F. *J. Phys. A.: Math. Gen.* 1978, 11, 1649.
(54) Deam, R. T.; Edwards, S. F. *Phil. Trans. Roy. Soc. London* 1976, 280, 317.
(55) Ullman, R. *Macromolecules* 1982, 15, 582.
(56) Oeser, R.; Picot, C.; Herz, J. In *Polymer Motion in Dense Systems*. Springer Verlag: Berlin. 1988.
(57) Vilgis, T. A.; Boué, F. *J. Polym. Sci., Polym. Phys. Ed.* 1988, 26, 2291.
(58) Edwards, S. F.; Vilgis, T. A. *Rep. Prog. Phys.* 1988, 51, 243.
(59) Warner, M. *J. Phys. C.* 1981, 14, 4985.
(60) Higgins, J. S. In *Treatise on Material Science and Technology–Neutron Scattering*. Academic Press: New York. 1979; p. 381.
(61) Higgins, J. S.; Nicholson, L. K.; Hayter, J. B. *Polymer* 1981, 22, 163.

15

Bioelastomers

There are a variety of biopolymeric materials which exhibit rubberlike elasticity[1-14]. This is perhaps to be expected when one recalls that most biopolymers are randomly coiled chains with considerable flexibility, and that they are frequently covalently cross-linked or have sufficient numbers of aggregated units to exist in network structures. One very large group of plant materials, the polysaccharides, are in this category, and they do require some elastomeric properties in their functioning. In many of these cases, however, the cross-linking is there primarily for a secondary purpose, such as preventing solubility. When swollen with water or aqueous solutions, such polysaccharides form gels which do exhibit the high deformability and recoverability that are the hallmarks of rubberlike elasticity. Not surprisingly, however, relatively few mechanical property measurements have been carried out to characterize the structures of these gels.

The bioelastomers occurring in animals, including vertebrates and mammals, however, are there *specifically* for their rubberlike elasticity. They are vital, for example, for the functioning of skin, arteries and veins, and much of the lung and heart tissue. Since they are produced by the ribosome "factories" in the body, they are proteins. Thus, the major focus of this chapter is on those proteins specifically designed to function as bioelastomers.

15.1 Some General Observations

It is useful to summarize some general information on bioelastomers that is presented elsewhere[15].

Even with the temporary restriction to bioelastomers which are proteins, there is an almost staggering variety of interesting materials. For example, there is elastin in vertebrates (including mammals)[2,5,10,16], resilin in insects[2,5], abductin in mollusks[5,17], arterial elastomer in octopuses[18-21], circulatory and locomotional proteins in cephalopods[22,23], and viscid silk in spider webs[24-26]. Since they are mammals, polymer scientists and engineers who are interested in bioelastomers have focused heavily on elastin! Any materials of this type, however, are worth studying in their own right, to learn more about rubberlike elasticity and biological function. Such studies should also provide guidance on how Nature might be mimicked by synthetic chemists[27], to produce better nonbiological elastomers[15].

Elastin is a chemical copolymer of α-amino acids which are placed along the chain backbone with sufficient irregularity to prevent crystallization entirely, no matter how low the temperature or how high the elongation. The chains are kept as flexible as possible in two ways. First, the α-amino acids for the repeat units are chosen so as to have side chains as small as possible, to avoid steric congestion and its associated stiffness[2,5]. The ones which predominate (with specification of their $-R$ side chains) are glycine ($-H$), alanine ($-CH_3$), serine ($-CH_2OH$), and valine [$-CH(CH_3)_2$]. Second, the side chains chosen are predominantly nonpolar. This was thought to have the advantage of minimizing interchain interactions that could interfere with the desired chain mobility and cause sluggishness. This now seems unlikely since the backbone is itself polar by virtue of the backbone in the ($-CHR-CO-NH-$) protein repeat unit, and this requires that the elastin be highly swollen with body fluids to take it from the glassy state into the elastomeric state. This swelling would increase the separation among the elastin chains and thus minimize any such undesirable interchain interactions.

The nonpolar side groups may have the purpose of enclosing or sheathing the polar peptide backbone. If so, it could be the origin of the unusual negative temperature coefficient of swelling, namely, the fact that elastin swells *more* as the temperature is decreased! This could be vital in the case of vertebrates (with variable body temperatures) whose elastin might otherwise become glassy at low temperatures[5].

Elastin chains synthesized in vivo are enzymatically cross-linked in a highly specific manner. The cross-linking occurs at lysine repeat units at carefully placed locations along the chain backbone. They are first oxidized, and then fused into an aromatic–heterocyclic cross-link of high stability[2,28]. This means that the cross-links are fixed in number and in location along the protein backbone. They might also be carefully positioned spatially, by virtue of the lysine repeat units being preceded and succeeded by alanine units, possibly in α-helical conformations.

As already mentioned, elastin is in the highly swollen state when it is functioning as a bioelastomer. Since it is bathed in excess body fluids, it is at swelling equilibrium and the system is semiopen, in the thermodynamic sense, in that it can exchange diluent (but not polymer) with the surroundings. In such an open system, stretching an elastomer causes it to swell more, and compressing it causes it to swell less[29]. It is therefore obviously important to take such changes into account when elasticity measurements are conducted under open conditions. Failure to do so[30] has caused considerable confusion in some of the literature on elastin[31]. Alternatively, the problems of a closed system can be avoided, either by removing the excess solvent and then immersing the swollen network in oil (to avoid evaporation of the solvent if it is volatile)[5], or by using a nonvolatile solvent as the swelling agent and then simply carrying out the measurements in the open atmosphere[32]. These experiments are described further below.

Meaningful stress–strain measurements on samples of elastin are difficult to conduct. Although elastin is totally amorphous, it is laid down along collagen fibers which subsequently provide reinforcement for it during its functioning in the body. When these collagen fibers are extracted as part of the typical elastin purification scheme, the elastin retains this fiberlike shape. These "fibers" are generally not in register with one another and thus experience different strains as the macroscopic deformation proceeds. The problem can be avoided by working with separate elastin fibers, but this involves rather painstaking manipulations under a microscope[33].

The question of the storage of elastic energy is of general importance in the area of rubberlike elasticity, and assumes particular importance in the case of bioelastomers. The issue is perhaps best illustrated by a simple example. Energy is expended in deforming an elastomer, and the elastomer stores some of this energy in the same way that a compressed gas or stretched spring stores energy. Both are capable of doing work on the surroundings. However, only a small fraction of the energy expended in deforming an elastomer will generally be recoverable during its retraction. This is because some of the energy of deformation is degraded into heat by viscous effects. Not only is this wasteful, but the heat generated has to be removed, whether the elastomer is part of an organism or part of an automobile tire. Much work has been carried out on bioelastomers to see whether Nature has been more successful than synthetic polymer chemists in preparing elastomers in which such "heat buildup" is minimized[5].

15.2 Chemical Aspects of Protein Bioelastomers

15.2.1 Overall Amino Acid Composition

Although all of the materials to be discussed in this section are proteins, they vary in their distribution in the animal kingdom, including both vertebrates and invertebrates. They can therefore be expected to vary considerably in their amino acid composition. The most important example, particularly to people, is elastin. It occurs only in vertebrates, including mammals, and is a major constituent in ligaments and arteries. In the case of ligaments and other connective tissue, the static (near-equilibrium) properties are of greatest importance. In the case of the circulatory system, dynamic properties come to the fore, particularly the ability of elastin to store and release elastic energy during the pulsating flow of the blood. Although most samples studied scientifically have been obtained from pigs or cows, the results obtained are thought to apply to elastin from any source, including human beings[2,5,34].

Resilin is the insect's answer to the elastin occurring in vertebrates[2]. It is found in the cuticle, where it can function either as part of the reciprocating wing system in the continuous flight of some insects, or as part of the more irregular jumping mechanism in others.

The analogous protein occurring in mollusks is called abductin[2]. It occurs at the hinges in their shells, where it counteracts the muscles holding the shell closed. The abductin, which is compressed when the shell is closed, provides the energy for opening it when these muscles relax. Some mollusks use this device cyclically as a means of locomotion. In the case of cephalopod mollusks, such as octopus, squid, and cuttlefish, elastomeric proteins occur in the arteries[18-21], where they serve a dynamic-mechanical purpose similar to that served by elastin in the arteries of vertebrates.

Some final examples are somewhat different from the others in that the elastomer functions outside of the body of the animal that produced it. They are several of the silks produced by spiders in making the webs they use to capture prey[24-26,35-38]. The two types of greatest interest are the viscid silk in the catching spiral, and the frame silk and dragline fibers, particularly when swollen with water.

In discussing the chemical nature or primary structure of some of these protein bioelastomers, it is important to first scrutinize the overall amino acid composition to look for general trends, and this is done in the remainder of this section. One can then examine the sequencing of these amino acids within the protein chain, as is done in the following section. The repercussions of these features on higher order structural aspects, including chain flexibility, are discussed whenever appropriate.

Some of the most relevant compositional information for representative protein rubbers are given in table 15.1. As can be seen from these results, glycine is particularly prevalent in elastin, resilin, and abductin. Since glycine has the smallest possible side group, the $-H$ atom, it was originally thought to be present because of the flexibility it would give the elastomeric network chains. This flexibility may not be of critical importance, in general, since glycine is present to only a small extent in octopus arterial elastomer (OAE), for example. Also of relevance is the fact that alanine has the $-CH_3$ side group, serine the $-CH_2OH$ group, and valine the $-CH(CH_3)_2$ group, all of which are small (at least when compared with the alternatives). Yet, in the proteins surveyed in table 15.1, alanine and valine are major components only in elastin, and serine is present in large amounts only in resilin. The same conclusion can be reached from the condensed data given in the first row of table 15.2. These results imply that Nature has several devices for giving protein chains the flexibility and mobility required for rubberlike elasticity, and only one of these involves the use of small side groups.

It was also thought that the nonpolarity of the side groups in elastin is important with regard to its elasticity, since the absence of strong coulombic interactions between the chains facilitate chain mobility. As can be seen from table 15.2,

Table 15.1 Some Amino Acid Compositions for Several Protein Bioelastomers, in Mole Percent[a]

Amino acid	Elastin	Resilin	Abductin	OAE[b]	Viscid silk
Glycine	31.3	42.2	58.4	8.5	44.2
Alanine	24.4	7.0	6.7	7.2	8.3
Serine	1.2	12.8	5.1	7.2	3.1
Valine	12.8	1.2	0.6	6.2	6.7
Proline	11.3	7.5	0.8	5.5	20.5

[a] Adapted from reference 5. [b] Octopus aterial elastomer.

Table 15.2 Summarized Compositional Information[a]

Side chain	Elastin	Resilin	Abductin	OAE[b]	Viscid silk
Small	54.9	62.0	70.2	22.9	55.6
Nonpolar	60.9	22.0	28.5	39.3	41.6
Polar	7.5	35.0	13.1	51.9	14.3

[a] Adpated from reference 5.   [b] Octopus arterial elastomer.

however, resilin and octopus bioelastomers, in particular, have high ratios of polar to nonpolar side groups and still exhibit good rubberlike elasticity. Perhaps this should have been expected, since these proteins are so highly swollen when they are elastomeric. The network chains would be well separated by the presence of solvent molecules, and would interact relatively little, even if their side groups were very polar. In fact, increased polarity could be one of the alternate mechanisms for *improving* rubberlike elasticity. The polarity would increase the swelling in water that is essential for any of these proteins to be elastomeric.

In any case, the conditions for rubberlike elasticity certainly seem to be met in these proteins. In elastin, for example, almost all the experimental evidence indicates a highly random, irregular network structure, consisting of chains of very high mobility. Also, the "proof is in the pudding," of course, and the rubberlike elasticity of these materials is obvious to those studying them in vitro (even if it is not always consciously appreciated by the animals using them in vivo).

15.2.2 Amino Acid Sequencing

The total sequencing is not known for any of the proteins mentioned in the previous sections, but enough of it is known in the case of elastin to detect some interesting trends[5,39].

Chain segments occurring near cross-links along the chain trajectory are rich in alanine, and these alanines are thought to occur in pairs or triplets on both sides of single lysine units. The grouping of alanines is possibly important since they could then form short α-helical stretches to which the lysine groups attach[5]. As described below, the lysine groups are used for the introduction of cross-links, and having them at the ends of rodlike helical stretches could be an important way of positioning them spatially for this cross-linking reaction.

The parts of the chains well away from the cross-links are rich in glycine (Gly), alanine (Ala), valine (Val), and proline (Pro), as shown in table 15.1. In many parts, these four α-amino acids are not randomly arranged, but appear in three repeat sequences. These sequences are the –Val–Pro–Gly–Gly– tetrapeptide, the –Val–Pro–Gly–Val–Gly– pentapeptide, and the –Ala–Pro–Gly–Val–Gly–Val– hexapeptide[5,39].

In all three of these repeat sequences, a proline is followed by a glycine, and this pairing could give rise to a series of conformations called "β turns," in which the chain almost doubles back on itself. It has been proposed that a number of such turns could be coiled to form a β helix or spiral. This has given rise to what is called the "librational" model for the elasticity exhibited in elastin[40-45]. In this model, the peptide units in these segments are visualized as undergoing rocking or librational motions of relatively large amplitude. Stretching the elastin stretches these β helices in a way that damps the amplitudes of these rocking motions. This constraint on the chain dynamics is viewed as the origin of the decrease in entropy that, in all other elastomers, has been attributed to the simple stretching out of chains in a totally random network structure[15]. It was thus proposed that in elastin the elastic force has a uniquely different molecular origin[40,41].

In spite of a large number of experimental studies and conformational calculations in its support, there is still great skepticism regarding this model. One objection involves the fact that only a relatively small portion of the elastin backbone consists of the repeat units cited. The rest seems totally random and therefore presumably incapable of forming any types of regular structures that would undergo the described librational motions. Also, NMR studies of elastin indicate that its backbone is highly mobile, with correlation times of the order of 0.1 µs, and this would seem to argue against the presence of helices of any significant stability or permanence. Also, it is possible to satisfactorily explain the thermoelastic[46] and elastic[47] properties of elastin using the standard rotational isomeric state representation[48] of elastin sequences, without the need for introducing additional structural features of greater complexity.

15.2.3 Cross-Linking Chemistry

The cross-linking of elastin, and presumably most of the other elastomeric proteins as well, occurs with much better control than is usually the case with nonbiopolymeric elastomers[5,49]. The cross-links generally appear to be covalent bonds, as evidenced by the fact that these materials do not dissolve in any solvents except those that are known to cleave peptide bonds. Some of the spider-web silks mentioned earlier are exceptions in that they do dissolve in nondegradative solvents. These cross-links are therefore physical and temporary, and are thought to be small crystallites[24-26,50-54]. If so, these elastomers would be interesting analogues to plasticized poly(vinyl chloride), which has some elasticity because of the small crystallites it contains. These are formed from chain sequences that have sufficient stereochemical regularity (specifically syndiotacticity) to crystallize[55]. It would be fascinating to obtain similarly detailed information on the regular sequences that crystallize in spider-web silk[36,37,56].

In both elastin and resilin, unique cross-links are formed by a series of steps, at least some of which are enzymatically controlled. In the specific case of elastin, the cross-links result from the oxidation and fusion of four lysine side chains to form an aromatic heterocyclic pyridinium ring of considerable stability. The entire cross-link structure consists of either of two isomers called desmosine and isodesmosine. They are shown in parts (A) and (B), respectively, of figure 15.1[5]. Each such structure has four attachment points, which means it could, in principle, become a cross-link of functionality eight. The attachment points act pairwise, however, with a given pair attaching to two relatively closely spaced repeat units on the same chain. The effective functionality may thus be only four. If the short

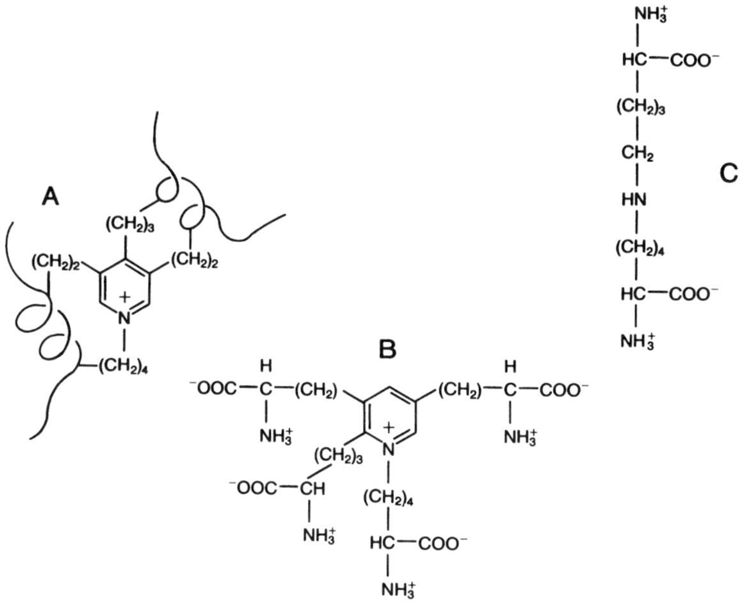

Figure 15.1 Structures of three types of cross-links occurring in elastin[5]. (A) is desmosine, (B) is isodesmosine, and (C) is dehydrolysino-nor-leucine.

segments between the closely spaced cross-links act as effective network chains, however, the network would have an interesting bimodal distribution of network chain lengths[15], as described in chapter 13. The mole fraction of the very short chains, however, would presumably be much too small to have a beneficial effect on the mechanical properties of the elastomer. There are thought to be other structures that possibly contribute to the cross-linking in elastin. One of these, formed from two lysine side chains, is shown in part (C) of figure 15.1.

The cross-links in resilin are similarly highly specific, but are formed from the phenolic side groups of tyrosine repeat units[5]. They are thought to be the dityrosine and trityrosine structures shown in parts (A) and (B), respectively, of figure 15.2. Details of their biosynthesis are essentially entirely unknown.

Even less is known about the nature of the cross-linking in the other protein elastomers[5], except for the already mentioned probable role of small amounts of crystallization in the case of the spider-web silks[36].

15.3 Network Thermoelasticity

15.3.1 General Relevance

In proposing models for the elastic deformation of a rubberlike material, the most important piece of experimental information is how much of the elastic force is entropic and how much is energetic[15,29]. As was described in chapter 9, this

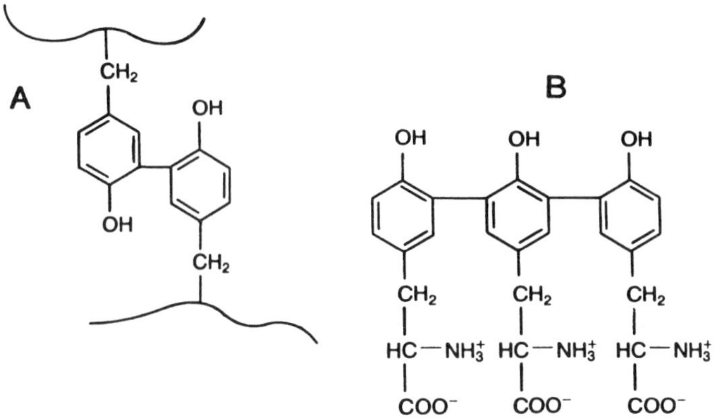

Figure 15.2 Structures of two types of cross-links occurring in resilin[5]. (A) is based on dityrosine, and (B) on trityrosine.

information is obtained from thermoelastic or force–temperature measurements. Since the bioelastomers differ so greatly in chemical structure from the commercial nonbiological elastomers which were used to establish the foundations of elasticity theories, they have been much studied with regard to their thermoelasticity. Elastin has been the most studied, resilin considerably less, and the others hardly at all.

As already mentioned, studies of this type are greatly complicated by the fact that these bioelastomers have to be highly swollen to be elastomeric. Water is the preferred diluent since aqueous solutions are used to obtain the desired swelling in vivo. Since water is relatively volatile, however, water-swollen samples are almost invariably studied in contact with excess water, which means the elastomer is at swelling equilibrium, in a semiopen system. Since the degree of swelling depends on the extent of deformation as well as on the temperature[29], it is essential to take into account such changes in interpreting the thermoelastic data. These corrections are, of course, elaborations of the usual ones which account for the fact that the measurements are carried out at constant pressure rather than constant volume[15,29]. Molecular models proposed to account for the deformation mechanism of the material studied would then have to be consistent with the thermoelastic results, particularly with regard to the relative importance of entropic and energetic effects.

15.3.2 Elastin

The earliest thermoelastic studies on elastin were carried out in 1934![57] They were conducted on water-swollen samples, in elongation, in a thermodynamically open system. The enthalpy of deformation was found to be negative and large, and this observation was correctly explained much later as arising from the effects of thermal expansion[58]. More specifically, elastin increases its degree of equilibrium swelling with water as its temperature is lowered. The correctness of this explanation was verified by doing measurements on elastin swollen in a water and ethylene glycol solution of a composition so chosen as to bring the thermal expansion coefficient of the swollen elastin network essentially to zero. Under these conditions, there were no large enthalpic changes and, in fact, the elastic force was found to be approximately 80% entropic in origin, that is, f_e/f was approximately 0.2.[58] Additional measurements carried out in open systems with sizable changes in volume were corrected for this through use of appropriate equations from Gaussian elasticity theory. These results were also consistent with $f_e/f \approx 0.2$.[5] Any molecular theory or picture for the deformation of elastin would have to be consistent with this result. This is, of course, a necessary but not sufficient requirement since any number of models could have this specific attribute.

Although deformation-induced changes in swelling complicate the quantitative analysis of thermoelastic data, they obviously occur without much concern for what experimentalists think when elastin is functioning in the body. They can therefore assuredly contribute to the elastic force and to the storage of elastic energy under these conditions[5]. More recent calorimetric experiments were consistent with the earliest work in showing that large amounts of heat were released when elastin was stretched while in swelling equilibrium with water. These studies suggested that the decrease in enthalpy was due to increased exposure of the numerous nonpolar side chains in elastin to the water environment. It was not realized at this stage, however, that the increased exposure was due simply to the increase in swelling induced by the imposed elongation. Failure to recognize this consequence of having an open system led to the construction of a model in which the extent of these nonpolar–polar or hydrophobic–hydrophilic interactions would depend on deformation, even at constant volume. The model proposed, consisting by necessity of two phases, was called the "oil-drop" model[30]. In it, the protein chains were viewed as being in spherical globules (or oil drops), surrounded by water in the spaces between them. The model is shown schematically in figure 15.3[5,30].

Since the relevant interactions would occur only at the interfaces between the globules and the surrounding water or aqueous solution, they would be at a minimum when the elastin was in the unstretched state. Deforming the elastin would change these spherical globules to ellipsoids of higher surface area. This would increase the magnitude of the interactions of the polar groups with the surrounding water, which would explain the large enthalpy changes observed experimentally.

The oil-drop model is radically different from the usual model of a one-phase network of random chains which greatly interpenetrate one another[15,29,59]. In the former case, the water is nonuniformly consigned to the spaces between the protein droplets, whereas in the latter it is uniformly dispersed throughout the system. As might be expected, the two models predict very different properties under various experimental conditions, and several of the appropriate experiments have been carried out.

The critical experiment involves carrying out thermoelasticity measurements in a thermodynamically closed system. The oil-drop model predicts that large

Figure 15.3 The "oil-drop" model for elastin[5]. The originally spherical globules of the protein shown would be deformed into prolate ellipsoids by the imposed elongation.

enthalpy changes should still be observed since an increased amount of water is not required for an increase in the level of the interactions. The random network model would predict disappearance of the enthalpy effects because its one-phase nature keeps the interactions constant at constant composition.

In one such experiment, the system was kept closed by swelling the elastin networks in a limited amount of a very nonvolatile solvent, triethylene glycol, and then carrying out the thermoelastic measurements on the swollen samples simply surrounded by an inert gas[32]. The anomalous enthalpy effects were found to be completely absent under these conditions.

In a related experiment, it was possible to use water as the swelling agent instead of the glycol, which is admittedly rather different from the usual environment for elastin. This was done by immersing the water-swollen samples of elastin in mineral oil, thus making the system closed in spite of the volatility of the water[5]. In this case, the enthalpy of stretching was measured directly by calorimetry. The excess enthalpy was found to be essentially zero, in good agreement with the thermoelastic results in the closed gylcol system.

Values of the thermoelastic quantity f_e/f can also be obtained by calculations of the unperturbed dimensions $\langle r^2 \rangle_0$ of elastinlike chains. Such theoretical studies were carried out for the elastin repeat sequences mentioned earlier, as modeled by rotational isomeric models[48]. Doing the calculations at two or more temperatures then gives $T d \ln \langle r^2 \rangle_0 / dT$, which is equal to f_e/f.[46]

Test calculations were carried out on homopolymers of glycine, valine, and alanine, and alternating copolymers of alanine and proline, and valine and proline. The model chains of greatest relevance to elastin consisted of the repeat sequences [–Val–Pro–Gly–Gly–]$_x$, [–Val–Pro–Gly–Val–Gly–]$_x$, [Val–Ala–Pro–Gly–Val–Gly–]$_x$. The first stage of the calculations involved semiempirical

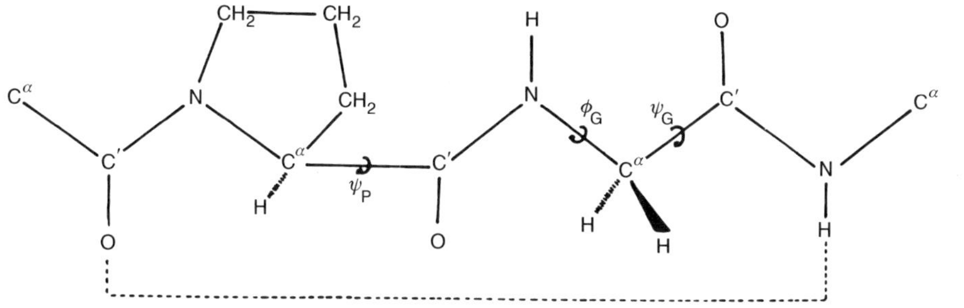

Figure 15.4 Schematic diagram of the proline–glycine–unspecified tripeptide unit used to incorporate the longer range hydrogen bonding (depicted by the broken line) into the conformational energy surface of the glycine residue[46]

molecular mechanics energy calculations on the amino acid residues listed. The calculations were carried out at three levels of approximation: the first with no account at all for hydrogen bonding, the second with account for local hydrogen bonding (between adjacent repeat units), and the last with account also for long-range hydrogen bonding. This type of long-range hydrogen bonding between the residues on either side of a Pro–Gly pair is thought to be important with regard to the β-turns in the Urry model for the deformation of elastin[40,41]. It is illustrated in figure 15.4[46].

Figure 15.5 shows a typical energy map, thus generated, for glycine units. In spite of the fact that long-range hydrogen bonding is taken into account, there is seen to be a great deal of the configuration space available to these residues. This is shown by the large areas corresponding to relatively low energies, that is, to less than 5 kcal mol^{-1}. This conformational versatility is, of course, consistent with the flexibilizing effects of the glycine units mentioned earlier. In the second stage, transformation matrices were defined so that the skeletal bonds, or the virtual bonds replacing them, could all be transformed into a single coordinate system for vector addition. These transformation matrices were Boltzmann averaged over each of the generated energy maps at two chosen temperatures. The last step introduced the transformation matrices into generator matrices for each residue at each of the two temperatures. The generator matrices are sequentially multiplied in the order of the residues in the primary sequence structure of the polypeptide being modeled. Extraction of the appropriate terms in the resulting product matrix then directly gives the unperturbed dimensions $\langle r^2 \rangle_0$.[48] Calculations were carried for the limit of long chains, that is, for large degree of polymerization x.

The values of f_e/f obtained for the homopolymeric polypeptides, such as polyalanine, were negative, as would perhaps be anticipated from the regularity of the chemical structure[46]. This regularity generally means that there is also enough regularity of conformational sequences so that relatively high spatial extension is obtained. Increase in temperature then causes departures from these regular sequences, which leads to a decrease in unperturbed dimensions. In the

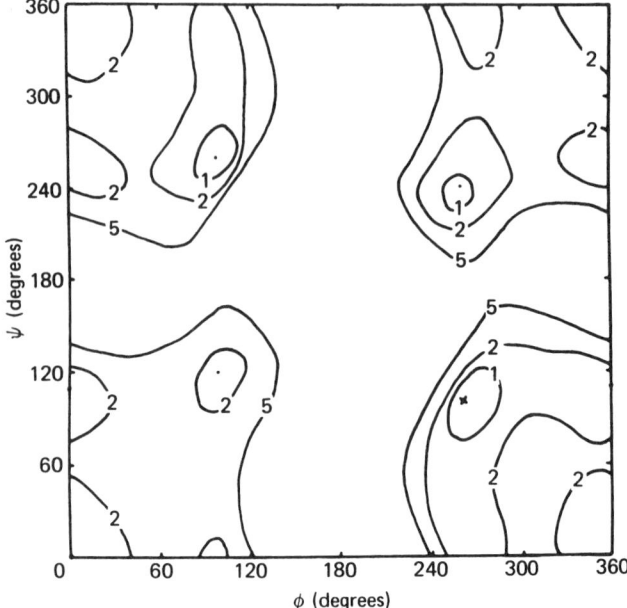

Figure 15.5 Conformational energy surface for a glycine unit in which long-range hydrogen bonding has been included, as described in figure 15.4[46].

case of these polypeptides, the regular conformation at the lower temperature could be extended β structures, some of which could change to more compact α helices as the temperature is increased. In any case, such changes give a negative value to both $d \ln \langle r^2 \rangle_0 / dT$ and f_e/f.

When the copolymeric chains consisting of the elastin repeat sequences were studied in this way, however, the values calculated for f_e/f were very different, presumably because sequence regularities were less pronounced. The values were now generally small and positive. In the most realistic case, where account was taken of even long-range hydrogen bonding, the values were in very good agreement with the experimental result $f_e/f = 0.26 \pm 0.09$.[46]

Thus, the thermoelastic results, both experimental and theoretical, support the random network model, and discredit the oil-drop model and, for that matter, any two-phase model for swollen elastin.

Additional evidence in support of this conclusion includes fluorescence probe measurements[60] and, more important, measurements of glass transition temperatures[61]. Two-phase models for swollen elastin would predict two glass transition temperatures (for the two phases), and these two temperatures would be independent of the relative amounts of the two phases (i.e., independent of composition). Instead, only one glass transition temperature is found for swollen elastin, and it decreases monotonically with increasing amounts of diluent. This is exactly what is to be expected from the one-phase random network model and, in essence, a parallel situation occurs whenever more plasticizer is added to a polymer to make it "more pliable."

Although the two-phase model for swollen elastin is incorrect, the "hydrophobic bonding" between the nonpolar elastin side chains and water does make an important contribution to the elastic force in elastin in vivo[5]. This bonding causes an ordering in the water structure near the nonpolar groups, and this decrease in entropy gives rise to an increase in the free energy of the system. Part of the retractive force in stretched elastin is therefore the attempt of the system to decrease its free energy by minimizing its hydrophobic bonding. It does this by retracting, since this causes a deswelling that decreases the amount of water available for interacting with the nonpolar side chains.

15.3.3 Resilin

Thermoelasticity measurements have been carried out on resilin tendons from a dragonfly[5,62]. These were heroic experiments because the minute size of the samples required that the entire series of measurements be carried out under a microscope! They were conducted in an open system, in swelling equilibrium with water. Quantitative corrections for changes in swelling were not carried out, the justification being that there was some evidence that they were small. The results obtained by analysis of the data were only semiquantitative, but clearly showed the same trends as elastin. More specifically, the elastic force was almost entirely entropic, and the swollen resilin was well represented by the standard, random network model[15,29,59]

15.3.4 Other Protein Elastomers

Abductin from a scallop was also studied with regard to its thermoelastic properties[5,17]. Again, only approximate data were obtained, but they supported the primarily entropic nature of the force and a random network structure.

Octopus arterial elastomer (OAE) was studied more quantitatively, with corrections for changes in swelling[5,20]. The force in this case is also almost entirely entropic, but there may now be an important difference relative to the other protein bioelastomers. Because of the large number of polar amino acids in this polymer, there may not be significant contributions from hydrophobic bonding.

There are no published thermoelastic results on any of the spider-web silks as yet. Any such results may be inherently unreliable, however. If these materials are partially crystalline, then swelling, deformation, and temperature, could change the nature and amount of the crystallinity. Structural changes such as these, occurring during the thermoelasticity measurements, could make the data essentially uninterpretable in the present context.

15.4 Stress–Strain Behavior

15.4.1 General Results

As is true for elastomers, in general, the moduli of the elastomeric proteins vary with deformation, and swelling[16]. The dependence on deformation is particularly strong in the case of elongation, which is almost always the deformation used with bioelastomers. Nonetheless, values of the elongation modulus that are relatively constant have been reported for these materials. This would be expected in view of the very specific way in which the cross-links are introduced, except possibly in the crystallite-cross-linked spider-web silks. In any case, these values are listed in the first row of table 15.3 for the protein elastomers being discussed. The values are remarkably similar in magnitude, considering the diversity of the structures in which these proteins appear. As a result, the values of the molecular weight between cross-links is also very similar, as can be seen from the second row of table 15.3. They are all well within the range $10^{-3}M_c = 5-15\,\mathrm{g\,mol^{-1}}$ which is characteristic of commercial nonbiological elastomers[15,29,59].

Some typical force–extension curves are presented in figure 15.6, along with the limiting Gaussian curve. Data were available for all of the protein elastomers under discussion, save for abductin (possibly because of the unusual shape of samples of this material). All values of the force have been normalized by dividing them by the limiting modulus νkT to facilitate comparisons. Since the magnitudes of the moduli are not very different for these materials, as documented previously, this permits focusing on the shapes of the isotherms. The most striking feature is the fact that all four of these elastomers show very significant non-Gaussian upturns in the force at high elongations. Also very notable are the relatively low values of the elongation at which this occurs.

One way of interpreting these data is in terms of *equivalent* freely jointed chains[59]. In this approach, a number n' of consecutive skeletal bonds are grouped into a single virtual bond. The number n' is made large enough so that the correlations between the first and last skeletal bonds in such a grouping becomes negligible. This makes consecutive virtual bonds uncorrelated by design, and thus the real chain is replaced by its freely jointed equivalent. This is a semiquantitative way of characterizing non-Gaussian behavior in terms of chain stiffness. The stiffer the chains, the larger the non-Gaussian effect at constant deformation and the larger the values of n' required to form the uncorrelated virtual bonds.

Table 15.3 Some Network Characteristics for several Protein Bioelastomers[a]

Property	Elastin	Resilin	Abductin	OAE[b]	Viscid silk
G (MPa)	0.41	0.65	≈ 0.70	0.47	≈ 0.30
$10^{-3}M_c$ (g mol^{-1})	6.7	5.1	≈ 5	6.7	≈ 9

[a] Adapted from reference 5.   [b] Octopus arterial elastomer.

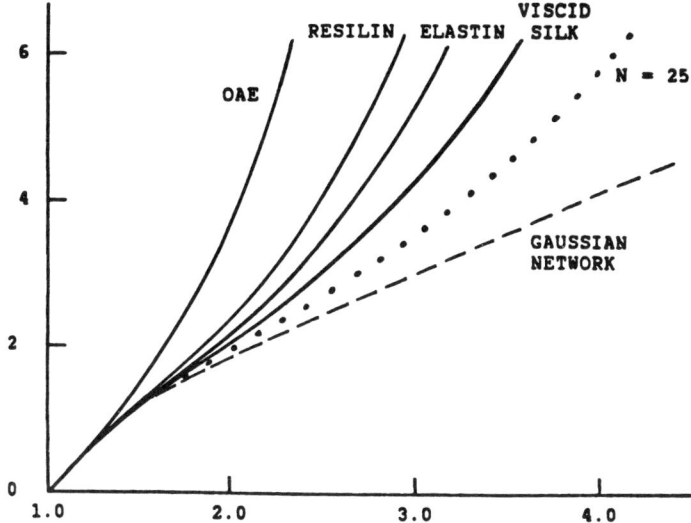

Figure 15.6 Stress–strain isotherms in elongation for bioelastomeric proteins, with the Gaussian limit being given by the dashed line. Curve OAE is for octopus arterial elastomer, and the curve labeled $N = 25$ is for a non-Gaussian network consisting of 25 ideal random links. Adapted from reference 5.

Figure 15.6 suggests that the viscid silk is the lowest in stiffness of the bioelastomers shown, and the octopus arterial elastomer is the highest. If each of the curves is fitted using the non-Gaussian isotherms generated for freely jointed equivalents, then the number of amino acid repeat units required per virtual bond is approximately 7 for the viscid silk and the resilin, 8 for the elastin, and 12–14 for the OAE[5]. Since the simple flexible polymer polyethylene $[-CH_2CH_2-]$ has approximately 6 repeat units per virtual bond in this approach, all four of the bioelastomers are seen to have significant chain stiffness. It is very interesting that the first three of these, with values of n' of approximately 7, are the ones with relatively large numbers of the amino acids having small side chains, as described in table 15.1. The number of repeat units per virtual bond is much higher in the case of the OAE. This is presumably due to its relatively high content of large, polar amino acids, as described in table 15.2.

15.4.2 Elastin

The equivalent freely jointed interpretation of any data of this type is severely limited, however, because it is not very molecular[48]. That is, there is no real direct connection between the extent of the non-Gaussian behavior and the detailed molecular structure of the chains. Such a connection has been made in the case of elastin[47]. Stress–strain isotherms were generated using an approach very similar to the theoretical calculations described previously for f_e/f. In this case, however, the "degree of polymerization" x is very important, and values were chosen that were comparable to those present between cross-links in native elastin. Specifically, these values were $x = 5$ and 10. Also, alanine units were placed at the ends of the chains to model the alanine sequences found both before and after the lysines used to generate the cross-links. In this approach, Monte Carlo methods were applied to the conformational energy maps to generate large numbers of chains having the desired conformational characteristics. The end-to-end distance r was then calculated for each of these conformations, and the various values of r placed into a histogram. The histograms were subsequently smoothed using the cubic spline method, thus producing a continuous distribution, as described in chapter 8. Derivatives were taken at various points along the distribution, and were used in the final step of the calculation, the generation of the theoretical stress–strain isotherms from the standard three-chain model. The calculations were carried out on the elastin repeat sequences described earlier in this chapter.

A typical distribution thus produced, and its Gaussian approximation, are shown in figure 15.7[47]. As expected, the Gaussian approximation is not very good for chains that are this short. Some of the calculated stress–strain results represented as the nominal stress against the elongation itself are shown in figure 15.8. The representation is the same as that used for the experimental data in figure 15.6. The same calculated results, now in the Mooney-Rivlin representation[15,29], are presented in figure 15.9. At least some of these theoretical isotherms did show upturns that were qualitatively similar to those observed in the experimental studies on elastin. The upturns in the theoretical isotherms are due to the limited extensibility of the elastin repeat sequences. Most notably, they occurred

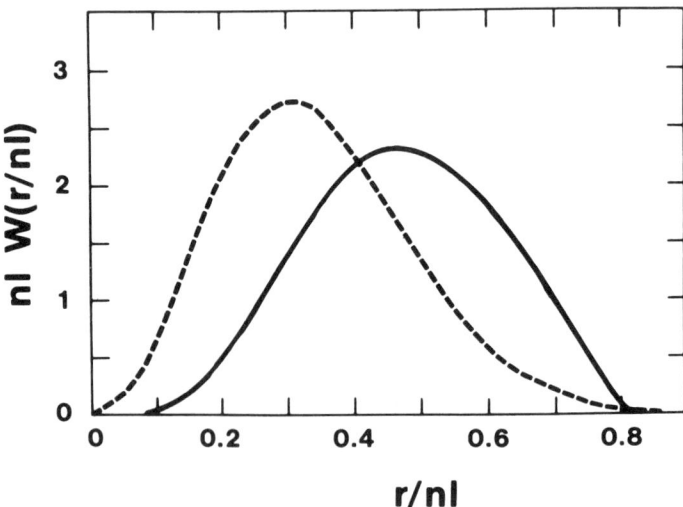

Figure 15.7 Monte Carlo generated normalized radial distribution function for an alanine-terminated tetrapeptide elastin model chain having a degree of polymerization x of 5 (———), and the Gaussian distribution having the same value of the unperturbed dimensions $\langle r^2 \rangle_0$ (- - - - -)[47].

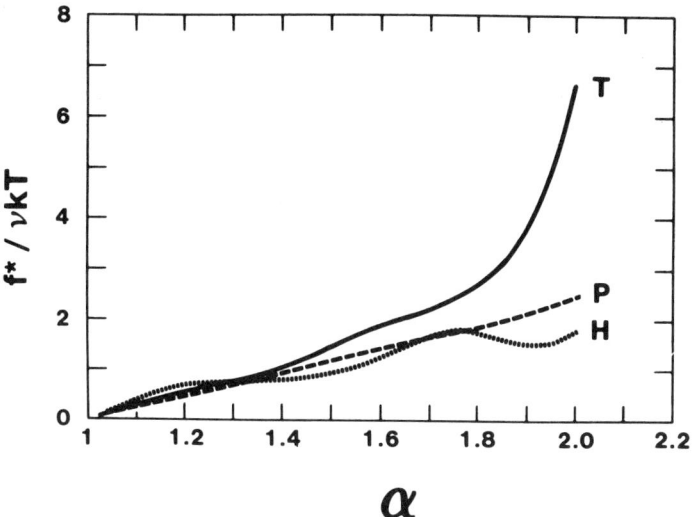

Figure 15.8 Theoretical nominal stress shown as a function of elongation for alanine-terminated elastin model chains with $x = 5$.[47] The repeat units consist of either the tetrapeptide (T), or the pentapeptide (P), or the hexapeptide (H).

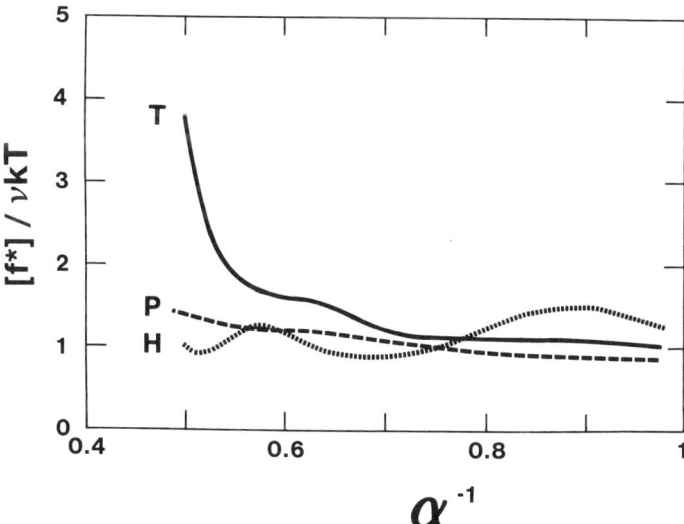

Figure 15.9 Theoretical modulus shown as a function of inverse elongation for the elastin chains described in figure 15.8[47].

at approximately the same elongation, ca. 1.8, at which the experimental isotherms showed this behavior, at least in the most reliable case of single elastin fibers (which avoid the problem of multiple fibers being out of register in larger samples). Unfortunately, it is very difficult to extend these simulations to higher elongations to characterize the upturns more thoroughly. This is because of the rapid decrease in the number of generated conformations at high elongations, and difficulties in evaluating derivatives of the distribution function[47].

These non-Gaussian upturns in the modulus could be extremely useful for the survival of the organism[2]. In the case of arteries, the increase in modulus could decrease the likelihood of ballooned-out areas called aneurisms, and the bursting of these aneurisms, or the rupture of the artery in general. This would be in addition to the usual reinforcement of these blood vessels by having the elastin extensively interpenetrated by strong, partially crystalline fibers of collagen.

15.4.3 Resilin

As shown in figure 15.6, the stress–strain isotherms for the resilin networks are significantly more non-Gaussian than those for elastin. It would therefore be of considerable interest to attempt to understand these differences through additional Monte Carlo simulations. Such studies await the required sequencing information on resilin.

15.4.4 Spider-Web Silk

The viscid silk fibers in the spider web have already been discussed with regard to their stress–strain isotherms. It is interesting, however, to compare some of mechanical properties of viscid silk with those of the frame silk occurring in the same web[26]. The frame silk fibers have a modulus approximately 3000 times higher than that of the viscid silk, and a strength approximately twice as high. They are not as extensible as the viscid fibers and, as a result, the energies required for rupture are nearly the same. These differences are presumably due to higher degrees of crystallinity in the frame silk.

In any case, the frame silk has a good balance of properties to resist rupture upon being impacted by a flying insect[36]. Also, these fibers are very "lossy" or hysteretic, which means they degrade much of this absorbed energy quickly into heat[26]. This helps prevent the insect from being shot back out of the web by the retractive force of the fibers.

15.4.5 Other Protein Elastomers

There are also interesting uniaxial and biaxial stress–strain data which have been reported for arteries from several cephalopods, namely a squid (*Nototodarus sloani*), a cuttlefish (*Sepia latimanus*), and a nautilius (*Nautilus polpilius*)[22]. There have also been studies of squid mantle wall in inflation, a deformation closely related to the inflation–expulsion cycle used by these animals for locomotion[23]. Most of these elastomeric materials are heavily reinforced with fibers and pervaded with muscle tissue, greatly complicating the interpretation of their mechanical properties in molecular terms.

15.5 Dynamic-Mechanical Properties

15.5.1 Viscoelastic Responses in General

The bioelastomers in vivo generally function in cyclic deformation–recovery processes, an example being the pulsatile flow of blood through the arteries. Thus, retarded responses due to viscous effects assume particular importance.

The viscoelastic experiments used to characterize such time-retarded behavior are generally carried out using cyclic, low-amplitude tensile deformations. In the case of elastin, measurements were generally made as a function of degree of hydration and deformation frequency, as well as of temperature[5,10,63]. Both thermodynamically open and thermodynamically closed systems have been investigated. Of course, the values of these variables occurring physiologically are of greatest relevance. As is typically the case, the effective frequency range is increased by the method of reduced variables. In this technique, data taken at one temperature are shifted along the frequency axis by an amount a_T that yields a smooth joining with data pertaining to another temperature. In this way, data from separate experiments at different temperatures can be combined into master curves.

The three viscoelastic quantities of greatest importance are the storage modulus E', the loss modulus E'', and the loss tangent tan δ (which is equal to E''/E')[64]. The loss tangent is of particular interest since it goes through a maximum at the glass transition temperature. In the most reliable investigations, the shift factors a_T are determined so as to give the best joinings for both the E' vs. frequency curves and the E'' vs. frequency curves. Some typical storage and loss moduli obtained on swollen elastin samples, in a closed system and at a series of temperatures, are shown in figure 15.10[10]. The combination of these data into master curves for E', E'', and tan δ are presented by the solid lines in figure 15.11. The values of the temperature shift factor required to do this are shown as a function of temperature in figure 15.12. The other set of master curves in figure 15.13 show the effects of increasing the degree of swelling on these viscoelastic properties.

There are a number of interesting features in these results. First, as is true, in general, for amorphous elastomers, decrease in temperature increases both the storage and loss modulus. That is, the material is approaching the temperature T_g at which it becomes glassy, rather than remaining elastomeric[64]. The same changes are seen to occur with decrease in swelling (removal of plasticizer) or with increase in frequency.

As mentioned earlier, decrease in the ambient temperature of an animal's body could be disastrous at constant composition since, if the temperature fell below the glass transition temperature of the swollen elastin, the elastin would lose its elastomeric properties. This could be a problem in the case of fish, amphibians, and reptiles, since their body temperatures can drop below 0°C. The effect is

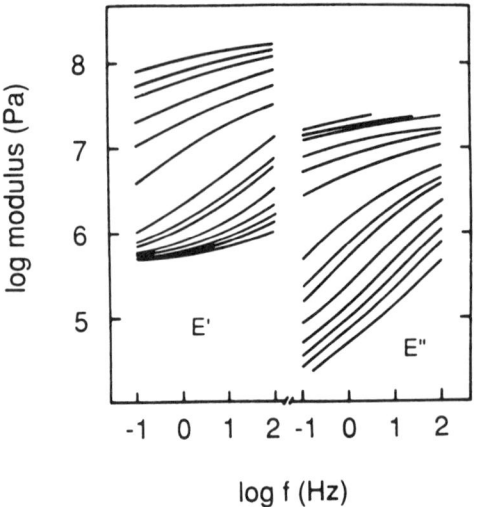

Figure 15.10 Storage moduli (E') and loss moduli (E'') for elastin at a water content of 28.1 wt% measured at the temperatures 0, 10, 15, 20, 23, 26, 33, 37, 42, 49, 56, 64, 73, and 79°C (from top to bottom)[10].

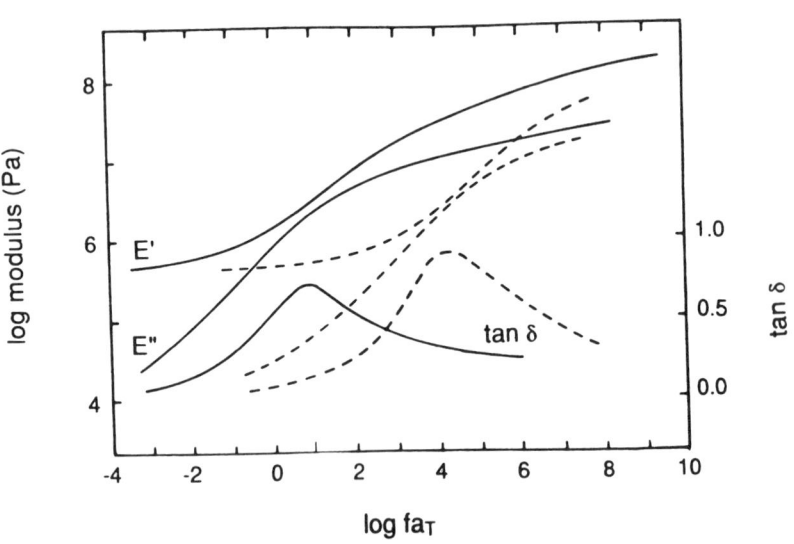

Figure 15.11 Master curves for elastin[10]. The solid lines, showing elastin at a water content of 28.1 wt%, were based on the data in figure 15.10. The broken lines show elastin at a water content of 43.3 wt%.

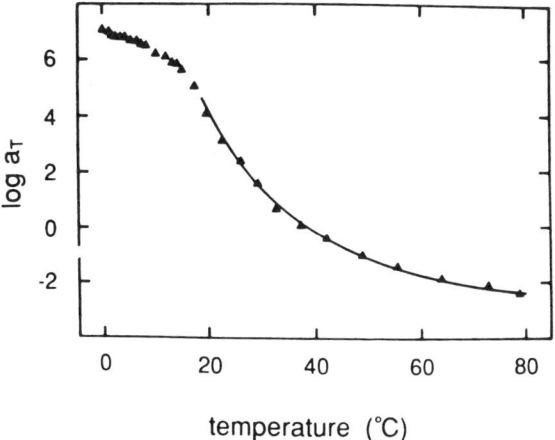

Figure 15.12 The logarithm of the temperature shift factor shown as a function of temperature for the data in figure 15.10[10].

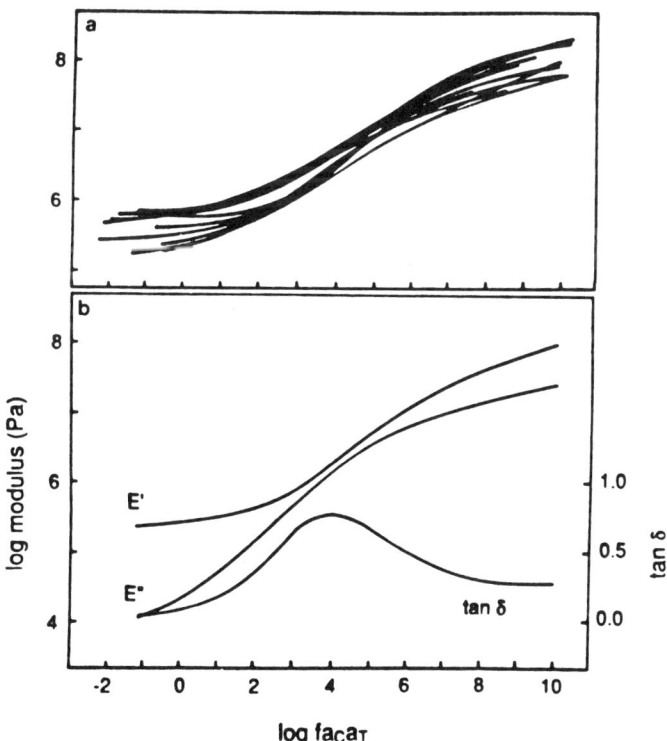

Figure 15.13 (a) The storage moduli for elastin samples at 12 water contents, corrected for the horizontal shifts, and (b) the master curves obtained by averaging data from all the water contents[10].

fortunately offset, however, by the fact that elastin in a living organism is an open system. In addition, the thermodynamics of the swelling of elastin is such that it swells *more* at the lowered temperature. As a result, the elastin remains elastomeric at the decreased temperature.

The effects of frequency, at constant temperature and constant degree of swelling, are even more intriguing[5,10]. At frequencies up to about 5 Hz, the range over which elastin functions in animals, elastin is an excellent elastomer. Its storage modulus is almost independent of frequency, and its loss modulus is quite low. Correspondingly, the loss tangent and damping effects are small, and the efficiency of elastic recovery (the "resilience") are large. However, at higher frequencies, both storage and loss moduli increase dramatically, indicating that the elastin is becoming glassy. As judged from the incipient maximum in tan δ, the glass transition at this composition and temperature should occur well above the frequencies normally encountered in a living organism. However, in the case of blood flow made turbulent by the presence of obstructions, high-frequency vibrations could be set up[10]. These vibrations could conceivably make the elastin sufficiently glasslike to lessen its resistance to damage by the usual fatigue mechanisms that eventually cause failure in any cyclically deformed material. As will be discussed further in the next section, loss of some of the plasticizing water by deswelling would lower the frequency at which this could become a problem.

There are a limited amount of data on resilin that indicate its viscoelastic properties are very similar to those of elastin[5]. Resilin is not used for circulatory systems, so some the problems cited above with regard to blood flow are not relevant. It is used, however, as parts of the reciprocating mechanisms in flying insects, presumably at frequencies below those at which the resilin would begin to show signs of glassiness. It is also used as part of the catapult mechanism used by jumping insects, such as the flea. In this application, a rapidly cycling deformation is not involved.

Abductin has been studied at only one frequency, approximately 3 Hz[5]. This frequency was chosen because it is very close to the frequency at which the scallop opens and closes its shell to propel itself through the water. About all that can be said from the results obtained is that the abductin is very similar to elastin in its viscoelastic properties at this relatively low frequency. It is a good elastomer, and there do not seem to be any circumstances that could place these properties in jeopardy.

15.5.2 Effects of Dehydration

Dehydration of elastin involves removing some of the water that plasticizes it, thereby bringing it closer to its glass transition. Since elastin in the body is at swelling equilibrium, it is fully hydrated. In spite of this, it is close to its glass transition and any deswelling could have serious effects on its elastomeric properties. The viscoelasticity curves under physiological conditions are very sensitive to composition, and relatively small decreases in water content can shift them to lower frequencies by one or two decades![10] Similar effects are found in analogous experiments in which very slow calorimetric or dilatometric measurements are used to determine the "equilibrium" (long-time) glass transition temperature at constant composition. The values are approximately 200°C at 0.00 g water per g elastin, 120°C at 0.05 g water per g elastin, and 15°C at 0.30 g water per g elastin.

This, of course, immediately raises the question of how some of the elastin in vivo might lose some of its hydration. The fact that elastin is very hydrophobic suggests that the water of hydration is not very tightly held, and this is confirmed in the absorption isotherms it exhibits[10]. At 100% relative humidity, swollen elastin contains approximately 0.6 g water per g elastin, but with only a 2% decrease, to 98%, there is only approximately 0.3 g water per g elastin left. This shows the ease with which swollen elastin can be dehydrated.

There are molecules in the extracellular matrix, such as chondritin sulfate, that are now known to dehydrate swollen elastin significantly at physiological concentrations. Other materials that might have similar effects are cholesterol and cholesterol esters. This could be highly relevant to changes in the functioning of elastin in disease states and as a result of aging[10].

15.6 Some Other Bioelastomeric Gels

15.6.1 General Properties

There are a wide variety of biopolymers that form gels in which the network structure results from physical cross-links, such as crystallites or helical "junction zones"[7,65-67]. These gels are typically formed by cooling aqueous solutions of the biopolymer to the point at which some of the chain segments aggregate sufficiently to form a macroscopic gel. Since the cross-link points in these materials are physical aggregates of one type or another, they are generally "thermoreversible," that is, they reliquefy upon increase in temperature. There are analogies in the realm of non-biological materials[68]; for example, the polyethylene gels[69] used in the spinning of ultra-high-strength fibers.

With regard to mechanical properties, both low-deformation and high-deformation experiments have been carried out. In the case of low deformations, the usual cyclic dynamic mechanical testing has been done to some extent[65]. Other, more applied approaches include compression under constant load[70], and penetration using standard probes.

One of the most important experiments involves monitoring the buildup of the modulus as the gelation reaction proceeds. Since these gels are quite weak, even when the gelation reaction is complete, such measurements are almost always carried out using one of the dynamic cycling techniques. Once the modulus of the gel has reached an apparent plateau, suggesting that the gelation reaction has neared completion, the frequency of the oscillation may be varied, thus obtaining E', E'', and tan δ over whatever frequency ranges are of interest. This type of experiment is particularly useful in identifying entanglement networks, in which only transient inter-chain entanglements hold the network structure together[7]. Obviously, in this case the moduli go to zero as the frequency of the deformation decreases.

There is another interesting aspect to the long-time behavior of gels of this type. Not only is there relatively little polymer present in these highly swollen materials, but what polymer there is, is frequently susceptible to bacterial attack[65]. This type of network degradation can be prevented by the addition of small amounts of bactericide, thus proving a nice parallel to the use of antioxidants and antiozonants to prevent network breakdown in commercial, nonbiological elastomers.

There have been very few attempts to characterize gels such as these with regard to their large-deformation, relatively static or equilibrium properties. Such data could be interpreted using the equilibrium molecular theories of rubberlike elasticity to obtain the usual information on network structure. A related subject in this area of relatively large deformations concerns the failure envelope[64]. In this case, the ultimate properties, namely the stress and strain to rupture, are plotted against each other for different concentrations of the solutions used to prepare the gels. A concentration shift factor is then used to construct a master curve called a "failure envelope." The method parallels the much-used time–temperature superposition used to extend the effective range of deformation rates in fracture experiments on nonbiological materials. The concentration dependence of the gel modulus is widely interpreted in terms of the currently accepted theories of gel formation, particularly the Flory and Stockmayer theory and the percolation theory[7,65,68].

In the case of thermally stable gels, small changes in temperature have only the usual, relatively small effect on the modulus. As described in chapter 9, the modulus is directly proportional to the absolute temperature in the case of an ideal elastomer, and deviations from this for real elastomers are not very great. For example, in this approximation a 10° increase in temperature in the vicinity of room temperature would cause an increase of only approximately 3% in the elastic force or modulus. The situation is completely different in the case of thermally reversible gels, where the aggregated segments acting as cross-links become less stable and disaggregate at the elevated temperature. A relatively small increase in temperature can therefore cause a very pronounced decrease in the effective degree of cross-linking, with a correspondingly large decrease in the modulus.

An example of a thermoreversible gel which could exhibit a pronounced decrease in modulus with increase in temperature would be one held into a network structure by small crystallites[68]. Another related but more specific example would be networks in which the cross-links are segments of double or triple helices, with the sequences involved coming from different polymer chains[65]. In this case, the changes in modulus can be interpreted using the now well-understood theory of the helix–coil transition in biopolymers.

A possible additional complication in biopolymer gels, in general, involves the interactions between the large amounts of aqueous electrolytes present in these systems, and the network chains, which can be highly polar or even ionic. For example, a correlated binding of the solvent with the polymeric solute would require reconsideration of results interpreted using a lattice theory. Account would have to be taken of the possibly nonrandom placing of solvent molecules and polymer segments on the lattice, a circumstance not anticipated in the simplest versions of the lattice theories.

Additional complications occur when the network chains are sufficiently rodlike to form anisotropic phases. This would occur only above a critical concentration, but for relatively long rods this concentration is quite small, frequently amounting to only a few percent polymer[71,72]. When this phase separation occurs, two phases are formed, the more dilute one remaining isotropic, while the more concentrated one consists of an anisotropic arrangement of the the aligned rodlike chains.

There have been calculations on the stress–strain behavior of isotropic systems of such rodlike segments[72]. If there are no permanent cross-links, the elastic response comes from the bending of the rodlike segments. The stress–strain isotherm in elongation is predicted to be S shaped; that is, the modulus should start at a low value, increase significantly, and then become small again. This is, of course, the opposite to the usual stress–strain isotherm exhibited by elastomers, which corresponds to the modulus starting at a relatively large value, decreasing significantly, and then increasing again. In the case of the rigid-rod gels, the hardening predicted is thought to arise from increases in the number of contact points between the rods, and the subsequent decrease to a saturation in the number of these contact points.

15.6.2 Some Specific Systems

15.6.2.1 Random-Coil Biopolymers Gelatin, which is probably the most important gel of this type, is a protein derived from collagen by hydrolytic degradation. It readily dissolves in warm water to form a solution of randomly coiled polymer. So long as the concentrations are not extremely low, for example, less than a few tenths of a percent, cooling such a solution results in the formation of a transparent elastic gel. These gels are, of course, well known in the food industry. The gelation process involves the peptide chains reforming some of the well-known collagen triple-helical sequences they had in the native state. Because of the dilute nature of the system, these regions are very limited in size and number[65]. They are sufficient, however, to link the chains into the observed network structure.

The gelation process itself has been studied by a variety of techniques, including measurements of optical rotation, mechanical properties, NMR spectra, dielectric relaxation, and radiation scattering. The studies are generally carried out as a function of reaction time, polymer concentration, temperature, and, in the case of the mechanical property measurements, deformation frequency. Different measurements can give very different types of information[65]. For example, optical rotation data can provide measures of the helical content of the gelatin, whereas mechanical property data can provide estimates of the effective cross-link density. The basic goal has been to understand the gelation process, particularly with regard to its mechanism. The two most important steps are obviously the conformational changes required to convert random-coil stretches along the chain into helices, and the association of these helices to form the physical cross-links. Several detailed mechanisms have been proposed for this process. There is also a limited amount of large-deformation data which has been used to estimate equilibrium moduli and ultimate properties.

One of the most interesting types of marine polysaccharides forming bioelastomeric gels are the carrageenans, derived from seaweed. They are partially sulfated polysaccharides having an AB disaccharide repeat unit, where the A unit is derived from a 1-3-linked β-D-galactose residue and the B unit from a 1-4-linked 3,6-anhydro-α-D-galactose residue. There are occasional departures from this simple alternating structure, in that some of the B units are modified sufficiently to cause kinking of the chains. Such kinking could well affect the packing of the chains and thus the formation of elastomeric gels. A variety of carrageenans exist, and differ from one another primarily in the number and location of the sulfate groups they contain[65]. Like gelatin, they dissolve at elevated temperature, and cooling of the solutions thus obtained results in the formation of thermoreversible gels. In this case, however, the cross-links are thought to consist of threefold right-handed helices stabilized by hydrogen bonding. Another difference is the fact that the sulfate groups can give the polymer considerable ionic charge, and this could certainly affect its properties. The associated counterions also have to be taken into account in detailed theories for the gelation process.

A related polysaccharide, agar, is a complex mixture of components also obtained from seaweeds. The most important component with regard to gelation is agarose, whose repeat unit is another disaccharide with A and B units of 1-3-linked β-D-galactose and 1-4-linked 3,6-anhydro-α-L-galactose, respectively[65]. Agars are quite similar to the carrageenans, except for the replacement of the D-galactose unit by its L counterpart, and by the near absence of ionizable units such as sulfate groups. Again, there are occasional departures from perfect structural regularity that can cause kinks in the chains, with attendant effects on the gelation process. There is some controversy regarding the exact molecular structure at the cross-links. There may not only be double helix formation, but also association of these helical sequences into microcrystalline regions. The mechanism for the gelation process may depend greatly on the rate at which the agarose solutions are cooled. It is thought that quenching gives rise to nonnucleated spinodal decomposition, whereas slower cooling gives nucleated network growth.

Also occuring in algae are the alginates, which are salts of alginic acid. These are apparently not alternating disaccharides, but are made up of varying amounts of 1-4-linked β-D-mannuronic acid and 1-4-linked α-L-guluronic acid[65]. The exact composition depends on the source of the alginate. In any case, the chains consist of homopolymeric blocks of the first species mixed with homopolymeric blocks of the second, and with blocks of irregular sequences of the two. The alginates gel upon the introduction of divalent cations such as Ca^{2+}, and the gels thus formed are not thermoreversible. It is thought that very specific ion binding occurs, with accompanying conformational changes in the chains. One model for the cross-link regions depicts them as crystalline aggregates in a twofold ribbon form, with the Ca^{2+} ions cooperatively bound among the chains. There have been a number of studies on how the rigidity of the alginate gel depends on molecular weight and concentration of the polymer and the concentration and type of the divalent cation. Dynamic-viscoelasticity studies showed viscoelastic effects to be small in these gels, which encouraged interpretation of their elastic moduli in terms of the molecular theories of rubberlike elasticity. Values of the molecular weight between cross-links obtained in this way seemed to be in satisfactory agreement with values obtained from swelling equilibrium measurements. The very complicated structure of these gels, however, suggests that the agreement may have been fortuitous.

Of the plant polysaccharides, cellulose is the most common and certainly the most important[65]. Since cellulose, itself, is not soluble in common solvents, it cannot be gelled into swollen networks in the way described for the other biopolymers. It can, however, be derivitized to esters and ethers, and these are readily soluble and do form gels, presumably thermoreversible ones. Hydroxypropyl cellulose is an example of such a system. Because of the derivatization, these materials are no longer purely biopolymeric, however.

Pectins are an example of polysaccharides that do form gels readily in their natural state[65]. They occur in the cell walls of certain plants and are important in the food industry. They are branched in their native state, but the extracted material is predominantly linear. The chains consist of α-(1-4)-linked D-galacturonic acid residues interrupted occasionally by L-rhamose units, and they have varying amounts of methoxy units. Gelation occurs under conditions where intermolecular electrostatic repulsions are reduced, for example at low pH. Addition of divalent ions, such as Ca^{2+}, facilitates the process, and at least some of the gels are thermoreversible. The cross-link points are rather ill-defined aggregates that are nonstoichiometric and are thought to involve noncovalent and nonelectrostatic bonding.

In the case of mammals, the most important polysaccharide is unquestionably starch, because of its edibility. Naturally occuring starch occurs in granules and consists of two fractions: the first is linear and called amylose and the second is branched and called amylopectin[65]. Both are based on glucose, with amylose being an α-(1-4)-linked D-glucose, while amylopectin has both α-(1-4) and α-(1-6) linkages. The starch is partially crystalline, largely because of the amylopectin, and this is rather surprising because of the branched nature of this component. Amylose is normally the minor component, but can also undergo crystallization. In the case of amylose, quenching solutions rapidly produced gelation, with considerable turbidity and a significant shear modulus. Crystallinity, however, as judged from wide-angle x-ray diffraction measurements, was very slow to develop. This suggests that gelation results from the formation of amylose double helices, which then slowly aggregate into crystallites capable of scattering x-rays. The turbidity is thought to arise from the separation of the system into polymer-rich and polymer-deficient regions. Although the gels are highly elastic, their complicated structures have discouraged interpretation of their properties in terms of molecular theories of rubberlike elasticity.

The amylopectin fraction of starch has also been studied with regard to its gelation[65]. The process is very slow, and thought to involve a slow association, with crystallization of short branches near the surfaces of the aggregates.

In addition to the algal and plant polysaccharides mentioned above, there are polysaccharides that occur outside the cells of certain bacteria and fungi. An example is xanthan, a well-known bacterial polysaccharide[65]. In some respects, these are similar to the other polysaccharides, but tend to form less stable gels, some of which are transient, entanglement-type networks.

There are also animal polysaccharides, most notably the proteoglycans (which consist of a protein backbone with attached linear polysaccharides called glycosaminoglycans) and glycoproteins (which consist of a central protein core surrounded by very short saccharide chains in a "bristle-brush" structure). These are used in living systems, more for their viscoelasticity, for example as mucus, rather than for their elastic properties. They are therefore not of direct interest in the present context.

There is also a variety of relatively solid, gel-like materials that are only transient network structures[7]. They exhibit this gel-like structure because of the extensive entangling of the biopolymer chains, and dilution, even at constant temperature, converts these materials into simple solutions. Guar galactomannan is an example of a material of this type[65]. These materials are also beyond the scope of the present treatment.

15.6.2.2 Globular and Rodlike Biopolymers Finally, there are globular biopolymers, particularly proteins, that can aggregate into rodlike structures which ultimately aggregate into fibrous or branched network structures[65]. For example, actin, tubulin, hemoglobin-S, insulin, and fibrin form fibrous assemblies; denatured proteins form branched networks; and casein forms micellar aggregates. The units undergoing aggregation can themselves be rodlike, as they are in the case of myosin and collagen. The network "chains" in these systems are thought to be quite stiff, and presumably they deform in ways rather different from the usual stretching out of random-coil chains. This could make the elastic force predominantly energetic or enthalpic, rather then entropic. Thermoelastic measurements on some of the more stable gels of this type could be highly informative.

References

(1) Vincent, J. F. V.; Currey, J. D., Eds. *The Mechanical Properties of Biological Materials*. Cambridge University Press: Cambridge. 1980; Vol. 34.

(2) Gosline, J. M. In *The Mechanical Properties of Biological Materials*; J. F. V. Vincent and J. D. Currey, Eds.; Cambridge University Press: Cambridge. 1980; p. 331.

(3) Vincent, J. F. V. *Structural Biomaterials*. Macmillan: London. 1982.

(4) Wainright, S. A.; Biggs, W. D.; Currey, J. D.; Gosline, J. M. *Mechanical Design in Organisms*. Princeton University Press: Princeton. 1986.

(5) Gosline, J. M. *Rubber Chem. Technol.* 1987, 60, 417.

(6) Kramer, O., Ed. *Biological and Synthetic Polymer Networks*. Elsevier: London. 1988.

(7) Burchard, W.; Ross-Murphy, S. B., Eds. *Physical Networks. Polymers and Gels.* Elsevier: London. 1990.

(8) Lillie, M. A.; Gosline, J. M. In *Physical Networks. Polymers and Gels*, W. Burchard and S. B. Ross-Murphy, Eds. Elsevier: London. 1990; p. 391.
(9) Vincent, J. F. V. *Structural Biomaterials*, revised ed. Princeton University Press: Princeton, NJ. 1990.
(10) Lillie, M. A.; Gosline, J. M. *Biopolymers* 1990, 29, 1147.
(11) Gosline, J. M. In *Concepts of Efficiency in Comparative Physiology*, R. W. Blake, Ed. Cambridge University Press: Cambridge. 1992.
(12) Lillie, M. A.; Chalmers, G. W. G.; Gosline, J. M. *Biopolymers* 1996, 39, 627.
(13) Lillie, M. A.; Gosline, J. M. *Biopolymers* 1996, 39, 641.
(14) Mann, S., Ed. *Biomimetic Materials Chemistry*. VCH Publishers: New York. 1996.
(15) Mark, J. E.; Erman, B. *Rubberlike Elasticity. A Molecular Primer*. Wiley-Interscience: New York. 1988.
(16) Bush, K.; McGarvey, K. A.; Gosline, J. M.; Aaron, B. B. *Conn. Tissue Res*. 1982, 9, 157.
(17) Alexander, R. M. *J. Exp. Biol.* 1966, 44, 119.
(18) Shadwick, R. E.; Gosline, J. M. *Science* 1981, 213, 759.
(19) Shadwick, R. E.; Gosline, J. M. *Can. J. Zool.* 1983, 61, 1866.
(20) Shadwick, R. E.; Gosline, J. M. *J. Exp. Biol.* 1985, 114, 239.
(21) Shadwick, R. E.; Gosline, J. M. *J. Exp. Biol.* 1985, 114, 259.
(22) Gosline, J. M.; Shadwick, R. E. *Pacific Sci.* 1982, 36, 283.
(23) Gosline, J. M.; DeMont, M. E. *Sci. Am.* 1985, 252, 96.
(24) Gosline, J. M.; Denny, M. W.; DeMont, M. E. *Nature* 1984, 309, 551.
(25) Calvert, P. *Nature* 1984, 309, 516.
(26) Gosline, J. M.; DeMont, M. E.; Denny, M. W. *Endeavor* 1986, 10, 37.
(27) Mark, J. E.; Calvert, P. D. *J. Mater. Sci., Part C* 1994, 1, 159.
(28) Ross, R.; Bornstein, P. *Sci. Am.* 1971, 224, 44.
(29) Treloar, L. R. G. *The Physics of Rubber Elasticity*, 3rd ed. Clarendon Press: Oxford. 1975.
(30) Weis-Fogh, T.; Andersen, S. O. *Nature* 1970, 227, 718.
(31) Hoeve, C. A. J.; Flory, P. J. *Biopolymers* 1974, 13, 677.
(32) Andrady, A. L.; Mark, J. E. *Biopolymers* 1980, 19, 849.
(33) Aaron, B. B.; Gosline, J. M. *Biopolymers* 1981, 20, 1247.
(34) Sage, E. H.; Gray, W. R. In *Elastin and Elastic Tissue*, L. B. Sandberg, W. R. Gray, and C. Franzblau, Eds. Plenum Press: New York. 1976.
(35) Kerkam, H.; Viney, C.; Kaplan, D.; Lombardi, S. *Nature* 1991, 349, 596.
(36) Vollrath, F. *Sci. Am.* 1992, 266(3), 70.
(37) Kaplan, D. L.; Adams, W. W.; Farmer, B.; Viney, C., Eds. *Silk Polymers. Materials Science and Biotechnology*. American Chemical Society: Washington, DC. 1994; Vol. 544.
(38) Gosline, J. M.; Nichols, C.; Guerette, P.; Cheng, A. In *Biomimetics. Design and Processing of Materials*, M. Sarikaya and I. A. Aksay, Eds. American Institute of Physics Press: Woodbury, NY. 1995; p. 237.
(39) Foster, J. A.; Rubin, C. B.; Kagan, H. M.; Franzblau, C.; Bruenger, E.; Sandberg, L. B. *J. Biol. Chem.* 1974, 249, 6191.
(40) Luan, C.-H.; Urry, D. W. *J. Phys. Chem.* 1991, 95, 7896.
(41) Luan, C.-H.; Parker, T. M.; Prasad, K. U.; Urry, D. W. *Biopolymers* 1991, 31, 465.
(42) Urry, D. W.; Peng, S. Q.; Parker, T. M. *Biopolymers* 1992, 32, 373.
(43) Sciortino, F.; Prasad, K. U.; Urry, D. W.; Palma, M. U. *Biopolymers* 1993, 33, 743.
(44) Urry, D. W. *Angew. Chem. Int. Ed. Engl.* 1993, 32, 819.
(45) Urry, D. W. *Sci. Am.* 1995, 272(1), 64.

(46) DeBolt, L. C.; Mark, J. E. *Polymer* 1987, 28, 416.
(47) DeBolt, L. C.; Mark, J. E. *J. Polym. Sci., Polym. Phys. Ed.* 1988, 26, 865.
(48) Flory, P. J. *Statistical Mechanics of Chain Molecules.* Interscience: New York. 1969.
(49) Gosline, J. M.; Rosenbloom, J. In *Extracellular Matrix Biochemistry*, K. A. Piez and A. H. Reddi, Eds. Elsevier: New York. 1984; p. 191.
(50) Thiel, B. L.; Kunkel, D. D.; Viney, C. *Biopolymers* 1994, 34, 1089.
(51) Anderson, J. P.; Cappello, J.; Martin, D. C. *Biopolymers* 1994, 34, 1049.
(52) Thiel, B. L.; Viney, C. *MRS Bull.* 1995, September, 52.
(53) Termonia, Y. *Macromol. Symp.* 1996, 102, 159.
(54) Simmons, A. H.; Michal, C. A.; Jedlinski, L. W. *Science* 1996, 271, 84.
(55) Billmeyer, F. W. *Textbook of Polymer Science*, 3rd ed. Wiley: New York. 1984.
(56) Kaplan, D. L.; Fossey, S.; Mello, C. M.; Arcidiacono, S.; Senecal, K.; Muller, W.; Stockwell, S.; Beckwitt, R.; Viney, C.; Kerkam, K. *MRS Bull.* 1992, October, 41.
(57) Weigand, W. B.; Snyder, J. W. *Trans. Inst. Rubber Ind.* 1934, 10, 234.
(58) Hoeve, C. A. J.; Flory, P. J. *J. Am. Chem. Soc.* 1958, 80, 6523.
(59) Flory, P. J. *Principles of Polymer Chemistry.* Cornell University Press: Ithaca, NY. 1953.
(60) Mark, J. E. *Biopolymers* 1976, 15, 1853.
(61) Hoeve, C. A. J.; Hoeve, M. B. J. A. *Polym. Eng. Sci.* 1980, 20, 290.
(62) Weis-Fogh, T. *J. Mol. Biol.* 1961, 3, 520.
(63) Andrady, A. L.; Mark, J. E. *Polym. Bull.* 1991, 27, 227.
(64) Ferry, J. D. *Viscoelastic Properties of Polymers*, 3rd ed. Wiley: New York. 1980.
(65) Clark, A. H.; Ross-Murphy, S. B. *Adv. Polym. Sci.* 1987, 83, 57.
(66) Ross-Murphy, S. B. *Polymer* 1992, 33, 2622.
(67) Ross-Murphy, S. B. In *Viscoelasticity of Biomaterials*. Washington, DC. 1992; Vol. 489; p. 2622.
(68) Russo, P. S., Ed. *Reversible Polymeric Gels and Related Systems.* American Chemical Society: Washington, DC. 1987.
(69) Yang, Y.; Ichise, N.; Li, Z.; Yuan, Q.; Mark, J. E.; Chan, E. K. M.; Alamo, R. G.; Mandelkern, L. In *Complex Fluids*, E. B. Sirota, D. Weitz, T. Witten, and J. Israelachvili, Eds. Materials Research Society: Pittsburgh, PA. 1992; p. 325.
(70) Li, Z.; Mark, J. E.; Chan, E. K. M.; Mandelkern, L. *Macromolecules* 1989, 22, 4273.
(71) Russo, P. S.; Siripanyo, S.; Saunders, M. J.; Karasz, F. E. *Macromolecules* 1986, 19, 2856.
(72) Aharoni, S. M., Ed. *Synthesis, Characterization, and Theory of Polymeric Networks and Gels.* Plenum Press: New York. 1992.

16

Multiphase Systems

One class of multiphase elastomers are those capable of undergoing strain-induced crystallization, as was discussed separately in chapter 12. In this case, the second phase is made up of the crystallites thus generated, which provide considerable reinforcement. Such reinforcement is only temporary, however, in that it may disappear upon removal of the strain, addition of a plasticizer, or increase in temperature. For this reason, many elastomers (particularly those which cannot undergo strain-induced crystallization) are generally compounded with a permanent reinforcing filler[1–12]. The two most important examples are the addition of carbon black to natural rubber and to some synthetic elastomers[3,5,13], and the addition of silica to siloxane rubbers[4]. In fact, the reinforcement of natural rubber and related materials is one of the most important processes in elastomer technology. It leads to increases in modulus at a given strain, and improvements of various technologically important properties, such as tear and abrasion resistance, resilience, extensibility, and tensile strength[3,5,7,11,14–17]. There are also disadvantages, however, including increases in hysteresis (and thus of heat buildup) and compression set (permanent deformation). Another problem in this area is the absence of a reliable molecular theory for filler reinforcement, in general, and even simple molecular pictures of the origin of the reinforcement are lacking. The subject is not even discussed in what has long been the standard reference book[18] on rubberlike elasticity!

On the other hand, there is an incredible amount of relevant experimental data available, with most of these data relating to reinforcement of natural rubber by carbon black[5,7,14]. Recently, however, other polymers such as poly(dimethylsiloxane), and other fillers, such as precipitated silica, metallic particles, and even glassy polymers, have become of interest[19–55].

These studies have shown that materials which act as fillers can vary substantially with respect to the chemical nature of their surfaces, and probably most solid, finely divided materials may advantageously be incorporated into an elastomer. In fact, this is one of the ways the crystallites discussed in chapter 12 improve the mechanical properties of an elastomer. Experimental evidence indicates that the extent of the reinforcement depends strongly on particle size. The maximum reinforcement is obtained for particles with diameters ranging from 10 to 100 nm, and larger particles can actually cause a detrimental effect. The reinforcement phenomenon pertains only to the rubbery state of the elastomer, with the same polymer in the glassy state frequently being essentially unaffected.

The most important unsolved problem in this area is the nature of the bonding between the filler particles and the polymer chains[17]. The network chains may adsorb strongly onto the particle surfaces, which would increase the effective degree of cross-linking. This effect will be especially strong if particles contain some reactive surface groups that may cross-link (or end-link) polymer chains. Chemisorption, with permanent chemical bonding between filler particles and polymer chains, can be dominant, particularly if the filler is precipitated into the elastomer during curing[22–24,42]. Another type of adsorption which can occur at a filler surface is physisorption, arising from long-range van der Waals forces between the surface and the polymer. Contrary to chemisorption, this physical adsorption does not severely restrict the movement of polymer chains relative to the filler surface when high stresses are applied. The available experimental data suggest that both chemisorption and physisorption contribute to reinforcement phenomena, and that the optimal degree of chemical bonding is quite low (of the order of 0.2 bonding sites per square nanometer of filler surface)[11]. Excessive covalent bonding, leading to immobilization of the polymer at the filler surface, is highly undesirable. A filler particle may thus be considered a cross-link of very high functionality, but transient in that it can participate in molecular rearrangements under strain.

There are probably numerous other ways in which a filler changes the mechanical properties of an elastomer, some of admittedly minor consequence[17]. For example, another factor involves changes in the distribution of end-to-end vectors of the chains due to the volume taken up by the filler[16,17,56]. This effect is obviously closely related to the adsorption of polymer chains onto filler surfaces, but the surface also effectively segregates the molecules in its vicinity and reduces entanglements. Another important aspect of filler reinforcement arises from the fact that the particles influence not only an elastomer's static properties (such as the distribution of its end-to-end vectors), but also its dynamic properties (such as network chain mobility). More specifically, the presence of fillers reduces the segmental mobility of the adsorbed polymer chains.

The most relevant physical factor in reinforcement by small filler particles, however, is probably the interfacial effect related to vacuole formation and the separation of the polymer from the surface of the particles[11]. The formation of vacuoles having small radii of curvature requires very high pressure, however, and is therefore unlikely. The effect of cavitation on the behavior of filled rubbers is probably of great importance[1,57]. The elastic modulus of the filler is much larger than the modulus of the surrounding rubber, and this leads to large strain gradients in the vicinity of the filler particle. This effect is very difficult to study from the molecular point of view, but has been investigated using various composite theories[9].

There have also been numerous studies characterizing the particles themselves. These have employed, for example, light microscopy and electron microscopy[58], scanning tunnel microscopy[59], x-ray and neutron scattering[22,24,60–63], and nuclear magnetic resonance spectroscopy[64–67]. Non-spherical particles are of growing interest[19,44,68–70], as are particles having regularly spacings within the elastomeric matrix[71–74].

As has been mentioned, the status of the theory for elastomer reinforcement is rather unsatisfactory at the present time[17]. Nonetheless, it will be useful to have an overview of what has been accomplished and what is presently in progress. This is provided in the following section.

16.1 Some Theoretical Approaches

The first theoretical attempt to explain the dependence of reinforcement on the concentration of filler was carried out by Guth and Gold[75]. These authors modified the Einstein viscosity equation for spherical particles in a viscous medium by adding a quadratic term to account for interactions between particles. They obtained the equation

$$\eta = \eta_0(1 + 2.5\phi + 14.1\phi^2) \quad (16.1)$$

where η and η_0 are the viscosities of the filled and unfilled rubber, respectively, and ϕ is the volume fraction of filler. Smallwood[76] assumed that a similar equation holds for changes in Young's modulus:

$$E = E_0(1 + 2.5\phi + 14.1\phi^2) \quad (16.2)$$

where E is the modulus for the filled elastomer, and E_0 for the unfilled elastomer. Equation (16.1) was generalized later, by Guth, to nonspherical particles[77]:

$$\eta = \eta_0(1 + 0.67\phi f + 1.62\phi^2 f^2) \quad (16.3)$$

where f is a shape factor used to estimate the average particle anisometry. Some authors[78] argue that the rubber adsorbed on the filler surface ("bound rubber")[79–81] should also be counted as a part of the filler particle when using the above equations in the analysis of experimental data.

There are numerous other models for reinforcement in polymer composites[68,69,82–91], for the types of interactions between filler and elastomer[92–95], and the effects of fillers on swelling[96,97]. A review of many of these theories has recently been given by Ahmed and Jones[9]. Most of them are not molecular, and there is a serious lack of rigorous extension of the statistical theory of rubber elasticity to filled elastomers[17]. Several attempts have been made to apply such theory to experimental data, however, in an attempt to estimate the number of filler–rubber attachments[98,99].

Bueche[100] modified the classical kinetic theory of rubber elasticity to account for additional cross-links produced by the filler. The equation obtained was

$$\sigma = (\nu_r + \nu_f)kT(\alpha - \alpha^{-2}) \tag{16.4}$$

where ν_r is the number of effective network chains in the unfilled rubber, and ν_f is the number of additional chains produced by the bonding to the filler. It was assumed that ν_f is proportional to the volume fraction of filler in the network v_f:

$$\nu_f = \frac{S}{s} v_f \tag{16.5}$$

where S is the surface area per unit volume of filler and s is the average area per attachment site. Equation (16.4), when written in terms of the reduced stress or the modulus $[f^*]$ for an unswollen network, becomes

$$[f^*] = (\nu_r + \nu_f)kT \tag{16.6}$$

The theory gives good results for swollen polymers, where entanglement effects may possibly be neglected. Otherwise, the Money-Rivlin equation should be applied [with the constant $2C_1$ set equal to $(\nu_r + \nu_f)kT$].

In more recent work, Heinrich and Vilgis[101] developed a statistical model for filled polymer networks based on the replica formalism. This study included the effect of entanglements on the mechanical properties of filled polymers. The van der Waals model of Kilian and coworkers has now also been successfully applied to filled systems[102], as has a double-network model[103]. Other analytical theories are being developed; for example, the treatment of reinforcement in terms of the lateral compression of the particles as represented by fractal aggregates[104]. Simulations are also being employed to obtain insight into the reinforcement. For example, attempts are being made to reformulate the network partition function so as to study the adsorption of chains on filler particles and the effect of this adsorption on network topology[105]. Finally, as described in chapter 8, it is also possible to carry out Monte Carlo simulations to determine the effect of the excluded volume of the filler particles on the network chain-length distributions and associated elastomeric properties[16,17,56].

As is obvious from this brief overview of filler theories[17], the mechanism of the reinforcement is only poorly understood. There are other, practical, difficulties in this area as well. Typically, such fillers are blended into a polymer of high molecular weight, prior to its being cross-linked into the final network structure. This is a difficult, tedious procedure since the polymer usually has a very high bulk viscosity and the filler is typically badly agglomerated[106]. It is not surprising, therefore, that good dispersions frequently require a great deal of time, energy, and patience. This also complicates, of course, the construction of realistic theoretical models for the elastomer–filler system. Some elucidation of the nature of the reinforcement obtained and simplification of the procedure itself might be obtained by focusing on model filled systems. One way of doing this is by precipitating the fillers into network structures rather than the usual blending of badly agglomerated fillers into polymers prior to their cross-linking. This in situ approach is the subject of several of the following sections.

16.2 In Situ Generation of Fillers in Elastomers

16.2.1 General Comments

The polymer which has been used most extensively in studies of this type is poly(dimethylsiloxane) (PDMS) $[-Si(CH_3)_2O-]_x$. It has been chosen in part because it is compatible with many of the organometallic materials used to generate ceramic-type phases, and in part because it is an elastomer which requires considerable reinforcement from silica or some other filler before it is useful in most industrial applications[3,4,10].

The idea of in situ generation of reinforcing fillers in polymers is based on an important advance in the ceramics area. It involves the use of preparative techniques heavily based on chemical reactions; for example, the generation of ceramic-type materials by the hydrolysis of an organometallic compound[48,107–117]. In the ceramics area, the advantages are the purity of the resulting products, the possibility of avoiding high temperatures (thereby permitting survival of organic guest molecules), the control of ultrastructure (at the nanometer level), the relative ease of forming ceramic "alloys," and the possibility of incorporating a different material into the resulting ceramic while it is still porous, prior to its densification. In the present application to elastomers, the new approach has the advantages of avoiding the difficult, time-consuming, and energy-intensive process of blending agglomerated filler into high molecular-weight polymers, and of simplifying the problem of obtaining good dispersions. The main disadvantage is the increased cost, and the reluctance of industry to carry out the major retooling required to introduce such a different technology. In any case, this approach has been investigated using a variety of fillers: for example, silica by hydrolysis of organosilicates, titania from titanates, alumina from aluminates, zirconia from zirconates, and so on[8,20–23,106,118–124].

The following sections consist, in large part, of descriptions of how some of these sol–gel ideas can be extended using concepts from the area of polymer chemistry[54]. The modifications are generally carried out by using elastomeric polymer chains terminated with functional groups that can participate in the ceramic-generating reaction, typically hydroxyl groups in the case of a hydrolysis reaction. This permits intimate bonding between the inorganic (ceramic) phase and the organic (polymer) phase, in a way that is most likely to give novel mechanical properties. The PDMS elastomers mentioned above have been most studied in such organic–inorganic hybrid composites.

Although the reinforcement of elastomers by dispersed ceramic phases is of primary interest in the present context, the case where the ceramic predominates and becomes the continuous phase[54] is also discussed. The elastomeric phase therefore becomes dispersed in it. Yet another approach to the novel reinforcement of elastomers involves monomers such as styrene or methyl methacrylate, polymerized in situ to give glassy polymeric domains, and some related systems in which there are either magnetically responsive particles or crystalline zeolites.

Of primary interest in this chapter is the reinforcement provided by these fillers. It is easy to switch the focus, however, so that the elastomer is viewed as only a matrix in which the ceramic materials are being generated. In this "matrix isolation" approach[125], x-ray and neutron scattering techniques, for example, can be used to obtain information that transcends these particular systems. It should be useful in a variety of areas, including the new sol–gel technique for preparing unmodified ceramics mentioned above.

16.2.2 Preparation

The most important reaction of this type involves the catalytic hydrolysis of tetraethoxysilane (TEOS)[20,22,23,35,106,118,119,121,126–131]:

$$Si(OC_2H_5)_4 + 2H_2O \rightarrow SiO_2 + 4C_2H_5OH \qquad (16.7)$$

In the original sol–gel technique, the gel thus formed is first dried to remove unreacted TEOS, ethanol, water, and catalyst (which is generally chosen to be volatile). It is then fired into a porous ceramic, which may then be densified into the final ceramic object. The kinetics of this reaction can be studied using air-pressure deformation measurements[132] on the gels as they evolve, in a manner similar to that used to characterize thermoreversible polyethylene gels[133,134].

In the simplest application of this approach for obtaining elastomer reinforcement, some of the organometallic material is absorbed into the cross-linked network, and the swollen sample is placed into water containing the catalyst (typically a volatile base such as ammonia or ethylamine). Hydrolysis to form the desired silicalike particles proceeds rapidly at room temperature to yield in the order of 50 wt% filler in less than an hour[8,21,23,135]. Alternatively, hydroxyl-terminated chains are blended with enough TEOS to both end-link them and to provide silica by the hydrolysis reaction. Thus, curing and filling take place simultaneously, in a one-step process[8]. In an alternative method, TEOS is blended into a polymer having end groups (e.g., vinyls) that are unreactive under the hydrolysis conditions. The silica is then formed in the usual manner [eq. (16.7)], and the product dried. The resulting slurry of polymer and silica is stable and can be cross-linked later using any of the standard cross-linking techniques, such as vinyl-silane coupling or peroxide thermolysis[8].

16.2.3 Electron Microscopy

Both transmission and scanning electron microscopy have been used to characterize these novel composite materials[8,20,21,136,137]. The information obtained in this way includes: (1) the nature of the precipitated phase (particulate or nonparticulate), (2) the average particle size, if particulate, (3) the distribution of particle sizes, (4) the degree to which the particles are well defined (smoothness of the interfaces), and (5) the degree of agglomeration of the particles[54].

A typical micrograph is shown in figure 16.1[138]. Such results show that the particles formed in this manner typically have a narrow distribution of sizes, with most diameters typically in the 200–250 Å range[136,139]. They are generally well dispersed and essentially unagglomerated, which suggests that the reaction may involve simple homogeneous nucleation. This is consistent with the fact that

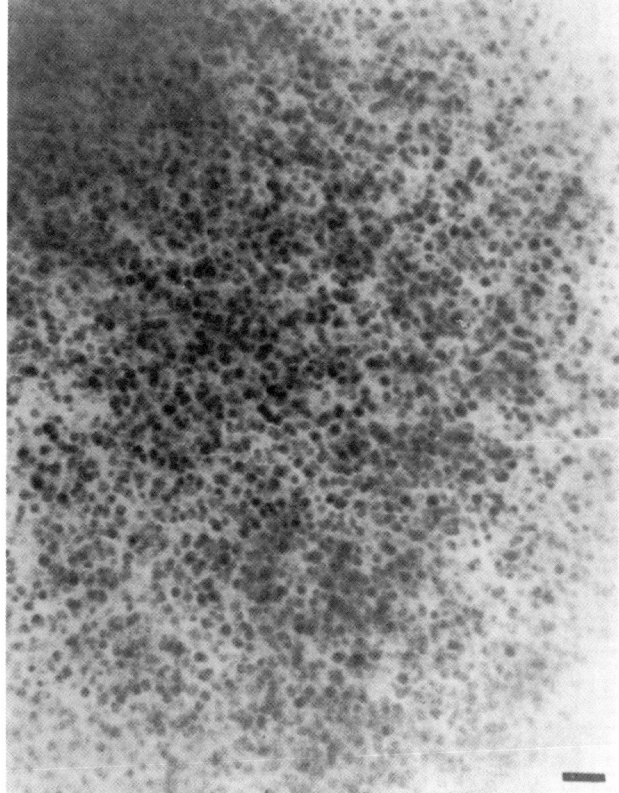

Figure 16.1 Transmission electron micrograph for a poly(dimethylsiloxane) (PDMS) network containing 34.4 wt% silica generated in situ[138]. The length of the bar in the figure corresponds to 1000 Å.

particles growing independently of one another and separated by cross-linked polymer would not agglomerate unless very high concentrations were reached. There is some evidence that, in at least some cases, the pore size of the network constrains the growth of these particles, in that the average diameters observed seem to increase with decrease in cross-link density[140].

One interesting result from characterization studies of this type is the conclusion that basic catalysts generally yield particles that are well defined, whereas acidic catalysts yield particles that are rather "fuzzy"[136]. This conclusion is in agreement with results obtained earlier in sol–gel ceramics investigations[141].

16.2.4 Scattering Techniques

A number of x-ray and neutron scattering studies have been carried out on these filled elastomers[8,20,21,24,54,61,142]. Some typical results, for systems in which the two

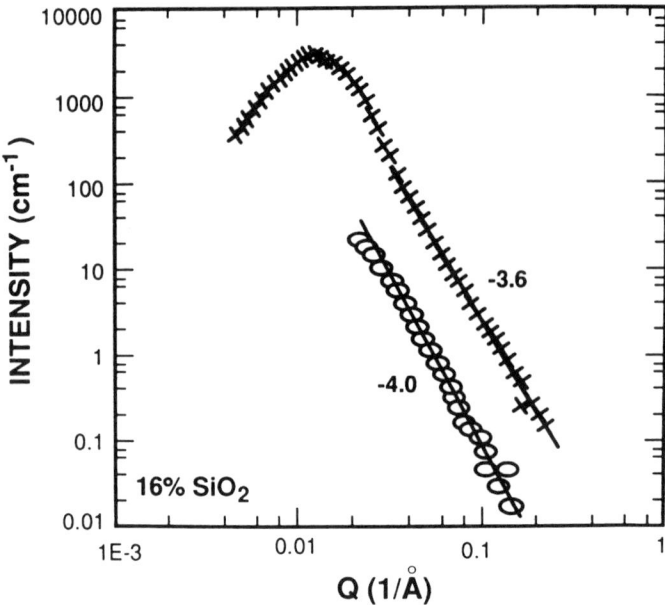

Figure 16.2 The SAXS and SANS scattering intensities for a SiO_2-PDMS composite thought be be bicontinuous, shown as a function of the scattering vector (in reciprocal Å)[142]. The weight percent of silica present is indicated in the figure, and the numbers attached to the two curves give the values of the terminal slope (which are of interest with regard to characterizing the roughness of the interfaces).

phases may be bicontinuous (interpenetrating), are shown in figure 16.2[142] and are discussed further in section 16.3. In general, such data can provide estimates of average particle size and possibly particle-size distribution. In addition, the terminal slopes give an indicate of the nature of the interfaces, with -3 corresponding to rough interfaces and -4 to smooth.

Although scattering results are generally consistent with those gotten by electron microscopy, there are some intriguing differences. Of particular interest is the observation that some fillers which appear to be particulate in electron microscopy, do not appear to be so in scattering studies[24,142]. Additional comparisons will be required to resolve this issue.

16.2.5 Nuclear Magnetic Resonance

Another way to characterize the in situ generated particles is by nuclear magnetic resonance (NMR) imaging, utilizing ^1H and ^{29}Si magic-angle spinning, with two-dimensional Fourier transform spin-echo techniques. One specific approach is to study ^1H spin–spin (T_2) relaxation times of the protons in a PDMS polymer as they are being constrained by silicalike material being generated in their vicinity[143]. Some illustrative results are given in figure 16.3, which shows ^1H NMR images of a elastomeric sample of PDMS containing in situ precipitated silica. For

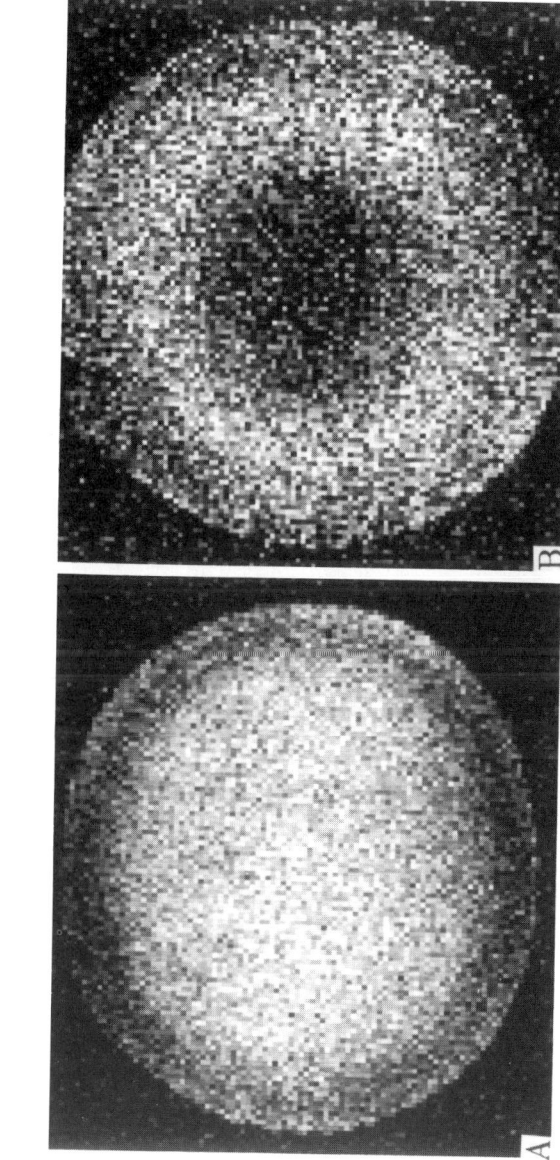

Figure 16.3 ^{1}H NMR images of a SiO$_2$-PDMS elastomeric sample obtained with a two-dimensional spin-echo sequence having an echo time of 3.3 ms (A) and 22.7 ms (B)[143]. The view is down the axis, and the resolution is 128× by 128× pixels of 211 and 236 mm, respectively (in A), and 211 mm in both axes (in B). Reprinted with permission from Garrido, L., et al. (1991), *Macromolecules*, **24**. Copyright 1997, American Chemical Society.

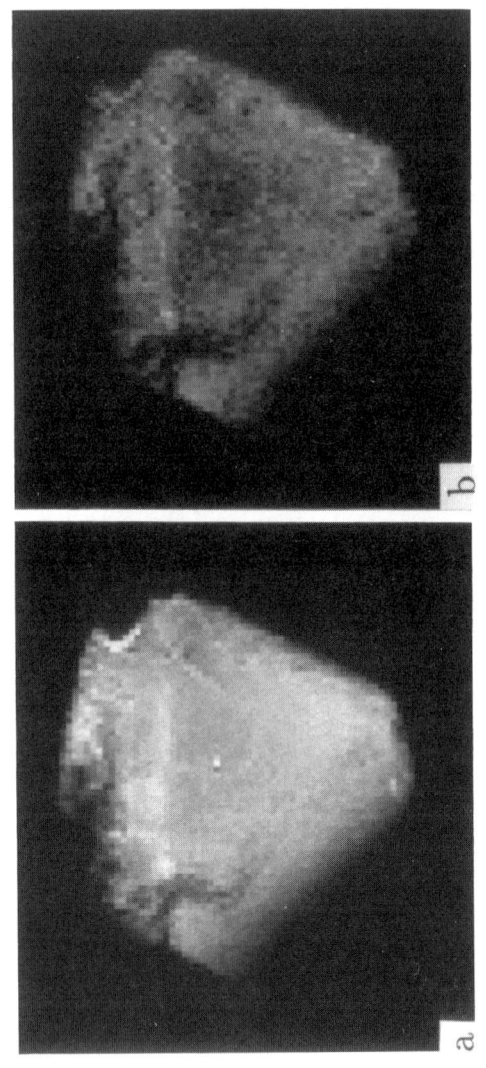

Figure 16.4 1H NMR images of an PDMS–aerogel composite obtained with a two-dimensional spin-echo sequence having an echo time of 2.82 ms (a) and 122.6 ms (b)[145].

testing purposes, this composite was intentionally made to be inhomogeneous, with much larger amounts of silica being on the surface. This was done by choosing a sample with relatively large dimensions, and carrying out the precipitation reaction for only a short period of time. The sample employed was cylindrical in shape, with a height of approximately 1 cm and a diameter of approximately 2 cm. The dark rim at the edge of the sample seen in figure 16.3 probably indicates a reduced mobility of the network chains due to the presence of the silica. Its change in location with time can help characterize the movement of the reaction front into the sample. This technique is obviously nondestructive, but if the sample can be sacrificed, then slices taken from the sample can be further studied in a gradient column with regard to density, by electron microscopy[144], or by x-ray or neutron scattering.

A PDMS elastomer which has been infiltrated into a silica aerogel can also be imaged in this way[145]. This is shown in figure 16.4, and will be discussed further in section 16.4.

16.2.6 Aging

Permitting precipitated silica particles to remain in contact with their aqueous catalyst solution can permit them to "age" or "digest"[136,139,146]. Electron microscopy results suggest that some reorganization is occurring, with the particles becoming better defined, more uniform in size, and possibly even less aggregated. There seem to be interesting parallels with "Ostwald ripening" in the area of colloid science[147].

16.2.7 Densities

Comparisons between the values of wt% filler obtained from density measurements and the values obtained directly from weight increases can give very useful information on the filler particles. For example, the fact that the former estimate is smaller than the latter in the case of silica-filled PDMS elastomers[140,148] indicates that there are probably either voids or unreacted organic groups in the filler particles, and possibly both. Incomplete conversion of tetraethoxysilane to silica is also indicated by ^{29}Si NMR measurements[140].

16.2.8 Calorimetry

Differential scanning calorimetry measurements at low temperatures have been carried out on PDMS elastomers containing silica precipitated in situ[149]. The presence of the silica was found to reduce both the extent of crystallization and the rate of crystallization when the elastomers were in the unstretched state. This is in interesting contrast to similar studies of PDMS in the stretched state, where the filler may facilitate the crystallization process[150].

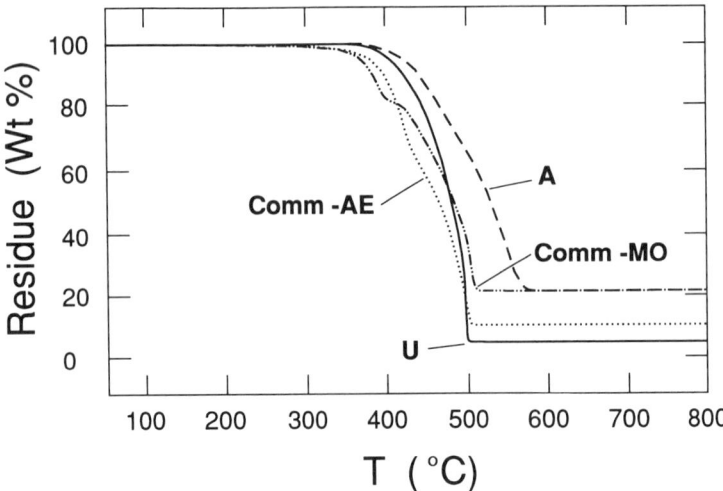

Figure 16.5 Comparison of TGA thermograms for PDMS networks that were unfilled (U), or contained either silica precipitated in situ (A) or commercial fume silica (COMM-MO and COMM-AE)[151]. The heating of the samples was carried out under nitrogen.

16.2.9 Thermogravimetric Analysis

In at least some cases, fillers precipitated in situ may have the advantage of increasing thermal stability. Typical results obtained by thermogravimetric analysis (TGA) of silica-filled PDMS elastomers are shown in figure 16.5[151]. The samples in which the SiO_2 was introduced in situ seem to have higher decomposition temperatures. A possible mechanism for this improvement would be increased capability of the silica produced in situ to tie up hydroxyl chain ends that participate in the degradation reaction.

16.2.10 Mechanical Properties and Equilibrium Swelling

16.2.10.1 Uniaxial Extension The reinforcing ability of such particles generated in situ has been amply demonstrated for a variety of deformations, including uniaxial extension (simple elongation), biaxial extension, shear, and torsion[8,54]. In the case of uniaxial extension, the modulus $[f^*]$ frequently increases by more than an order of magnitude, with the isotherms generally showing the upturns at high elongation that are the signature of good reinforcement[42,152]. Typical results are shown in figure 16.6[152]. As is generally the case in filled elastomers, there is seen to be considerable irreversibility in the isotherms, which is thought to be due to irrecoverable sliding of the chains over the surfaces of the filler particles. Some fillers other than silica, for example, titania (TiO_2), do give stress–strain isotherms

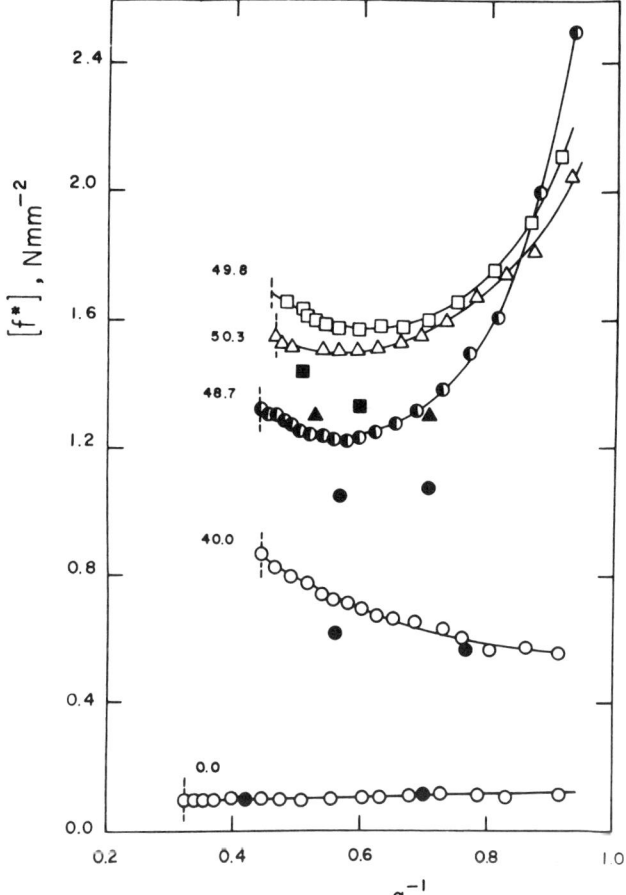

Figure 16.6 Reduced stress shown as a function of reciprocal elongation for PDMS networks that are either unfilled or filled with silica in situ[152]. Filled symbols are for results obtained out of sequence to test for reversibility, and each curve is labeled with the weight percent of filler that had been incorporated.

that are reversible, indicating interesting differences in surface chemistry[153] Titania fillers, and other nonsilica fillers, are discussed further in section 16.2.13.

16.2.10.2 Adhesion Between Phases It is possible to interpret equilibrium swelling measurements obtained on unfilled and filled elastomers to estimate the degree of adhesion between elastomer and filler particles[15,154,155]. Some typical results for fillers generated in situ are presented in figure 16.7[154]. The results differ greatly from those to be expected for nonadhering fillers, indicating good bonding between the two phases. Resistance to separation from the surface in such swelling tests, however, does not necessarily mean that the chains have no mobility *along* the surface, as was described in the preceding section.

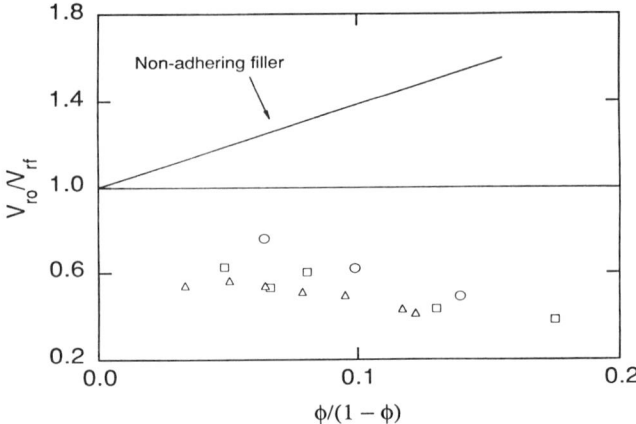

Figure 16.7 Plot of volume fraction ratio V_{r0}/V_{rf}, characterizing the swelling of an unfilled PDMS network relative to that of a filled PDMS network, against filler loading expressed as volume ratio of filler to rubber $\phi/(1-\phi)$ (where ϕ is the volume fraction of filler)[154]. Types of filler were silica–titania mixed oxides (□), silica (○), and titania (△).

16.2.10.3 Biaxial Extension As already mentioned, there have been numerous experiments carried out on such in-situ-filled PDMS networks in uniaxial extension, and they have shown that these elastomers have very good mechanical properties[8,156,157]. However, there is relatively little analogous data available for the other types of deformation, such as biaxial extension (compression), shear, and torsion. These types of deformation are important in the characterization of elastomeric materials, but are not much studied because they are more difficult to impose than simple elongation. This is true even for elastomers filled in the usual manner, by blending of particles prior to cross-linking[18]. It is obviously interesting and important to use a variety of deformations to characterize the effects of particles produced in situ on the mechanical properties of the filled elastomers[158].

Biaxial extension is one of the most important of these other mechanical deformations. The equations for it are identical to those for uniaxial elongation and, thus, the same elastomeric quantities pertain here as well[158,159]. The most convenient method for producing equibiaxial extension is inflation of a circular sheet, clamped around its circumference, into the form of part of a spherical balloon[18,158–160]. The strain can be obtained from the vertical distances between the points marked on the inflated sheet (as measured with a cathetometer), and from the horizontal distances (measured with a traveling microscope). The stress is then directly obtained from a manometer measuring the pressure on the deforming fluid.

Since biaxial extension is equivalent to uniaxial compression, both types of data can be combined in the same plot. In this way, a full spectrum of stress–strain data for both elongation and compression can be viewed, and the behavior of networks in the two deformation regions can be compared directly. Results

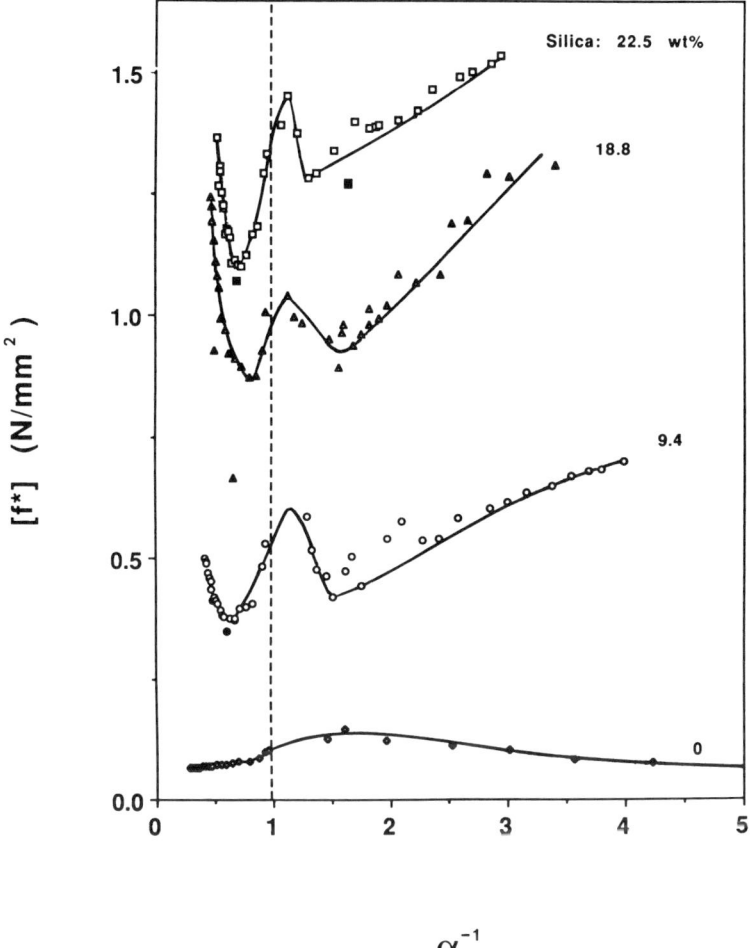

Figure 16.8 Stress–strain isotherms for PDMS–silica elastomers, reinforced in situ, in elongation (region to the left of the vertical dashed line, with $\alpha^{-1} < 1$) and in biaxial extension (compression) (to the right, with $\alpha^{-1} > 1$)[158]. The filled points represent the data used to test for reversibility.

obtained for both elongation (reciprocal elongation $\alpha^{-1} < 1$) and compression ($\alpha^{-1} > 1$) are therefore depicted in figure 16.8[158], where the modulus is plotted against α^{-1}. It is obvious that very strong reinforcing effects from the precipitated silica occur for biaxial extension, as well as for elongation. This is evident both from the large upward shifts of the isotherms as a whole, and from the pronounced upturns both at high elongations ($\alpha^{-1} < 1$) and at high compressions (biaxial extensions) ($\alpha^{-1} > 1$). As can also be seen, such desirable upturns do not occur for the reference (unfilled) samples.

It is interesting to note that the extent of the reinforcement, as gauged by the magnitudes of the upturns, is approximately the same in uniaxial and biaxial extension. The range of deformation over which it occurs, however, seems to be larger in the case of the biaxial extension. The pronounced maxima and minima in the isotherms for the filled elastomers in biaxial extension are not understood at present[158], and represent challenges to theories on the origin of reinforcing effects in elastomers, in general.

The effect of filler content on the rupture moduli $[f^*]_r$ in elongation and in compression is shown in figure 16.9. The dependences are seen to be quite similar, with the curves for uniaxial and biaxial extension being very nearly parallel to one another. Figure 16.10 shows the effects on the deformation at rupture α_r in elongation and in compression. The two behaviors are again quite similar, in that increase in filler content changes these values of α_r in the direction of smaller deformations ($\alpha \to 1$).

Similarly, values of the energy E_r required for rupture in both uniaxial and biaxial (compression) extension can be obtained by direct integration of $\int f^* d\alpha$. For both deformations, the increases in $[f^*]_r$ predominate over the decreases in α_r, and, as a result, E_r increases substantially with increase in filler content[158]. The two curves lie nearly parallel to one another, in a way similar to that shown in figure 16.9.

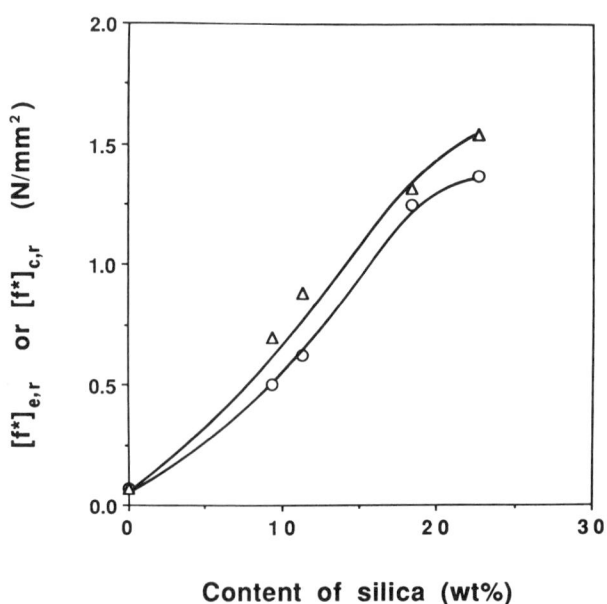

Figure 16.9 Effect of filler content on the rupture modulus in elongation $[f^*]_{e,r}$ (○), and in compression $[f^*]_{c,r}$ (△)[158]. (See figure 16.8.)

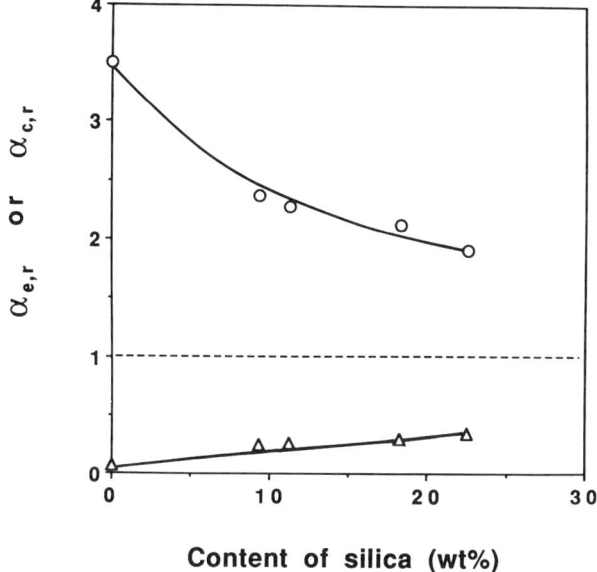

Figure 16.10 Effect of filler content on the deformation at rupture in elongation $\alpha_{e,r}$ and in compression $\alpha_{c,r}$[158]. (See figures 16.8 and 16.9.)

16.2.10.4 Shear As described in appendix C, pure shear involves extension in three perpendicular directions without rotation of the principal axes of the strain. It is conveniently imposed by stretching a wide rectangular sheet in a direction perpendicular to its width[18]. The extension ratio in this direction is α, and the perpendicular or transverse direction remains unchanged to keep the other elongation ratio α_2 at unity. This is achieved by having the width of the test sheet very much greater than its length, making changes in the width negligible. The stress t_2 is automatically generated as a result of the restraints introduced by the clamps. The principal stress t_1 and its extension ratio α can then be measured[18,158].

The stress-strain isotherms obtained for both unfilled and in-situ-filled PDMS networks, represented in terms of the pure shear modulus G and the principal extension ratio α, are shown in figure 16.11. The results in simple shear, in terms of the shear strain γ, were essentially identical and are presented elsewhere[161]. As in the case of uniaxial extension and compression, the addition of filler shifts the isotherms upward to significantly higher values of the modulus. In the case of the filled networks, there is an initial decrease in the modulus with increase in deformation. Also, the larger the amount of filler present, the more pronounced the decrease. This may be due to stress-induced rearrangements of the chains in the vicinity of the filler particles[158]. Also of possible relevance is the fact that an increase in the amount of filler decreases the number of load-bearing chains passing through the unit cross-sectional area, and changes the distribution of their end-to-end distances[16,17,56].

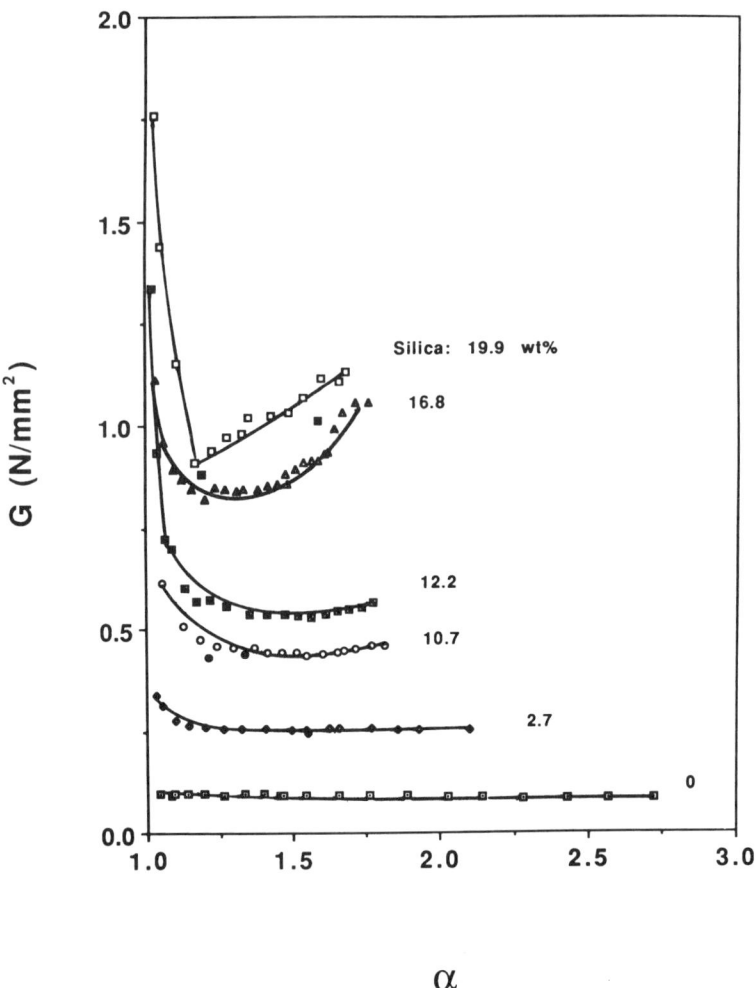

Figure 16.11 Stress–strain isotherms for PDMS–silica elastomers, reinforced in situ, in pure shear[158]. The filled points represent the data used to test for reversibility.

From these results, it can readily be seen that the filled samples have much higher shear moduli than the corresponding unfilled sample. Also, at high deformation, there are pronounced upturns in the reduced stress or modulus in the case of the filled samples, which indicates very good reinforcement. The magnitudes of the upturns are much less pronounced, however, than they were found to be in uniaxial extension and compression. In any case, the extension ratio α or shear strain γ at which the upturn appears is seen to decrease with increase in weight percent of filler. As is usually the case[3,4,10], the maximum deformability decreases as well. Clearly, the behavior observed here is very similar to that described previously for uniaxial and biaxial extension.

Also of interest is the dependence of the ultimate properties on the silica content[158]. It was observed that α_r and γ_r decrease with increase in silica content, while the values of the modulus G_r and stress f_r^* at rupture increase at lower contents, but appear to be leveling off or even decreasing at higher contents. This is presumably due to decreasing extensibility. The rupture energy E_r depends on silica content in a way very similar to that shown by G_r and f_r^*.

16.2.10.5 Torsion The apparatus for the torsion measurements was based on Gent's modification[162] of the original equipment described by Treloar[18]. For this type of deformation, the shear modulus can be calculated from $G = 2M/\pi\psi a^4$, where M is the torsion couple, ψ is the torsion in radians per unit length of the strained axial length, and a is the unstrained radius. Wide ranges of rotational deformation can be investigated, for example, up to $1500°$[163].

Some isotherms obtained on PDMS-SiO$_2$ elastomers filled in situ, in torsion,[163] are shown in figure 16.12. As was the case for the other deformations, increase in filler content shifts the isotherms upward, to significantly higher values of the shear modulus. These curves are significantly different from the ones for the other types of deformation, however, in that there is little evidence for upturns in the modulus at high deformations. This could be due either to the nature of the deformation or to the inability, to date, of reaching sufficiently high torsional deformation.

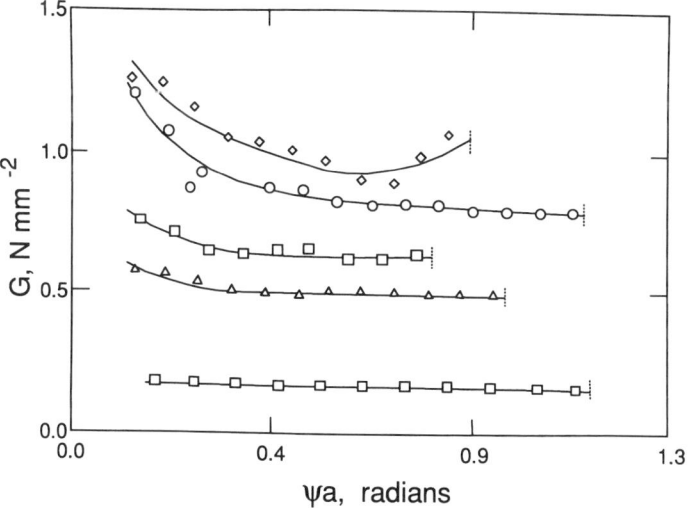

Figure 16.12 Shear moduli for PDMS–silica elastomers, reinforced in situ, shown as a function of the torsional strain expressed as ψa (radians), where ψ is the twist in radians per unit length of the strained axis and a is the unstrained radius of the (cylindrical) elastomer sample[163]. The modulus increases continuously as the amount of filler is increased in the sequence 0.0, 15.5, 17.3, 19.8, and 26.4 wt%. The vertical dashed lines locate the rupture points.

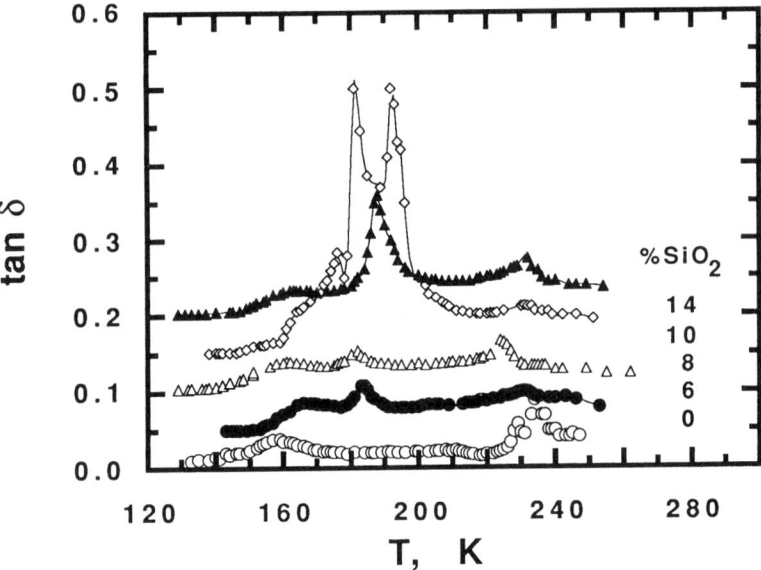

Figure 16.13 Temperature dependence of the dynamic-mechanical loss tangent (tan δ) for PDMS–silica elastomers, reinforced in situ[165]. The measurements were carried out at 11 Hz, and each curve is identified by the weight percent of SiO_2 present. For the purposes of clarity, the curves have been shifted arbitrarily along the ordinate.

16.2.10.6 Dynamic Mechanical Properties Measurement of dynamic mechanical losses can provide important information on transitions occurring in polymers[164], including filled elastomers. Typical values of the loss tangent obtained on PDMS–SiO_2 elastomers filled in situ are shown in figure 16.13[165]. Such results document the effect of filler on the glass transition temperature (in the vicinity of 160 K), and on the melting point (in the vicinity of 230 K) of the PDMS.

16.2.10.7 Cyclic Deformations It is also of considerable interest to see how filled elastomeric materials perform in cyclic deformations, in order to characterize their resistance to creep or to failure by fatigue. In a typical experiment of this type, silica-filled PDMS elastomers are subjected to cyclic compressions, with the modulus and sample dimensions monitored as a function of the number of cycles, the time period of cycling, or the power consumed[166].

Some cyclic compression experiments have been carried out on unfilled PDMS and on PDMS–SiO_2 systems[166]. Of particular interest in such experiments is the change in fractional length of a sample with time, since this characterizes the amount of creep or compression set resulting from the cyclic deformation. Typical results obtained on unfilled PDMS and on PDMS filled with a commercial fume silica are presented in figure 16.14. Both types of samples are seen to show considerable compression set. The corresponding results for PDMS samples containing silica precipitated in situ are given in figure 16.15. The filled samples in

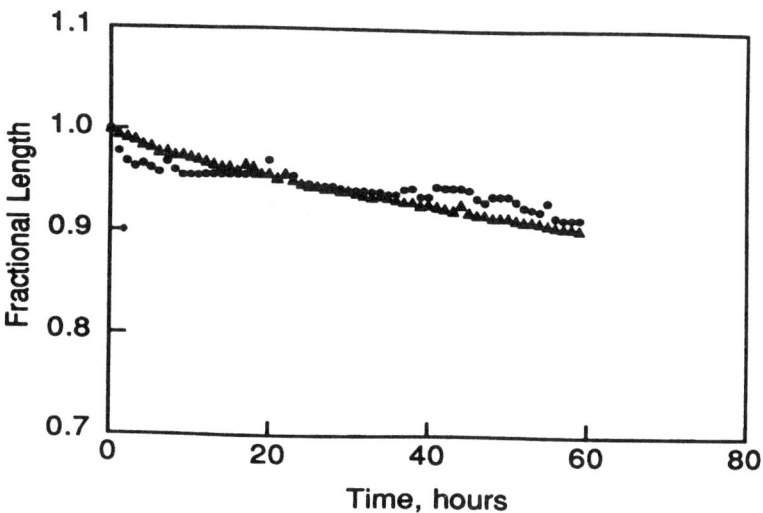

Figure 16.14 Effect of cyclic stress on unfilled PDMS and PDMS elastomer filled with fume silica, represented as the dependence of fractional length on time: unfilled (▲), 14.4 wt% silica (●)[166].

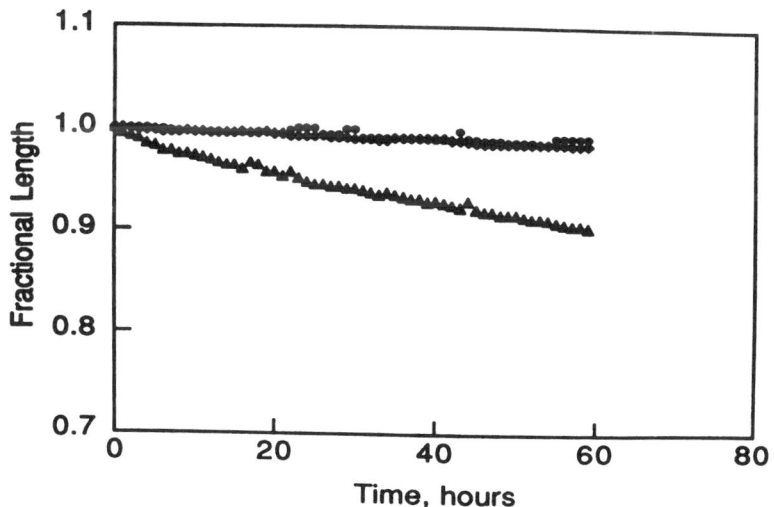

Figure 16.15 Effect of cyclic stress on unfilled PDMS and PDMS elastomers filled with silica precipitated in situ: unfilled (▲), 12.0 wt% silica (◆), 19.4 wt% silica (●)[166]. Reprinted from Wen, J., Mark, J. E., Fitzgerald, J. J. (1994), *J. Macromol. Sci., Macromol. Rep.*, **A31**, by courtesy of Marcel Dekker, Inc.

this case are seen to show very little compression set[166]. This suggests that the in situ silica can retard the chemical changes caused by cyclic stresses, as well as those caused by high temperatures, and is thus consistent with the TGA results[151] already cited.

16.2.10.8 Impact Resistance Impact strengths of some PDMS–SiO$_2$ samples have been determined by the Charpy pendulum impact test and by the falling-weight impact test[167]. In the former, the impact strength is determined from the energy given up by a pendulum striking the test sample. In the latter, the impact of a dart hitting the sample is directly measured by a quartz piezoelectric transducer. The samples investigated were PDMS-modified SiO$_2$ and SiO$_2$/TiO$_2$ glasses with PDMS contents ranging from 0 to 65 wt%. As expected, only samples with relatively high ceramic contents were sufficiently brittle to be studied in this manner. The influence of PDMS content and PDMS molecular weight on the impact strength is shown in figures 16.16 and 16.17, respectively. The larger the amount of PDMS used, or the higher its molecular weight, the higher the impact strength.

For PDMS-modified SiO$_2$ glasses, structural analysis shows that this hybrid material has some degree of localized phase separation of the PDMS component, even though OH-terminated PDMS can be successfully incorporated into the SiO$_2$ network by chemical bonding. The PDMS component can behave as an elastomeric phase because the glass transition temperature of PDMS is far below room temperature[168]. When the material is subjected to an impact test, the PDMS component can absorb a great deal of energy by motions of the PDMS chains, and can ameliorate the growth of cracks and fracture. Therefore, considerable toughening of the glass can be achieved by increasing the amount of PDMS introduced. The effect of increase in PDMS molecular weight may be to increase the phase separation which leads to the energy-absorbing domains[167].

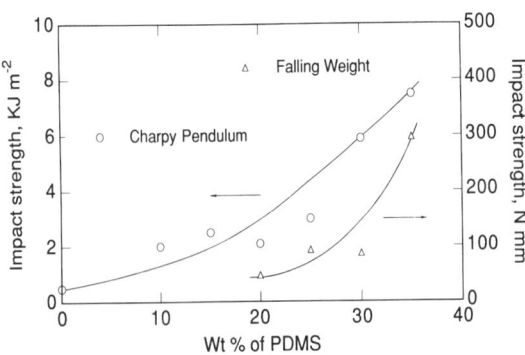

Figure 16.16 Dependence of two estimates of the impact strength on the amount of PDMS in PDMS-modified SiO$_2$ glasses[167]. The impact strengths were obtained from (○) the Charpy pendulum impact test, (△) the falling-weight impact test.

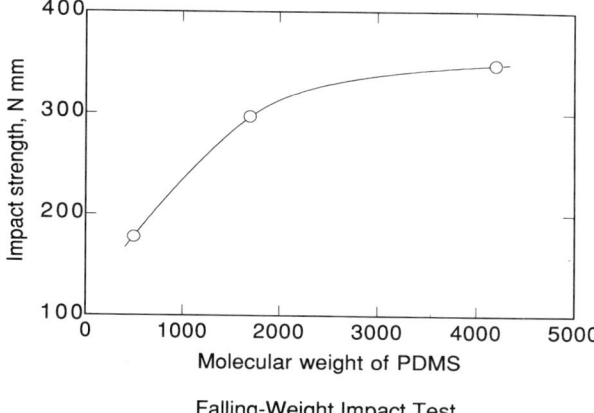

Figure 16.17 Dependence of impact strength (as obtained from the falling-weight impact test) on molecular weight of the PDMS in PDMS-modified SiO$_2$ glasses[167].

Microscopic observation of fracture surfaces can also provide useful information for interpreting impact strengths in terms of the physical processes involved[169]. Specifically, smooth fracture surfaces are associated with completely brittle failure[170], with little effective resistance to either initiation or propagation of cracks. This was observed in the case of samples having relatively low PDMS contents. In contrast, samples with high PDMS content had fracture surfaces showing some degree of "whitening" or shearing. This suggests a ductile, energy-absorbing response to the impacts, with increased resistance to crack propagation.

16.2.10.9 Equilibrium Swelling In a swelling experiment, the network is typically placed into an excess of solvent, which it imbibes until the dilational stretching of the chains prevents further absorption of solvent[8,18,171]. General aspects of this nonmechanical type of deformation are described in chapter 6. Adsorption of some network chains onto the particles of a reinforced elastomer would, of course, reduce the extent of swelling at equilibrium and lead to inhomogeneous strain fields[96]. Characterization of such equilibrium swelling as a function of filler constitution, filler amount, and the method of generating and incorporating it into an elastomer are the issues of greatest interest with regard to this type of deformation. An example of such an investigation was described in figure 16.7.

Although this is a potentially interesting area, results to date are rather limited.

16.2.11 Comparisons among Various Silica-Based Fillers

As is now obvious, there are a variety of ways to generate silica-type fillers useful for reinforcing PDMS networks, and the extent to which such fillers provide reinforcement was characterized in a study by Sun and Mark[172]. The materials and techniques employed were: (1) incorporating a commercial silica which has been treated with hexamethyldisilazane, (2) incorporating silica which had been precipitated from a silicate in an aqueous dispersion with PDMS, (3) precipitating silica directly into PDMS during its curing, (4) precipitating silica directly into a swollen PDMS network after it was cured, (5) incorporating silica prepared from tetraethoxysilane (TEOS) and containing some PDMS, and (6) incorporating silica prepared from partially hydrolyzed TEOS and also containing some PDMS. The resulting filled elastomers showed the largest values of the ultimate strength in the cases of (4) and (6), and the largest value of the rupture energy for (4). A better molecular understanding of filler reinforcement could provide a detailed interpretation of these results.

16.2.12 Other Polymers

As already mentioned, most of the studies to date on elastomer–ceramic materials have involved PDMS, primarily because of its great miscibility with TEOS. Similar studies[173] on the related polymer, poly(methylphenylsiloxane), however, are also of considerable importance. This is because of the stereochemically irregular structure of this polymer, which prevents it from undergoing strain-induced crystallization[8,174]. Good reinforcement generated in situ was also achieved in this polymer, suggesting that such crystallization is not important for this type of reinforcement[173].

A large number of other polymers have been treated in this way, as is described elsewhere[54,55]. For example, the same techniques have also been shown to give good reinforcement in polyisobutylene elastomers[175], and in poly(ethyl acrylate)[176]. The results for polyisobutylene are shown in figure 16.18. In the case of some polymers, it is useful to choose R groups in the alkoxy silane $Si(OR)_4$ so as to make the molecule more hydrophobic than TEOS. In the case of the poly(ethyl acrylate), it appears that the silica precipitation can be carried out during an emulsion polymerization[176].

16.2.13 Other Ceramic-Type Fillers

Silica particles in PDMS elastomers can be a problem at high temperature, since the silanol groups on their surfaces can cause degradation of the polymer[177]. For this reason, and others, a variety of other fillers have been precipitated into PDMS and other elastomers[54,55]. Examples are titania (TiO_2)[153,154,167,178–181], alumina (Al_2O_3)[180–184], and zirconia (ZrO_2)[120,181]. These nonsilica fillers also provide good reinforcement. One interesting difference, however, is the observation that the stress–strain isotherms in these cases frequently have much better reversibility[179]. This is illustrated for PDMS–TiO_2 systems in figure 16.19[153]. The differences are presumably due to different interactions between the groups present on these particle surfaces and the PDMS elastomeric matrix.

There is now increasing interest in introducing fillers consisting of more than one oxide[54,55,154,167,181,185]. One of the hopes is that there will be an advantageous "delegation of responsibilities" in these mixed oxides, with one type of material improving one set of properties and another improving a different set.

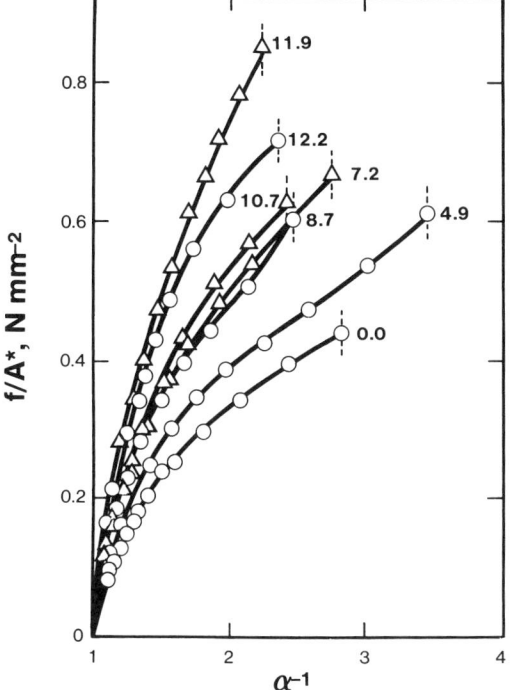

Figure 16.18 Stress–strain isotherms for polyisobutylene networks reinforced with silica generated in situ[175]. Each curve is labeled with the weight percent of filler introduced. Circles represent results obtained using tetraethoxysilane and triangles represent results using phenyltriethoxysilane (which can possibly introduce organic phenyl groups into the filler particles).

16.3 Preparation of Bicontinuous Systems

At some compositions and under some hydrolysis conditions, bicontinuous phases can be obtained (with the silica and polymer phases interpenetrating one another). Some of the evidence for such interpenetration is from small-angle x-ray scattering (SAXS) and small-angle neutron scattering (SANS) profiles[24,142], such as were illustrated in figure 16.2. The observed peak suggests that the SiO_2 cannot consist of individual particles, but rather forms a continuous network.

Other evidence is from stress–strain isotherms in elongation. Since the two networks interpenetrate one another, the mechanical properties exhibited by the material can be very peculiar. In the first deformation of the virgin material, the silica network can give a very high initial modulus, but once this structure is broken, additional deformation cycles can indicate much lower values of the modulus. The mechanism for the generation of bicontinuous materials may be spinodal decomposition, occurring either before or after the polymerization.

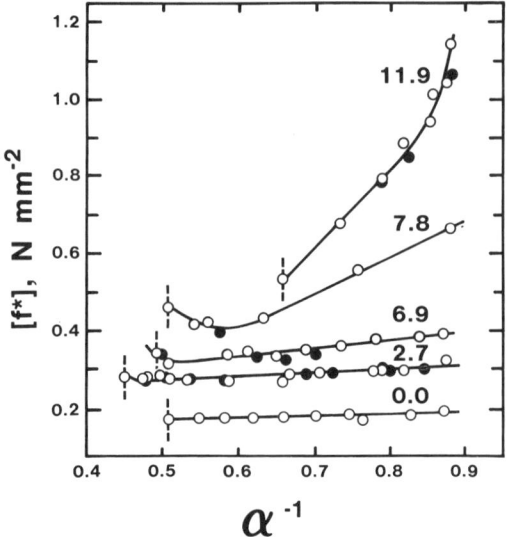

Figure 16.19 Stress–strain isotherms for PDMS networks reinforced with titania particles generated in situ[153]. Each curve is labeled with the weight percent of filler introduced, and filled circles locate results obtained out of sequence to test for reversibility.

16.4 In Situ Generation of Elastomers in Ceramics

If the hydrolyses in organosilicate–polymer systems are carried out with sufficiently large amounts of the silicate, then the silica generated can become the continuous phase, with the polymer dispersed in it[45,119,127,185–197]. The result is a polymer-modified ceramic, variously called an "ORMOCER"[119,186,187], "CERAMER"[188–190], or "POLYCERAM"[114,127,194,195]. It is obviously of considerable interest to determine how the polymeric phase, often elastomeric, modifies the ceramic in which it is dispersed.

One illustrative property of importance in such hybrid organic–inorganic composites is hardness. The case of the PDMS–SiO$_2$ system is illustrated in figure 16.20[196]. The hardness could be varied greatly by changing the ratio of organic-to-inorganic character, as measured by the molar ratio of organic R groups (here, CH_3 side groups) to Si atoms. Low values of the R/Si ratio yield a brittle ceramic, and high values yield an elastomer reinforced in situ. The most interesting range of values, R/Si $\sim$ 1, can give a hybrid material that can be viewed as a ceramic of reduced brittleness or an elastomer of increased hardness, depending on one's point of view. The impact measurements already described[167] are also very useful for characterizing materials of this type.

It is also possible to use the NMR technique described earlier to characterize composites in this category, as was shown in figures 16.3 and 16.4. Of particular interest is the one pertaining to aerogels, which are obtained by drying silica gels supercritically to wispy, foamlike structures. It is possible to vacuum impregnate

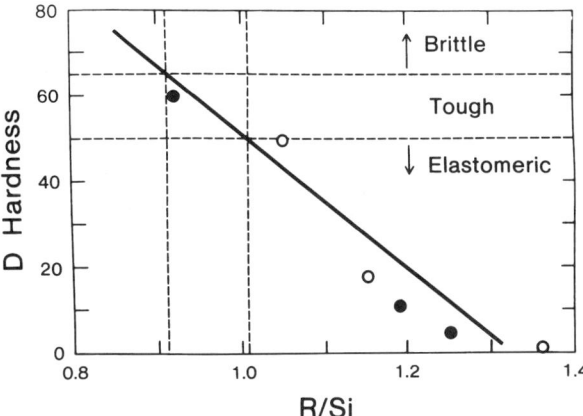

Figure 16.20 Dependence of the D-scale hardness of PDMS composites on the ratio of alkyl groups to silicon atoms[196]. The open circles correspond to bimodal PDMS, and the filled circles to unimodal PDMS.

such an aerogel with PDMS containing some vinyl groups to facilitate subsequent cross-linking with radiation[145]. Figure 16.4 shows two NMR images obtained on such a material. The variation of signal intensity across the sample can be used to obtain information on the distribution of cross-links and on the polymer–silica interactions.

16.5 In Situ Generation of Catalysts in Polymers

In this type of organic–inorganic composite, a relatively unreactive substance is transformed into a catalytically active one within a polymer which serves as a protective organic environment. An illustration of the techique involves nickel formate dissolved into a solution of poly(ethylene oxide) (PEO) in ethylene glycol, and then thermally decomposed in situ to form finely divided (high surface area) nickel particles[198]. The decomposition reaction is

$$\text{Ni(HCOO)}_2 \xrightarrow{200°C} \text{Ni} + \text{CO}\uparrow + \text{H}_2\text{O}\uparrow + \text{H}_2\uparrow + \text{CO}_2\uparrow \qquad (16.8)$$

When the reacted mixture is subsquently dried, the nickel particles are protected by their being encapsulated in the PEO matrix. A scanning electron micrograph of such a material is shown in figure 16.21. The nickel particles (at a loading of approximately 7.6 wt%) are clearly in evidence; they seem to be well dispersed, and appear to have diameters of the order of 0.4 μm. The bonding between the particles and the polymer matrix is seen to be relatively poor, but this should not be a disadvantage in the present (nonmechanical) application. Energy-dispersive x-ray analysis (EDXA) indicated that the only elements present were nickel (in the particles), and gold (in the coatings). The Ni distribution map of the same area, at

292 STRUCTURES AND PROPERTIES OF RUBBERLIKE NETWORKS

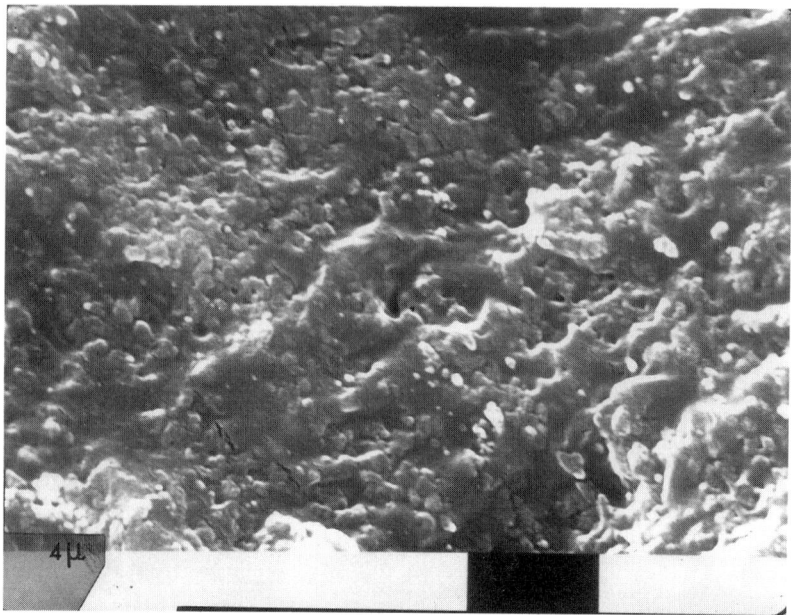

Figure 16.21 Scanning electron micrograph of a PEO–Ni sample (7.6% Ni) at an original magnification of 5000×[198]. The length of the bar corresponds to 4 μm.

the same resolution as figure 16.21, is shown in figure 16.22. The particles are seen to be well dispersed, in agreement with the electron microscopy results.

The protected catalyst sample can then be utilized by adding portions of it to a reaction medium, which dissolves the protective polymer and releases the nickel particles for catalyzing the reaction. For example, the Ni–PEO sample was very effective in the hydrogenation of an n-olefin, and should have greater convenience and extended shelf life. The PEO matrix was chosen because of its solubility in a wide variety of solvents, but it has the disadvantage of high permability to various gases (which could partly inactivate the catalyst). Other polymers[199,200], particularly elastomers such as polyisobutylene[201], with its unusually low permeability, would probably be better suited for this unusual type of application.

16.6 In Situ Polymerizations of Glassy Polymers

16.6.1 Isotropic Systems

It is also possible to obtain reinforcement by polymerizing a monomer, such as styrene, to yield hard glassy domains within an elastomer[202,203]. In PDMS, low concentrations of styrene give low-molecular-weight polymer that acts more like a soft, plasticizing liquid than as reinforcing particles[202]. The absence of evidence for polystyrene (PS) particles at lower styrene concentrations is demonstrated by the electron micrographs presented in figure 16.23. The plasticizing effect is

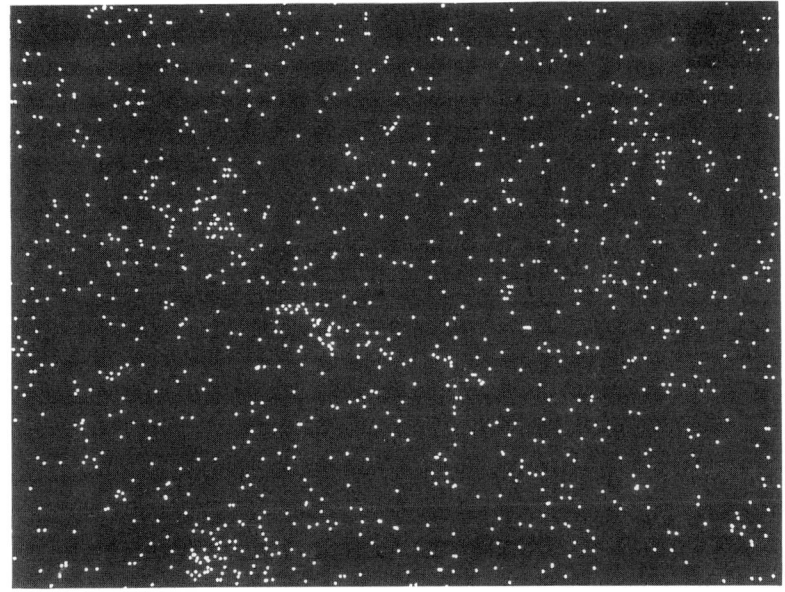

Figure 16.22 The EDXA nickel distribution map[198] of the same sample area shown in figure 16.21.

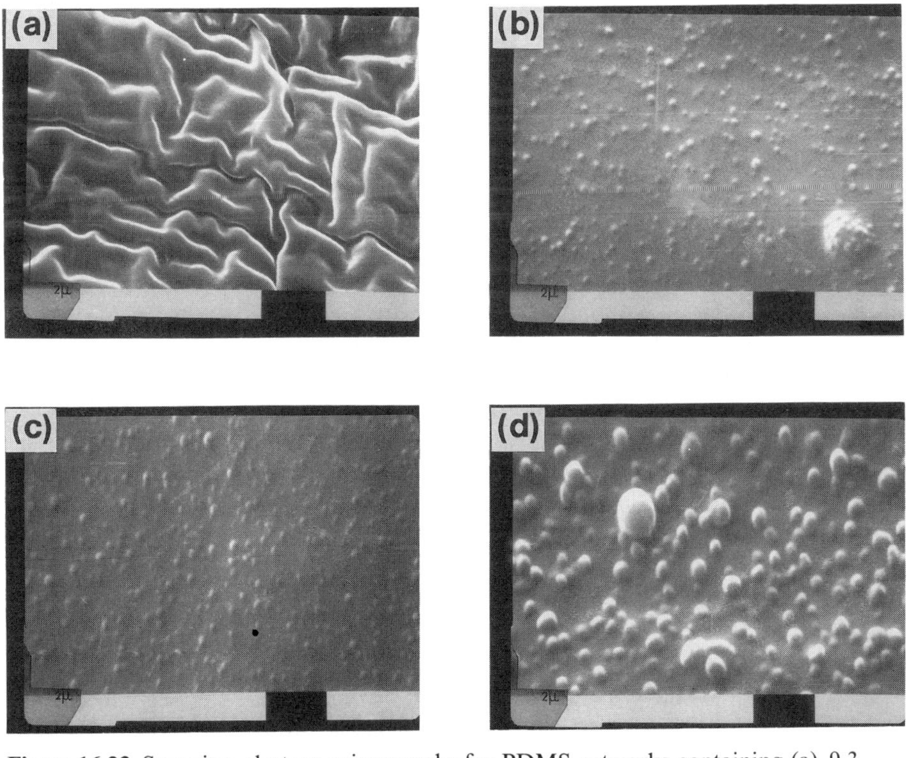

Figure 16.23 Scanning electron micrographs for PDMS networks containing (a) 9.3, (b) 20.2, (c) 32.0, and (d) 43.0 wt% polystyrene (PS)[202]. The widths of the rectangles at the bottom correspond to 2 μm.

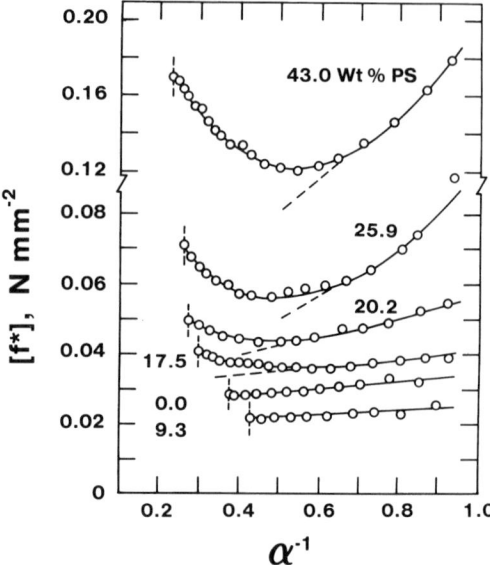

Figure 16.24 Stress–strain isotherms for the PDMS–PS composites[202]. Each curve is labeled with the weight percent of PS present in the composite, and the dashed lines locate the relatively linear portions of the curves characterized by the Mooney-Rivlin constants[18].

indicated by the stress–strain results obtained; for example, by the initial decrease in the modulus and ultimate strength, as is illustrated in figure 16.24. Higher styrene concentrations give pronounced reinforcement. Of particular interest is the fact that in addition to the increases in modulus and ultimate strength, there are increases in maximum elongation, instead of the usual decreases. Polyisobutylene has also been reinforced in this manner[203].

The glassy particles thus generated are relatively easy to extract from the elastomeric matrix, which means that there is little effective bonding between the two phases. It is possible, however, to get excellent bonding onto the filler particles. One way to do this is to include some $R'Si(OC_2H_5)_3$ in the hydrolysis, where R' is an unsaturated group. The R' groups on the surfaces of the particles then participate in the polymerization, thereby bonding the elastomer chains to the reinforcing particles[204]. Alternatively, the $R'Si(OC_2H_5)_3$ can be used as one of the end-linking agents, placing unsaturated groups at the cross-links[205]. Their participation in the polymerization would now tie the PS domains to the elastomer's network structure.

The PS domains have the disadvantage of having a relatively low glass transition temperature T_g ($\sim 100°C$)[168] and being totally amorphous. Above T_g, they would therefore soften and presumably lose their reinforcing capability. For this reason, similar studies have been carried out using poly(diphenylsiloxane) as the reinforcing phase[206]. It has a relatively high T_g of 49°C, is crystalline, and has an extraordinarily high melting point of 550°C![207,208]

The particles of glassy polymer generated by such in situ polymerizations are generally approximately spherical, and the systems are isotropic. The following section shows how these two constraints can be relaxed.

16.6.2 Anisotropic Systems

It is possible to convert the essentially spherical PS domains described above to rodlike ellipsoidal particles[209–213]. First, the PS–elastomer composite is raised to a temperature well above the T_g of the PS. It is then stretched uniaxially, and cooled while in the stretched state. The particles are deformed into prolate ellipsoids, and retain this shape when cooled. When the deforming force is removed, the elastomer is observed to retract, but only part of the way back to its original dimensions. The particles themselves can be characterized using scanning and transmission electron microscopy, as illustrated in figure 16.25. They were found to have axial ratios of approximately 2, and to have their axes preferentially oriented in the direction of the high-temperature stretching[210]. The reinforcement they provided was characterized using stress–strain measurements in elongation at room temperature. The results are given in figure 16.26. In these anisotropic materials, the modulus in the direction parallel to the original stretching direction was found to be significantly higher than that of the untreated (isotropic) PS–PDMS elastomer, whereas in the perpendicular direction it was significantly lower. A variation of this technique has been used to generate elongated TiO_2 particles by stretching a polymer film containing a titanate prior to its hydrolysis[51,214].

It is also possible to generate oblate ellipsoids by stretching such a PS–PDMS elastomer *biaxially*; for example, by inflation of a similarly softened sheet of the material[215].

16.7 Fillers Responding to Magnetic Fields

It is possible to prepare particles of metals or metal oxides by related techniques. Examples are the in situ generation of hydrated iron oxide by the hydrolysis of ferric chloride, and iron and iron oxides by the thermolysis of iron carbonyls. Such particles can, of course, also be prepared separately and then blended into a polymer. They can then be locked into position either by cooling the polymer into the crystalline or glassy state, or by cross-linking it in the case of an elasomer[121,216,217].

Of particular interest in this category are particles that respond to an external magnetic field. For example, magnetic ferrite particles dispersed in PDMS can be aligned in a magnetic field during cross-linking. In this way, anisotropic mechanical properties can be obtained, even from essentially spherical particles. The reinforcement is found to be significantly higher in the direction parallel to the magnetic lines of force[121,218], as is illustrated in figure 16.27.

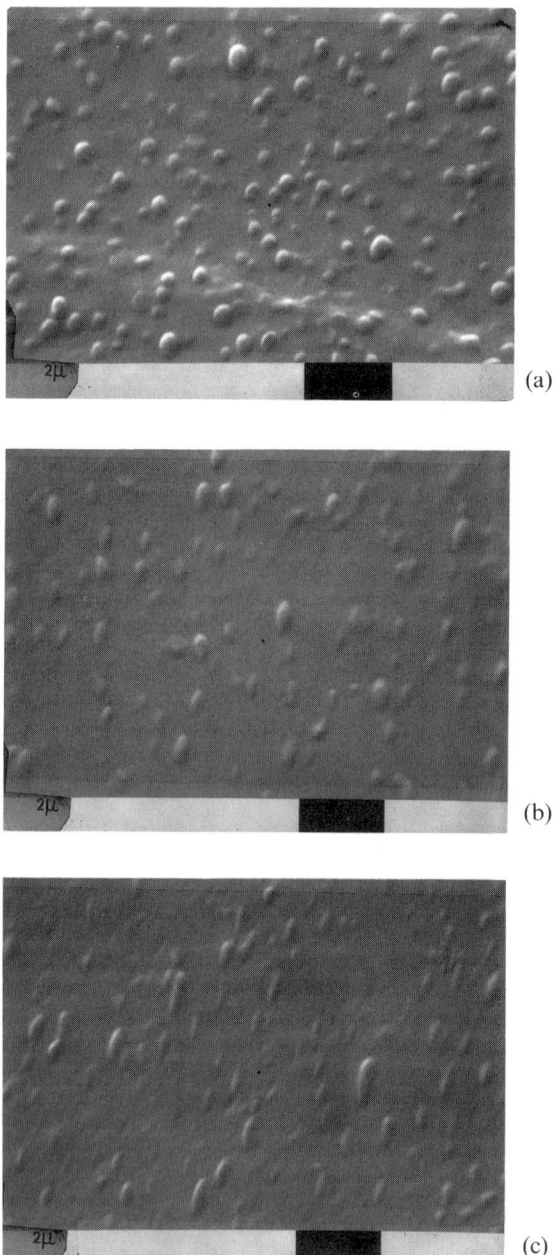

Figure 16.25 Scanning electron micrographs of a PDMS–PS composite stretched at high temperature and cooled, at different draw ratios: (a) unstretched sample, (b) elongation of 1.8, (c) elongation of 2.2[210]. The widths of the black rectangles correspond to 2 μm. Reprinted with permission from Wang, S. and Mark, J. E. (1990), *Macromolecules*, **23**. Copyright 1997, American Chemical Society.

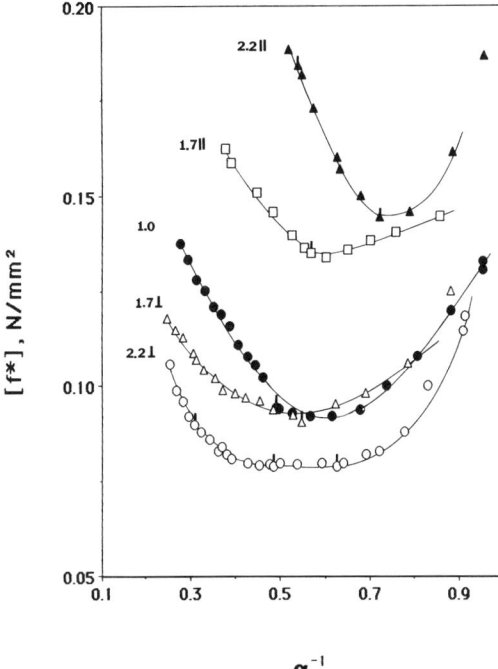

Figure 16.26 Stress–strain isotherms of the PDMS–PS composites described in figure 16.25[210]. Values of the draw ratio and testing directions are indicated on each curve. The symbols with small tabs attached represent data used to test for reversibility. Reprinted with permission from Wang, S. and Mark, J. E. (1990), *Macromolecules*, **23**. Copyright 1997, American Chemical Society.

This technique could be combined with the in situ approach by generating metal or metal oxide magnetic particles in a magnetic field[217,219], for example, by the thermolysis or photolysis of a metal carbonyl.

16.8 Fillers of Controlled Crystalline Structure

Commercially, the reinforcement of PDMS has concentrated on the use of high-surface-area silica, which is much used because of the excellent reinforcement it provides. This material is amorphous, whether generated separately or in situ, as is the carbon black used in the reinforcement of other classes of elastomers. In order to obtain a better molecular understanding of reinforcement mechanisms, however, it may also be useful to study crystalline fillers, preferably of known structures. An example of a class of such inorganic substances are the zeolites, which are hydrated silicates of aluminum and sodium of the general formula $Na_2O \cdot Al_2O_3 \cdot nSiO_2 \cdot xH_2O$. Their structures have been extensively studied, largely because they are novel (with one or more cavities), and they have important

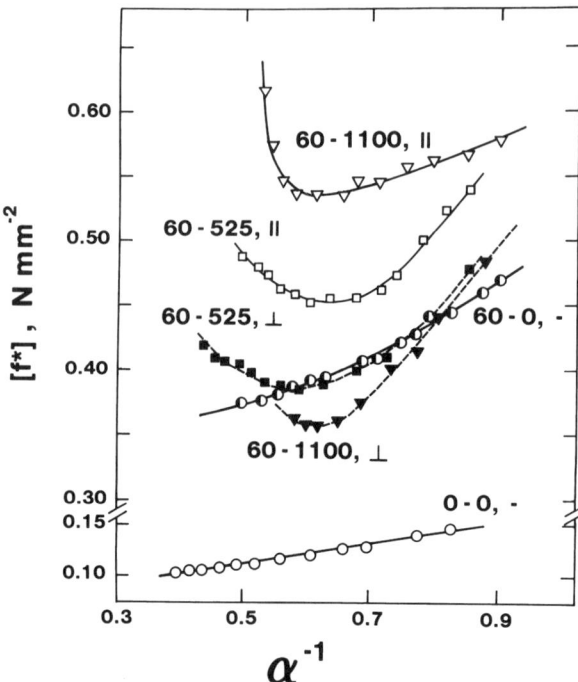

Figure 16.27 Stress–strain isotherms for PDMS composites containing magnetically responsive particles[218]. Each curve is identified by the weight percent of filler present, the strength of the magnetic field (in gauss) applied during cross-linking, and the orientation (parallel or perpendicular) of the sample strip relative to the lines of force of the magnetic field.

applications as catalysts, sorption agents, ion exchange resins, and so on[220–222]. Two studies have investigated the use of zeolites for reinforcement of PDMS elastomers[223,224]. Two types of zeolites, having cavity sizes of 4 and 10 Å, respectively, were employed, and the reinforcement they provided was characterized by stress–strain measurements on the elastomers in elongation. Some of the results are shown in figure 16.28, and demonstrate the good reinforcement provided by these types of fillers.

It was hoped that the zeolite with the larger pore size would be better penetrated by the polymer chains, thus giving significantly better reinforcement. Small-angle neutron scattering profiles for the PDMS elastomers filled with the two types of zeolite were nearly the same[224], however. This argues against chain interpenetration occurring in the large-pore zeolite and not in the small-pore one. Since, statistically, penetration by *loops* would be much more likely than by ends, it may be necessary to have zeolite pore diameters more than twice the diameter of the potentially penetrating chain. Even then, there may be insufficient driving force for the chains to go into and through such cavities.

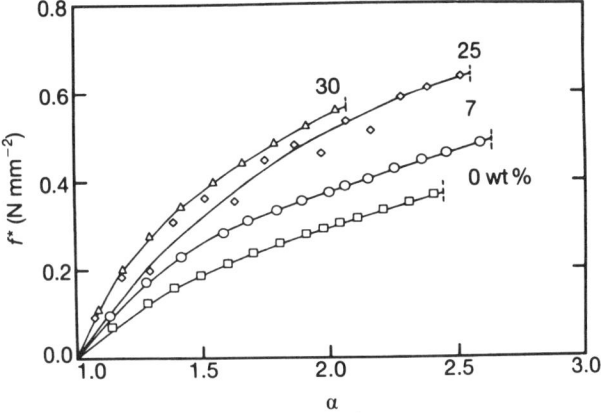

Figure 16.28 Stress–strain isotherms for some PDMS networks filled with a zeolite (13x), with each curve labeled with the weight percent of filler present in the network[224].

It is possible, however, to design other important experiments on the use of zeolites in filler-reinforced elastomers. For example, the likelihood of chain interpenetration should be greatly increased if *monomer* is absorbed into the zeolite, or into a related structure such as a nanotube, and then polymerized[225–229]. This virtually ensures chain penetration of the cavities, with unusually intimate filler–elastomer interactions.

References

(1) Oberth, A. E. *Rubber Chem. Technol.* 1967, 40, 1337.
(2) Boonstra, B. B. In *Rubber Technology*, M. Morton, Ed. Van Nostrand Reinhold: New York. 1973; p. 51.
(3) Boonstra, B. B. *Polymer* 1979, 20, 691.
(4) Warrick, E. L.; Pierce, O. R.; Polmanteer, K. E.; Saam, J. C. *Rubber Chem. Technol.* 1979, 52, 437.
(5) Rigbi, Z. *Adv. Polym. Sci.* 1980, 36, 21.
(6) Queslel, J. P.; Mark, J. E. In *Encyclopedia of Polymer Science and Engineering, Second Edition*, R. A. Meyers, Ed. Wiley-Interscience: New York. 1986; p. 365.
(7) Donnet, J.-B.; Vidal, A. *Adv. Polym. Sci.* 1986, 76, 103.
(8) Mark, J. E.; Erman, B. *Rubberlike Elasticity. A Molecular Primer*. Wiley-Interscience: New York. 1988.
(9) Ahmed, S.; Jones, F. R. *J. Mater. Sci.* 1990, 25, 4933.
(10) Enikolopyan, N. S.; Fridman, M. L.; Stalnova, I. O.; Popov, V. L. *Adv. Polym. Sci.* 1990, 96, 1.
(11) Edwards, D. C. *J. Mater. Sci.* 1990, 25, 4175.
(12) Medalia, A. I.; Kraus, G. In *Science and Technology of Rubber*, 2nd ed., J. E. Mark, B. Erman, and F. R. Eirich, Eds. Academic Press: New York. 1994; p. 387.
(13) Karasek, L.; Sumita, M. *J. Mater. Sci.* 1996, 31, 281.
(14) Kraus, G. *Adv. Polym. Sci.* 1971, 8, 155.
(15) Kraus, G., Ed. *Reinforcement of Elastomers*. Interscience: New York. 1965.

(16) Kloczkowski, A.; Sharaf, M. A.; Mark, J. E. *Comput. Polym. Sci.* 1993, 3, 39.
(17) Kloczkowski, A.; Sharaf, M. A.; Mark, J. E. *Chem. Eng. Sci.* 1994, 49, 2889.
(18) Treloar, L. R. G. *The Physics of Rubber Elasticity*, 3rd ed. Clarendon Press: Oxford. 1975.
(19) Matijevic, E.; Scheiner, P. *J. Coll. Interfac. Sci.* 1978, 63, 509.
(20) Mark, J. E. *Kautschuk + Gummi Kunstoffe* 1989, 42, 191.
(21) Mark, J. E. *CHEMTECH* 1989, 19, 230.
(22) Schaefer, D. W.; Mark, J. E., Eds. *Polymer-Based Molecular Composites*. Materials Research Society: Pittsburgh, PA. 1990; Vol. 171.
(23) Mark, J. E.; Schaefer, D. W. In *Polymer-Based Molecular Composites*, D. W. Schaefer and J. E. Mark, Eds. Materials Research Society: Pittsburgh, PA. 1990; Vol. 171; p. 51.
(24) Schaefer, D. W.; Mark, J. E.; McCarthy, D.; Jian, L.; Sun, C.-C.; Farago, B. In *Polymer-Based Molecular Composites*, D. W. Schaefer and J. E. Mark, Eds. Materials Research Society: Pittsburgh, PA. 1990; Vol. 171; p. 57.
(25) Yasrebi, M.; Kim, G. H.; Gunnison, K. E.; Milius, D. L.; Sarikaya, M.; Aksay, I. A. In *Better Ceramics Through Chemistry IV*, B. J. J. Zelinski, C. J. Brinker, D. E. Clark, and D. R. Ulrich, Eds. Materials Research Society: Pittsburgh, PA. 1990; Vol. 180; p. 625.
(26) Chung, Y. J.; Ting, S.-J.; Mackenzie, J. D. In *Better Ceramics Through Chemistry IV*, B. J. J. Zelinski, C. J. Brinker, D. E. Clark, and D. R. Ulrich, Eds. Materials Research Society: Pittsburgh, PA. 1990; Vol. 180; p. 981.
(27) Saegusa, T.; Chujo, Y. *J. Macromol. Sci.—Chem.* 1990, A27, 1603.
(28) Mauritz, K. A.; Jones, C. K. *J. Appl. Polym. Sci.* 1990, 40, 1401.
(29) Mauritz, K. A.; Scheetz, R. W.; Pope, R. K.; Stefanithis, I. D.; Wilkes, G. L.; Huang, H.-H. *Preprints, Div. Polym. Chem., Inc., Am. Chem. Soc.* 1991, 32(3), 528.
(30) Bianconi, P. A.; Lin, J.; Strzelecki, A. R. *Nature* 1991, 349, 315.
(31) Okada, A.; Fukumori, K.; Usuki, A.; Kojima, Y.; Sato, N.; Kurauchi, T.; Kamigaito, O. *Preprints, Div. Polym. Chem., Inc., Am. Chem. Soc.* 1991, 32(3), 540.
(32) Yano, K.; Usuki, A.; Okada, A.; Kurauchi, T.; Kamigaito, O. *Preprints, Div. Polym. Chem., Inc., Am. Chem. Soc.* 1991, 32(1), 65.
(33) Calvert, P. In *U.S.-Japan Workshop on Smart/Intelligent Materials and Systems*, I. Ahmad, A. Crowson, C. A. Rogers, and M. Aizawa, Eds. Technomic: Lancaster. 1991; p. 162.
(34) Calvert, P. In *Ultrastructure Processing of Advanced Materials*, D. R. Uhlmann and D. R. Ulrich, Eds. Wiley: New York. 1992; p. 149.
(35) Mackenzie, J. D.; Chung, Y. J.; Hu, Y. *J. Non.-Cryst. Solids* 1992, 147&148, 271.
(36) Hu, Y.; Mackenzie, J. D. *J. Mater. Sci.* 1992, 27, 4415.
(37) Brennan, A. B.; Rodrigues, D. E.; Wang, B.; Wilkes, G. L. In *Chemical Processing of Advanced Materials*, L. L. Hench and J. K. West, Eds. Wiley: New York. 1992; p. 807.
(38) Huang, H.; Glaser, R. H.; Brennan, A. B.; Rodigues, D.; Wilkes, G. L. In *Ultrastructure Processing of Advanced Materials*, D. R. Uhlmann and D. R. Ulrich, Eds. Wiley: New York. 1992; p. 425.
(39) Schmidt, H. In *Ultrastructure Processing of Advanced Materials*, D. R. Uhlmann and D. R. Ulrich, Eds. Wiley: New York. 1992; p. 409.
(40) Schmidt, H. K. In *Submicron Multiphase Materials*, R. H. Baney, L. R. Gilliom, S.-I. Hirano, and H. K. Schmidt, Eds. Materials Research Society: Pittsburgh, PA. 1992; Vol. 274; p. 121.

(41) Ellsworth, M. K.; Novak, B. M. In *Submicron Multiphase Materials*, R. H. Baney, L. R. Gilliom, S.-I. Hirano, and H. K. Schmidt, Eds. Materials Research Society: Pittsburgh, PA. 1992; Vol. 274; p. 67.
(42) Mark, J. E.; Wang, S.; Xu, P.; Wen, J. In *Submicron Multiphase Materials*, R. H. Baney, L. R. Gilliom, S.-I. Hirano, and H. K. Schmidt, Eds. Materials Research Society: Pittsburgh, PA. 1992; Vol. 274; p. 77.
(43) Sun, L.; Aklonis, J. J.; Salovey, R. *Polym. Eng. Sci.* 1993, 33, 1308.
(44) Matijevic, E. *Chem. Mater.* 1993, 5, 412.
(45) Novak, B. M. *Adv. Mater.* 1993, 5, 422.
(46) Mark, J. E.; Calvert, P. D. *J. Mater. Sci., Part C* 1994, 1, 159.
(47) Mark, J. E. In *Frontiers of Polymers and Advanced Materials*, P. N. Prasad, Ed. Plenum Press: New York. 1994; p. 403.
(48) Mark, J. E.; Lee, C. Y-C; Bianconi, P. A., Eds. *Hybrid Organic-Inorganic Composites*. American Chemical Society: Washington, DC. 1995; Vol. 585.
(49) Mark, J. E. In *Diversity into the Next Century*, R. J. Martinez, H. Arris, J. A. Emerson, and G. Pike, Eds. SAMPE: Covina, CA. 1995; Vol. 27.
(50) Schmidt, H. K. *Macromol. Symp.* 1996, 101, 333.
(51) Calvert, P. In *Biomimetic Materials Chemistry*, S. Mann, Ed. VCH Publishers: New York. 1996; p. 315.
(52) Giannelis, E. P. In *Biomimetic Materials Chemistry*, S. Mann, Ed. VCH Publishers: New York. 1996; p. 337.
(53) Wen, J.; Wilkes, G. L. In *Polymeric Materials Encyclopedia: Synthesis, Properties, and Applications*, J. C. Salamone, Ed. CRC Press: Boca Raton, FL. 1996.
(54) Mark, J. E. *Hetero. Chem. Rev.* 1996, 3, 307.
(55) Mark, J. E., *Polym. Eng. Sci.* 1996, 36, 2905.
(56) Yuan, Q. W.; Kloczkowski, A.; Mark, J. E.; Sharaf, M. A. *J. Polym. Sci., Polym. Phys. Ed.* 1996, 34, 1674.
(57) Gent, A. N.; Lindley, P. *Proc. Roy. Soc. London* 1959, A249, 195.
(58) Hess, W. M.; Ford, F. P. *Rubber Chem. Technol.* 1963, 36, 1175.
(59) Kim, S. J.; Reneker, D. H. *Rubber Chem. Technol.* 1993, 66, 559.
(60) Young, R. J.; Al-Khudhairy, D. H. A.; Thomas, A. G. *J. Mater. Sci.* 1986, 21, 1211.
(61) Landry, M. R.; Coltrain, B. K.; Landry, C. J. T.; O'Reilly, J. M. *J. Polym. Sci., Polym. Phys. Ed.* 1995, 33, 637.
(62) McCarthy, D. W.; Mark, J. E.; Schaefer, D. W. *Polym. J.*, submitted.
(63) McCarthy, D. W.; Mark, J. E.; Clarson, S. J.; Schaefer, D. W. *Polym. J.*, submitted.
(64) Garrido, L.; Ackerman, J. L.; Mark, J. E. In *Polymer-Based Molecular Composites*, D. W. Schaefer and J. E. Mark, Eds. Materials Research Society: Pittsburgh, PA. 1990; Vol. 171; p. 65.
(65) Simon, G. *Polym. Bull.* 1991, 25, 365.
(66) Legrand, A. P.; Lecomte, N.; Vidal, A.; Haidar, B.; Papirer, E.; Donnet, J. B. *J. Appl. Polym. Sci.* 1992, 46, 2223.
(67) Litvinov, V. M.; Spiess, H. W. *Macromol. Chem.* 1991, 192, 3005.
(68) Chow, T. S. *J. Polym. Sci., Polym. Phys. Ed.* 1978, 16, 959.
(69) Chow, T. S. *J. Polym. Sci., Polym. Phys. Ed.* 1978, 16, 967.
(70) Younan, A. F.; Ismail, M. N.; Yehia, A. A. *J. Appl. Polym. Sci.* 1992, 45, 1967.
(71) Sunkara, H. B.; Jethmalani, J. M.; Ford, W. T. *Chem. Mater.* 1994, 6, 362.

(72) Sunkara, H. B.; Jethmalani, J. M.; Ford, W. T. In *Hybrid Organic-Inorganic Composites*, J. E. Mark, C. Y-C Lee, and P. A. Bianconi, Eds. American Chemical Society: Washington, DC. 1995; Vol. 585; p. 181.
(73) Pu, Z.; Mark, J. E.; Jethmalani, J. M.; Ford, W. T. *Polym. Bull.* 1996, 37, 545.
(74) Jethmalani, J. M.; Ford, W. T. *Chem. Mater.* 1996, 8, 2138.
(75) Guth, E.; Gold, O. *Phys. Rev.* 1938, 53, 322.
(76) Smallwood, H. M. *J. Appl. Phys.* 1944, 15, 758.
(77) Guth, E. *J. Appl. Phys.* 1945, 16, 20.
(78) Brennan, J. J.; Jermyn, T. E. *J. Appl. Polym. Sci.* 1965, 9, 2749.
(79) Kendall, K.; Sherliker, F. R. *Brit. Polym. J.* 1980, 12, 85.
(80) Vidal, A.; Donnet, J.-B. *Prog. Coll. Polym. Sci.* 1987, 75, 201.
(81) Meissner, B. *J. Appl. Polym. Sci.* 1993, 50, 285.
(82) Kerner, E. H. *Proc. Phys. Soc.* 1956, B69, 808.
(83) Hashin, Z.; Shtrikman, S. *J. Mech. Phys. Solids* 1963, 11, 127.
(84) Hirsch, T. J. *J. Am. Conc. Inst.* 1962, 59, 427.
(85) Takayanagi, M.; Nemura, S.; Minami, S. *J. Polym. Sci.* 1964, C5, 113.
(86) Charlesby, A.; Morris, J.; Montague, P. *J. Polym. Sci., Part C* 1969, 16, 4505.
(87) Kraus, G.; Rollmann, K. W. In *Multi-Component Systems*. American Chemical Society: Washington, DC. 1971; p. 189.
(88) Chow, T. S. *J. Appl. Phys.* 1977, 48, 4072.
(89) Kontou, E.; Spathis, G. *J. Appl. Polym. Sci.* 1990, 39, 649.
(90) Wolff, S.; Donnet, J.-B. *Rubber Chem. Technol.* 1990, 63, 32.
(91) Berlin, A. A.; Zhuk, A. V.; Knunjantz, N. N.; Oshmjan, V. G.; Topolkarev, V. A. *Macromol. Chem., Macromol. Symp.* 1993, 44, 295.
(92) Berrod, G.; Vidal, A.; Papirer, E.; Donnet, J. B. *J. Appl. Polym. Sci.* 1981, 26, 833.
(93) Berrod, G.; Vidal, A.; Papirer, E.; Donnet, J. B. *J. Appl. Polym. Sci.* 1981, 26, 1015.
(94) Rigbi, Z. *Kautschuk + Gummi Kunstoffe* 1989, 42, 1107.
(95) Vondracek, P.; Pouchelon, A. *Rubber Chem. Technol.* 1990, 63, 202.
(96) Kotani, T.; Sternstein, S. S. In *Polymer Networks: Structure and Mechanical Properties*, A. J. Chompff and S. Newman, Eds. Plenum Press: New York. 1971; p. 273.
(97) Horkay, F.; Zrinyi, M.; Geissler, E.; Hecht, A.-M.; Pruvost, P. *Polymer* 1991, 32, 835.
(98) Bueche, F. *Rubber Chem. Technol.* 1965, 38, 1070.
(99) Kraus, G. *J. Appl. Polym. Sci.* 1963, 7, 1165.
(100) Bueche, A. M. *J. Polym. Sci.* 1957, 25, 139.
(101) Heinrich, G.; Vilgis, T. A. *Macromolecules* 1993, 26, 1109.
(102) Strauss, M.; Pieper, T.; Peng, W.; Kilian, H.-G. *Makromol. Chem., Macromol. Symp.* 1993, 76, 131.
(103) Reichert, W. F.; Goritz, D.; Duschl, E. J. *Polymer* 1993, 324, 1216.
(104) Witten, T. A.; Rubinstein, M.; Colby, R. H. *J. Phys. II France* 1993, 3, 367.
(105) Galiatsatos, V. private communications.
(106) Mark, J. E.; Pan, S.-J. *Makromol. Chemie, Rapid Commun.* 1982, 3, 681.
(107) Hench, L. L.; Ulrich, D. R., Eds. *Ultrastructure Processing of Ceramics, Glasses, and Composites*. Wiley: New York. 1984.
(108) Mackenzie, J. D.; Ulrich, D. R., Eds. *Ultrastructure Processing of Advanced Ceramics*. Wiley: New York. 1988.
(109) Ulrich, D. R. *J. Non-Cryst. Solids* 1988, 100, 174.

(110) Ulrich, D. R. *CHEMTECH* 1988, 18, 242.
(111) Mackenzie, J. D. *J. Non-Cryst. Solids* 1988, 100, 162.
(112) Ulrich, D. R. *J. Non-Cryst. Solids* 1990, 121, 465.
(113) Brinker, C. J.; Scherer, G. W. *Sol-Gel Science*. Academic Press: New York. 1990.
(114) Uhlmann, D. R.; Ulrich, D. R., Eds. *Ultrastructure Processing of Advanced Materials*. Wiley: New York. 1992.
(115) Baney, R. H.; Gilliom, L. R.; Hirano, S.-I.; Schmidt, H. K., Eds. *Submicron Multiphase Materials*. Materials Research Society: Pittsburgh, PA. 1992; Vol. 274.
(116) Hampden-Smith, M. J.; Klemperer, W. G.; Brinker, C. J., Eds. *Better Ceramics Through Chemistry V*. Materials Research Society: Pittsburgh, PA. 1992; Vol. 271.
(117) Cheetham, A. K.; Brinker, C. J.; Mecartney, M. L.; Sanchez, C., Eds. *Better Ceramics Through Chemistry VI*. Materials Research Society: Pittsburgh, PA. 1994; Vol. 346.
(118) Wilkes, G. L.; Brennan, A. B.; Huang, H.-H.; Rodrigues, D.; Wang, B. In *Polymer-Based Molecular Composites*, D. W. Schaefer and J. E. Mark, Eds. Materials Research Society: Pittsburgh, PA. 1990; Vol. 171; p. 15.
(119) Schmidt, H.; Wolter, H. *J. Non-Cryst. Solids* 1990, 121, 428.
(120) Wang, S. B.; Mark, J. E. *J. Macromol. Sci., Macromol. Rep.* 1991, A28, 185.
(121) Mark, J. E. *J. Appl. Polym. Sci., Appl. Polym. Symp.* 1992, 50, 273.
(122) Rodrigues, D. E.; Wilkes, G. L. *J. Inorg. Organomet. Polym.* 1993, 3, 197.
(123) Clarson, S. J.; Mark, J. E. In *Siloxane Polymers*, S. J. Clarson and J. A. Semlyen, Eds. Prentice Hall: Englewood Cliffs, NJ. 1993; p. 616.
(124) Wang, B.; Wilkes, G. L. *J. Macromol. Sci., Pure Appl. Chem.* 1994, A31, 249.
(125) Craddock, S.; Hinchliffe, A. *Matrix Isolation*. Cambridge University Press: New York. 1975.
(126) Kovar, R. F.; Lusignea, R. W. In *Ultrastructure Processing of Advanced Ceramics*, J. D. Mackenzie and D. R. Ulrich, Eds. Wiley-Interscience: New York. 1988; p. 715.
(127) Doyle, W. F.; Uhlmann, D. R. In *Ultrastructure Processing of Advanced Ceramics*, J. D. Mackenzie and D. R. Ulrich, Eds. Wiley-Interscience: New York. 1988; p. 795.
(128) McGrath, J. E.; Pullockaren, J. P.; Riffle, J. S.; Kilic, S.; Elsbernd, C. S. In *Ultrastructure Processing of Advanced Ceramics*, J. D. Mackenzie and D. R. Ulrich, Eds. Wiley-Interscience: New York. 1988; p. 55.
(129) Prasad, P. N. In *Sol-Gel Optics*, J. D. Mackenzie and D. R. Ulrich, Eds. SPIE—The International Society for Optical Engineering: Bellingham, WA. 1990; Vol. 1328; p. 168.
(130) Morita, K.; Hu, Y.; Mackenzie, J. D. In *Better Ceramics Through Chemistry V*, M. J. Hampden-Smith, W. G. Klemperer, and C. J. Brinker, Eds. Materials Research Society: Pittsburgh, PA. 1992; Vol. 271; p. 693.
(131) Mark, J. E.; Eisenberg, A.; Graessley, W. W.; Mandelkern, L.; Samulski, E. T.; Koenig, J. L.; Wignall, G. D. *Physical Properties of Polymers*, 2nd ed. American Chemical Society: Washington, DC. 1993.
(132) Saunders, P. R.; Ward, A. G. In *Proceedings of the Second International Congress of Rheology*. Butterworth Scientific: London. 1953.
(133) Li, Z.; Mark, J. E.; Chan, E. K. M.; Mandelkern, L. *Macromolecules* 1989, 22, 4273.

(134) Yang, Y.; Ichise, N.; Li, Z.; Yuan, Q.; Mark, J. E.; Chan, E. K. M.; Alamo, R. G.; Mandelkern, L. In *Complex Fluids*, E. B. Sirota, D. Weitz, T. Witten, and J. Israelachvili, Eds. Materials Research Society: Pittsburgh, PA. 1992; p. 325.
(135) Ning, Y. P.; Zhao, M. X.; Mark, J. E. In *Chemical Processing of Advanced Materials*, L. L. Hench and J. K. West, Eds. Wiley: New York. 1992; p. 745.
(136) Mark, J. E.; Ning, Y.-P.; Jiang, C.-Y.; Tang, M.-Y.; Roth, W. C. *Polymer* 1985, 26, 2069.
(137) Mark, J. E. In *Frontiers of Macromolecular Science*, T. Saegusa, T. Higashimura, and A. Abe, Eds. Blackwell Scientific: Oxford. 1989; p. 289.
(138) Ning, Y.-P.; Tang, M.-Y.; Jiang, C.-Y.; Mark, J. E.; Roth, W. C. *J. Appl. Polym. Sci.* 1984, 29, 3209.
(139) Xu, P.; Wang, S.; Mark, J. E. In *Better Ceramics Through Chemistry IV*, B. J. J. Zelinski, C. J. Brinker, D. E. Clark, and D. R. Ulrich, Eds. Materials Research Society: Pittsburgh, PA. 1990; Vol. 180; p. 445.
(140) Ulibarri, T. A.; Beaucage, G.; Schaefer, D. W.; Olivier, B. J.; Assink, R. A. In *Submicron Multiphase Materials*, R. H. Baney, L. R. Gilliom, S.-I. Hirano, and H. K. Schmidt, Eds. Materials Research Society: Pittsburgh, PA. 1992; Vol. 274; p. 85.
(141) Schaefer, D. W.; Keefer, K. D. *Phys. Rev. Lett.* 1984, 53, 1383.
(142) Schaefer, D. W.; Jian, L.; Sun, C.-C.; McCarthy, D. W.; Jiang, C.-Y.; Ning, Y.-P.; Mark, J. E.; Spooner, S. In *Ultrastructure Processing of Advanced Materials*, D. R. Uhlmann and D. R. Ulrich, Eds. Wiley: New York. 1992; p. 361.
(143) Garrido, L.; Mark, J. E.; Sun, C. C.; Ackerman, J. L.; Chang, C. *Macromolecules* 1991, 24, 4067.
(144) Wang, S.; Mark, J. E. *J. Macromol. Sci., Macromol. Rep.* 1994, A31, 253.
(145) Garrido, L.; Mark, J. E.; Wang, S.; Ackerman, J. L.; Vevea, J. M. *Polymer* 1992, 33, 1826.
(146) Bartlett, J. R.; Woolfrey, J. L. In *Chemical Processing of Advanced Materials*, L. L. Hench and J. K. West, Eds. Wiley: New York. 1992; p. 247.
(147) Ostwald, W. *Z. Physik. Chem.* 1900, 34, 495.
(148) Mark, J. E.; Jiang, C.-Y.; Tang, M.-Y. *Macromolecules* 1984, 17, 2613.
(149) Clarson, S. J.; Mark, J. E.; Dodgson, K. *Polym. Commun.* 1988, 29, 208.
(150) Levin, V. Y.; Solonimski, G. L.; Andrianov, K. A.; Zhdanov, A. A.; Godovski, Y. A.; Papkov, V. S.; Lyubavskaya, A. Y. *Polym. Sci., U.S.S.R.* 1963, 15, 256.
(151) Sohoni, G. B.; Mark, J. E. *J. Appl. Polym. Sci.* 1992, 45, 1763.
(152) Mark, J. E.; Ning, Y.-P. *Polym. Bull.* 1984, 12, 413.
(153) Wang, S.-B.; Mark, J. E. *Polym. Bull.* 1987, 17, 271.
(154) Wen, J.; Mark, J. E. *Rubber Chem. Technol.* 1994, 67, 806.
(155) Burnside, S. D.; Giannelis, E. P. *Chem. Mater.* 1995, 7, 1597.
(156) Mark, J. E. *Brit. Polym. J.* 1985, 17, 144.
(157) Ning, Y.-P.; Mark, J. E. *Polym. Eng. Sci.* 1986, 26, 167.
(158) Wang, S.; Xu, P.; Mark, J. E. *Rubber Chem. Technol.* 1991, 64, 746.
(159) Xu, P.; Mark, J. E. *Rubber Chem. Technol.* 1990, 63, 276.
(160) Pak, H.; Flory, P. J. *J. Polym. Sci., Polym. Phys. Ed.* 1979, 17, 1845.
(161) Wang, S. Ph. D. in Chemistry, University of Cincinnati, OH. 1991.
(162) Gent, A. N.; Kuan, T. H. *J. Polym. Sci., Polym. Phys. Ed.* 1973, 11, 1723.
(163) Wen, J.; Mark, J. E. *Polym. J.* 1994, 26, 151.
(164) Ferry, J. D. *Viscoelastic Properties of Polymers*, 3rd ed. Wiley: New York. 1980.
(165) Sharaf, M. A.; Kloczkowski, A.; Mark, J. E. *Rubber Chem. Technol.* 1995, 68, 601.

(166) Wen, J.; Mark, J. E.; Fitzgerald, J. J. *J. Macromol. Sci., Macromol. Rep.* 1994, A31, 429.
(167) Wen, J.; Mark, J. E. *Polym. J.* 1995, 27, 492.
(168) Brandrup, J.; Immergut, E., Eds. *Handbook of Polymer Science.* Wiley: New York. 1975.
(169) Vincent, P. I. *Impact Tests and Service Performance of Thermoplastics.* Plastics Institute: London. 1971.
(170) Reed, P. E. In *Developments in Polymer Fracture—1*, E. H. Andrews, Ed. Applied Science: London. 1979.
(171) Erman, B.; Mark, J. E. *Ann. Rev. Phys. Chem.* 1989, 40, 351.
(172) Sun, C.-C.; Mark, J. E. *Polymer* 1989, 30, 104.
(173) Clarson, S. J.; Mark, J. E. *Polym. Commun.* 1987, 28, 249.
(174) Mark, J. E. *Polym. Eng. Sci.* 1979, 19, 409.
(175) Sun, C.-C.; Mark, J. E. *J. Polym. Sci., Polym. Phys. Ed.* 1987, 25, 1561.
(176) Qu, W.; Mark, J. E. unpublished results.
(177) Thómas, D. K. *Polymer* 1966, 7, 99.
(178) Sur, G. S.; Mark, J. E. *Eur. Polym. J.* 1985, 21, 1051.
(179) Clarson, S. J.; Mark, J. E. *Polym. Commun.* 1989, 30, 275.
(180) Wang, B.; Brennan, A. B.; Huang, H.-H.; Wilkes, G. L. *J. Macromol. Sci.—Chem.* 1990, A27, 1447.
(181) Wen, J.; Mark, J. E. *J. Appl. Polym. Sci.* 1995, 58, 1135.
(182) Mark, J. E.; Wang, S.-B. *Polym. Bull.* 1988, 20, 443.
(183) Wang, B.; Huang, H.-H.; Brennan, A. B.; Wilkes, G. L. *Preprints, Div. Polym. Chem., Inc. Am. Chem. Soc.* 1989, 30(2), 146.
(184) Nass, R.; Schmidt, H. *J. Non-Cryst. Solids* 1990, 121, 329.
(185) Ning, Y. P.; Zhao, M. X.; Mark, J. E. In *Frontiers of Polymer Research*, P. N. Prasad and J. K. Nigam, Eds. Plenum Press: New York. 1991; p. 479.
(186) Schmidt, H. In *Inorganic and Organometallic Polymers*, M. Zeldin, K. J. Wynne, and H. R. Allcock, Eds. American Chemical Society: Washington, DC. 1988; p. 333.
(187) Nass, R.; Arpac, E.; Glaubitt, W.; Schmidt, H. *J. Non-Cryst. Solids* 1990, 121, 370.
(188) Wang, B.; Wilkes, G. L. *J. Polym. Sci., Polym. Chem. Ed.* 1991, 29, 905.
(189) Wilkes, G. L.; Huang, H.-H.; Glaser, R. H. In *Silicon-Based Polymer Science*, J. M. Zeigler and F. W. G. Fearon, Eds. American Chemical Society: Washington, DC. 1990; Vol. 224; p. 207.
(190) Brennan, A. B.; Wang, B.; Rodrigues, D. E.; Wilkes, G. L. *J. Inorg. Organomet. Polym.* 1991, 1, 167.
(191) Sobon, C. A.; Bowen, H. K.; Broad, A.; Calvert, P. D. *J. Mater. Sci. Lett.* 1987, 6, 901.
(192) Calvert, P.; Mann, S. *J. Mater. Sci.* 1988, 23, 3801.
(193) Azoz, A.; Calvert, P. D.; Kadim, M.; McCaffery, A. J.; Seddon, K. R. *Nature* 1990, 344, 49.
(194) Doyle, W. F.; Fabes, B. D.; Root, J. C.; Simmons, K. D.; Chiang, Y. M.; Uhlmann, D. R. In *Ultrastructure Processing of Advanced Ceramics*, J. D. Mackenzie and D. R. Ulrich, Eds. Wiley-Interscience: New York. 1988; p. 953.
(195) Boulton, J. M.; Fox, H. H.; Neilson, G. F.; Uhlmann, D. R. In *Better Ceramics Through Chemistry IV*, B. J. J. Zelinski, C. J. Brinker, D. E. Clark, and D. R. Ulrich, Eds. Materials Research Society: Pittsburgh, PA. 1990; Vol. 180; p. 773.
(196) Mark, J. E.; Sun, C.-C. *Polym. Bull.* 1987, 18, 259.

(197) Zhao, M. X.; Ning, Y. P.; Mark, J. E. In *Advanced Composite Materials*, M. D. Sacks, Eds. American Ceramics Society: Westerville, OH. 1993; p. 891.
(198) Wang, S.; Mark, J. E. *Polym. Bull.* 1992, 29, 343.
(199) Mayer, A. B. R.; Mark, J. E. In *Nanostructured Materials*, G.-M. Chow and K. E. Gonsalves, Eds. American Chemical Society: Washington, DC. 1996; p. 137.
(200) Mayer, A. B. R.; Mark, J. E. *Polym. Bull.* 1996, 37, 683.
(201) Morton, M., Ed.; *Rubber Technology*. Van Nostrand Reinhold: New York. 1973.
(202) Fu, F.-S.; Mark, J. E. *J. Polym. Sci., Polym. Phys. Ed.* 1988, 26, 2229.
(203) Fu, F.-S.; Mark, J. E. *J. Appl. Polym. Sci.* 1989, 37, 2757.
(204) Sur, G. S.; Mark, J. E. *Polym. Bull.* 1988, 20, 131.
(205) Sur, G. S.; Mark, J. E. *Eur. Polym. J.* 1988, 24, 913.
(206) Wang, S.; Mark, J. E. *J. Mater. Sci.* 1990, 25, 65.
(207) Ibemesi, J.; Gvozdic, N.; Keumin, M.; Lynch, M. J.; Meier, D. J. *Preprints, Div. Polym. Chem., Inc., Am. Chem. Soc.* 1985, 26(2), 18.
(208) Ibemesi, J.; Gvozdic, N.; Keumin, M.; Tarshiani, Y.; Meier, D. J. In *Polymer-Based Molecular Composites*. D. W. Schaefer and J. E. Mark, Eds. Materials Research Society: Pittsburgh, PA. 1990; Vol. 171; p. 105.
(209) Nagy, M.; Keller, A. *Polymer Commun.* 1989, 30, 130.
(210) Wang, S.; Mark, J. E. *Macromolecules* 1990, 23, 4288.
(211) Ho, C. C.; Hill, M. J.; Odell, J. A. *Polymer* 1993, 34, 2019.
(212) Ho, C. C.; Keller, A.; Odell, J. A.; Ottewill, R. H. *Polym. Int.* 1993, 30, 207.
(213) Ho, C. C.; Keller, A.; Odell, J. A.; Ottewill, R. H. *Coll. Polym. Sci.* 1993, 271, 469.
(214) Burdon, J. W.; Calvert, P. In *Hierarchically Structured Materials*, I. A. Aksay, E. Baer, M. Sarakaya, and D. A. Tirrell, Eds. Materials Research Society: Pittsburgh, PA. 1992; Vol. 255; p. 375.
(215) Wang, S.; Xu, P.; Mark, J. E. *Macromolecules* 1991, 24, 6037.
(216) Rigbi, Z.; Mark, J. E. *J. Polym. Sci., Polym. Phys. Ed.* 1985, 23, 1267.
(217) Liu, S.; Mark, J. E. *Polym. Bull.* 1987, 18, 33.
(218) Sohoni, G. B.; Mark, J. E. *J. Appl. Polym. Sci.* 1987, 34, 2853.
(219) Sur, G. S.; Mark, J. E. *Polym. Bull.* 1987, 18, 369.
(220) Flank, W. H.; Whyte, T. E., Jr., Eds. *Perspectives in Molecular Sieve Science*. American Chemical Society: Washington, DC. 1988.
(221) Thomas, J. M. *Sci. Am.* 1992, 266(4), 112.
(222) Bonneviot, L.; Kaliaguine, S., Eds. *Zeolites: A Refined Tool for Designing Catalytic Sites*. Elsevier: Amsterdam. 1995.
(223) Al-Ghamdi, A. M. S.; Mark, J. E. *Polym. Bull.* 1988, 20, 537.
(224) Wen, J.; Mark, J. E. *J. Mater. Sci.* 1994, 29, 499.
(225) Frisch, H. L.; Xue, Y. *J. Polym. Sci., Polym. Chem. Ed.* 1995, 33, 1979.
(226) Frisch, H. L.; Maaref, S.; Xue, Y.; Beaucage, G.; Pu, Z.; Mark, J. E. *J. Polym. Sci., Polym. Chem. Ed.* 1996, 34, 673.
(227) Frisch, H. L.; Xue, Y.; Maaref, S.; Beaucage, G.; Pu, Z.; Mark, J. E. *Macromol. Symp.* 1996, 106, 147.
(228) Frisch, H. L.; West, J. M.; Goltner, C. G.; Attard, G. S. *J. Polym. Sci., Polym. Chem. Ed.* 1996, 34, 1823.
(229) Frisch, H. L.; Mark, J. E. *Chem. Mater.* 1996, 8, 1735.

Appendixes

A. Network Structural Parameters

The conventional way of forming a network is by means of cross-linking a long "primary" polymer chain, at random points, to neighboring chains[1,2]. This is referred to as random cross-linking, and the most important example is the vulcanization of natural rubber by sulfur. Some of the time-honored concepts in this area are described in appendix B. Recent progress in synthetic polymer chemistry, however, makes it possible to form elastomeric networks by more controlled reactions, for example, by end-linking functionally terminated chains as described in chapters 10 and 13. Another way of forming such "model" networks is by linking chains together at well-defined points, such as reactive side chains, along their lengths. These more specific and more controlled methods of network formation have the advantage of providing information on the structure of the resulting network, particularly the molecular weight between cross-links and thus a measure of the degree of cross-linking. Describing the different ways of representing such structural information is the main object of this appendix.

To begin, a chain extending between two linkages or junctions is called a "network chain," and the total number of chains and junctions in a network is denoted by ν and μ, respectively. The molecular weight of a network chain is denoted by M_c, and, since network chains generally exhibit a distribution of molecular weights, M_c represents the average of this distribution. The number of chains meeting at a junction is called the "functionality" ϕ of that junction. A network having different functionalities at different junctions may be characterized by an average functionality. A chain connected to a junction at only one end is called a "dangling chain," and one that is attached to the same junction at both of its ends is called a "loop." A dangling chain does not contribute to the elastic free energy of a network at equilibrium; neither does a loop that is not penetrated by another network chain that is itself elastically effective. A network with no dangling chains or loops, and no junctions having a functionality less than three, is called a "perfect network."

A network may be visualized as being formed in two hypothetical steps[3]. In the first step, all chains are united at multifunctional junctions (including bifunctional

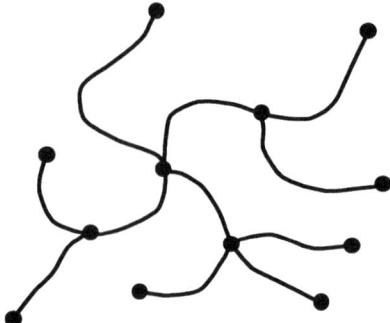

Figure A.1 A tree with filled circles showing the labeled points forming the junctions.

ones, if present) to form a molecule in the form of a "tree" (which by definition contains no cycles). A simple example is shown in figure A.1. The junctions, which are referred to as the "labeled" points, are indicated with filled circles, some of which may contain reactive groups that could further react with one another to form the final network structure. There are $\nu + 1 \approx \nu$ junctions in such a tree. In the second step, ξ pairs of these junctions are combined. The chain ends may be included in this joining process, following which the tree is converted into a network (which may contain dangling chains). The process reduces the number of labeled points to $\nu - \xi + 1 \approx \nu - \xi$, and introduces exactly ξ independent cyclic paths. This quantity ξ is referred to as the "cycle rank" of the network, and is the number of chains that must be cut to reduce the network back to a tree. In phantomlike theories of rubber elasticity, described in chapter 2, the cycle rank is the only structural parameter required for describing the elasticity of a network, irrespective of whether the network is perfect or imperfect[3–6].

There are thus five parameters used to characterize network structure; specifically, M_c, ν, μ, ϕ, and ξ. For a perfect network, only two of these are independent. The remaining three parameters may be obtained from the equations

$$\mu = 2\nu/\phi$$
$$\xi = (1 - 2/\phi)\nu \quad \text{(A.1)}$$
$$\xi/V^0 = (1 - 2/\phi)\rho N_A/M_c$$

where V^0 is the volume of the network in the state of formation, ρ is the corresponding density, and N_A is Avogadro's number. A derivation of the relations given by eq. (A.1) can be found elsewhere[7]. An additional expression which is useful in interpeting experimental data on swollen networks is

$$n = x_c V_1 \rho_d / M_l \quad \text{(A.2)}$$

where n is the number of bonds in the expression $\langle r^2 \rangle_0 = C_\infty n l^2$, x_c is the number of segments (the portion of the chain whose volume is equal to that of the solvent volume), V_1 is the molar volume of the solvent, ρ_d is the density of the unswollen polymer, and M_l is its molecular weight per bond. In terms of these variables, the third part of eq. (A.1) becomes

$$\xi/V^0 = (1 - 2/\phi)v_2^0 N_A / x_c V_1 \qquad (A.3)$$

where v_2^0 is the volume fraction of polymer during formation of the network.

For imperfect networks, the expressions given by eq. (A.1) require modifications. The presence of dangling chains, loops, or bifunctional junctions requires the identification of those chains and junctions that are "active" or "effective." An active junction is defined by Scanlan[8] and Case[9] as one joined to the gel network by at least three paths. An active chain, on the other hand, is one terminated by active junctions at both its ends. Pearson and Graessley[10] have shown that for a randomly interconnected network whose junctions are of even functionality,

$$\xi = \nu_a - \mu_a \qquad (A.4)$$

where ν_a and μ_a are the numbers of active chains and active junctions, respectively, as defined by Scanlan[8] and Case[9]. Flory[11] later showed that this equation holds for networks of any functionality, and that the networks do not have to be random. Flory also showed[11] that the definition of active junctions and chains by Scanlan[8] and Case[9] depends on the local topology of a network and hence leads to ambiguities. Instead, Flory defined the term "effective" chains to be those that effectively contribute to the elasticity of the network, and related their number ν_{eff} to the cycle rank by

$$\nu_{\text{eff}} = 2\xi \qquad (A.5)$$

General expressions relating the effective number of chains to other network parameters in imperfect networks are not available at the present time. For an imperfect tetrafunctional network, however, the number of effective chains is given by[1,12]

$$\nu_{\text{eff}} = \nu_0 (1 - 2M_c / \bar{M}_n) \qquad (A.6)$$

where ν_0 is the total number of chains in the network. The quantity $\bar{M}_n$ is the average molecular weight of the primary molecules (the primary or precursor chains being cross-linked into the network structure). Expressions relating the cycle rank and the number of active junctions to M_c have been given by Queslel and Mark[13]. For imperfect networks, the values of M_c may be determined from gel curves and sol fraction measurements, making use of the Pearson-Graessley[10] or Charlesby[14] equations.

The number of skeletal bonds in typical network chains ranges between 100 and 1000, which conforms to the Gaussian picture of the single chain[3]. Networks consisting of chains having much fewer than 100 bonds are more like hard thermosets and do not show the typical large extensibility of an elastomeric network. Those having chains with more than 1000 bonds typically take very long times to reach equilibrium under load, and the gels they form when swollen by solvents are extremely fragile. In a typical network of flexible chains, the root-mean-squared distance $\langle r^2 \rangle_0^{1/2}$ of a chain having about 500 skeletal bonds is of the order of 70–

80 Å. A spherical domain of this radius contains about 40–50 cross-links and their pendent chains. Thus, a network chain shares its space with many other network chains, which topologically may be either close or distant. This high degree of interpenetration of network chains results in entanglements that are permanently trapped into the network structure as a result of cross-linking. Such entanglements are discussed in chapters 3, 4, and 10.

References

(1) Flory, P. J. *Principles of Polymer Chemistry*. Cornell University Press: Ithaca, NY. 1953.
(2) Coran, A. Y. In *Science and Technology of Rubber*, J. E. Mark, B. Erman, and F. R. Eirich, Eds. Academic Press: San Diego, CA. 1994; p. 339.
(3) Flory, P. J. *Proc. Roy. Soc. London, A* 1976, 351, 351.
(4) Duiser, J. A.; Staverman, A. J. In *Physics of Noncrystalline Solids*, J. A. Prins, Ed. North Holland: Amsterdam. 1965.
(5) Graessley, W. W. *Macromolecules* 1975, 8, 186.
(6) Graessley, W. W. *Macromolecules* 1975, 8, 865.
(7) Mark, J. E.; Erman, B. *Rubberlike Elasticity. A Molecular Primer*. Wiley-Interscience: New York. 1988.
(8) Scanlan, J. J. *J. Polym. Sci.* 1960, 43, 501.
(9) Case, L. C. *J. Polym. Sci.* 1960, 45, 397.
(10) Pearson, D. S.; Graessley, W. W. *Macromolecules* 1978, 11, 528.
(11) Flory, P. J. *Macromolecules* 1982, 15, 99.
(12) Flory, P. J. *Chem. Rev.* 1944, 35, 51.
(13) Queslel, J. P.; Mark, J. E. *J. Chem. Phys.* 1985, 82, 3449.
(14) Charlesby, A.; Pinner, S. H. *Proc. Roy. Soc. Series A* 1959, 249, 367.

B. Definitions in the Area of Rubber Technology

The technology of rubber primarily involves the more applied aspects of the preparation of cross-linked, filler-reinforced elastomeric materials. In many cases, there is not much scientific understanding for many of the procedures used in the industry, since they have apparently been developed over the decades by trial-and-error methods. Nonetheless, some familiarity with these ideas could be useful to scientists working in the area of rubberlike elasticity, particularly with regard to the preparation, processing, and utilization of these materials. This section gives some of the more important definitions in the area. More detailed definitions and discussions of these and related concepts are readily available in the literature[1-8].

B.1 Basic Definitions

Accelerator: A substance that accelerates the curing of an elastomer in the case of sulfur vulcanization. Many of these are sulfur-containing compounds.

Activator: A substance that increases the efficiency of an accelerator. There are some parallels here with catalysts and cocatalysts in small-molecule chemistry.

Antidegradant: Any substance that suppresses or slows down the degradation of an elastomer, which typically occurs from reactions with oxygen, ozone, and so on.

Antioxidant: An additive that protects an elastomer from oxidative degradation under use conditions. Such protection is particularly important in the case of elastomers having unsaturation, for example, the diene elastomers.

Antiozonant: An additive that protects an elastomer, frequently of the unsaturated type, from degradative reaction with ozone during service. Some accomplish this by migrating to the surface and forming a barrier layer, while others chemically react with the ozone.

Blooming: Migration of additives in elastomers to the surface to minimize surface free energy, thereby frequently giving the material a chalky or discolored appearance.

Bonding agent: Substance used to improve the adhesion or bonding between an elastomer and a fiber, film, fabric, metallic surface, and so on.

BuNa rubbers: Elastomers produced from butadiene (Bu) that was polymerized using a sodium (Na) catalyst.

Butyl rubber: Polyisobutylene containing a few mole percent unsaturated comonomeric units to permit cross-linking.

Cavitation: Mode of failure when a rubber vulcanizate is subjected to hydrostatic tension, with all three principal stresses equal.

Chlorinated or chlorosulfonated polyethylene: Elastomer obtained from this normally thermoplastic polymer by substitution of a large enough number of chlorine atoms or chlorosulfonate groups for hydrogen atoms that crystallinity is suppressed.

Coagulation: Precipitation of an elastomer from its latex.

Cold rubber: Styrene–butadiene rubber made by emulsion polymerizations carried out at low temperatures in order to improve its properties.

Compounding: Incorporation of additives into an elastomer, typically by a milling or other blending technique.

Compression set: Deformation that remains after an imposed compressive force has been removed.

Crepe rubber: Natural rubber containing only a small amount of sodium bisulfite in order to bleach it.

Cure: Synonym for cross-linking.

Cure time: Amount of time required for a polymer to reach its maximum level of cross-linking.

Durometer: Device used to measure the hardness of an elastomer.

EPDM: Elastomer based on ethylene, propylene, and a diene comonomer (to facilitate cross-linking).

Extenders: Materials added during compounding to decrease the cost of an elastomer.

Fatigue life: Number of deformation cycles an elastomer can undergo before failure.

Green strength: Strength of an uncured sample of rubber, of importance with regard to processing techniques.

Guayule rubber: Natural rubber from the guayule shrub.

GR-S: Synonym for SBR (styrene–butadiene rubber).

Hard rubber: Rubber so heavily cross-linked that it is more suitable for molding as a thermoset, than as an elastomer.

Hardness: Resistance to indentation.

Hevea rubber: Natural rubber from the *Hevea brasiliensis* tree.

Hysteresis: Failure of a property that has been changed to return to its initial value after removal of the modifying condition. In elastomers, the relevant property is the stress, and the corresponding energy losses appear as heat buildup.

Knotty tearing: Tearing in which the tear tip circles around on itself to form new tear breaks in an irregular pattern.

Latex: Stabilized emulsion of elastomer particles in water, frequently dried so as to produce an elastomeric coating on a surface.

Lubricant: An additive that simplifies processing by facilitating the flow of an elastomer.

Masterbatching: Introducing filler into an elastomer by having it present as a slurry in a latex while the polymer is being coagulated.

Mastication: Processing step in which the elastomer chains are broken by shear and by chemical reaction, during milling, so as to intentionally reduce the viscosity of the elastomer mixture.

Mullins effect: Decrease in modulus of an elastomer, especially when filled, after deformation (particularly after the first deformation–retraction cycle).

Nonreinforcing filler: Material, typically inorganic, such as clay, talcum, finely divided metal, and so on, added to an elastomer to lower its cost, change its appearance, or increase its thermal or electrical conductivity.

Oil-extension: Addition of an oil, typically a hydrocarbon, to dilute an elastomer, using it to serve as either a plasticizer or liquid-type of nonreinforcing filler, or both.

Peptizer: Additive that promotes the chemical breakdown of the elastomer chains during mastication.

Phr: Measure of composition, specifically the parts by weight of a substance per 100 parts of elastomer.

Plasticizer: Low-molecular-weight material added to a polymer, typically to "soften" it (lower its glass transition temperature).

Promoter: Synonym for activator.

Pure-gum vulcanizate: Cross-linked but unfilled elastomer.

Reinforcing filler: Material, typically inorganic, added to an elastomer to improve its mechanical properties, particularly to increase its modulus. The most important examples are carbon black and silica.

Resilience: Ratio of the energy returned upon recovery from deformation to that giving rise to the deformation.

Retarder: Substance that slows or inhibits the cross-linking process.

Reversion: Decrease in the modulus of an elastomer during curing, presumably due to the opposing effect of chain degradation.

SBR: Styrene–butadiene rubber.

Scorch: Unfortunate term used for premature cross-linking, which interferes with the subsequent processing of the elastomer.

Scorch time: Amount of time a rubber mixture can be worked before it becomes unprocessable.

Silicones: Misnomer for siloxanes, relevant here with regard to polysiloxane elastomers.

Smoked sheet: Natural rubber that has been dried in the smoke from burning wood or coconut shells.

Soft rubber: A material having a degree of cross-linking low enough to give it the high extensibility normally associated with elastomers, that is, well below that used to produce hard rubber.

Solution SBR: Styrene–butadiene rubber produced by anionic polymerization in solutions, primarily to give higher molecular weights and narrower molecular weight distributions.

Stabilizer: An antidegradant that protects a polymer, particularly before and during the cross-linking step.

Staining: Discoloration of an elastomer surface, typically due to oxidation of additives.

Structure index: Measure of the extent of agglomeration or chainlike structure of the particles in a filler.

Sulfur donor: An organic polysulfide that supplies some of the sulfur required in the cure of an elastomer by elemental sulfur.

Tack strength: Force required to separate two uncured polymer surfaces after they have been brought into contact.

Tackifiers: Materials added to an elastomer to improve its tack strength.

Thermoset: Any material that becomes a solid because of cross-linking (thus includes elastomers). Its more specific use is to materials so heavily cross-linked (such as hard rubber, epoxy resins, and phenol–formaldehyde resins) that they are used primarily to mold rigid objects rather than elastomeric ones.

Vulcanization: Cross-linking an elastomer by heating a mixture of polymer and elemental sulfur, frequently with the addition of accelerators and activators. Also used loosely for any type of cross-linking process.

References

(1) Morton, M., Ed. *Rubber Technology*. Van Nostrand Reinhold: New York. 1973.
(2) Eirich, F. R., Ed. *Science and Technology of Rubber*. Academic Press: New York. 1978.
(3) Billmeyer, F. W. *Textbook of Polymer Science*, 3rd ed. Wiley: New York. 1984.
(4) Rodriguez, F. *Principles of Polymer Engineering*, 2nd ed. McGraw-Hill: New York. 1982.
(5) Alger, M. S. M. *Polymer Science Dictionary*. Elsevier: London. 1989.
(6) Stinson, S. *Chem. Eng. News* 1990, May 21, 45.
(7) Gent, A. N., Ed. *Engineering with Rubber. How to Design Rubber Components*. Hanser: New York. 1992.
(8) Mark, J. E.; Erman, B.; Eirich, F. R., Eds. *Science and Technology of Rubber*; 2nd ed. Academic: New York. 1994.

C. Deformation and Stress

The state of deformation and stress in an elastic body may be described in various ways, but this appendix summarizes the most commonly adopted definitions in the area of rubber elasticity. Other definitions of stress and deformation in non-linear elastic bodies may be found in several more general treatments in the literature[1-4].

C.1 Deformation

C.1.1 General Aspects

The deformation of a body may be described by the displacement gradient tensor λ, defined as

$$\lambda = \begin{vmatrix} \dfrac{\partial x_1}{\partial X_1} & \dfrac{\partial x_1}{\partial X_2} & \dfrac{\partial x_1}{\partial X_3} \\ \dfrac{\partial x_2}{\partial X_1} & \dfrac{\partial x_2}{\partial X_2} & \dfrac{\partial x_2}{\partial X_3} \\ \dfrac{\partial x_3}{\partial X_1} & \dfrac{\partial x_3}{\partial X_2} & \dfrac{\partial x_3}{\partial X_3} \end{vmatrix} \tag{C.1}$$

where x_i and X_i are the coordinates of a point in the deformed and the undeformed states of the body, respectively. This form of the displacement gradient matrix describes the deformation of a body in the most general state of non-homogeneous strain. In some treatments, x_1 is replaced by x, x_2 by y, and x_3 by z. The latter notation is also used frequently in this book.

Theories of rubber elasticity are often presented in terms of homogeneous principal deformations, in which case eq. (C.1) simplifies to

$$\lambda \equiv \begin{vmatrix} \dfrac{X_1}{X_{10}} & 0 & 0 \\ 0 & \dfrac{X_2}{X_{20}} & 0 \\ 0 & 0 & \dfrac{X_3}{X_{30}} \end{vmatrix} = \begin{vmatrix} \lambda_1 & 0 & 0 \\ 0 & \lambda_2 & 0 \\ 0 & 0 & \lambda_3 \end{vmatrix} \qquad (C.2)$$

The state of deformation represented by eq. (C.2) is shown in figure C.1 for a network in the form of a prism with sides X_{10}, X_{20}, and X_{30} in the reference state (defined as the state in which the network was formed). This state may be isotropic or anisotropic, and the chains may be in the bulk state or may be diluted by a solvent. The dimensions of the prism in the deformed state become X_1, X_2, and X_3. The determinant of λ is equal to the ratio of the final volume V of the network to the reference volume V_0 during formation, that is,

$$\det \lambda = V/V^0 = v_{2C}/v_2 \qquad (C.3)$$

Here, v_{2C} is the volume fraction of polymer present during the formation of the network, and v_2 is the volume fraction at the beginning of the mechanical property measurements. It is often convenient to decompose the displacement gradient tensor into the product of a dilation part and a pure distortion part:

$$\lambda = \left(\frac{V}{V^0}\right)^{1/3} \alpha = (v_{2C}/v_2)^{1/3} \begin{vmatrix} \alpha_1 & 0 & 0 \\ 0 & \alpha_2 & 0 \\ 0 & 0 & \alpha_3 \end{vmatrix} \qquad (C.4)$$

where α_i is the ratio of the length along the ith axis to the length in the initial undistorted state (in which the polymer may be swollen with solvent). Defined in this manner, α represents the distortion of the sample, and is referred to as the "distortion tensor." This definition conveniently separates the deformation into two parts. First, the sample is visualized as being swollen to its final equilibrium volume V, followed by its distortion under the applied stress. During the application of the stress, the volume of the sample is usually assumed to be constant, as a result of which the determinant of α equates to unity, that is, $\alpha_1\alpha_2\alpha_3 = 1$.

Some simple examples of deformations used in theoretical and experimental studies of rubber elasticity are given below[5]. These examples hold for networks

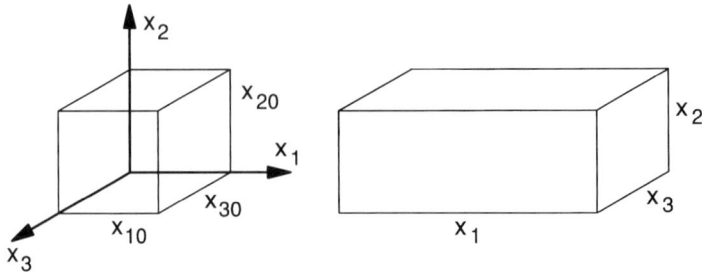

Figure C.1 An undeformed and a deformed rectangular prism.

cross-linked in the isotropic state. For networks cross-linked in the anisotropic state, further care must be taken in describing the state of the deformation and its relationship to stress. Treatments of such networks may be found elsewhere[6-8].

C.1.2 Isotropic Swelling

The components of the tensor α that measure the degree of distortion equate to unity, that is, $\alpha = \mathbf{E}$ ($\mathbf{E}$ being the 3×3 identity matrix), and λ is given by

$$\lambda = \begin{vmatrix} (v_{2C}/v_2)^{1/3} & 0 & 0 \\ 0 & (v_{2C}/v_2)^{1/3} & 0 \\ 0 & 0 & (v_{2C}/v_2)^{1/3} \end{vmatrix} = (v_{2C}/v_2)^{1/3} \mathbf{E} \qquad (C.5)$$

C.1.3 Simple Tension

The change in volume upon stretching a network is of second order compared with changes in the linear dimensions, and is therefore generally taken to be zero. Taking the direction of stretch along the x_1 axis, and letting $\lambda_1 = \lambda$ leads to

$$\lambda = \begin{vmatrix} \lambda & 0 & 0 \\ 0 & (V/\lambda V^0)^{1/2} & 0 \\ 0 & 0 & (V/\lambda V^0)^{1/2} \end{vmatrix} \qquad (C.6)$$

where $V/V^0 = v_{2C}/v_2$. Letting $\alpha_1 = \alpha$, $\alpha_2 = \alpha_3 = \alpha^{-1/2}$ and using eq. (C.4), the components of λ will be related to those of α by

$$\lambda = (v_{2C}/v_2)^{1/3} \begin{vmatrix} \alpha & 0 & 0 \\ 0 & \alpha^{-1/2} & 0 \\ 0 & 0 & \alpha^{-1/2} \end{vmatrix} \qquad (C.7)$$

C.1.4 Biaxial Extension

When the dimensions of a rubber sheet are changed independently by the factor α_1 along the x_1 axis and by α_2 along the x_2 axis, then the tensor λ is given by

$$\lambda = \begin{vmatrix} \lambda_1 & 0 & 0 \\ 0 & \lambda_2 & 0 \\ 0 & 0 & (v_{2C}/v_2)(\lambda_1 \lambda_2)^{-1} \end{vmatrix}$$

$$= \begin{vmatrix} (v_{2C}/v_2)^{1/3} \alpha_1 & 0 & 0 \\ 0 & (v_{2C}/v_2)^{1/3} \alpha_2 & 0 \\ 0 & 0 & (v_{2C}/v_2)^{1/3}(\alpha_1 \alpha_2)^{-1} \end{vmatrix} \qquad (C.8)$$

C.1.5 Pure Shear

The displacement gradient tensor λ for the state of pure shear is obtained[5] by setting $\alpha_1 = \alpha$ and $\alpha_2 = 1$ in eq. (C.4). The result is

$$\lambda = \begin{vmatrix} (v_{2C}/v_2)^{1/3}\alpha & 0 & 0 \\ 0 & (v_{2C}/v_2)^{1/3} & 0 \\ 0 & 0 & (v_{2C}/v_2)^{1/3}\alpha^{-1} \end{vmatrix} \tag{C.9}$$

C.2 Stress

In general, the state of stress at a point is given by the tensor

$$\mathbf{T} = \begin{vmatrix} T_{11} & T_{12} & T_{13} \\ T_{21} & T_{22} & T_{23} \\ T_{31} & T_{32} & T_{33} \end{vmatrix} \tag{C.10}$$

where the first index indicates the direction of the force and the second index identifies the direction of the normal of the plane on which the force is acting. The stress tensor may conveniently be taken as symmetric in the study of rubber elasticity. A homogeneous state of stress is obtained if the stress is the same at all points of the body. A coordinate system for which $\mathbf{T}$ is diagonal is referred to as the principal coordinate system, and the corresponding stresses are called the principal stresses. In this book, the notation adopted by Flory[9] is used, according to which the principal stress tensor $\mathbf{t}$ is given by

$$\mathbf{t} = \begin{vmatrix} t_1 & 0 & 0 \\ 0 & t_2 & 0 \\ 0 & 0 & t_3 \end{vmatrix} \tag{C.11}$$

The principal directions of stress and strain are identical. Thus, the three principal stress components of eq. (C.11) are associated with the three extension ratios of eq. (C.2).

The components of $\mathbf{t}$ may be obtained from the change in the Helmholtz (or elastic) free energy ΔA, by writing the infinitesimal work $-dW$ done under the forces $f_1, f_2,$ and f_3 acting on the faces of the prism shown in figure C.1, and then equating it to the change in the Helmholtz free energy. Thus,

$$-dW = (d\Delta A)_T = f_1 dL_1 + f_2 dL_2 + f_3 dL_3 \tag{C.12}$$

where $dL_1, dL_2,$ and dL_3 represent the respective changes in the sides of the prism. Work done against volume change as well as the distortion of the body is included in eq. (C.12) inasmuch as it is written for constant temperature. The relation of $t_1, t_2,$ and t_3 to $(d\Delta A)_T$ is obtained after performing the following identity operations on eq. (C.12):

$$(d\Delta A)_T = \frac{f_1}{L_2 L_3} L_2 L_3 dL_1 + \frac{f_2}{L_1 L_3} L_1 L_3 dL_2 + \frac{f_3}{L_1 L_2} L_1 L_2 dL_3$$

$$= t_1 L_2 L_3 dL_1 + t_2 L_1 L_3 dL_2 + t_3 L_1 L_2 dL_3$$

$$= t_1 \frac{L_{10}}{L_1} L_1 L_2 L_3 d \frac{L_1}{L_{10}} + t_2 \frac{L_{20}}{L_2} L_2 L_1 L_3 d \frac{L_2}{L_{20}} + t_3 \frac{L_{30}}{L_3} L_3 L_1 L_2 d \frac{L_3}{L_{30}}$$
(C.13)

Here, the component of stress t_i is defined as the force divided by the deformed area on which it acts. The last equality of eq. (C.13) is written as

$$(d\Delta A)_T = V \sum_{i=1}^{3} t_i \lambda_i^{-1} d\lambda_i$$
(C.14)

The definition of stress then follows from eq. (C.14) as

$$t_i = \frac{\lambda_i}{V} \left(\frac{\partial \Delta A}{\partial \lambda_i} \right)_T = \frac{\alpha_i}{V} \left(\frac{\partial \Delta A}{\partial \alpha_i} \right)_{T,V}$$

$$= \frac{2\lambda_i^2}{V} \left(\frac{\partial \Delta A}{\partial \lambda_i^2} \right)_T = \frac{2\alpha_i^2}{V} \left(\frac{\partial \Delta A}{\partial \alpha_i^2} \right)_{T,V}$$
(C.15)

where the second line of eq. (C.15) is obtained by a trivial algebraic transformation. Taking the differential of eq. (C.4) and substituting into eq. (C.14) leads to

$$(d\Delta A)_T = \left(\frac{1}{3} \sum_{i=1}^{3} t_i \right) dV + V \sum_{i=1}^{3} t_i (\alpha_i^{-1} d\alpha_i)$$
(C.16)

The term in the first parentheses on the right-hand-side of eq. (C.16) represents the hydrostatic pressure, which leads to work against the volume change. The second term represents the work done in distorting the elastic body.

The change in the Helmholtz (or elastic) free energy is given[9] as the sum of two terms:

$$\Delta A = \Delta A^*(T, V) + \Delta A_{el}(T, \lambda)$$
(C.17)

where $\Delta A^*(T, V)$ is a function of temperature T and volume V, and denotes the contribution from intermolecular forces such as those in simple liquids. The quantity $\Delta A_{el}(T, \lambda)$ is the elastic free energy arising from the elasticity of the network chains.

Substituting eq. (C.17) into eq. (C.15) leads to the following expression for the stress:

$$t_i = p^* + \frac{2\alpha_i^2}{V} \left(\frac{\partial \Delta A_{el}}{\partial \alpha_i^2} \right)_{T,V}$$
(C.18)

where $p^* = -(\partial \Delta A^* / \partial V)_T$ is the contribution to the pressure from intermolecular interactions such those occurring in simple liquids. One may eliminate p^* from eq. (C.14) by taking the difference between values of t_i along two directions. This leads to the Treloar relations[5], expressed as

$$t_i - t_k = \frac{2}{V}\left[\alpha_i^2\left(\frac{\partial \Delta A_{el}}{\partial \alpha_i^2}\right) - \alpha_k^2\left(\frac{\partial \Delta A_{el}}{\partial \alpha_k^2}\right)\right]_{T,V} \quad (\text{C.19})$$

The above description may be generalized to nonprincipal stresses and strains by defining the strain tensor Λ as

$$\Lambda = \lambda^T \lambda \quad (\text{C.20})$$

where the superscript T denotes the transpose.

Letting A be a function of Λ transforms eq. (C.13) into

$$(d\Delta A)_T = Tr\left[\frac{\partial \Delta A}{\partial \Lambda} d\Lambda\right] \quad (\text{C.21})$$

Here, Tr denotes the trace operator, that is, the sum of the diagonal elements of the tensor given in the square brackets. The definition of the stress tensor $\mathbf{T}$ follows, by the generalization of the expression given in eq. (C.15), as

$$\mathbf{T} = \frac{2}{V}\lambda\left(\frac{\partial \Delta A}{\partial \Lambda}\right)_T \lambda^T \quad (\text{C.22})$$

References

(1) Murnaghan, F. D. *Finite Deformations of an Elastic Solid*. John Wiley & Sons: New York. 1951.
(2) Truesdell, C.; Toupin, R., Eds. *The Classical Field Theories*. Springer-Verlag: Berlin. 1960; Vol. III/1.
(3) Eringen, A. C. *Nonlinear Theory of Continuous Media*. McGraw-Hill: New York. 1962.
(4) Ogden, R. W. In *Mechanics of Solids*, H. G. Hopkins and M. J. Sewell, Eds. Pergamon Press: Oxford, 1982.
(5) Treloar, L. R. G. *The Physics of Rubber Elasticity*, 3rd ed. Clarendon Press: Oxford. 1975.
(6) Flory, P. J. *Trans. Faraday Soc.* 1960, 56, 722.
(7) Erman, B. In *Networks, 86. 8th Polymer Networks Group Meeting*. Elsevier: Elsinore, Denmark. 1988; p. 497.
(8) Batsberg, W.; Hvidt, S.; Kramer, O.; Fetters, L. J. In *Networks, 86. 8th Polymer Networks Group Meeting*, Elsevier: Elsinore, Denmark. 1988; p. 509.
(9) Flory, P. J. *Trans. Faraday Soc.* 1961, 57, 829.

D. Summary of Thermodynamics and Statistical Mechanics

Consider a system of N molecules at fixed temperature T and volume V. At any given instant, this system will be in an energy state E_i. Placing together a large number n of these systems at temperature T, and subsequently insulating them from their surroundings, establishes the canonical ensemble of classical statistical mechanics. The canonical ensemble partition function is defined as

$$Z(N, V, T) = \sum_i \exp(-E_i/kT) \tag{D.1}$$

where the summation is carried over all n systems of the ensemble. The partition function obtained from statistical mechanical considerations serves as a bridge between statistical mechanics and thermodynamics, as outlined in textbooks of statistical mechanics (see, e.g., Hill[1], or Chandler[2]).

The probability p_i of the system to be in the state where its energy is E_i will be

$$p_i = \frac{\exp(-E_i/kT)}{\sum_i \exp(-E_i/kT)} \tag{D.2}$$

The correspondence between statistical mechanics and thermodynamics is established through the Helmholtz free energy $A(N, V, T)$ by the relation

$$A(N, V, T) = -kT \ln Z(N, V, T) \tag{D.3}$$

The infinitesimal work dW done on the system is

$$dW = -pdV + fdL \tag{D.4}$$

where p and f are the pressure and force acting on the system, and V and L are its volume and its length in the direction along which the force is applied. Denoting the internal energy of the system by E and the heat evolved by dQ, and using the thermodynamic relations

$$dE = dQ + dW$$
$$dQ = TdS \qquad \text{(D.5)}$$
$$A = E - TS$$

one obtains the expression for dA in terms of dT, dV, and dL as

$$dA = -SdT - pdV + fdL \qquad \text{(D.6)}$$

Comparing eq. (D.6) with the differential form,

$$dA = \left(\frac{\partial A}{\partial T}\right)_{V,L} dT + \left(\frac{\partial A}{\partial V}\right)_{T,L} dV + \left(\frac{\partial A}{\partial L}\right)_{T,V} dL \qquad \text{(D.7)}$$

and using eq. (D.3), leads to the the entropy (S), pressure (p), force (f), and energy (E) of the system:

$$
\begin{aligned}
S &= -\left(\frac{\partial A}{\partial T}\right)_{V,L} = kT\left(\frac{\partial \ln Z}{\partial T}\right)_{V,L} + k\ln Z \\
p &= -\left(\frac{\partial A}{\partial V}\right)_{T,L} = kT\left(\frac{\partial \ln Z}{\partial V}\right)_{T,L} \\
f &= \left(\frac{\partial A}{\partial L}\right)_{T,V} = -kT\left(\frac{\partial \ln Z}{\partial L}\right)_{T,V} \\
E &= -T^2\left(\frac{\partial A/T}{\partial T}\right)_{V,L} = kT^2\left(\frac{\partial \ln Z}{\partial T}\right)_{V,L}
\end{aligned}
\qquad \text{(D.8)}
$$

The third of the above expressions is the one used for obtaining the macroscopic force required to keep a network at constant length L.

The single-chain configuration partition function may be written down in analogy with the partition function for the macroscopic system described above[3,4]. Replacing the number of particles N of the above system with the number of skeletal bonds n of the chain, and the length L by the chain end-to-end vector $\mathbf{r}$, one obtains expressions equivalent to eqs. (D.3) and (D.8):

$$A(\mathbf{r}) = A(T) - kT \ln Z_\mathbf{r}$$

$$
\begin{aligned}
S &= -\left(\frac{\partial A}{\partial T}\right)_\mathbf{r} = kT\left(\frac{\partial \ln Z_\mathbf{r}}{\partial T}\right)_\mathbf{r} = k\ln Z_\mathbf{r} \\
f &= \left(\frac{\partial A}{\partial \mathbf{r}}\right)_T = -kT\left(\frac{\partial \ln Z_\mathbf{r}}{\partial \mathbf{r}}\right)_T \\
E &= -T^2\left(\frac{\partial A/T}{\partial T}\right)_\mathbf{r} = kT^2\left(\frac{\partial \ln Z_\mathbf{r}}{\partial T}\right)_\mathbf{r}
\end{aligned}
\qquad \text{(D.9)}
$$

where $A(T)$ is a function of temperature. The single-chain configuration partition function $Z_\mathbf{r}$ at fixed temperature is defined as the sum over the energies of all available isomeric states of the chain at fixed end-to-end vector $\mathbf{r}$:

$$Z_r = \sum_i \exp(-E_i/kT) \tag{D.10}$$

For continuously changing rotational angles, eq. (D.10) may be written in integral form as

$$Z_r\, d\mathbf{r} = (8\pi^2)^{-1} \int_r \cdots \int \exp(-E/kT)\, d\{\mathbf{I}\} \tag{D.11}$$

where the front factor $(8\pi^2)^{-1}$ normalizes the partition function with respect to the spatial orientations of the chain as a rigid body. The integration therefore is carried out over all internal configurations $\{\mathbf{I}\}$ of the chain subject to the constraint that the end-to-end vector is fixed. Here, $\{\mathbf{I}\}$ indicates the set of bond vectors obtained under the condition that the end-to-end vector lies between $\mathbf{r}$ and $\mathbf{r} + d\mathbf{r}$.

The distribution function for the chain vector $\mathbf{r}$ is[3,4]

$$W(\mathbf{r}) = \frac{Z_r}{Z} \tag{D.12}$$

This equation is important because it relates the partition function to the distribution function for which expressions of different degrees of approximation exist in the literature, the simplest of which is the Gaussian. Various approximations to $W(\mathbf{r})$ are given in appendix F. Finally, substituting eq. (D.12) into eq. (D.9) results in the expressions for A, S, f, and E in terms of the distribution function.

Alternative derivations of the statistical mechanical distribution laws have been described in the literature[5].

References

(1) Hill, T. L. *An Introduction to Statistical Thermodynamics*. Addison-Wesley: Reading, MA. 1960; pp. 18 & 19.
(2) Chandler, D. *Introduction to Modern Statistical Mechanics*. Oxford University Press: New York. 1987.
(3) Flory, P. J. *Statistical Mechanics of Chain Molecules*. Interscience: New York. 1969.
(4) Volkenstein, M. *Configurational Statistics of Polymer Chains*. Interscience: New York. 1963.
(5) Wall, F. T. *Proc. Nat. Acad. Sci.* 1971, 68, 1720.

E. Fluctuations in Phantom Networks

Fluctuations of junction positions and chain dimensions in a phantom network may be deduced once the Γ_τ matrix given in chapter 2 is known. In this appendix we characterize this matrix and its inverse, and derive important relations involving fluctuations. For additional information, the reader is referred to the paper by Kloczkowski et al.[1], which is a concise presentation in matrix form of previous work by Eichinger[2], Graessley[3,4], Flory[5], and Pearson[6].

E.1 The Matrix Γ and its Inverse

All quantities in this appendix are related exclusively to free junctions described in the phantom network model of chapter 2. The index τ will therefore be dropped in Γ and $\mathbf{R}$. The matrix Γ is known as the Kirchoff or valency–adjacency matrix[2] in graph theory, and its form depends on the topological structure and connectivity of the network. Flory[5] and Pearson[6] have considered networks with equal-length chains forming a treelike structure, and we follow their model. An example of such a structure is shown in figure E.1[1]. The network, which is tetrafunctional,

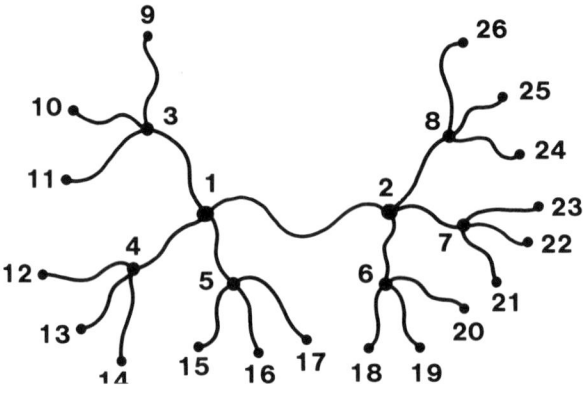

Figure E.1 A treelike structure consisting of equal-length chains. Reprinted with permission from Kloczkowski, A., Mark, J. E., Erman, B. (1989), *Macromolecules*, **22**. Copyright 1997, American Chemical Society.

grows symmetrically from the central chain between junctions 1 and 2, and these two junctions constitute the first tier of junctions. The second tier is made up of the six junctions numbered 3 to 8 in the figure. The network is assumed to grow in this manner without forming any short-circuited cyclic connections. The matrix Γ for this tetrafunctional structure is given by

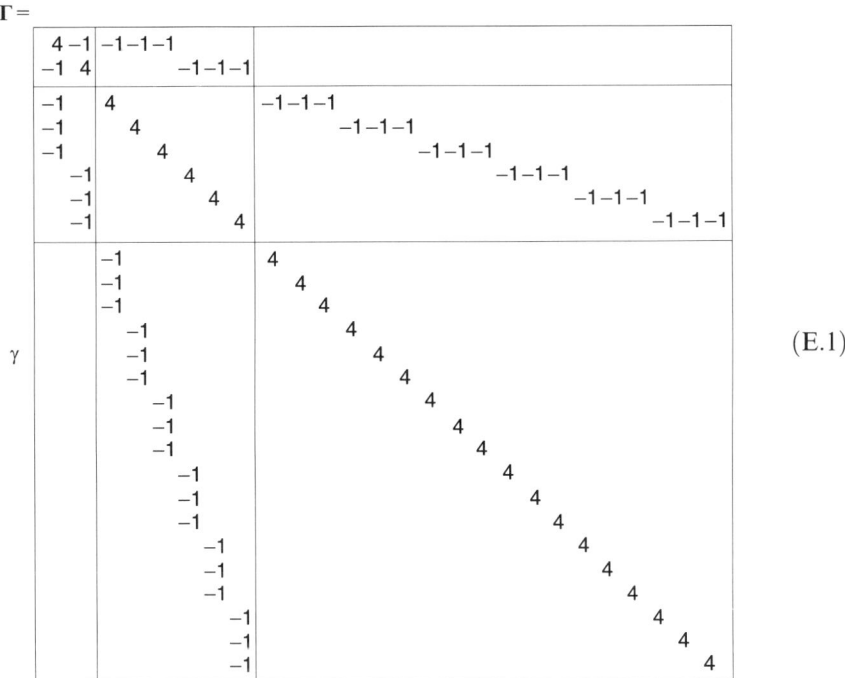

(E.1)

The inverse of Γ is required for calculating the average fluctuations in the phantom network. As described earlier[1], Γ may be cast in a pseudo-diagonal form:

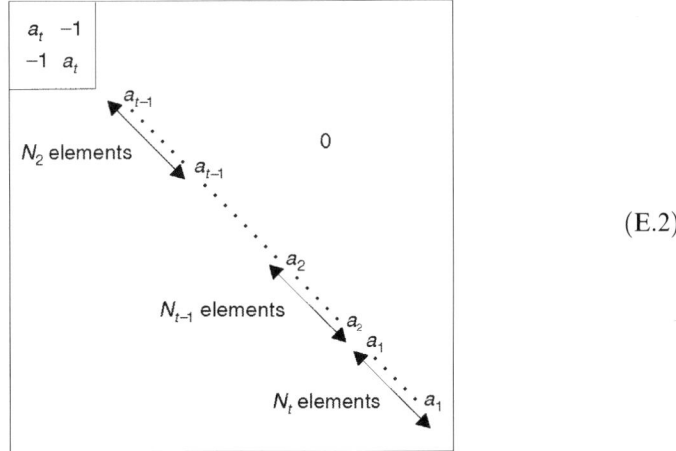

(E.2)

where all off-diagonal elements in the upper right part of the matrix are zero, except Γ_{12} (which is equal to -1). The a terms are indexed starting from the lower right corner, and the subscript on a indicates the tier to which it belongs. The recurrence relations for the a_i terms are obtained from eq. (E.2) as

$$a_1 = \phi$$
$$a_n = \phi - \frac{(\phi - 1)}{a_{n-1}} \tag{E.3}$$

and the solution of eq. (E.3) is

$$a_n = \frac{[(\phi - 1)^{n+1} - 1]}{[(\phi - 1)^n - 1]} \tag{E.4}$$

In the limit where the number of tiers goes to infinity, a_n is independent of n:

$$\lim_{n \to \infty} a_n = \phi - 1 \tag{E.5}$$

In the following, an infinite network is considered.

The fluctuations in chain dimensions are obtained from fluctuations of junctions by the relation

$$\langle (\Delta r_{ij})^2 \rangle = \langle (\Delta \mathbf{R}_i - \Delta \mathbf{R}_j)^2 \rangle = \langle (\Delta R_i)^2 \rangle + \langle (\Delta R_j)^2 \rangle - 2\langle \Delta \mathbf{R}_i \cdot \Delta \mathbf{R}_j \rangle$$
$$= (3/2)[(\Gamma_\tau^{-1})_{ii} + (\Gamma_\tau^{-1})_{jj} - 2(\Gamma_\tau^{-1})_{ij}] \tag{E.6}$$

where r_{ij} is the magnitude of the vector joining two junctions i and j. The latter are not necessarily at the ends of a single chain, and may be separated by several chains of the network.

E.2 Expressions for Various Fluctuations

E.2.1 Two Junctions Joined by a Single Chain

The first four elements of Γ^{-1} follow from eqs. (E.5) and (2.26):

$$\begin{bmatrix} \langle (\Delta R_1)^2 \rangle & \langle \Delta \mathbf{R}_1 \cdot \Delta \mathbf{R}_2 \rangle \\ \langle \Delta \mathbf{R}_1 \cdot \Delta \mathbf{R}_2 \rangle & \langle (\Delta R_2)^2 \rangle \end{bmatrix} = \frac{3}{2} \begin{bmatrix} (\Gamma^{-1})_{11} & (\Gamma^{-1})_{12} \\ (\Gamma^{-1})_{21} & (\Gamma^{-1})_{22} \end{bmatrix}$$
$$= \frac{3}{2\gamma} \begin{bmatrix} \dfrac{\phi - 1}{\phi(\phi - 2)} & \dfrac{1}{\phi(\phi - 2)} \\ \dfrac{1}{\phi(\phi - 2)} & \dfrac{\phi - 1}{\phi(\phi - 2)} \end{bmatrix} \tag{E.7}$$

Substituting from eq. (E.7) into eq. (E.6) leads to the fluctuations in the dimensions of a phantom network chain being given by

$$\langle (\Delta r_{ij})^2 \rangle = \frac{3}{2\gamma} \frac{2}{\phi} = \frac{2}{\phi} \langle r_{ij}^2 \rangle_0 \tag{E.8}$$

E.2.2 Two Junctions Separated by Several Chains

When two junctions m and n are separated by d other junctions, their average fluctuations are given by the matrix

$$\begin{bmatrix} \langle(\Delta R_m)^2\rangle & \langle\Delta \mathbf{R}_m \cdot \Delta \mathbf{R}_n\rangle \\ \langle\Delta \mathbf{R}_n \cdot \Delta \mathbf{R}_m\rangle & \langle(\Delta R_n)^2\rangle \end{bmatrix} = \frac{3}{2}\begin{bmatrix} (\Gamma^{-1})_{mm} & (\Gamma^{-1})_{mn} \\ (\Gamma^{-1})_{nm} & (\Gamma^{-1})_{nn} \end{bmatrix}$$

$$= \frac{3}{2\gamma}\begin{bmatrix} \dfrac{\phi-1}{\phi(\phi-2)} & \dfrac{1}{\phi(\phi-2)(\phi-1)^d} \\ \dfrac{1}{\phi(\phi-2)(\phi-1)^d} & \dfrac{\phi-1}{\phi(\phi-2)} \end{bmatrix}$$

(E.9)

The derivation of the off-diagonal terms of the matrix in eq. (E.9) is somewhat lengthy and is given in its most general form in the appendix of the paper by Kloczkowski et al.[1].

The mean-square fluctuations in the dimensions of the path between junctions m and n is obtained by substitution from eq. (E.9) into eq. (E.6). The result is

$$\langle(\Delta r_{mn})^2\rangle = \frac{2}{\phi(\phi-2)(d+1)}\frac{(\phi-1)^{d+1}-1}{(\phi-1)^d}\langle r_{mn}^2\rangle_0$$

$$= \frac{2}{\phi(\phi-2)}\frac{(\phi-1)^{d+1}-1}{(\phi-1)^d}\langle r_{12}^2\rangle_0$$

(E.10)

where the second line follows by the use of $\langle r_{mn}^2\rangle_0 = (3/2\gamma)(d+1) = (d+1)\langle r_{12}^2\rangle_0$. In the limit $d \to \infty$, eq. (E.10) reduces to

$$\frac{\langle(\Delta r_{mn})^2\rangle_{d\to\infty}}{\langle r_{12}^2\rangle_0} = \frac{2(\phi-1)}{\phi(\phi-2)}$$

(E.11)

Equation (E.11) shows that the ratio of the mean-square fluctuations of the dimensions of an infinite path in a network to those of the single network chain is constant.

E.2.3 Points on a Network Chain

The network chain is assumed to consist of n Kuhn or freely jointed segments and to be attached to two ϕ-functional junctions at its extremities. A point on the chain is defined as the junction of two Kuhn segments. The fractional distances ζ and θ of the ith and jth points, respectively, are defined as

$$\zeta = \frac{i-1}{n} \qquad 0 \leq \zeta \leq 1$$
$$\theta = \frac{j-1}{n} \qquad 0 \leq \theta \leq 1$$

(E.12)

The fluctuations of the two points i and j on the chain have been derived[1] to be

$$\begin{bmatrix} \langle(\Delta R_i)^2\rangle & \langle\Delta \mathbf{R}_i \cdot \Delta \mathbf{R}_j\rangle \\ \langle\Delta \mathbf{R}_i \cdot \Delta \mathbf{R}_j\rangle & \langle(\Delta R_j)^2\rangle \end{bmatrix} = \frac{3n}{2\gamma_0} \times \begin{bmatrix} \frac{\phi-1}{\phi(\phi-2)} + \frac{\zeta(1-\zeta)(\phi-2)}{\phi} & \frac{\phi-1}{\phi(\phi-2)} + \frac{(\phi-2)[\min(\zeta,\theta)-\zeta\theta]-\eta}{\phi} \\ \frac{\phi-1}{\phi(\phi-2)} + \frac{(\phi-2)[\min(\zeta,\theta)-\zeta\theta]-\eta}{\phi} & \frac{\phi-1}{\phi(\phi-2)} + \frac{\theta(1-\theta)(\phi-2)}{\phi} \end{bmatrix}$$

(E.13)

Here,

$$\eta = |\zeta - \theta| \tag{E.14}$$

$$\gamma_0 = 3/2\langle r_1^2\rangle_0 \tag{E.15}$$

where $\langle r_1^2\rangle_0$ is the mean-square end-to-end distance of a single segment such that $\langle r^2\rangle_0 = n\langle r_1^2\rangle_0$. The $\min(\zeta,\theta)$ function in eq. (E.13) may be written as[6]

$$\min(\zeta,\theta) = (\zeta + \theta - |\zeta - \theta|)/2 \tag{E.16}$$

The fluctuations in the dimensions of a chain joining the points i and j are obtained by using eq. (E.13) in eq. (E.6):

$$\langle(\Delta r_{ij})^2\rangle = \frac{3}{2\gamma}\left[\eta - \frac{(\phi-2)}{\phi}\eta^2\right] \tag{E.17}$$

E.2.4 Points on Network Chains that are Separated by Several Junctions

The fluctuation matrix for the case where points i and j are on two chains separated by d junctions is given by[1]

$$\begin{bmatrix} \langle(\Delta R_i)^2\rangle & \langle \Delta \mathbf{R}_i \cdot \Delta \mathbf{R}_j \rangle \\ \langle \Delta \mathbf{R}_i \cdot \Delta \mathbf{R}_j \rangle & \langle(\Delta R_j)^2\rangle \end{bmatrix} = \frac{3n}{2\gamma} \times$$

$$\begin{bmatrix} \dfrac{\phi-1}{\phi(\phi-2)} + \dfrac{\zeta(1-\zeta)(\phi-2)}{\phi} & \dfrac{\phi-1}{\phi(\phi-2)} + \dfrac{[1+\zeta(\phi-2)][(\phi-1)-\theta(\phi-2)]}{\phi(\phi-2)(\phi-1)^d} \\ \dfrac{\phi-1}{\phi(\phi-2)} + \dfrac{[1+\zeta(\phi-2)][(\phi-1)-\theta(\phi-2)]}{\phi(\phi-2)(\phi-1)^d} & \dfrac{\phi-1}{\phi(\phi-2)} + \dfrac{\theta(1-\theta)(\phi-2)}{\phi} \end{bmatrix}$$

(E.18)

where ζ and θ are measured relative to the junctions to the left of points i and j, respectively.

The mean-square fluctuation in $\mathbf{r}_{ij}$ is given by the relation

$$\langle(\Delta r_{ij})^2\rangle = \frac{3}{2\gamma}\left\{\frac{2(\phi-1)}{\phi(\phi-2)}\left[1 - \frac{1}{(\phi-1)^d}\right]\right.$$
$$\left. + \frac{\phi-2}{\phi}\left[\zeta(1-\zeta) + \theta(1-\theta) - \frac{\zeta+\theta-2\theta\zeta}{(\phi-1)^d}\right] + \frac{\eta-d}{(\phi-1)^d}\right\}$$

(E.19)

where the distance between i and j is $\eta = d + \theta - \zeta$ if point i is on the left side of point j along the contour. Otherwise, $\eta = d + \zeta - \theta$.

Fluctuations, correlations, and small-angle neutron scattering from end-linked Gaussian chains in regular bimodal networks have also been derived along the lines of the above presentation[7].

References

(1) Kloczkowski, A.; Mark, J. E.; Erman, B. *Macromolecules* 1989, 22, 1423.
(2) Eichinger, B. *Macromolecules* 1972, 5, 496.
(3) Graessley, W. W. *Macromolecules* 1975, 8, 186.
(4) Graessley, W. W. *Macromolecules* 1975, 8, 865.
(5) Flory, P. J. *Proc. Roy. Soc. London, A* 1976, 351, 351.
(6) Pearson, D. S. *Macromolecules* 1977, 10, 696.
(7) Kloczkowski, A.; Mark, J. E.; Erman, B. *Macromolecules* 1991, 24, 3266.

F. Distributions of the Chain End-to-End Vector

F.1 Examples of Distribution Functions

F.1.1 The Freely Jointed Chain

The single chain may be modeled at different levels of complexity. The simplest model is the freely jointed chain, in which the bonds are connected by universal joints such that two consecutive bonds may adopt all directions relative to each other with equal probability. Thus, bond angles are not preserved and the orientations of different bonds are uncorrelated in this hypothetical model. The distribution function or the probability density $W(r)$ for having one end of a chain at the origin of a coordinate system, and the other end at a distance r from the origin, serves as a useful quantity for describing the statistical properties of the chain[1-6].

F.1.1.1 The Exact Expression For a chain of n bonds each of length l, the exact expression for $W(r)$ is

$$W(r) = (8\pi r l^2)^{-1} n(n-1) \sum_{t=0}^{\tau} \frac{(-1)^t}{t!(n-t)!} \left[\frac{n - (r/l) - 2t}{2}\right]^{n-2} \tag{F.1}$$

where τ is the integer in the range specified by

$$[(n - r/l)/2] - 1 \leq \tau < (n - r/l)/2 \tag{F.2}$$

Equation (F.1) was obtained by Treloar[3], but the form given above is from Flory[2].

For a chain with a sufficiently large number of bonds, an approximate expression for $W(r)$ given by Kuhn and Grün[7] is

$$\ln W(r) = \text{constant} - n\left(\frac{r}{nl}\beta + \ln \frac{\beta}{\sinh \beta}\right) \tag{F.3}$$

where β is the inverse Langevin function $L^{-1}(r/nl)$ obtained from the solution of the expression

$$r/nl = \coth\beta - (1/\beta) = L(\beta) \tag{F.4}$$

The function $L(\beta)$ is known as the Langevin function.

F.1.1.2 The Gaussian Approximation The series expansion of the right-hand side of eq. (F.3), as done by Kuhn and Kuhn[8], leads to

$$\ln W(r) = \text{constant} - n\left\{\frac{3}{2}\left(\frac{r}{nl}\right)^2 + \frac{9}{20}\left(\frac{r}{nl}\right)^4 + \frac{99}{350}\left(\frac{r}{nl}\right)^6 + \cdots\right\} \tag{F.5}$$

When the end-to-end distance of the chain is much smaller than the fully stretched length (i.e., when $r/nl \ll 1$), the higher order terms in eq. (F.5) may be neglected, and the distribution function reduces to the Gaussian form,

$$W(r) = (3/2\pi nl^2)^{3/2} \exp(-3r^2/2nl^2) \tag{F.6}$$

Equation (F.6) obeys the normalization condition

$$\int_0^\infty W(r)4\pi r^2 dr = 1 \tag{F.7}$$

The mean-square end-to-end distance $\langle r^2 \rangle$ of the freely jointed chain is obtained as

$$\langle r^2 \rangle = \int_0^\infty r^2 W(r)4\pi r^2 dr = nl^2 \tag{F.8}$$

A more widely used quantity in rubber elasticity, and in chain statistics in general, is the characteristic ratio C_n obtained by dividing the mean-square end-to-end vector by nl^2:[2,9]

$$C_n = \langle r^2 \rangle / nl^2 \tag{F.9}$$

From eq. (F.8), the characteristic ratio for the freely jointed chain equates to unity, irrespective of the number of bonds. Thus, the freely jointed chain is the simplest model, consisting of bonds that are devoid of any correlations with one another, and all spatial arrangements correspond to the same energy. The magnitude of the characteristic ratio generally increases with increasing correlations among bonds of the chain.

F.1.2 More Realistic Pictures of the Network Chain

F.1.2.1 General Comments The freely jointed chain discussed in the preceding sections is characterized by only two constants—the number of bonds and the bond length—both of which are geometrical in nature. For this reason, this model does not reflect the consequences of the local chemical structure of the chain. A more realistic expression for the end-to-end chain-length distribution is the Hermite series, expressed in terms of the even moments $\langle r^n \rangle$ of the end-to-end vector. The rotational isomeric state (RIS) scheme allows for easy calculation of the even moments for any given chain of known chemical structure. The distribution function is[2]:

$$W(r) = \left(\frac{2\langle r^2\rangle_0}{3\pi}\right)^{-3/2} \exp\left(-\frac{3r^2}{2\langle r^2\rangle_0}\right)$$

$$\left[(1 + 3\cdot 5g_4 + 3\cdot 5\cdot 7g_6 + 3\cdot 5\cdot 7\cdot 9g_8 + \cdots)\right.$$

$$- (30g_4 + 315g_6 + 3780g_8 + 51975g_{10} + \cdots)\frac{r^2}{\langle r^2\rangle_0} \quad \text{(F.10)}$$

$$+ (9g_4 + 189g_6 + 3402g_8 + \cdots)\frac{r^4}{\langle r^2\rangle_0^2}$$

$$\left. - (27g_6 + 972g_8 + \cdots)\frac{r^6}{\langle r^2\rangle_0^3} + \cdots\right]$$

where

$$g_{2p} = 2^{-p}\left\{-\frac{\Delta_4}{2!(p-2)!} + \frac{\Delta_6}{3!(p-3)!} - \cdots + \frac{(-1)^{p-1}}{p!}\Delta_{2p}\right\} \quad \text{(F.11)}$$

with

$$\Delta_{2p} = 1 - \left[\frac{3^p}{3\cdot 5\cdot 7\cdots(2p+1)}\right]\frac{\langle r^{2p}\rangle_0}{\langle r^2\rangle_0^p} \qquad p > 0 \quad \text{(F.12)}$$

F.1.2.2 The Long-chain Approximation: The Gaussian Chain Calculations show that for sufficiently long chains, the ratios $\langle r^{2p}\rangle_0/\langle r^2\rangle_0^p$ in eq. (F.12) become negligibly small compared with unity. Thus, all the g_{2p} terms vanish in eq. (F.11) and the distribution function reduces to the well-known Gaussian form:

$$W(\mathbf{r}) = \left(\frac{2\langle r^2\rangle_0}{3\pi}\right)^{-3/2} \exp\left(-\frac{3r^2}{2\langle r^2\rangle_0}\right) \quad \text{(F.13)}$$

$$A(\mathbf{r}) = A(T) + \frac{3kT}{2\langle r^2\rangle_0}r^2 \quad \text{(F.14)}$$

In the case of the freely jointed chain, eq. (F.13) reduces to eq. (F.6) by virtue of eq. (F.9). Shorter chains require the use of eq. (F.10), in which the number of required terms containing higher order moments increases as the length decreases.

F.1.2.3 The Fixman-Alben Distribution[10] A complete series expression for the distribution function becomes prohibitively complicated as the chain length decreases. The following approximate but simple expression[11] for short chains was originally given by Fixman and Alben[10]:

$$W(\mathbf{r}) \propto \exp\left[-\frac{3}{2}\frac{r^2}{\langle r^2\rangle_0} - \left(\frac{3br^2}{\langle r^2\rangle_0}\right)^q\right] \tag{F.15}$$

The quantities b and q are parameters to be adjusted to give the best fit to actual distribution functions for short chains, for example, those given by Monte Carlo simulations.

F.2 Transformation of Distribution Functions under Deformation

Let the deformation be represented by the deformation gradient matrix, relating the end-to-end vector $\mathbf{r}_0$ of chain vectors in the undeformed state to $\mathbf{r}$ in the deformed state, by $\mathbf{r} = \lambda \mathbf{r}_0$. The probability distribution of $\mathbf{r}$ is denoted by $W_0(\mathbf{r}_0)d\mathbf{r}_0$ in the undeformed state and by $W(\mathbf{r})d\mathbf{r}$ in the deformed state. The differentials $d\mathbf{r}_0$ and $d\mathbf{r}$ represent the volume elements $dx_0 dy_0 dz_0$ and $dxdydz$, respectively, and are related to each other by

$$d\mathbf{r} = \det(\lambda)d\mathbf{r}_0 \tag{F.16}$$

where det is the determinant of the deformation gradient tensor, and is equal to the ratio of the volume element in the deformed state to that in the undeformed state. The probability distributions are transformed by the deformation as

$$W_0(\mathbf{r}_0)d\mathbf{r}_0 \xrightarrow{\lambda} W(\mathbf{r})d\mathbf{r} \tag{F.17}$$

The distribution function in the final state may be written in terms of the distribution function in the undeformed state as

$$W(\mathbf{r})d\mathbf{r} = (\det\lambda)^{-1} W^0(\lambda^{-1}\mathbf{r})d\mathbf{r} \tag{F.18}$$

where λ^{-1} is the inverse of λ. In terms of the original coordinates, the final form of the distribution function is

$$W(\mathbf{r})d\mathbf{r} = (\det\lambda) W(\lambda \mathbf{r}_0)d\mathbf{r}_0 \tag{F.19}$$

References

(1) Volkenstein, M. V. *Configurational Statistics of Polymer Chains*. Interscience: New York. 1963.
(2) Flory, P. J. *Statistical Mechanics of Chain Molecules*. Interscience: New York. 1969.
(3) Treloar, L. R. G. *The Physics of Rubber Elasticity*, 3rd ed. Clarendon Press: Oxford. 1975.
(4) Mark, J. E.; Erman, B. *Rubberlike Elasticity. A Molecular Primer*. Wiley-Interscience: New York. 1988, p. 196.
(5) Boyd, R. H.; Phillips, P. J. *The Science of Polymer Molecules*. Cambridge University Press: Cambridge. 1993.
(6) Grosberg, A. Y.; Khokhlov, A. R. *Statistical Physics of Macromolecules*. American Institute of Physics: Woodbury, NY. 1994.
(7) Kuhn, W.; Grün, F. *Kolloid-Z.* 1942, 101, 248.

(8) Kuhn, W.; Kuhn, H. *Helv. Chim. Acta* 1943, 26, 1394.
(9) Mattice, W. L.; Suter, U. W. *Conformational Theory of Large Molecules.* Interscience: New York. 1994.
(10) Fixman, M.; Alben, R. *J. Chem. Phys.* 1973, 58, 1553.
(11) Erman, B.; Mark, J. E. *J. Chem. Phys.* 1988, 89, 3314.

G. Fortran Program for Monte Carlo Calculations

G.1 Program (Calculation of Persistence Lengths and Mean-Square End-To-End Distances)

```
      program monte carlo
      dimension e(3,3),el(3),fi(200),rtr(200,3)
      dimension fi(200),t(200,3,3),prod(200,3,3)
      dimension tt(200,3,3)
      dimension delr(3)
```
c***
c   NOTATION FROM:"FOUNDATIONS OF ROTATIONAL ISOMERIC STATE THEORY
c   AND GENERAL METHODS FOR GENERATING CONFIGURATIONAL AVERAGES",
c   Flory, P. J. *Macromolecules* **1974** 7, 381.
c
c   description of variables
c
c   maximum number of bonds that may be treated: 200
c
c   e(3,3):3x3 unit matrix
c   el(3): column vector, showing the x, y and z components of the bond vector
c   in the local coordinate system
c   fi(200): torsional angle for the bonds
c   rtr(200,3): the end to end vector between the zeroth atom and each atom
c   on the chain. First index is the atom number, second is the
c   component
c   fi1,fi2,fi3: three isomeric state angles
c   p(i): equilibrium probabilities of the three states
c   t(200,3,3): transformation matrix, separately identified for each bond
c   tt(i,j,k): product of transformation matrices, ending at the i'th bond
c   mc: number of Monte Carlo runs
c   idum: any positive integer for seed for generating a random number
c   theta: complementary bond angle

```
c      bond: length of bond
c      n: number of bonds (odd number in order to define the middle bond)
c      the chain starts with zeroth atom and ends with n'th atom.
c      pt, pgp, pgm: probabilities of trans, gauche+ and gauche- states
c      delr(j): components of the middle bond in the reference coordinate system
c*****************************************************************************
c      opens the file named MC.dat to read the required variables
       open (1,file ='MC.dat')
       read (1,*) idum,mc
       read (1,*) n,theta,el(1),el(2),el(3)
       read (1,*) fi1,fi2,fi3
       read (1,*) pt,pgp,pgm
       close (1)

       zzz=mc
c      opens the file named out MC to read the required variables
       open (2,file ='out MC', status='new')
       write (2,*)' '
       bond=(el(1)**2+el(2)**2+el(3)**2)**.5
       write(2,*) 'number of bonds  = ',n
       write(2,*) 'bond angle  = ',theta
       write(2,*) 'bond length  =',bond
       write(2,*) 'locations of isomeric minima =',fi1, fi2, fi3
       write(2,*) 'trans probability  = ',pt
       write(2,*) 'gauche+ probability  = ',pgp
       write(2,*) 'gauche- probability  = ',pgm
       write(2,*) 'averages are performed over ',mc,' Monte Carlo chains'

       pi=3.141592654
       theta=theta*pi/180.
       alfa=cos(theta)
       beta=sin(theta)
       fi1=fi1*pi/180.
       fi2=fi2*pi/180.
       fi3=fi3*pi/180.

       mid=(n+1)/2
       sumpr=0.
       rsumx=0.
       rsumy=0.
       rsumz=0.
       rsq=0.

       do 4321 ijk=1,mc
       do 10 i=1,3
       do 1156 j=1,3
1156      e(i,j)=0.
10     e(i,i)=1.
       fi(1)=0.
       pttt=pt+pgp
```

```
        do 40 i=2,n-1
        call ran2(idum,ran)
        fi(i)=fi3
        if(ran.lt.pt) fi(i)=fi1
40      if(ran.gt.pt.and.ran.lt.pttt) fi(i)=fi2
c       the state of each bond is generated at the end of this step
c***************************************************************************
c       generation of the transformation matrix for each bond
        do 750 i=1,n-1
        fii=fi(i)
        t(i,1,1)=alfa
        t(i,1,2)=beta
        t(i,1,3)=0.
        t(i,2,1)=beta*cos(fii)
        t(i,2,2)=-alfa*cos(fii)
        t(i,2,3)=sin(fii)
        t(i,3,1)=beta*sin(fii)
        t(i,3,2)=-alfa*sin(fii)
        t(i,3,3)=-cos(fii)
750     continue
c***************************************************************************
c       forming the matrix products
c       prod(i,j,k) is the product of transformation matrices in eq 14
c       up to i'th term
        do 5000 k=1,3
        do 5000 l=1,3
5000       prod(1,k,l)=t(1,k,l)
        do 5010   i=2,n-1
        do 5020   k=1,3
        do 5020   l=1,3
        qq=0.
        do 5030   m=1,3
5030       qq=qq+prod(i-1,k,m)*t(i,m,l)
5020       prod(i,k,l)=qq
5010       continue
c***************************************************************************
c       calculating the position vector of each backbone atom relative to
c       the laboratory fixed coordinate system affixed to the first bond
        rtr(1,1)=el(1)
        rtr(1,2)=el(2)
        rtr(1,3)=el(3)
        do 5040 i=2,n
        do 5040 k=1,3
        rtr(i,k)=rtr(i-1,k)+prod(i-1,k,1)*el(1)+prod(i-1,k,2)*el(2)
5040       rtr(i,k)=rtr(i,k)+prod(i-1,k,3)*el(3)
```

```
c***********************************************************************
      c     calculating the middle bond vector in terms of the laboratory frame
            do 9223 j=1,3
9223        delr(j) = rtr(mid,j)-rtr(mid-1,j)

c***********************************************************************
      c     calculating the (r^2 cos^2 FI) where FI is the angle between the
      c     middle bond and the chain end-to-end vector
            sum = 0.
            do 3343 j=1,3
3343        sum = sum + delr(j)*rtr(n,j)
            sumpr = sumpr + sum**2/bond**2
            sumpr = sumpr/zzz

      c     sumpr = summation of r^2*cos^2
c***********************************************************************
      c     calculating the x, y and z components of the persistence vector and
      c     the mean square end-to-end vector
            rsumx = rsumx + rtr(n,1)
            rsumy = rsumy + rtr(n,2)
            rsumz = rsumz + rtr(n,3)
            rsq = rsq + rtr(n,1)**2 + rtr(n,2)**2 + rtr(n,3)**2
4321        continue
            a1 = rsumx/zzz
            a2 = rsumy/zzz
            a3 = rsumz/zzz
            rsq = rsq/zzz
            cr = rsq/n/bond**2

c***********************************************************************
      c     calculating the front term D0 for orientation function
            d0 = (3.*sumpr/rsq-1.)/10.

            write(2,*) ' '
            write(2,*) 'x-component of the persistence vector =',a1
            write(2,*) 'y-component of the persistence vector =',a2
            write(2,*) 'z-component of the persistence vector = ',a3
            write(2,*) 'mean square end-to-end vector         =',rsq
            write(2,*) 'characteristic  ratio                 =',cr
            write(2,*) 'average of r^2 cos^2(FI)              =',sumpr
            write(2,*) 'D0                                    =', d0
4922        format(5f15.5)
            close (2)
            stop
            end
```

```
      subroutine ran2(idum,ran)
      parameter (m=714025,ia=1366,ic=150889,rm=1./m)
      dimension ir(97)
      data iff/0/
      if(idum.lt.0.or.iff.eq.0) then
      iff=1
      idum=mod(ic-idum,m)
      do 11 j=1,97
      idum=mod(ia*idum+ic,m)
      ir(j)=idum
11    continue
      idum=mod(ia*idum+ic,m)
      iy=idum
      endif
      j=1+(97*iy)/m
      if(j.gt.97.or.j.lt.1)pause
      iy=ir(j)
      ran=iy*rm
      idum=mod(ia*idum+ic,m)
      ir(j)=idum
      return
      end
```

G.2 Form of the Data Set

a seed number for random number generator number of MC trials
bonds in each chain, complementary bond angle, x, y, z components of bond vector
torsional angles for trans, gauche$^+$ and gauche$^-$
probabilities for trans, gauche$^+$ and gauche$^-$ states
53 10
51   70.53   1.   0.0   0.0
0.   120.   -120.
.4808   .2596   .2596

G.3 Output of the Program

number of bonds = 51
bond angle = 70.53
bond length = 1.000
locations of isomeric minima = .0000 120.0 -120.0
trans probability = .4808
gauche+ probability = .2596
gauche- probability = .2596
averages are performed over 10 Monte Carlo chains

x-component of the persistence vector = 3.267
y-component of the persistence vector = 1.320
z-component of the persistence vector = .4897

mean square end-to-end vector = 172.2
characteristic ratio = 3.376
average of r^2 cos^2(FI) = 26.87
D0 = -5.3185E-02

H. Some Historical Aspects

The first material ever used for its elastomeric properties was natural rubber, obtained from the latex of the tree *Hevea brasiliensis*[1-6]. This tree is indigenous to South America, particularly the Amazon valley in Brazil. Because of its susceptibility to plant disease endemic to this region, commercial production is somewhat limited. However, similar climatic conditions occur in other tropical regions, particularly in South-East Asia, where most of the world's supply of natural rubber is now obtained. The rubber particles are of micron size and are dispersed in an aqueous medium. The material obtained directly from the latex contains a number of contaminants, including carbohydrates, proteins, mineral salts, and fatty acids, some of which stabilize the rubber suspension[1]. These are removed in the various purification steps carried out after the rubber is coagulated, or precipitated, from the latex after it is harvested. This is typically done by the addition of acid to the latex.

The remainder of this appendix gives a brief overview of some of the historical milestones in the discovery and utilization of natural rubber, and of some of the synthetic elastomers[7] developed to replace it.

H.1 The Earliest History

The earliest known record of natural rubber being used because of its elastomeric properties pertains to games played during the Mayan civilization, possibly back as far as the 11th century![2,3] Such games were observed by the crew of Columbus' voyages to the New World in 1496, specifically on the islands of Trinidad and Hispaniola. In what must certainly have been one of its very first applications, it was used by the Aztec indians to form rubber balls that were then propelled by various means through a hoop or between posts at opposite ends of a playing field. From available descriptions, it was a game not unlike modern-day basketball. Natural rubber was also used by the Indian tribes and some of the Spanish colonists to waterproof articles of clothing, other fabrics, and footware. It was even used to make airtight bags for preserving food during wars or while traveling. News of this fascinating material caused considerable interest in Europe. Around 1500, Columbus himself brought samples of these bouncing balls back to the court of Queen Isabella of Spain[1]. On the basis of these dates, it appears that the events leading to the extensive awareness of natural rubber are approaching their 500th anniversary.

H.2 Natural Rubber in the Un-Cross-Linked State

In spite of these early "discoveries," these materials were treated as curiosities and there were essentially no commercial developments for nearly 250 years. In 1745, two Frenchmen named F. Fresneau and C. de la Condamine reported to the Paris Academy of Sciences on their discoveries in the Amazon valley[1-4]. They had obtained considerable information on the hevea tree, and brought back several samples of dried latex, which the natives called cauchu, or "weeping wood." This apparently revived interest in using the material, particularly as a coating for waterproofing fabrics. This application required solutions of the rubber, so considerable effort was expended in finding suitable solvents. Both turpentine and ether were used for this purpose. Because the rubber was, of course, un-cross-linked, the resulting materials had very poor properties. They were, for example, sticky in warm weather and hard and brittle when cold.

In England in 1770, J. Priestley remarked on one of the secondary but important uses of the dried latex[1-4,8]. It was found to be capable of rubbing pencil marks off of paper, a purpose for which some elastomers are still used today. Because of this capability, the material gradually became known as "rubber," a name which has also persisted to the present.

One of the earliest quantitative studies on natural rubber as an elastomer as carried out by J. Gough around this time, in 1805[9-11]. Gough, an Englishman with wide-ranging interests, carried out a variety of experiments, some of which involved the characterization of naturally occurring materials. Incredibly, these experiments were not thwarted by the fact that he had been blind since youth from an attack of smallpox. For example, he used his highly developed sense of touch to identify species of plants by touching them to his lips and tongue. He used the same technique in his studies of elastomers. He first showed that rapidly stretching a strip of rubber significantly increases its temperature, while letting it snap back decreases it. In the case of natural rubber, part of the increase in temperature was undoubtedly due to the latent heat accompanying the strain-induced crystallization, and part of the decrease was due to the latent heat of melting. The part that is more significant with regard to the mechanism of the deformation is that portion due to the decrease in the entropy of the chains caused by the stretching. In his second series of experiments, Gough demonstrated that increasing the temperature of a rubber sample increases its "stiffness," that is, increases its modulus. It thus contracted, reversibly, when heated in the stretched state under constant load. Its behavior is thus much closer to that of a gas than that of most solids. In his third series of experiments, Gough found that the density of natural rubber increased significantly when it was cooled while in the highly stretched state. He also observed that such a strip of rubber lost a good part of its ability to contract, until it was warmed. Both observations are now known to be due to the crystallization induced by the stretching process, which so decreases the configurational entropy of the polymer chains that their melting point is considerably elevated. Some related experiments were carried out by Page, in 1847[11,12]. Such investigations involving stress, strain, and temperature are called "thermoelasticity" experiments, and are discussed in detail, in the light of modern thermodynamics, by Flory[10].

Approximately 50 years later, Lord Kelvin showed that, according to thermodynamics, the results of either of Gough's first and second series of experiments preordained the results of the other. Gough's experiments were repeated in 1859 by J. P. Joule, who was able to use thermocouples to measure the temperature changes more accurately[10,11,13]. Since this was after the invention of vulcanization, described below, he was also able to use cross-linked samples, which gave more reproducible results. This work was carried out just a few years after entropy was introduced as a thermodynamic variable. Considerably later, more comprehensive stress–temperature studies were carried out on natural rubber by K. H. Meyer and C. Ferri, in 1935[14,15], and by R. L. Anthony, R. H. Caston, and E. Guth, in 1942[15,16]. In spite of the length of time that has elapsed, such "thermoelastic" experiments are still being carried out, the current interest being characterization of the thermodynamic nonideality of elastomers[10,15,17]. A less serious view of some of these early experiments is depicted in the cartoon shown in figure H.1.

In spite of the shortcomings of rubber-coated fabrics, the first patent involving rubber was issued in England in 1791, and the first rubber factory established in Vienna in 1811[1]. Other important early milestones were an 1813 U.S. patent for waterproofing shoes, and an 1823 patent to C. Macintosh in Glasgow for covering two pieces of cloth held together with a layer of solvent-deposited rubber[1] Around 1819, an English coach maker and inventor named T. Hancock began producing rubber-coated articles and invented the process of mastication, in which the molecular weight of the rubber is reduced to make it more tractable[1,2,4].

An important breakthrough on the synthesis side occurred in 1826, in England, when M. Faraday completed one of the first analyses of natural rubber. His results, upon subsequent correction for errors in the scale of atomic weights, came very close to the correct formula (C_5H_8) for the material now known to be the monomeric species[1,18]. Additional, related information was obtained by F. C. Himly, in Germany, in 1938[18]. He was able to obtain a volatile distillate from natural rubber, which was presumably the monomer or something closely related to it. A more definitive separation was obtained by an English chemist, C. G. Williams, who distilled natural rubber into three fractions, one of which had the composition C_5H_8.[1,18] Williams was the first to call this material, the monomer of natural rubber, "isoprene."

Not surprisingly, these developments were soon followed by attempts to reverse the process, to prepare the polymer from the monomer or from other low molecular-weight species. C. G. Williams was at least partially successful with his monomeric fractions in this regard. Also, in France in 1875, G. Bouchardat was able to convert isoprene in the presence of hydrogen chloride

On Oct. 23, 1927, three days after its invention,
the first rubber band is tested.

Figure H.1 "The Far Side" by Gary Larson is reprinted by permission of Chronicle Features, San Francisco, CA. All rights reserved.

gas to a rubberlike substance[1,18,19]. A few years later, W. A. Tilden in England produced isoprene by the destructive distillation of rubber. Upon storage, some of his monomer also converted into a high-molecular-weight material with some elastic properties[18,19] These were the earliest beginnings of still-continuing efforts to synthesize elastomers independently of Nature.

Some of the earliest x-ray investigations on the crystallization of polymers were carried out on natural rubber, in the 1920s[11,20]. This was presumably due not only to the commercial importance of this particular elastomer, but also to its ability to readily and conveniently undergo strain-induced crystallization. Investigations of this type gave results that transcended the field of rubberlike elasticity, and greatly helped to elucidate the nature of the crystalline state in polymers in general.

An important scientific advance with regard to the theories of rubberlike elasticity was the demonstration by H. Staudinger in the 1920s that polymers were covalently bonded long-chain molecules[21]. This was followed, on the experimental side by the demonstration that mechanical deformations of elastomers occurred at very nearly constant volume[10,15]. The very earliest suggested mechanisms for the elastic deformation were due to Ostwald, Fikentscher, and Mark, and to Mack[15]. Although discussed by Treloar for historical reasons, they were not correct and were quickly abandoned. However, K. H. Meyer, G. von Susich, and E. Valko did discover for the first time, in 1932, the correct qualitative picture for the deformation mechanism[10,15,21]. They proposed that the elastic force was due to the decreased entropy of the oriented and stretched chains, thereby explaining the thermoelastic results of Gough and Joule. They also concluded that the elastic force should be proportional to the absolute temperature. The same ideas were used by E. Karrer in 1933 to interpret the properties of muscle. W. Kuhn, in 1936, then argued that the elastic force should be proportion to the number of polymer "molecules" in the elastomer. The stage was thus set for W. Kuhn, E. Guth, and H. F. Mark to develop the first molecular theories of rubberlike elasticity. The elastic equations of state resulting from these theories were in at least approximate agreement with much of the experimental data available at that time. Also, quite remarkably, they had some features that were very similar to the equations obtained later using much more sophisticated approaches[10,15,17].

H.3 Cross-Linked Natural Rubber

In 1939, C. Goodyear in the United States sought ways to improve the properties of natural rubber, particularly its stickiness and liquidlike behavior[1,2]. In his studies, he accidently found that heating a mixture of natural rubber, sulfur, and lead carbonate converted it into an insoluble mass. It is now known that the sulfur atoms bonded or "cross-linked" the rubber chains into a network structure. The resulting material had properties that made it a vast improvement over the native un-cross-linked rubber. For this breakthrough, Goodyear was granted a U.S. patent in 1844. Hancock, in England, independently made a similar discovery, for which he coined the term "vulcanization" (after Vulcan, the Roman god of fire)[1,2,4]. Although Hancock's patent came after Goodyear's, he was much more successful in exploiting this new technology. Goodyear began to publicize and create interest in his invention by putting on rubber "shows," in London in 1851 and in Paris in 1855. He also accumulated over 200 rubber patents in his name, but had to spend much of his time and energies dealing with infringements on them. In one of the most famous of these, he was represented by the legendary D. Webster, who was the U.S. Secretary of State at the time. In spite of these efforts, he died heavily in debt, in 1860[2].

Hancock and Macintosh, on the other hand, continued finding successful applications for this new vulcanization technique. For example, they turned out rubber fire hose, which was first used in the great fire of London in 1827. In one extremely important event, in 1846, Hancock built solid rubber tires for the carriage of Queen Victoria[1]. These were clearly the predecessors of the pneumatic tires that are the basis of the entire automobile industry.

During this period, up to around 1870, there was increasing demand for more rubber for these growing industries. As a result, there was considerable expansion in the growing and harvesting of natural rubber in South America[1]. The increasing demands increased the price of the rubber, and this led to considerable economic development of the Amazon valley. It also led, however, to brutal exploitation of the natives[22]. A similarly tragic situation arose in the rubber-growing regions of Africa, particularly the Belgian Congo. It is said that the costs were one human life for every 4 kg of rubber, which is approximately the amount that goes into an automobile tire! Many rubber trees were also lost in these regions from improper tapping techniques[22].

H.4 The Plantation Movement East

There had long been interest in planting and cultivating the hevea rubber tree in places other than South America and Africa. The first attempts took place in 1873, when a London curator was able to bring some hevea seeds from Brazil back to England. A few did produce plants, but all died before they could be planted elsewhere. In 1876, however, an English coffee planter named H. Wickham was able to transport tens of thousands of the seeds to Kew Gardens in London, where many were successfully germinated into viable rubber plants[1,4]. They were sent to Ceylon, where they were used to establish the first rubber plantations in the Far East. These were then expanded to Malaya, Singapore, Indonesia, India, Java, and other countries. This permanently broke the South American and African monopoly. By 1912, the South-East Asian plantations matched those of Brazil for the first time, and 2 years later exceeded them by a factor of two. By 1922, they produced 93% of the world supply, and by 1932, 98%[1].

During the period 1880–1890, there were also important advances in the science of rubber tree cultivation. For example, the planting of the trees was carried out more methodically, and tapping methods and routines were improved to the point where trees could produce the desired latex efficiently for several decades[1].

H.5 Some Additional Scientific Developments

In 1888, a Belfast veterinarian by the name of J. Dunlop made the first pneumatic tires, for bicycles[1,4]. It is hard to overestimate the importance of this development, since the bicycle was then, and still is, the most common means of transportation in the world. The new types of tires made the bicycle even more attractive, and thus encouraged vast improvements in the road systems everywhere. Even more important is the fact that automobiles were under rapid development around then, just in time to take advantage of the improved road systems. Not surprisingly, therefore, the pneumatic tire became of central importance to the development of the automobile industry, as well as the rubber industry.

By 1888, the highly experimental motorized bicycles and tricycles had evolved into four-wheeled automobiles, which were now on sale. The first of these to be fitted with pneumatic tires was a Peugeot entered in a race in France by the Michelin brothers. Although many of these early developments of pneumatic tires took place in Europe, leadership in this field soon moved to the United States, particularly to Akron, Ohio.

In 1906, an Akron chemist named G. Oensager discovered the first of a long series of compounds that accelerate the vulcanization process[4]. Around this time, important additional work was being carried out on the synthesis of new elastomers, that is, materials with chemical structures different from that of natural rubber. For example, in 1901, I. Kondakov in Russia was able to produce a rubberlike material by heating 2,3-dimethylbutadiene in the presence of potash[18]. His compatriot S. V. Lebedev had some success in polymerizing butadiene in 1910. Also in 1910, University of Kiel chemistry professor C. D. Harries prepared some synthetic rubber by polymerizing 2,3-dimethylbutadiene[18,19]. Both a hard rubber type, called "Methyl H," and a soft rubber type, called "Methyl W," were produced. Similar work was being carried out in the United States, at the Hood Rubber Company and the Diamond Rubber Company. Several million tons of these materials were produced in Germany by the end of World War I.

In related work, in 1927, K. Ziegler and chemists at the Bayer company in Germany developed another synthetic rubber called BuNa–S (for a butadiene–styrene copolymer polymerized using sodium, Na)[7,19]. There was also a BuNa–N rubber which contained acrylonitrile instead of styrene, making it more polar and therefore more resistant to oils and other hydrocarbons. This material was introduced in the United States in 1937. Around this time, inorganic materials such as carbon black were first used to provide reinforcement for elastomers.

Also in the United States, in 1929, polysulfide rubber was prepared accidently by the hydrolysis of ethylene chloride in a sodium polysulfide solution. Two years later, Du Pont announced the commercial availability of a polychloroprene rubber called "Neoprene." The synthesis of this material was the result of efforts by their polymer chemists W. H. Carrothers, J. A. Nieuwland, and E. K. Bolton[4,7,18,19]. The same elastomer was produced in the former Soviet Union under the name "Sovprene."

Polyisobutylene was first prepared by R. M. Thomas and W. J. Sparks of the Standard Oil Company, in 1937[7,19,23]. The modification prepared by copolymerizing isobutylene with a comonomer providing unsaturated groups for cross-linking was called "butyl rubber," and was first marketed in 1943. Its most remarkable characteristic is its extremely low permeability to gases. It was therefore much used for inner tubes in tires, and is now used as one of the layers in tubeless tires. Two other elastomers should be mentioned, both being important because their essentially unsaturated structures make them much less susceptible to degradation than the diene elastomers. The first class is the polysiloxanes, with Si — O backbones, which give them superb thermal stability[24,25]. The other type is the ethylene–propylene terpolymers, which consist primarily of ethylene and propylene with just enough of a third comonomer to provide some unsaturated sites to facilitate cross-linking[7,19,26].

During the 1940s, there were also important new developments with regard to the theories of rubberlike elasticity[10,15,17,27]. H. M. James and E. Guth developed what has become known as the theory of "phantom networks," so named because the chains were visualized as being able to pass through one another. Soon thereafter, the alternative "affine network" theory was proposed by P. J. Flory, J. Rehner, and F. T. Wall. Both theories are discussed in detail in chapter 2. Attempts to reconcile and significantly extend these theories were not very successful until the 1970s. The descriptions of these modern theories make up a very significant part of this book.

H.6 The Effects of World War II

By 1940, war already seemed imminent and the United States was stockpiling natural rubber, effectively doubling its normal imported amount of around a half-million tons a year[1,4]. In 1941, Japan occupied South-East Asia, cutting off supplies of natural rubber to the United States. In its first response to this supply crisis, the U.S. government ordered the planting of tens of thousands of acres of guayule. This shrub, which thrives in the western parts of the United States and in Mexico, also contains rubber latex. It has the disadvantage of yielding the rubber only with difficulty; the plant must be ground up and extracted, thus requiring a constant supply of new plants.

In its second response, the government orchestrated the gearing up of U.S. industry to produce various synthetic rubber substitutes[4,7,28]. The impetus for this colossal project came from B. Baruch, J. B. Conant, and K. Compton, who convinced President Roosevelt of the desperate nature of the situation. In one of the most remarkable success stories in modern industry, a capacity of 1.5 billion pounds a year was reached in 1945!

H.7 The Postwar Period

At this point, the South-East Asian rubber plantations were recovered from the Japanese[1]. Although considerable rehabilitation work was required, the plantations were eventually producing at their normal levels, dominating the world market in natural rubber. One additional scientific development was the discovery of the extent to which elastomeric materials can be degraded by reaction with substances such as ozone. This led directly to the development of organic antidegradants, particularly antiozonants[4].

One of the greatest challenges in synthesizing natural rubber itself, in the laboratory, is to have all of the monomer units add to the chain in exactly the same way. Specifically, both double bonds in the monomer have to be opened ("1,4-addition"), and the new double bond thus formed in the repeat unit has to have the "*cis*" configuration[7]. Natural rubber is essentially pure *cis*-1,4-polyisoprene, and, because of this structural regularity, it can undergo strain-induced crystallization. This is a very important advantage with regard to mechanical properties, since the crystallites thus produced greatly reinforce the elastomer[10,15,17], as described in chapter 12. Polyisoprenes that have monomer additions other than the 1,4 variety, or have mixtures of *trans* and *cis* placements, do not have this capability for self-reinforcement. One of the triumphs of modern synthetic polymer chemistry is the capability, demonstrated relatively recently, of being able to produce diene polymers having structures rivaling those of natural rubber in their regularity.

Shortly after the war, the threat to natural rubber posed by the synthetic elastomers quickly became more pronounced[1]. Some of these synthetic materials have novel and very attractive properties. For example, triblock copolymers were developed in which the end blocks aggregated into hard glassy domains that could function as temporary cross-links for the elastomeric central block[29-31] These "thermoplastic elastomers" have the tremendous advantage of reprocessability.

By 1973, U.S. consumption of natural rubber had reached a low of 22% of all rubber used. Nonetheless, it has always been the elastomer of choice for some, highly specialized applications. A modern-day example is its very extensive use in airplane tires, and it is also much used in blends with other elastomers. The competitive situation with regard to natural rubber started to reverse, however, because of the petroleum shortage, concerns about environmental pollution from the oil industry, and worries about the depletion of nonrenewable resources[1]. For example, it takes 3–6 tons of crude oil to produce 1 ton of synthetic rubber. The efficiency of producing natural rubber has also been greatly increased, particularly by horticultural breeding techniques. There has even been development of artificial stimulants to enhance the flow of latex during the tapping procedure[1]. Rapid strides are also being made in understanding the enzymatic biosynthesis of natural rubber in the *Hevea braziliensis* tree, and in the application of modern techniques of genetic engineering[32]. Another intriguing experiment is the separation of enzymes from the natural rubber latex, and using them in the in vitro synthesis of natural rubber[18]. For a variety of reasons, such as these, natural rubber made up, in 1990, 37% of the world production of elastomers of all types. Of this total production (25 billion pounds), the United States consumes 6.4 billion pounds, 27% of which is natural rubber.

Much hinges on the fortunes of the natural rubber industry[4,33-35]. It both provides a critically important raw material to the industrially advanced countries and represents a major element in the economies of a number of developing countries. How much of its market will be permanently lost to synthetic materials is not obvious at this point, and remains a question of considerable importance.

REFERENCES

(1) Hurley, P. E. *J. Macromol. Sci., Chem.* 1981, A15, 1279.
(2) Moynihan, J. *Today's Chemist* 1989, October, 14.
(3) Kauffman, G. B.; Seymour, R. B. *J. Chem. Ed.* 1990, 67, 422.
(4) Stinson, S. *Chem. Eng. News* 1990, May 21, 45.
(5) Semegen, S. T. In *Rubber Technology*, M. Morton, Ed. Van Nostrand Reinhold: New York. 1973; p. 152.
(6) Brydson, J. A. *Rubber Chemistry*. Applied Science: London. 1978.
(7) Lal, J. In *Encyclopedia of Polymer Science and Engineering*. Wiley: New York. 1989; p. 106.
(8) Reese, K. M.; Eichinger, B. E. *Chem. Eng. News* 1983, December 19, 78.
(9) Gough, J. *Proc. Lit. Philos. Soc., Manchester, 2nd Ser.* 1805, 1, 288.
(10) Flory, P. J. *Principles of Polymer Chemistry*. Cornell University Press: Ithaca, NY. 1953.
(11) Mandelkern, L. *Rubber Chem. Technol.* 1993, 66, G61.
(12) Page, C. G. *Am. J. Sci.* 1847, 54, 341.
(13) Joule, J. P. *Trans. Roy. Soc. (London)* 1859, A149, 91.
(14) Meyer, K. H.; Ferri, C. *Helv. Chim. Acta* 1935, 18, 570.

(15) Treloar, L. R. G. *The Physics of Rubber Elasticity*, 3rd ed. Clarendon Press: Oxford. 1975.
(16) Anthony, R. L.; Caston, R. H.; Guth, E. *J. Phys. Chem.* 1942, 46, 826.
(17) Mark, J. E.; Erman, B. *Rubberlike Elasticity. A Molecular Primer.* Wiley-Interscience: New York. 1988.
(18) Anderson, K. J. *MRS Bull.* 1990, June, 71.
(19) Kauffman, G. B.; Seymour, R. B. *J. Chem. Ed.* 1991, 68, 217.
(20) Katz, J. R. *Naturwissenshaften* 1925, 13, 900.
(21) Morawetz, H. *Polymers: The Origins and Growth of a Science.* Wiley-Interscience: New York. 1985.
(22) Mason, P. *Cauchu. The Weeping Wood.* Australian Broadcasting Commission: Sydney. 1979.
(23) Zapp, R. L.; Hous, P. In *Rubber Technology*, M. Morton, Ed. Van Nostrand Reinhold: New York. 1973; p. 249.
(24) Zeigler, J. M.; Fearon, F. W. G., Eds. *Silicon-Based Polymer Science. A Comprehensive Resource.* American Chemical Society: Washington, DC. 1990.
(25) Warrick, E. L. *Forty Years of Firsts. The Recollections of a Dow Corning Pioneer.* McGraw-Hill: New York. 1990.
(26) Borg, E. L. In *Rubber Technology*, M. Morton, Ed. Van Nostrand Reinhold: New York. 1973; p. 220.
(27) Mark, J. E.; Erman, B., Eds. *Elastomeric Polymer Networks.* Prentice Hall: Englewood Cliffs, NJ. 1992.
(28) Morris, P. J. T. *The American Synthetic Rubber Research Program.* University of Pennsylvania Press: Philadelphia, PA. 1989.
(29) Morton, M., Ed. *Rubber Technology.* Van Nostrand Reinhold: New York. 1973.
(30) Legge, N. R.; Holden, G.; Schroeder, H. E., Eds. *Thermoplastic Elastomers. A Comprehensive Review.* Oxford University Press: New York. 1988.
(31) Seymour, R. B.; Kauffman, G. B. *J. Chem. Ed.* 1992, 69, 967.
(32) Dennis, M. S.; Henzel, W. J.; Bell, J.; Kohr, W.; Light, D. R. *J. Biol. Chem.* 1989, 264, 18618.
(33) Greek, B. F. *Chem. Eng. News* 1991, May 13, 37.
(34) Reisch, M. S. *Chem. Eng. News* 1993, May 10, 24.
(35) Stein, R. S., Ed. *Polymer Science and Engineering. The Shifting Research Frontiers*, National Academy Press: Washington, DC. 1994.

Selected General Bibliography

The following is a chronological list of some important publications regarding rubberlike networks.

(1) Flory, P. J. *Principles of Polymer Chemistry*. Cornell University Press: Ithaca, NY. 1953. One of the most authoritative texts on polymers, in general, with extensive coverage of rubberlike elasticity. A classic which is considered the bible of polymer science by many, and is still in print despite its age. Rigorous, with derivations, and much physical insight.

(2) Meares, P. *Polymers: Structure and Bulk Properties*. Van Nostrand: New York. 1965. Good coverage of structure-property relationships of polymers in the bulk state. Exceptionally good for mechanical properties, including rubberlike elasticity. Does include some derivations.

(3) Dusek, K.; Prins, W. *Adv. Polym. Sci.* 1969, 6, 1. A review article that is really a short book on the structure of networks and rubberlike elasticity. Very comprehensive.

(4) Chompff, A. J.; Newman, S., Eds. *Polymer Networks: Structure and Mechanical Properties*. Plenum Press: New York. 1971. Proceedings of an ACS Symposium on "Highly Cross-Linked Polymer Networks," and thus covers thermosets as well as elastomers.

(5) Smith, K. J., Jr. In *Polymer Science*, A. D. Jenkins, Ed. North-Holland: Amsterdam. 1972. Book chapter on rubberlike elasticity that is obviously out of date, but still very useful.

(6) Morton, M., Ed. *Rubber Technology*. Van Nostrand Reinhold: New York. 1973. Several general chapters on applied subjects, followed by more detailed chapters on a number of important commercial elastomers.

(7) Wall, F. T. *Chemical Thermodynamics*, 3rd ed. Freeman: San Francisco, CA. 1974. The chapter on "Statistical Thermodynamics of Rubber" is excellent from the pedagogic point of view.

(8) Dunn, A. S., Ed. *Rubber and Rubber Elasticity*. Wiley-Interscience: New York. 1974; Vol. Polymer Symposium 48. Proceedings of a Manchester Symposium honoring the late L. R. G. Treloar. Twelve articles on various aspects of rubberlike elasticity.

(9) Treloar, L. R. G. *The Physics of Rubber Elasticity*, 3rd ed. Clarendon Press: Oxford. 1975. One of the standard books on rubberlike elasticity for the last several decades. Earlier editions have interesting information on crystallization in elastomers.

(10) Labana, S. S., Ed. *Chemistry and Properties of Crosslinked Polymers*. Academic Press: New York. 1977. Proceedings of an ACS Symposium, similar to the book cited in number 4.

(11) Eirich, F. R., Ed. *Science and Technology of Rubber*. Academic Press: New York. 1978. Wide coverage of elastomers, from polymerization and vulcanization, through rubberlike elasticity and dynamical mechanical properties, to the manufacture of tires.

(12) Brydson, J. A. *Rubber Chemistry*. Applied Science: London. 1978. Very broad survey, as in the book cited in number 11, but has much more information on specific elastomers.

(13) Nash, L. K. *J. Chem. Ed.* 1979, 56, 363. Very detailed derivation of an elastic equation of state, using the simplest possible approach.

(14) Mark, J. E. *J. Chem. Ed.* 1981, 898. Semiquantitative discussion of basic topics, plus some more advanced areas that were current at that time.

(15) Mark, J. E.; Lal, J., Eds. *Elastomers and Rubber Elasticity*. American Chemical Society: Washington, DC. 1982; Vol. 193. Proceedings of an ACS Symposium, partly on the preparation of elastomers, but mostly on their characterization.

(16) Eichinger, B. E. *Ann. Rev. Phys. Chem.* 1983, 34, 359. Thermodynamics and continuum mechanics used to cover the subject "The Theory of High Elasticity." Very useful descriptions of unsolved problems, particularly in the area of theory.

(17) Labana, S. S.; Dickie, R. A., Eds. *Characterization of Highly Cross-Linked Polymers*, American Chemical Society: Washington, DC. 1984. An ACS Symposium proceedings that are a sequel to the book cited in number 10.

(18) Furukawa, J. *Makromol. Chem. Suppl.* 1985, 14, 3. Review of interpretative work carried out using a "pseudo cross-link" model.

(19) Kilian, H. G.; Enderle, H. F.; Unseld, K. *Coll. Polym. Sci.* 1986, 264, 866. Review of interpretative work carried out using a van der Waals model.

(20) Lal, J.; Mark, J. E., Eds. *Advances in Elastomers and Rubber Elasticity*. Plenum Press: New York. 1986. Proceedings of an ACS Symposium that is a sequel to the book cited in number 15.

(21) Queslel, J. P.; Mark, J. E. In *Encyclopedia of Polymer Science and Engineering, Second Edition*, R. A. Meyers, Ed. Wiley-Interscience: New York. 1986; p. 365. Coverage of a wide variety of topics, including some ASTM test methods.

(22) Singler, R. E.; Byrne, C. A., Eds. *Elastomers and Rubber Technology*. U.S. Government Printing Office: Washington, DC. 1987. Proceedings of a Sagamore Army Materials Research Conference which emphasized some of the more applied topics involving rubberlike elasticity.

(23) Queslel, J. P.; Mark, J. E. In *Encyclopedia of Physical Science and Technology*, R. A. Meyers, Ed. Academic Press: New York. 1987. Extensive survey, with considerable coverage of relevant chain statistics.

(24) Queslel, J. P.; Mark, J. E. *J. Chem. Ed.* 1987, 64, 491. Pedagogic treatment that is a sequel to the review article cited in number 14.

(25) Edwards, S. F.; Vilgis, T. A. *Rep. Prog. Phys.* 1988, 51, 243. Sophisticated review of the tube model of rubberlike elasticity.

(26) Kramer, O., Ed. *Biological and Synthetic Polymer Networks*. Elsevier: London. 1988. Proceedings of the Nertworks 86 Meeting held in Elsinore, Denmark, with some emphasis on biological networks and on swelling.

(27) Mark, J. E.; Erman, B. *Rubberlike Elasticity. A Molecular Primer*. Wiley-Interscience: New York. 1988. Introductory textbook, giving a broad overview and emphasizing physical concepts.

(28) Heinrich, G.; Straube, E.; Helmis, G. *Adv. Polym. Sci.* 1988, 85, 33. Review and analysis of the state of the statistical mechanics of rubberlike elasticity in 1988.

(29) Erman, B.; Mark, J. E. *Ann. Rev. Phys. Chem.* 1989, 40, 351. Aspects of elastomers and elasticity likely to be of interest to physical chemists desiring a brief overview.

(30) Queslel, J. P.; Mark, J. E. In *Comprehensive Polymer Science*, G. Allen, Ed. Pergamon: Oxford. 1989; p. 271. Review article on the physical properties of networks, written in a concise, encyclopedic style.

(31) Mark, J. E. *Kautschuk + Gummi Kunstoffe* 1989, 42, 191. Semiquantitative discussion of some of the molecular theories of rubberlike elasticity.

(32) Queslel, J. P.; Mark, J. E. In *Determination of Molecular Weight*, A. R. Cooper, Ed. Wiley-Interscience: New York. 1989; p. 487. Mechanical and solvent-swelling methods for characterizing insoluble polymers.

(33) Mark, J. E. In *Frontiers of Macromolecular Science*, T. Saegusa, T. Higashimura, and A. Abe, Eds. Blackwell Scientific: Oxford. 1989; p. 289. Review article emphasizing model networks and novel reinforcement techniques.

(34) White, J. L.; Murakami, K., Eds. *International Seminar on Elastomers, Proceedings.* Wiley: New York. 1989. Wide variety of topics covered at an international seminar.

(35) Baumgartner, A.; Picot, C. E. *Springer Proceedings in Physics.* Springer: Berlin. 1989; Vol. 42. Proceedings of a workshop held in Julich, Germany, which was arranged to provide for discussions between theoreticians and experimentalists interested in the physics of polymer networks.

(36) Burchard, W.; Ross-Murphy, S. B., Eds. *Physical Networks. Polymers and Gels.* Elsevier: London, 1990. Proceedings of the Networks 88 Meeting held in Freiburg, Germany, with considerable coverage of temporary, reversible gels.

(37) Queslel, J. P.; Mark, J. E. In *Encyclopedia of Physical Science and Technology*, 2nd ed. R. A. Meyers, Ed. Academic Press: New York. 1991. Updating of the article cited in number 23.

(38) Mark, J. E.; Erman, B., Eds. *Elastomeric Polymer Networks*, Prentice Hall: Englewood Cliffs, NJ, 1992. Book dedicated to the memory of Eugene Guth, and thus with an emphasis on contributions from many of his colleagues and areas in which he contributed.

(39) Mark, J. E. *Comput. Polym. Sci.* 1992, 2, 135. Survey of applications of molecular modeling to problems in the area of elastomeric polymer networks.

(40) Mark, J. E. *Angew. Makromol. Chemie* 1992, 202/203, 1. Recent advances in both the theoretical and experimental sides of rubberlike elasticity.

(41) Gent, A. N., Ed. *Engineering with Rubber. How to Design Rubber Components.* Hanser: New York. 1992. A book intended to teach the beginning engineer the principles of rubber science and technology, with a great deal of practical information.

(42) Aharoni, S. M., Ed. *Synthesis, Characterization, and Theory of Polymeric Networks and Gels.* Plenum Press: New York. 1992. Proceedings of an ACS Symposium, with some sections on unusual topics, such as fractal aspects of networks and gels, and networks consisting of rigid or semiflexible chains.

(43) Mark, J. E.; Eisenberg, A.; Graessley, W. W.; Mandelkern, L.; Samulski, E. T.; Koenig, J. L.; Wignall, G. D. *Physical Properties of Polymers*, 2nd ed. American Chemical Society: Washington, DC. 1993. Textbook-type treatment of a variety of topics, including rubberlike elasticity and viscoelasticity.

(44) Clarson, S. J.; Mark, J. E. In *Siloxane Polymers*, S. J. Clarson and J. A. Semlyen, Eds. Prentice Hall: Englewood Cliffs, NJ. 1993; p. 616. Review of polysiloxanes as elastomers.

(45) Mark, J. E.; Erman, B.; Eirich, F. R., Eds. *Science and Technology of Rubber*, 2nd ed. Academic Press: New York. 1994. Update of the book cited in number 11.

(46) Mark, J. E.; Erman, B. In *Encyclopedia of Advanced Materials*, D. Bloor, Ed. Pergamon: Oxford. 1994; p. 1739. Descriptions of some types of elastomers in the category of advanced materials.

(47) Mark, J. E.; Erman, B. In *Polymer Networks*, R. F. T. Stepto, Ed. Blackie Academic, Chapman & Hall: Glasgow, 1997; in press. Essentially an update of the review cited in number 29.

(48) Mark, J. E.; Erman, B. In *Performance of Plastics*, W. Brostow, Ed. Hanser: Cincinnati, 1997; in press. Similar to the article cited in number 47, but with more of an emphasis on practical aspects of rubberlike elasticity.

Author Index

A selected listing of authors most directly involved with elastomers and rubberlike elasticity

Abe, A. 270, 304, 353
Abe, Y. 127, 128, 132, 133
Aharoni, S. M. 86, 165, 185, 259, 264, 353
Aklonis, J. J. 265, 301
Aksay, I. A. 263, 265, 295, 300, 306
Alamo, R. G. 85, 264, 270, 303
Alben, R. 100, 105, 199, 218, 332, 334
Alfrey, T. 179, 187
Allcock, H. R. 189, 217, 290, 305
Allegra, G. 23, 32, 180, 187
Allen, G. 21, 50–52, 109, 114, 115, 117, 121, 131, 166, 186, 223, 234
Amis, E. J. 21, 223, 231, 234
Andrady, A. L. 134, 141–145, 147, 149, 182, 183, 187, 188, 192, 193, 196, 199, 207, 210, 216, 217, 218, 236, 253, 263

Bahar, I. 58–60, 70, 104, 105, 134, 148, 153, 154–157, 159, 160, 164, 166, 168, 173, 175–178, 186, 189, 213, 217
Ball, R. C. 33–35, 37, 42, 61, 70, 224, 230, 234
Bastide, J. 62, 70, 223, 231–234
Baumgaertner, A. 234, 353
Beltzung, M. 21, 223, 233, 234
Benoit, H. 8, 21, 134, 146, 149, 220, 223, 224, 233
Bianconi, P. A. 105, 265, 267, 269, 300, 301, 302
Biggs, W. D. 262
Binder, K. 104, 106
Boerio, F. J. 134, 147, 189, 198, 217

Bokobza, L. 134, 148, 153, 157–161, 164, 189, 217
Boonstra, B. B. 265, 266, 299
Boue, F. 62, 70, 223, 231–233, 234
Boyer, R. F. 108, 131
Brennan, A. B. 265, 267, 269, 270, 288, 290, 300, 302, 303, 305
Brinker, C. J. 265, 268, 270, 290, 300, 303–305
Brochard, F. 62, 71
Brostow, W. 354
Brotzman, R. W. 61, 70
Bruzzone, M. 187
Brydson, J. A. 341, 349, 352
Buckley, G. S. 134, 147, 189, 212, 217
Bucknall, C. B. 216, 219
Bueche, F. 188, 216, 267, 302
Burchard, W. 62, 71, 235, 257, 258, 262, 263, 353

Calvert, P. 235, 238, 240, 263, 265, 290, 300, 301, 305
Candau, S. 62, 70, 161, 164, 233
Candia, F. 116, 120, 122, 132
Case, L. C. 309, 310
Cassasa, E. F. 68, 72
Chompff, A. J. 267, 302, 351
Ciferri, A. 108, 110, 118, 120, 121, 123, 124, 131–133, 187
Clarson, S. J. 66, 70, 72, 134, 142, 147, 149, 189, 195, 211, 217, 218, 267, 269, 275, 288, 303, 304, 305, 353

AUTHOR INDEX

Cohen, C. 83, 134, 137, 148
Colby, R. H. 38, 42, 134, 137, 148, 268, 302
Coran, A. Y. 307, 310
Cotton, J. P. 8, 21, 220, 223, 233
Curro, J. G. 55, 70, 94, 95, 96, 97, 100, 102, 105, 200, 218

de Gennes, P. G. 55, 62, 70, 141, 149, 166, 186, 223, 233
Deam, R. T. 8, 16, 19, 21, 34, 35, 37, 42, 224, 234
DeBolt, L. C. 66–69, 72, 94, 105, 126, 129, 133, 240, 244–246, 251, 252, 263, 264
Deloche, B. 157, 161, 163, 164
des Cloizeaux 8, 21, 223, 233
Dickie, R. A. 94, 105, 134, 145, 149, 352
DiMarzio, E. A. 166, 186
Doi, M. 33–35, 37, 42, 85
Donald, A. M. 86, 219
Dondos, A. 134, 146, 149
Donnet, J. B. 265, 267, 299, 301, 302
Dubault, A. 157, 161, 163, 164
Duiser, J. 21, 308, 310
Duplessix, R. 62, 70, 220, 224, 233
Dusek, K. 73, 83, 118, 132, 137, 148, 351

Edwards, S. F. 8, 10, 16, 18, 21, 23, 32–35, 37, 42, 61, 70, 137, 148, 165, 185, 224, 234, 352
Eichinger, B. E. 12, 16, 21, 23, 32, 61, 68, 70, 72, 85, 134, 137, 148, 324, 329, 342, 349, 352
Eirich, F. R. 265, 299, 307, 310, 311, 314, 352, 354
Eisenbach, C. D. 147
Eisenberg, A. 353
Erman, B. 7, 13, 20, 21, 23, 24, 29, 32, 33, 38, 40, 42, 49–52, 54, 57–60, 62, 65, 68, 70–72, 74, 83, 100, 101, 104, 107, 108, 109, 117, 119, 121, 131, 134, 137, 139, 146–149, 152, 154–160, 164, 166, 168, 173, 175–178, 182, 186–189, 191, 193, 195, 199, 200, 205, 213, 216, 218, 223, 224, 228, 230–232, 234, 240–243, 247
Ewen, B. 223, 234

Falender, J. R. 134, 144, 145, 148, 149
Farnoux, B. 8, 21, 220, 223, 233, 234
Fearon, F. W. G. 149, 290, 305, 347, 350

Ferry, J. D. 38, 42, 137, 148, 149, 216, 219, 253, 258, 264, 284, 304
Fetters, L. J. 38, 42, 317, 320
Finkelmann, H. 135, 148, 165, 166, 185, 186
Fitzgerald, J. J. 207, 210, 218, 284–286, 304
Fixman, M. 100, 105, 199, 218, 332, 334
Flory, P. J. 7, 8, 10, 11, 16, 18, 20, 23, 24, 29, 32, 39, 42, 44, 48, 49, 50, 52–55, 57, 58, 61, 62, 67, 70, 72–74, 85, 87, 89, 92, 93, 95, 96, 104, 105, 107–110, 116, 122, 123, 125–129, 131–133, 134, 139, 146, 147, 149, 151, 152, 154, 155, 157, 160–164, 166–168, 180, 186–188, 199, 203, 216, 218, 223, 233, 236, 242, 243, 247, 263, 264, 307–310, 317–320, 322, 323, 324, 329, 330, 331, 333, 335, 342–345, 348, 349, 351
Ford, W. T. 99, 105, 267, 301, 302
Freed, K. 10, 21, 37, 42
Freire, J. J. 101, 105, 199, 218
Frisch, H. L. 29, 66, 70, 72, 306
Furukawa, J. 352

Galiatsatos, V. 68, 72, 134, 142, 146, 148, 149, 189, 195, 209, 211, 217, 218
Gao, J. 104, 105
Garrido, L. 63–66, 70–72, 195, 217, 272, 304
Gaylord, R. J. 42, 43, 180, 187
Gee, G. 53, 61, 70, 108, 114, 131
Geissler, E. 62, 71, 267, 302
Gent, A. N. 132, 134, 137, 146, 148, 149, 178, 186, 206, 218, 267, 283, 301, 304, 311, 314, 353
Godovski, Y. K. 108, 121, 131, 166, 186, 275, 304
Goritz, D. 132, 134, 148, 268, 302
Gosline, J. M. 235–244, 247–249, 252, 253, 256, 262–264
Gottlieb, M. 134, 147, 189, 216
Gough, J. 342, 349
Graessley, W. W. 14, 16, 21, 33, 42, 43, 65, 71, 105, 134, 137, 141, 143, 145, 146, 148, 308, 310, 324, 329, 353
Grest, G. S. 104, 105, 106
Gronski, W. 134, 147
Grun, F. 163, 167, 169, 170, 186, 330, 331, 333
Guenet, J. M. 85, 86
Guth, E. 7, 10, 18, 20, 267, 302, 343, 350

AUTHOR INDEX

Haidar, B. 134, 147, 217
Haliloglu, T. 104, 105
Han, C. C. 8, 21, 223, 233, 234
Hedrick, J. L. 134, 147, 217
Heinrich, G. 21, 23, 32, 34, 42, 43, 134, 147, 268, 302, 353
Helmis, G. 21, 23, 32, 34, 42, 353
Herz, J. E. 21, 134, 147, 157, 159–161, 163, 164, 231, 233, 234
Higgins, J. 8, 21, 220, 221, 223, 233, 234
Higgs, P. G. 224, 230, 234
Hill, T. L. 59, 70, 321, 323
Hoeve, C. A. J. 8, 21, 108–110, 131, 146, 149, 236, 242, 246, 263, 264
Hoffman, A. S. 84
Holden, G. 349, 350
Horkay, F. 59, 60, 62, 70, 71, 73, 83, 267, 302
Hu, Y. 84, 265, 279, 301, 303
Hvidt, S. 73, 83, 317, 320

Ilavski, M. 74, 84, 85
Iwata, K. 70, 72

James, H. M. 7, 10, 19, 20
Jannink, G. 8, 21, 220, 223, 224, 233
Jarry, J. P. 154, 163, 166, 173, 186
Joule, J. P. 343, 349

Kaplan, D. L. 238, 240, 263, 264
Kawabata, S. 204, 218
Keller, A. 85, 86, 295, 306
Kelley, F. N. 134, 147, 189, 206, 207, 216
Kennedy, J. P. 134, 147, 148
Khokhlov, A. R. 62, 71, 85
Kilian, H. G. 62, 71, 108, 131, 199, 203, 218, 268, 302, 352
Kirkham, M. J. 21, 50–52, 109, 114–116, 120, 131, 132
Kirste, R. G. 223, 233
Kloczkowski, A. 7, 10, 13, 20, 21, 40, 42, 87, 99, 100, 101, 103, 105, 166, 168, 173, 175–178, 180, 181, 186, 187, 189, 212, 217, 224, 228, 230, 234, 265, 266, 281, 284, 300, 301, 304, 324–329
Koenig, J. L. 65, 71, 105, 158, 164, 353
Kramer, O. 73, 83, 134, 137, 148, 235, 262, 317, 320, 352
Kraus, G. 265, 267, 299, 302
Kremer, K. 101, 105

Krigbaum, W. R. 163
Kuan, T. H. 112, 132, 146, 149, 218, 283, 304
Kuhn, W. 7, 18, 20, 167, 169, 170, 186, 209, 218, 330, 331, 333

Labana, S. S. 94, 105, 352
Lal, J. 21, 147, 189, 216, 341, 347, 348, 349, 352
Landry, M. R. 21, 234
Langley, N. R. 33, 42, 137, 145, 148, 149
Legge, N. R. 349, 350
Leibler, L. 223, 231, 233, 234
Lemstra, P. J. 85, 86
Llorente, M. A. 101, 105, 134, 137, 141–147, 149, 182, 183, 187, 188, 192, 193, 195, 197, 199, 200, 216, 217, 218
Lohse, D. J. 38, 42, 220, 233

Macconnachi, A. 220, 233, 234
Macosko, C. W. 134, 137, 147, 189, 216
Madkour, T. M. 189, 212, 213, 217
Mandelkern, L. 65, 71, 85, 105, 134, 141, 143, 145, 146, 149, 165, 185, 257, 264, 342, 343, 349, 353
Mann, S. 235, 263, 265, 290, 301, 305
Mark, H. F. 7, 20, 179, 187
Mark, J. E. 7, 13, 20, 21, 23, 32, 40, 42, 63–72, 85, 86, 87, 90, 94, 95, 96, 97, 100, 101, 102, 104, 105, 107–137, 139–149, 166, 168, 173–178, 180–184, 186–188, 216–218, 219, 228, 230, 234–236, 240–248, 250, 253, 257, 263, 265–308, 311–320, 332, 333, 343, 345, 347, 348, 350, 352, 353, 354
Marucci, G. 42, 43
Mattice, W. L. 67, 72, 87, 89, 94, 104, 107, 108, 110, 127, 128, 129, 131, 332, 334
McCrum, N. G. 126, 133
McGill, W. J. 209, 218
McGrath, J. E. 270, 303
McKenna, G. B. 73, 83, 85, 209, 218
Meares, P. 351
Medalia, A. I. 265, 299
Menduina, C. 101, 105, 199, 218
Merrill, E. W. 134, 137, 147, 189, 216
Meyer, K. H. 343, 349
Mitchell, G. R. 165, 166, 185
Moller, M. 134, 148, 186

Monnerie, L. 24, 32, 33, 38, 40, 42, 51, 52, 134, 148, 153, 154, 157–161, 164, 166, 173, 186, 189, 217
Mooney, M. 114, 132
Morawetz, H. 345, 350
Morton, M. 265, 292, 299, 306, 311, 314, 349, 350, 351
Murnaghan, F. D. 44, 45, 52, 315, 320
Muthukumar, M. 85

Nagai, K. 146, 149, 150, 155, 163
Nakajima, A. 114, 115, 124, 132, 133
Nishi, T. 85
Noda, I. 157, 163
Noel, C. 52
Nomura, S. 85, 134, 148
Novak, B. M. 265, 301

Obata, Y. 204, 218
Ober, C. K. 185
Ober, R. 8, 21, 233
Oberth, A. E. 265, 266, 299
Oeser, R. 234
Ogden, R. W. 108, 131, 315, 320
Okamoto, T. 218, 219
Oppermann, W. 134, 137, 148, 149
Opshcoor, A. 117, 132
Orofino, T. A. 118, 121, 132

Pak, H. 49, 52, 278, 304
Pearson, D. S. 13, 21, 224–226, 228, 234, 309, 310, 324, 328, 329
Pechold, W. 165, 166, 185, 186
Peterlin, E. 209, 218
Picot, J. 8, 21, 62, 70, 220, 223, 231, 233, 234, 353
Polmanteer, K. E. 42, 265, 269, 282, 299
Prauznitz, J. M. 62, 70, 84
Price, C. 21, 50–52, 107, 109, 114–118, 120–122, 131, 132
Prins, W. 73, 83, 117, 132, 351
Prud'homme, R. K. 62, 71

Queslel, J. P. 155, 157, 159, 160, 163, 189, 216, 265, 299, 309, 310, 352, 353

Rehage, G. 62, 71, 117, 132, 185
Rehner, J. 53, 57, 70
Rempp, P. 66, 72, 134, 147, 233
Reneker, D. H. 267, 301

Riande, E. 209, 218
Richards, R. W. 220, 233
Richter, D. 223, 234
Rigbi, Z. 70, 72, 265, 267, 295, 299, 302, 306
Ringsdorff, H. 135, 148, 186
Rivlin, R. S. 49, 50–52, 114, 132, 205, 218
Roe, R. J. 87, 104
Roland, C. M. 134, 147, 189, 212, 217
Ronca, G. 23, 32, 166, 167, 186, 187, 189
Ross-Murphy, S. B. 85, 235, 257, 258, 262, 263, 353
Russo, P. S. 257–259, 264

Saam, J. C. 265, 269, 282, 299
Saeguso, T. 265, 270, 300, 304
Saiz, E. 129, 133, 209, 218
Sakurada, I. 114, 132
Samulski, E. T. 65, 71, 105, 164, 270, 303, 353
Sanchez, C. 269, 303
Sarikaya, M. 263, 265, 295, 300, 306
Saunders, D. W. 49, 50–52, 205, 218
Scanlan, J. J. 309, 310
Schaefer, D. W. 265–267, 269, 270–272, 300, 303, 304
Schaeffer, J. 42, 43
Schmidt, H. K. 265, 266, 269, 271, 275, 276, 288, 290, 300, 301, 303, 304, 305
Schwarz, J. 55, 62, 71, 117, 132, 166, 186
Semegen, S. T. 341, 349
Semlyen, J. A. 66, 70, 72, 218, 269, 303, 353
Sharaf, M. A. 99, 100, 101, 102, 103, 105, 134, 135, 137, 144, 145, 147, 148, 149, 180, 181, 187, 189, 214, 215, 217, 219, 265, 266, 281, 300, 301, 304
Shen, M. 109, 116, 118, 120, 121, 131, 133
Siessler, H. W. 186
Singler, R. E. 352
Smith, K. J., Jr. 108, 114–116, 120, 121, 131–133, 180, 187, 351
Smith, T. L. 134, 147, 189, 206–209, 217
Sperling, L. H. 220, 233
Spiess, H. W. 134, 147
Stadler, R. 134, 147
Staverman, A. J. 21, 109, 131, 308, 310
Stein, R. S. 134, 137, 146, 147–149, 158, 164, 189, 217, 218, 349, 350
Stepto, R. F. T. 147, 354
Straube, E. 21, 23, 32, 34, 42, 353

AUTHOR INDEX

Sullivan, J. L. 131, 134–137, 144, 145, 147, 149
Sundararajan, P. R. 130, 133
Suter, U. W. 67, 72, 87, 89, 94, 104, 107, 108, 110, 127–129, 131, 133, 331, 333

Tanaka, T. 74, 83–85, 186
Tatara, Y. 58, 70
Termonia, Y. 217, 240, 264
Thirion, P. 145, 149
Tomalia, D. A. 65, 71
Tonelli, A. 157, 163
Treloar, L. R. G. 7, 8, 18, 20, 44, 45, 51, 52, 53, 70, 100, 105, 107, 108, 111–113, 117, 119, 121, 131, 132, 134, 141, 147, 157, 160–163, 200, 204, 206, 216, 236, 241–243, 247, 248, 250, 263, 265, 278, 281, 283, 287, 294, 300, 316, 320, 330, 333, 343–345, 348, 349, 350, 351
Truesdell, C. 44, 45, 52, 315, 320

Uhlmann, D. R. 265, 269, 270, 290, 300, 303–305
Ullman, R. 220, 225, 233, 234
Ulrich, D. R. 265, 269, 290, 300, 302–305
Urry, D. W. 240, 245, 263

Veiss-Fogh, T. 236, 243, 263
Vilgis, T. A. 23, 32–37, 42, 43, 101, 105, 108, 131, 137, 148, 186, 199, 218, 231, 234, 268, 302, 352

Vincent, J. F. V. 235, 262, 263
Viney, C. 263, 264
Volkenstein, M. 8, 21, 89, 105, 322, 323, 330, 333

Wagner, M. H. 42, 43
Wall, F. T. 7, 18, 20, 167, 186, 323, 351
Wang, S. 277, 279–283, 288, 290–292, 295–297, 305, 306
Warner, M. 33–35, 37, 42, 165, 166, 185, 186, 229, 231, 234
Warrick, E. L. 265, 269, 282, 299, 347, 350
Wasserman, E. 70, 72
Wiener, J. H. 104, 105
Wignall, G. D. 65, 71, 105, 220–223, 233, 303, 353
Wilkes, G. L. 265, 272, 275, 288, 290, 300, 301, 303, 305
Witten, T. A. 264, 268, 302, 303
Wood, L. A. 115, 133

Xu, P. 218, 265, 266, 270, 275, 278–283, 295, 299, 301, 304, 306

Yu, H. 8, 21, 223, 230, 231, 233, 234

Zentel, R. 135, 148, 165, 166, 185, 186
Zimm, B. 224, 234
Zrinyi, M. 59, 60, 62, 70, 71, 267, 302

Subject Index

Abductin 235, 238, 247, 248, 256
Abrasion resistance 265
Absorbance 158
Accelerator 311
Actin 262
Activator 311
Activity of solvent 55, 56, 61, 76
Additivity of free energies 10
Aerogels 274, 275, 290
Affine deformation 8, 10, 156, 167, 171, 179, 192, 213, 224
Affine limit 140, 182
Affine network 3, 7, 16, 19, 23, 29, 32, 37, 44, 46-48, 54, 77, 138, 151, 154, 155, 159, 180, 181, 223, 228–232
Agar 260
Agarose 74, 260
Aging 275
Alanine 236, 238, 239, 244, 250, 251
Alginates 260
Alumina 288
Amino acid composition 238, 245, 249
Amylopectin 261
Amylose 261
Aneurism 252
Annealed system 36
Antidegradant 311
Antioxidant 258, 311
Antiozonant 311

Bacterial attack 258
Biaxial compression 280

Biaxial extension 4, 203, 204, 276, 278, 279, 282, 295, 317
Bicontinuous systems 289
Bifunctional systems 213
Bimodal distributions 117, 142, 182, 188, 190, 195, 196, 241
Bimodal networks 5, 189, 190, 192–216, 224
Bimodality, effects of 201
Bimodals 291
Bioelastomers 5, 110, 235–269
Biomimetic materials 263, 301
Birefringence, strain 40, 87, 124, 146, 155, 157, 209, 218
Birefringence-temperature relations 195, 197
Blooming 311
Bonding agent 312
Bound rubber 267
Brittle failure 287
Bulk compressibility 57
BuNa rubbers 312
Butterfly shapes 223
Butyl rubber 312

c^* theorem 62
Calorimetry 121, 122, 211, 242, 244, 257
Carbon black 6, 265
Carrageenans 260
Cavitation 266, 312
Cellulose 261
Center of constraints 25, 26, 27, 38, 39
Center of entanglements 26, 27

SUBJECT INDEX

Ceramics 269
Chain flexibility 238
Chain folding 180
Chain interpenetration 139, 140
Chain-junction entangling 145
Chain length distributions 188, 190, 201, 207, 214, 268
　bimodal 117, 142, 182, 188, 190, 195, 196, 241
　unimodal 188, 190, 192
Chain scission 144
Characteristic dimensions 221
Characteristic ratios 96, 139
Charpy pendulum (impact) test 286
Chemical potential
　of solvent 4, 53, 55, 56, 59, 75, 76
　reduced 76
Chemisorption 266
Chi parameter 75, 76, 123
Chlorinated polyethylene 312
Chlorosulfonated polyethylene 312
Classical theories 3, 7, 22
Closed systems 53, 58, 115, 236, 242, 253
Coagulation 312
Coefficient of dilation 58
Coherent scattering 222
Coil–globule transition 74
Cold rubber 312
Collagen 79, 236, 252, 259, 262
Compounding 312
Compressibility 58
Compression 3, 23, 44, 45, 46, 49, 108, 109, 112, 118, 154, 203, 257
　set 265, 284, 312
Conditional probabilities 28
Configuration partition function 10, 14, 17
Constrained-chain model 24, 33, 38–42, 48, 49, 51, 52, 54, 57, 152, 180, 230
Constrained-junction model 3, 22–33, 37, 39–41, 44, 46–50, 54, 59, 61, 62, 76, 77, 100, 140, 151, 152, 155, 160, 161, 162, 180, 181, 199, 205, 232
Constraint models 137
Constraints 139
Coulombic interactions 146
Crack propagation 287
Creep 284
Crepe rubber 312
Critical phenomena 53, 54, 73–86
Critical point 75

Cross-linking 116, 145, 188, 190, 235, 236, 278
　by radiation 65
　chemistry 240
　degree of 116, 117, 119, 139, 140, 145
　density 57, 59, 60, 271
　effective degree of 266
　in solution 145, 155, 162
　inhomogeneous 195
　light 195
Cross-links 34, 35, 57, 73, 80, 83, 136, 239, 241, 242, 248, 250, 258–260, 294, 307–311
　helical junction 257
　permanent 259
　physical 79, 257, 260
　temporary 80
Crystalline fibers 252
Crystalline materials 150, 295
Crystallinity, degree of 115, 207, 247, 252
Crystallite nuclei 179
Crystallites 79, 80, 83, 141, 165, 178, 180, 182, 183, 240, 257, 258, 261, 265
Crystallization 80, 99, 101, 116, 129, 145, 149, 195, 209, 212, 236, 241, 275
　incipient 179
　strain-induced 5, 23, 49, 119, 141, 142, 178–187, 189, 193, 197, 199–201, 213, 265, 288
　temperature, incipient 180, 211
Cure 312
　time 312
Cycle rank 17, 29, 37, 307–310
Cyclic deformations 5, 207, 209, 253, 284
Cyclic stress 284, 285

Dangling chains 63, 83, 142–145, 230, 307–310
Deam–Edwards theory 8, 19
Decomposition temperature 276
Deformation 315–320
　tensor 9, 54, 150
　uniaxial (tension/compression) 9, 45, 54, 57, 108, 109, 111, 118, 119, 120, 150, 153-155, 157, 171, 172, 174, 205, 222, 223, 230–232, 252, 276, 278–282, 295
Deformation–recovery process 253
Degree of crystallinity 180, 181, 307–310
Degree of entanglement 22

SUBJECT INDEX 363

Degree of interpenetration 22, 24
Dehydration 256
Dehydrolysino-nor-leucine 241
Dendrimers 65
Desmosine 240, 241
Deuterated chains 124
Dichroic ratio 158
Dichroism, strain 87
Dielectric constant, solvent 146
Dielectric relaxation 259
Differential scanning calorimetry 209, 212, 275
Differential sorption measurements 61
Diffused constraints 33, 40–42
Diffusion coefficient 63, 65, 66
Dilation 45
Dilatometry 257
Diluent
  branched 65
  linear 63
  oligomeric 63
  polymeric 63
  reptating 63
Diluents, cyclic 53, 65, 66
Dipole moments 218
Disaccharides 260
Disinterspersion of network chains 62
Disorientation index 167, 168
Displacement gradient tensor 26, 45, 315–320
Distortion 45
Distribution 6, 10, 11, 27–30, 39, 100, 101, 250
  particle size 102, 272
Distributions
  bimodal 98
  end-to-end vector 330–334
  multimodal 98
Dityrosine 241, 242
DNA gels 74
Domains of constraint 30
Double-network model 268
Drug release 65
Ductile behavior 287
Durometer 312
Dynamic mechanical properties 253–257, 284
Dynamic mechanical tests 257
Dynamics, chain 90

Effects of World War II 348
Elastic contribution to swelling 60
Elastic deformation energy 107
Elastic energy 242
Elastic equation of state 9, 146
Elastic force 56, 108, 110, 146, 241, 247, 258
Elastic free energy 7, 8, 9, 13, 16–19, 22, 29–33, 37, 39, 40, 46, 47, 51, 53, 54, 81, 87, 104, 174, 203, 318
Elastic stress 107
Elastically effective chains 143
Elastin 112, 126, 235, 236, 238–257
Electric field 74
Electrolytes 258
Electron microscopy 79, 267, 270, 272, 275
  scanning 270, 291–293, 295, 296
  transmission 270, 271, 295
Electron scattering 221
Elongation 4, 22, 23, 40, 49, 100, 101, 103, 107–109, 112, 114, 117–120, 122, 136, 143, 248
Emulsion polymerization 288
End groups 82
End-labeled polymer 124
End-linking 4, 16, 63, 64, 66, 69, 134, 136, 143, 188–190, 195, 212, 215, 294
  in solution 205
End-to-end dimension (vector) 9, 10, 11, 87, 94–100, 124, 127, 139, 141, 151, 199, 200, 250, 266, 281, 330–334
Energetic contributions 4, 108
Energy dispersive x-ray analysis (EDXA) 291, 293
Engineering stress 46
Entangled chains 63
Entanglement
  chain-junction 206
  loci 139, 140, 160
  networks 262
  phenomena 139
Entanglements 3, 8, 20, 22–33, 38, 104, 116, 136, 137, 140, 146, 152, 195, 257, 266, 267
  localized 140
  permanent 139
  trapped 139, 161
Enthalpic component of the force 241, 262
Enthalpy 122
Entropic contributions 4, 8, 107, 108, 241, 262

SUBJECT INDEX

Entropic force 56, 247
Entropy decrease 247
Entropy of fusion 180
EPDM 312
Equation of state 9, 107, 109, 145
  approach 55
Equibiaxial extension 204, 205, 278
Ethylene hexene-1 copolymers 79, 80, 81
Ethylene–propylene copolymers 117, 126, 129
Euler angles 167
Excluded volume 89, 99, 123, 124, 268
Extenders 312
Extensibility, finite (limited) 87, 96, 101, 141-143, 193, 199
Extension ratios 23, 45, 51, 52, 54, 57, 151, 154, 155, 159, 171, 175, 176, 228, 281, 282
Extent of reaction 63
Extraction 53, 63–70
  time 64

Failure envelope 258
Falling weight (impact) test 286, 287
Fatigue failure 209, 284
Fatigue life 312
Fibrin 262
Filled networks 99, 276
Filler content 280, 281, 283, 298
Fillers 4, 101, 102, 104, 206, 265–299
First order volume transition 74
Flory number 24, 139, 140
Flory–Rehner equation 57
Fluctuations 6, 8, 10, 12, 13, 18, 23, 24, 26–30, 38, 39, 42, 48, 88, 151, 152, 160, 180, 222, 223, 225, 226, 233, 324–329
Fluorescence measurements 246
Force, reduced 46, 47, 51, 52, 123
Force–extension curve 248
Force–temperature relations 109, 110, 111, 119, 242
Fractal aggregates 268
Fractional precipitation 79
Free energy 176, 177
  increase 247
  of mixing 53, 61, 74, 168, 169
Freely-jointed chain 8, 34, 35, 87, 154, 156, 157, 166, 171, 179, 225, 226, 248, 249
Freezing point 209
Frequency 253, 256

Front factor 7, 20
Functional groups 144
Functionality of junctions (cross-links) 5, 7, 15, 16, 19, 20, 24, 25, 29, 40, 42, 57, 61, 63, 69, 134, 136, 137, 189, 195, 228, 240, 307–310

Galactose 260
Gamma radiation 144, 145
Gas–liquid chromatography 123
Gaussian elasticity theory 242
Gaussian limit 3, 8, 9, 10, 11, 14, 26–28, 35, 39, 67, 87, 94–99, 101, 141, 155, 165, 199, 213, 225, 228, 248–250
Gel, elastic 259
Gel permeation chromatography 79
Gelatin 74, 79, 259, 260
Gelation 79, 82, 257, 261
  point 79
Gels 4, 62, 66, 71, 73–86, 235, 257, 258, 260–263
  biological 77
  polyelectrolyte 62, 74, 79
  temperature sensitive 77, 78
  thermoreversible 74, 79–83
Gibbs free energy 54
Glass transition temperature 90, 246, 253, 256, 257, 284, 294
Glassy polymers 150, 292, 295
Glassy state 236, 256, 295
Globular proteins 262
Glucose 261
Glycine 236, 238, 239, 244–246
Glycoproteins 262
Glycosaminoglycans 262
Green strength 312
GR-S rubber 312
Guar galactomannan 262
Guayule rubber 312
Gutta percha 126

Hard rubber 312
Hardness 291, 312
Heat of deformation 121
Heat of fusion 180
Helix–coil transition 258
Helmholtz free energy 3, 8, 36, 44, 54, 170, 176, 213, 318, 321
Hemoglobin-S 262

Hevea rubber 312
Hexapeptide 251
High energy electrons 144
High-temperature stretching 295
Historical information 340–350
Hooke's law 46
Hybrid inorganic-organic composites 290, 302
Hydrogen bonding 245
Hydrolysis 269, 270
Hydrophobic bonding 247, 257
Hydrophobic–hydrophilic interactions 242
Hydroxypropyl cellulose 261
Hysteresis 265, 312
Hysteretic effects 252

Ideal mixing law 54
Impact resistance 202, 204, 286
Impact strength 286, 287
Imperfect networks 19
In situ, fillers 265–299
Incoherent scattering 222, 223, 233
Incompressibility 19
Infrared dichroism, Fourier transform 153, 157, 159
Infrared, polarized Fourier transform (FTIR) 157, 159, 160, 161, 162
Infrared spectroscopy 150
Insulin 262
Interaction parameter 54, 55, 73, 74
Interchain interactions 139, 224, 236
Intermolecular contributions to orientation 161, 162, 173
Intermolecular correlations 166, 170, 231
Intermolecular coupling 150
Intermolecular effects 162, 197
Intermolecular interactions 4, 8, 9, 10, 20, 22–45, 88, 104, 119, 123, 125, 157
Intermolecular orientational correlations 157
Interpenetrating network structure 195, 196
Interpenetrating phases 289
Intramolecular interactions 107, 110, 119, 129, 150, 197
Ion concentrations 78
Ionic charge 260
Ionic groups 74, 75
Ionic networks 62, 78
Ionic potential 75
Ionizable groups 74, 75, 260

Ionizable networks 74, 76
Ionized gels 74
Isodesmosine 240, 241
Isotropic–nematic phase transitions 175-178
Isotropic solutions 176
Isotropic swelling 317

James–Guth theory 7, 10, 23, 26, 233
Joint probabilities 28
Junction functionality 14, 24, 40, 135, 136, 155, 160, 307–310

Kappa parameter 29, 32, 39, 47, 77, 100, 199
Kirchoff adjacency matrix 11
Knotty tearing 312
Kratky plot 231, 232
Kuhn length (of segment) 34, 160, 166, 167, 176
Kuhn model 160

Lame constants 19
Lamellar structure 82
Latexes 312
Lattice
  model 168, 170, 178
  theory 54, 166, 258, 259
Legendre polynomial, second 154
Lennard–Jones potential 88, 102
Librational motions 240
Light microscopy 267
Light scattering
  depolarization 124
  dynamic 212
  spectroscopy 74
Limited chain extensibility 23, 119, 182, 184, 188, 195, 198, 200, 209, 250
Linear spring law 25, 27
Liquid crystalline, main-chain 165
Liquid crystalline networks 5, 135
  side-chain 165
Local chain axis 157
Loss tangent 210, 253, 256, 284
Lozenge shapes 223
Lubricant 312
Lysine 236, 239, 240, 250

Macroscopic deformation 151, 155
Magnetic fields 295

SUBJECT INDEX

Magnetic particles 298
Master curves 253–255, 258
Masterbatching 312
Mastication 312
Matrix isolation approach 270
Matrix multiplication methods 67
Maximum extensibility 63, 144, 145, 213
Mean-square end-to-end length 6, 8
Melting points 180, 189, 284, 295
Methyl methacrylate 269
Micronetworks 14
Microphase separation 79
Microscopic deformation 151, 152, 230
Mobility
  chain 238, 266, 277
  segmental 87
Model elastomers 5, 134–149
Model networks 63, 134, 136, 141, 144, 189
Modulus 80, 82, 101, 135, 136–140, 145, 178, 182, 192, 193, 198, 211, 248, 250, 257–259, 261, 265, 266, 276, 283, 284, 289, 294
  bulk 58, 59
  elastic 3, 4, 8, 22, 23, 46, 63, 66, 104, 213
  equilibrium 79
  filled elastomer 267
  loss 253, 254, 256
  orientational 154, 159
  phantom 139
  plateau 137, 140
  shear 46, 47, 101, 137, 261, 281, 282, 283
  storage 253–255
  upturn 100, 101, 141, 142
Molecular deformation 5, 150, 152
Molecular dynamics 3, 90, 104, 106
Molecular weight 5, 6, 29, 56, 57, 63–65, 80, 82, 83, 207, 212, 214
  between cross-links 134, 143, 159, 189, 204, 231, 261, 307–310
  between entanglements 38, 139
  distribution 5, 63, 134
  network chains 136, 139
  number average 135
Monodisperse distributions 188
Monte Carlo method 6, 14, 68, 91, 93–100, 104, 154, 156, 199, 250, 252, 268, 335
Mooney–Rivlin equation 51, 100, 101, 114, 116, 142, 145, 194, 195, 200, 201, 250, 268, 294
Mullins effect 313

Multiaxial stress 51
Multimodal chain length distributions 188, 213
Multimodal networks 188–219
Multiphase systems 265–306
Myosin 262

Natural rubber 49–52, 61, 101, 102, 111, 112, 114, 115, 117, 118, 120–122, 126, 141, 142, 183, 195, 205, 265
  cross-linked state 345
  oriented 116
  un-cross-linked state 342
  unoriented 116
N'-diethylacrylamide 74
Nematic 62, 165, 176, 177
Network
  chains 307–310
  imperfections 63, 143
  inhomogeneities 223
  structure 6, 11, 14, 134, 307–310
  topology 87, 145, 146
  unfolding 223
Networks, heterogeneous 195, 197
Neutron scattering 5, 8, 10, 40, 62, 124, 212, 267, 270, 271, 275
  elastic 221
  inelastic 220
  small angle (SANS) 220–234, 272
Neutron spin echo experiments 38, 223, 233, 272–274
Neutron spin-echo technique 222
n-functional junctions 213
N-isopropylacrylamide 74
Nonaffine deformations 13, 135, 136, 191, 192, 199
Nonaffineness 136, 180
Noncrystallizable networks 142
Non-Gaussian distributions 6, 67, 94, 99, 100, 104, 117, 142, 161, 180, 184, 188–200, 209, 248, 249, 252
Nonionic networks 53, 73, 74–76, 78
Nonphantomlike networks 161
Nonpolar-polar interactions 242
Nonpolar side groups, bioelastomers 238
Nonreinforcing effects 313
Nuclear magnetic resonance (NMR) 65, 79, 161, 240, 267, 272–274, 290
  deuterium 157

SUBJECT INDEX 367

Octopus arterial elastomer 238, 247–249
Oil-drop model, elastin 242, 243, 246
Oil-extension 313
Olympic networks 69
Open systems 115, 242, 247, 253
Optical anisotropy 124, 174
Optical properties 146
Optical rotation 259
Order parameter 177
Organic–inorganic hybrid composites 269
Organosilicate-polymer systems 290
Orientation-deformation relation 175
Orientation function 94, 154, 155, 158, 173, 174, 177
Orientational distribution 169, 172
Osmotic compressibility 57–60
Osmotic pressure 59, 75
Osmotic swelling and deswelling 62

Pair correlation functions, intramolecular 224
Particle size 270, 272
Partition function 10, 11, 12, 34, 167, 168, 174, 268, 321–323
Pectin 261
Penetration 257
Pentapeptide 251
Peptizer 313
Percolation theory 258
Perfect networks 19, 145, 307–311
Permeation chromatography 66
Peroxide curing 144
Peroxide thermolysis 188, 270
Persistence length 6, 94
Phantom-like chains 165, 173
Phantom limit 140, 182
Phantom network 3, 7, 8, 10–19, 22–25, 28–31, 37, 39, 44, 46–48, 51, 54, 57, 60, 62, 138, 151, 155, 156, 159, 160, 166, 213, 223–228, 232
Phantom structures 139
Phase separation 259
Phase transitions
 isotropic–nematic 166
 orientational 175
 volume 4, 62, 73–86
Phenylethoxysilane 289
phr (parts per hundred notation) 313
Physisorption 266
Plantation movement 346

Plasticizer 313
Plasticizing effect 66, 292
Poisson's ratio 62
Polar polymers 146
Polarity, bioelastomers 239
polarized fluorescence 162
Polarized light microscopy 79
Poly(2-hydroxyethyl methacrylate) 126
Poly(acrylamide) 74
Polyalanine 245
Polyamide 165
Polybutadiene 138
 cis-1,4 116, 117, 118, 120–122, 126, 127, 128, 141, 142, 182–184, 195
 $trans$-1,4 126, 128
Poly(butene-1) 126, 128
Poly(dimethylsilmethylene) 126, 130
Poly(dimethylsiloxane) (PDMS) 61, 64, 65, 66, 68, 94–97, 99–101, 103, 104, 114, 115, 117, 118, 121, 124, 126, 127, 134–136, 138, 139, 142, 144, 145, 157, 159–162, 184, 185, 189, 193–212, 223, 265, 269, 271–279
 unoriented 116, 120
Poly(diphenylsiloxane) 294
Poly(ethyl acrylate) 288
Polyethylene 79, 80, 81, 83, 87, 88, 90, 96–98, 101, 102, 104, 110, 111, 116, 117, 120, 121, 123, 124, 126–128, 249, 257
 oriented 116
 unoriented 116
Poly(ethylene oxide) 204, 211, 291
Poly(isobutylene) 75, 114, 115, 120, 121, 124, 126, 129, 130, 288, 289, 292, 294
Polyisoprene 162, 230–232
 cis-1,4 117, 127, 128
 $trans$-1,4 118, 121, 128
Polymer modified ceramics 290
Polymer–solvent interaction 73, 109, 123, 146
Poly(methyl hydrogen siloxane) 136, 215
Poly(methyl methacrylate) 79, 117, 126
Poly(methylphenylsiloxane) 288
Poly(n-pentene-1)
 atactic 126, 128
 isotactic 124, 126
Polyoxyethylene 104, 111, 113, 118, 124, 126, 128, 201

Poly(oxymethylene) 98, 99
Polysaccharides 235, 260, 261, 262
Polystyrene 74, 75, 79, 118, 121, 292, 294–297
  deuterated 223
Poly(tetrahydrofuran) 202, 212
Poly(tetramethylene oxide) 126, 128
Poly(trimethylene oxide) 126, 130
Polyurethane 194, 209, 211
Poly(vinyl acetate) 59, 60
Poly(vinyl alcohol) 114, 115
Poly(vinyl chloride) 126, 129, 240
Pore size 63
Postwar period 348
Pressure pulsing 79
Principal axis of deformation 27
Principal components of deformation 45
Principal components of stress 45
Principal deformations 315–320
Proline 239, 244, 245
Promoter 313
Proteins, bioelastomeric 235
Proteoglycans 262
Pure gum vulcanizates 313

Quasi-elastic scattering 222, 223
Quenched systems 36

Radiation, high energy 116
Raman spectroscopy 79
Random cross-linking 16
Random network model, elastin 242, 243, 246, 247
Random phase approximation 55
Rate of absorption 63
Reduced chemical potential 55
Reduced forces 46–52, 123, 159
Reduced modulus 101
Reduced orientation 159, 161
Reference interaction site model 55
Reference state of a network 26
Reinforcement 5, 265
Reinforcing filler 265, 313
Reinforcing particle size 266
Repeat units 56
Replica approach 36
Replica formalism 35, 61, 229, 268
Reptating chains 145
Resilience 256, 265, 313
Resilin 235, 238–242, 247–249, 252, 256

Retarder 313
Reversion 313
Ring-chain cyclization 124
Ring opening polymerization 189
Rotational isomeric state theory 4, 6, 67, 89, 93, 94, 95, 96, 104, 108, 110, 125, 126, 131, 154, 156, 157, 199, 240, 244
Rotational isomerization 87
Rotational potentials 88
Rotational state 107
Rouse model 223
Rupture 143, 190, 193, 194, 200, 204, 205, 258, 281
  energy 193, 283, 288
  modulus 280

SBR 313
Scanning calorimetry 79
Scanning tunnel microscopy 267
Scattering cross-section 222, 224
Scattering form factor 230
Scattering law 224, 226
Scattering length 222
Scorch 313
  time 313
Segmental orientation 5, 87, 104, 150–164, 166, 168, 178, 200, 213
Semiflexible chains 5, 165–178
Semi-open system 44, 53, 236
Semirigid chains 5, 176
Serine 236, 238
Shear 4, 51, 118, 203, 205, 276, 278, 281
  modulus 46, 175
  pure 318
Silica 6, 265–299
Silicon-based polymers 149
Silicones 313
Simple elongation 4
Simple tension 317
Simulations, computer 87–106, 148, 212, 213
Slip-links 3, 22, 33-35, 37, 42, 44, 48, 49, 54, 57, 137
Smoked sheet 313
Soft rubber 313
Solution SBR 313
Sorption 53, 63-70
Spatial neighbors 25
Specific solvent effects 146
Spider web silk 235, 238, 240, 241, 247, 248, 252

SUBJECT INDEX 369

Spinodal decomposition 260, 289
Spinodal point 59
Stabilizer 313
Staining 313
Starch 261
Strain invariants 18
Strain-orientation data 159
Stress 4–6, 58, 63, 73, 101, 103, 111, 134, 174, 315–320
  at rupture 283
  nominal 193, 250, 251
  reduced nominal (modulus) 181, 182, 199, 268, 277
  relaxation 145
  uniaxial 46, 47
Stress–deformation relations 175
Stress–elongation curves 141
Stress–optical coefficients 209, 211
Stress–strain
  behavior 248, 259
  data 278
  isotherms 23, 40, 92, 100–102, 109, 111, 113, 141, 145, 182, 183, 191, 193–195, 198–200, 203, 205, 206, 249, 250, 252, 259, 276, 279, 281, 282, 289, 290, 294, 297, 298, 299
  measurements 142, 236, 295
  properties in elongation 145
  relationships 3, 4, 42–53, 119
Stress–temperature relations 108, 110, 111, 112, 195
Structure index 313
Structure–property relations 5
Styrene 269
Sulfur donors 313
Superabsorbent polymers 83
Swelling 4, 22, 44, 46, 47, 50, 51, 53–72, 73–86, 109, 111, 112, 114, 115, 118, 119, 121, 137, 139, 141, 142, 145, 146, 155, 161, 162, 182, 195, 212, 223, 230, 232, 235, 242-244, 246, 247, 253, 256
  high degree 146

Tack strength 313
Tackifiers 313
Tear 5, 206
  energy 206-208
  resistance 265
Temperature coefficients 125, 236
  chain dimensions 122, 123

intrinsic viscosity 122
Temperature shift factors 255
Tensile load 154
Tensile strength 209, 213, 265
Tension
  simple 179, 180
  uniaxial 9, 51, 52
Tetraethoxysilane (TEOS) 270, 288, 289
Tetraethyl orthosilicate 134, 189
Tetrafunctional networks 7, 19, 20, 24, 25, 29, 57, 61, 135, 144, 153, 166, 167, 210, 224, 229
Tetrapeptide 250, 251
Thermal expansion coefficients 109, 114, 242
Thermal stability 276
Thermoelasticity 87, 107–133, 206, 241–253, 262
Thermogravimetric analysis 276
Thermoplastics, rubber toughened 216
Thermoreversible gels 257, 258, 260, 261, 270
Thermosets 202, 313
Thermotropic interactions 174, 177
Three-chain model 200, 250
Time-averaged position of a junction 26
Time–temperature superposition 258
Titania 269, 276–278, 288, 290, 295
Topological neighbors 25
Torsion 4, 111, 112, 113, 117, 118, 122, 203, 206, 276, 278, 283
Toughening 141
Toughness 5, 193
Transition moments 157
Trapped entanglements 33
Trapping
  factor 140
  of cyclics 53, 66–68
Tree 17
Treloar relations 45
Trifunctional networks 135, 189, 210, 213
Trimodals 189, 212, 213
Triphasic equilibrium 76, 77
Trityrosine 241, 242
Trouser-tear method 206
True stress 47
Tube model 42
Tubulin 262
Two phase model, swollen elastin 247
Tyrosine 241

Ultimate properties 5, 141, 143, 144, 184, 185, 188, 189, 193, 204, 213, 258, 260, 283
Ultimate strength 63, 141, 144, 145, 198, 288
Unimodal networks 14, 19, 117, 188, 189, 190–192, 202–211, 214, 224, 229, 291
Unperturbed configurations 125
Unperturbed dimensions 89, 96, 110, 122, 123, 125, 126, 146, 151, 199, 223, 244, 245
Unperturbed radius of gyration 67–69
Unperturbed state 8, 124, 125, 139
Upturns, in stress-strain isotherms 141, 143, 161, 182, 183, 192, 193, 195–197, 199, 205, 248, 250, 252, 276, 279, 283
UV radiation 144

Vacuole formation 266
Valine 236, 238, 239, 244
Van der Waals theory of rubber elasticity 199, 268
Vapor pressure measurements 123
Vinyl-terminated chains 144
Vinyl/vinylidene polymers 79
Viscoelastic effects 138, 207

Viscoelastic properties 148, 256
Viscoelastic responses 253
Viscoelasticity 207, 253, 261, 262
Viscometry 79, 124
Viscosity 122, 123
 bulk 268
Viscosity–temperature coefficients 123
Viscosity–temperature measurements 122, 123
Volume phase transitions 73, 74
Vulcanization 6, 188, 314

Wall–Flory model 8, 19, 23
Wave propagation vector 220, 221
Weakest link theory 188, 190–192
Wormlike chain 166

Xantan 262
X-ray diffraction 178, 261
X-ray scattering 62, 124, 221, 222, 267, 270, 271, 275
 small angle (SAXS) 79, 272, 289

Zeolites 269, 298, 299
Zirconia 288